Communications in Computer and Information Science

2482

Series Editors

Gang Li, *School of Information Technology, Deakin University, Burwood, VIC, Australia*

Joaquim Filipe, *Polytechnic Institute of Setúbal, Setúbal, Portugal*

Zhiwei Xu, *Chinese Academy of Sciences, Beijing, China*

Rationale

The CCIS series is devoted to the publication of proceedings of computer science conferences. Its aim is to efficiently disseminate original research results in informatics in printed and electronic form. While the focus is on publication of peer-reviewed full papers presenting mature work, inclusion of reviewed short papers reporting on work in progress is welcome, too. Besides globally relevant meetings with internationally representative program committees guaranteeing a strict peer-reviewing and paper selection process, conferences run by societies or of high regional or national relevance are also considered for publication.

Topics

The topical scope of CCIS spans the entire spectrum of informatics ranging from foundational topics in the theory of computing to information and communications science and technology and a broad variety of interdisciplinary application fields.

Information for Volume Editors and Authors

Publication in CCIS is free of charge. No royalties are paid, however, we offer registered conference participants temporary free access to the online version of the conference proceedings on SpringerLink (http://link.springer.com) by means of an http referrer from the conference website and/or a number of complimentary printed copies, as specified in the official acceptance email of the event.

CCIS proceedings can be published in time for distribution at conferences or as postproceedings, and delivered in the form of printed books and/or electronically as USBs and/or e-content licenses for accessing proceedings at SpringerLink. Furthermore, CCIS proceedings are included in the CCIS electronic book series hosted in the SpringerLink digital library at http://link.springer.com/bookseries/7899. Conferences publishing in CCIS are allowed to use Online Conference Service (OCS) for managing the whole proceedings lifecycle (from submission and reviewing to preparing for publication) free of charge.

Publication process

The language of publication is exclusively English. Authors publishing in CCIS have to sign the Springer CCIS copyright transfer form, however, they are free to use their material published in CCIS for substantially changed, more elaborate subsequent publications elsewhere. For the preparation of the camera-ready papers/files, authors have to strictly adhere to the Springer CCIS Authors' Instructions and are strongly encouraged to use the CCIS LaTeX style files or templates.

Abstracting/Indexing

CCIS is abstracted/indexed in DBLP, Google Scholar, EI-Compendex, Mathematical Reviews, SCImago, Scopus. CCIS volumes are also submitted for the inclusion in ISI Proceedings.

How to start

To start the evaluation of your proposal for inclusion in the CCIS series, please send an e-mail to ccis@springer.com

A. Mirzazadeh · Zohreh Molamohamadi ·
Erfan Babaee Tirkolaee ·
Gerhard-Wilhelm Weber · Kathryn E. Stecke
Editors

Optimization and Data Science in Industrial Engineering

Second International Conference, ODSIE 2024
Istanbul, Turkey, November 7–8, 2024
Proceedings

 Springer

Editors
A. Mirzazadeh
Kharazmi University
Tehran, Iran

Zohreh Molamohamadi
Kharazmi University
Tehran, Iran

Erfan Babaee Tirkolaee
Istinye University
Istanbul, Türkiye

Gerhard-Wilhelm Weber
Poznań University of Technology
Poznań, Poland

Kathryn E. Stecke
University of Texas
Dallas, TX, USA

ISSN 1865-0929   ISSN 1865-0937 (electronic)
Communications in Computer and Information Science
ISBN 978-3-031-93600-5   ISBN 978-3-031-93601-2 (eBook)
https://doi.org/10.1007/978-3-031-93601-2

This Springer imprint is published by the registered company Springer Nature Switzerland AG
The registered company address is: Gewerbestrasse 11, 6330 Cham, Switzerland

If disposing of this product, please recycle the paper.

Preface

The Second International Conference on Optimization and Data Science in Industrial Engineering (ODSIE 2024) was held online in Istanbul, Turkey on November 7–8, 2024 by International Scientific Services (RefConf) and İstinye University. It provided an energetic knowledge-transferring atmosphere for participants (as several comments revealed).

ODSIE 2024 attracted the attention of students and professionals internationally. The subjects covered included, but were not limited to, "Industry 4.0, IoT and smart manufacturing", "Digital twins and virtual commissioning", "Sustainable and smart cities", "Artificial intelligence and expert systems", "Metaheuristic algorithms with applications in IE", "Machine learning algorithms", "Big data analytics and data mining", "Robotic process automation", "Decision support systems", "E-Government, E-Commerce and E-Learning", "Supply chain design and logistics", "Optimization and Data Science based case studies of manufacturing/service industries", "Quantitative finance and risk modelling", and "Other fields of study related to Optimization and Data Science with applications in IE".

ODSIE 2024 was honored to be enriched by outstanding keynote speakers and workshop organizers from the USA, Mexico, Turkey, Saudi Arabia, India, and Iran. In this event, nine universities from the USA, UK, Czech Republic, Tunisia, Chile, Pakistan, and Turkey were present as scientific sponsors.

The conference included participants from 34 countries. The geographical diversity of international committee members was from 22 countries.

The review criteria in this step were: content; originality; relevance; contribution to the professional literature; significance and potential impact of the paper; language accuracy; study validity; accuracy of methodology and analysis; paper organization, required relevant data, citations, and references; adequate referring of the background information; consistency of references, symbols, and units throughout the paper; quality and clarity of tables and figures.

The papers in ODSIE 2024 were presented in 10 panel sessions: "Data & AI for Sustainable Development", "E-Government, E-Commerce, and E-Learning", "Biomedical Informatics: Towards a Connected Healthcare", "Artificial Intelligence, Transportation Manufacturing Systems, Scheduling", "Artificial Intelligence Solutions for Engineering Problems", "AI, Data Science, and Optimization Applications for Smart Transportation Systems", "Industry 4.0 in Supply Chain Management", "AI Transformations: Driving Efficiency and Security in Modern Business", "Contextualized Knowledge Representation and Analysis", and "Integrating Data Science and Optimization for Efficient Engineering Management".

ODSIE 2024 included seven outstanding keynote speeches and applied workshops by outstanding lecturers with good-sized audiences. The subjects were very well received by the participants and were entitled: "Recent Investment Opportunities and Sustainability Concerns in North America"; "A Decision-Making Framework for Determinants of an Organization's Readiness for Smart Warehouse"; "Embracing Artificial Intelligence in Modern Education: Opportunities & Challenges"; "From Data to Decisions: Hybrid Deep Learning Models in the Field of Health"; "Integrating Data Science and Optimization for Efficient Engineering Management"; "Statistics vs. Machine Learning: Approaches, Applications, and Decision-Making"; and "Innovation Labs: An Approach to Innovation Process Digitalization".

ODSIE 2024 recognized and awarded a prize for the best doctoral thesis in Industrial Engineering and related fields to encourage young scholars in research on Optimization and Data Science. Given the interdisciplinary nature of these fields, Ph.D. graduates from various disciplines and universities worldwide were invited to participate.

December 2024

A. Mirzazadeh
Zohreh Molamohamadi
Erfan Babaee Tirkolaee
Gerhard-Wilhelm Weber
Kathryn E. Stecke

Organization

Conference Chairs

Abolfazl Mirzazadeh	Kharazmi University, Iran/International Scientific Services (RefConf), Oman
Erfan Babaee Tirkolaee	İstinye University, Turkey
Saliha Karadayi-Usta (Co-chair)	İstinye University, Turkey

İstinye University

Erkan Ibiş	Rector
Hatice Gülen	Vice Rector
Mehmet Alper Tunga	Dean of the Faculty of Engineering and Natural Sciences

Conference Coordinators

Leila Chehreghani	Ulster University, UK/International Scientific Services (RefConf), Oman
Zohreh Molamohamadi	Kharazmi University, Iran/International Scientific Services (RefConf), Oman
Saina Sahragard	International Scientific Services (RefConf), Oman
Elahe Reeyazati	International Scientific Services (RefConf), Oman

Technical Program Committee Chairs

Gerhard-Wilhelm Weber	Poznań University of Technology, Poland
Kathryn Stecke	University of Texas at Dallas, Naveen Jindal School of Management, USA
Zohreh Molamohamadi	Kharazmi University, Iran/International Scientific Services (RefConf), Oman

Steering Committee Members

A. Mirzazadeh	Kharazmi Uniersity/International Scientific Services (RefConf), Iran
Erfan Babaee Tirkolaee	İstinye University, Turkey
Saliha Karadayi-Usta	İstinye University, Turkey
Janny M. Y. Leung	University of Macao, China
Tatiana Tchemisova	University of Aveiro, Portugal
Chefi Triki	University of Kent, UK
Mehmet Alper Tunga	İstinye University, Turkey

Ph.D. Thesis Competition Jury

Saliha Karadayı Üsta (Chair)	İstinye University, Turkey
Emre Çakmak (Co-chair)	İstinye University, Turkey
Noyan Sebla Sezer (Member)	İstinye University, Turkey

Technical Program Committee Members

Maria Grazia Speranza	University of Brescia, Italy
Alexandre Dolgui	IMT Atlantique, France
Michael Pecht	University of Maryland, USA
Josef Jablonsky	Prague University of Economics and Business, Czech Republic
Ruben Ruiz Garcia	Polytechnic University of Valencia, Spain
Michael G. Kay	North Carolina State University, USA
Sırma Zeynep Alparslan Gök	Süleyman Demirel University, Turkey
Elif Kılıç Delice	Ataturk University, Turkey
Leopoldo Eduardo Cárdenas-Barrón	Tecnológico de Monterrey, Mexico
Paulina Golinska	Poznań University of Technology, Poland
Sankar Kumar Roy	Vidyasagar University, India
Sadia Samar Ali	King Abdulaziz University, Saudi Arabia
Serap Ergun	Isparta University of Applied Sciences, Turkey
Eloisa Macedo	University of Aveiro, Portugal
Aybike Özyüksel Çiftçioğlu	Manisa Celal Bayar University, Turkey
Dinh Tran Ngoc Huy	International University of Japan, Japan
Svetlana Rastvortseva	HSE University Moscow, Russia
Hakan Gultekin	Sultan Qabus University, Oman
Taicir Moalla Loukil	University of Sfax, Tunisia

Safa Bhar Layeb	University of Tunis El Manar, Tunisia
Marwa Hasni	University of Sfax, Tunisia
Adeyinka Peter Ajayi	Redeemer's University, Nigeria
İlkay Saraçoğlu	Halic University, Turkey
Alireza Goli	University of Isfahan, Iran
Nazanin Pilevari	Islamic Azad University, West Tehran Branch, Iran
Mostafa Hajiaghaei-Keshteli	Tecnológico de Monterrey, Mexico
Sarfaraz Hashemkhani Zolfani	Universidad Católica del Norte, Chile
M. Irfan Uddin	Kohat University of Science and Technology, Pakistan
Alireza Amırteımoorı	İstinye University, Turkey
Amin Hosseinian Far	University of Hertfordshire, UK
Saliha Karadayi-Usta	İstinye University, Turkey
Tofigh Allahvıranloo	İstinye University, Turkey
Vladimir Simic	University of Belgrade, Serbia
Majid Tavana	La Salle University, USA
Murat Yeşilkaya	Tokat Gaziosmanpaşa University, Turkey
Ezgi Özer	Piri Reis University, Turkey

Keynote Speakers

Mostafa Hajiaghaei-Keshteli	Tecnológico de Monterrey, Mexico
Sadia Samar Ali	King Abdulaziz University, Saudi Arabia

Workshop Organizers

Meenakshi Kaushik	TIIPS, Guru Gobind Singh University, India
Ezgi Özer	Piri Reis University, Turkey
Vuppulapati Chandra Sekhar Naidu	Coforge, USA
Vasanthi Govindaraj	National General (An Allstate Company), USA
Ali Bagheri	SRBIU, Iran
Sanjib Biswas	Amity University Kolkata, India

Executive Committee Members

Amirhossein Sabri	Kharazmi University, Iran
Şura Canik	İstinye University, Turkey

İrem Yılmaz   İstinye University, Turkey
Alper Koçak   İstinye University, Turkey
Merve Taştan   İstinye University, Turkey

Reviewers

Aybike Özyüksel Çiftçioğlu   Manisa Celal Bayar University, Turkey
Alireza Goli   University of Isfahan, Iran
Harshavardhan Yedla   Persistent Systems, USA
Aditi Srivastava   ABES Business School, India
Sonia Nasri   LARODEC, Higher Institute of Management of
                                     Tunis/Manouba University, Tunisia
Ammar Odeh   Princess Sumaya University for Technology,
                                     Jordan
Rachid Ouchekh   Lund University, Sweden
Binnur Demir Erdem   Near East University, Cyprus
Seyed Ahmad Ghasemi   Tehran Institute of Humanities and Social Studies
                                     of ACECR, Iran
Vasanthi Govindaraj   National General (An Allstate Company), USA
Alper Bora Koçak   İstinye University, Turkey
Hela Limam   University of Tunis El Manar, Tunisia
Anak Agung Gde Satia Utama   Universitas Airlangga, Indonesia
Shahid Amin   Institute of Technology and Management,
                                     Gwalior, India
Firas Ali   Northern Technical University, Iraq
Sajid Ali   Quaid-i-Azam University, Pakistan
Sita Ram Sharma   Chitkara University, India
Wided Oueslati   Ecole Supérieure de Commerce de
                                     Tunis/Université de La Manouba, Laboratoire
                                     BESTMOD, Tunisia
Mehmet Kabak   Gazi University, Turkey
Serap Ergün   Isparta University of Applied Sciences, Turkey
Sapthami Iraganaboina   Jawaharlal Nehru Technological University
                                     Hyderabad, India
Farnaz Javadi Gargari   Alzahra University, Iran
Osama Hashmi   Shaheed Zulficar Ali Bhutto University, Pakistan
Nguyễn Thành Luân   Ho Chi Minh City University of Foreign
                                     Languages-Information Technology, Vietnam
Ali Asghar Rahmani Hosseinabadi   University of Regina, Canada
Marilisa Botte   Federico II University of Naples, Italy
Vuppulapati Chandra Sekhar   Coforge, USA
    Naidu

Saptadeep Biswas	National Institute of Technology Agartala, India
Hamed Shakerian	Islamic Azad University, , Yazd Science and Research Branch, Iran
Vasudeva G	Dayananda Sagar Academy of Technology and Management, India
Meenakshi Kaushik	TIIPS, Guru Gobind Singh University, India
Gerhard-Wilhelm Weber	Poznań University of Technology, Poland
Mohamed El Merouani	Abdelmalek Essaadi University, Morocco
Sherzod Ibodulloev	TIIAME National Research University, Uzbekistan
Sohit Reddy Kalluru	Buffalo State University, USA
Md. Ashraful Islam	University of Information Technology & Sciences, Bangladesh
Simerjeet Singh Bawa	Chitkara Business School, Chitkara University, India
Viraj Lele	DHL Supply Chain, Canada
Fatereh Sadat Moosavi	Islamic Azad University, Iran
Mouad El Mesoudy	Sidi Mohamed Ben Abdellah University, Morocco
Gopalakrishnan Chinnasamy	Jain Deemed to be University, India
Zohreh Molamohamadi	Kharazmi University, Iran
Erfan Babaee Tirkolaee	İstinye University, Turkey
Alexandre Dolgui	IMT Atlantique, France
Josef Jablonsky	Prague University of Economics and Business, Czech Republic
Ruben Ruiz Garcia	Polytechnic University of Valencia, Spain
Michael Kay	North Carolina State University, USA
Maria Grazia Speranza	University of Brescia, Italy
Michael Pecht	University of Maryland, USA
Sırma Zeynep Alparslan Gok	Suleyman Demirel University, Turkey
Elif Kılıç Delice	Ataturk University, Turkey
Leopoldo Eduardo Cárdenas-Barrón	Tecnológico de Monterrey, Mexico
Tatiana Tchemisova	University of Aveiro, Portugal
Paulina Golinska	Poznań University of Technology, Poland
Sankar Kumar Roy	Vidyasagar University, India
Sadia Samar Ali	King Abdulaziz University, Saudi Arabia
Serap Ergun	Isparta University of Applied Sciences, Turkey
Eloisa Macedo	University of Aveiro, Portugal
Dinh Tran Ngoc Huy	International University of Japan, Japan
Svetlana Rastvortseva	National Research University Higher School of Economics, Moscow, Russia
Hakan Gultekin	Sultan Qabus University, Oman
Taicir Moalla Loukil	University of Sfax, Tunisia

Safa Bhar Layeb	University of Tunis El Manar, Tunisia
Marwa Hasni	University of Tunis El Manar, Tunisia
Adeyinka Peter Ajayi	Redeemer's University, Nigeria
İlkay Saraçoğlu	Halic University, Turkey
Nazanin Pilevari	Islamic Azad University, West Tehran Branch, Iran
Mostafa Hajiaghaei-Keshteli	Tecnológico de Monterrey, Mexico
Safarnaz Hashemkhani	Universidad Católica del Norte, Chile
M. Irfan Uddin	Kohat University of Science and Technology, Pakistan
Eren Özceylan	Gaziantep University, Turkey
Erfan Babaee Tirkolaei	İstinye University, Turkey
Reza Kiani Mavi	Edith Cowan University, Australia
Sara Ahmed	Concordia University, Canada
Dalia Streimikiene	Vilnius University, Lithuania
Mamdouh El Haj Assad	University of Sharjah, UAE
Nancy-Paulina Carreón	Webasto Group, USA
Muhammad Muzammal	Bahria University, Pakistan
Mazdak Khodadadi-Karimvand	University of Science and Culture, Iran
Mehdi Gheisari	Southern University of Technology and Science, China
Nadi Serhan Aydin	İstinye University, Turkey
Siamak Naderi	University of Warwick, UK
Hamiden Khalifa	Cairo University, Egypt
Mohammad Shamsuddoha	University of Chittagong, Bangladesh
Stephen Cahoon	University of Tasmania, Australia
Manvinder Pahwa	Manipal University Jaipur, India
Shalendra Rao	Mohanlal Sukhadia University, India
Chanicha Moryadee	Suan Sunandha Rajabhat University, Thailand
Jan Medlock	Oregon State University, USA
Qimin Huang	Case Western Reserve University, USA
Parag Siddique	University of Louisville, USA
Yulin Sun	Southwestern University of Finance and Economics, China
Agnes Nalini Vincent	AMITY Global Business School, Mauritius

Scientific Sponsors

Kent Business School, UK

Prague University of Economics and Business, Czech Republic

The University of Texas at Dallas, USA

Universidad Católica del Norte Chile

Isparta University of Applied Sciences, Turkey

University of Sfax, Tunisia

Kohat University of Science and Technology, Pakistan

Manisa Celal Bayar University, Turkey

Piri Reis University, Turkey

Contents

Artificial Intelligence and Expert Systems in Engineering

Evaluation of Large Language Models in Annotating Metaphorical
Sentences 3
    Figen Eğin, Aytuğ Onan, and Hatice Yıldız Durak

Machine Learning-Based Predictive Approach for the Vessel's Trajectory
Using Automatic Identification System and Meteorological Data 17
    Houyem Mjadri, Wided Oueslati, and Afef Bahri

Harnessing Artificial Intelligence for Proactive Cybersecurity
and Operational Optimization in Business Processes 36
    *Ammar Odeh, Walid Salameh, Anas Abu Taleb, Qasem S. Abu Al-Haija,
    and Tareq Alhajahjeh*

Digital Technologies in Healthcare Systems

Integrative Review of Machine Learning and Deep Learning Approaches
for Cardiovascular Disease Detection, Classification, and Prediction 51
    Muhammad Anas, Saeid Nahavandi, and Jingxin Zhang

Psychographic Profiles of Healthcare Professionals and Their Impact
on Prescribing Behaviour 69
    Kashif Pervaiz and Simerjeet Singh Bawa

The Impact of Artificial Intelligence-Driven Digital Marketing Strategies
on Pharmaceutical Consumer Behavior 91
    Kashif Pervaiz and Simerjeet Singh Bawa

The Impact of Artificial Intelligence on Medical Professional Language
and Doctor-Patient Interactions 106
    Albena Dobreva

A New Hybrid Algorithm to Diagnose MS Using MRI Image Processing 119
    *Maryam Oghbaei, Ali Asghar Rahmani Hosseinabadi,
    and SeyedSaeid Mirkamali*

Comparative Approach of Deep Learning Methods for Predicting
Functional Outcomes After Stroke 137
    Hela Limam and Eya Jouini

Smart Supply Chain and Manufacturing Systems

Leveraging Supply Chain Analytics in the Era of Industry
4.0: A Comprehensive Guide to Data-Driven Optimization
and Decision-Making 153
    Oussama Zabraoui, Anas Chafi, and Salaheddine Kammouri Alami

A Case Study on Traffic Forecasting for Supply Chain Management:
Optimization of the Long Short-Term Memory Model 169
    Vasanthi Govindaraj and Prakash Periyasamy

Defining Key Competencies for Smart and Sustainable Humanitarian
Logistics 179
    Zeynep Yüksel, Dursun Emre Epcim, Süleyman Mete, and Eren Özceylan

Smart Transportation and Logistics Systems

Integrating Pedestrian and Scooter Traffic: A Model for Safe Urban
Mobility 197
    Serap Ergün

A Study on Optimizing Training Time for Traffic Speed Estimation Using
Graph Neural Networks 217
    Serap Ergün

Optimization of Signalized Intersections: Analyzing Autonomous Vehicle
Behaviors Through Data-Driven Simulations 232
    Syed Shah Sultan Mohiuddin Qadri, Mustafa Albdairi, Ali Almusawi,
    Ahmet Kabarcik, and H. S. Abdulrahman

Integrating Machine Learning with Optimization to Solve Time-Dependent
Cash in Transit Vehicle Routing Problem with Time Windows 245
    Alev Taskin, Aslihan Sagiroglu, Ezgi Zehra Seker,
    Melisa Caliskan Demir, and Khaled Dandis

Creation of a Fixed Wing UAV Based on Ardupilot Firmware
with Elements of Radio Interference Countermeasure System 265
    Serhii Lienkov, Alexander Myasischev, Vadym Ovcharuk,
    Oleksandr Sieliukov, Nataliia Lytvynenko, and Vitalii Tarhonskyi

Optimizing Vehicle Routing in the Dial-a-Ride Problem Using Deep
Q-Networks 281
    Mariem Ayari, Sonia Nasri, Hend Bouziri, and Wassila Aggoune-Mtalaa

Data Analytics and Advanced Optimization

Computational Approach for Solving Fredholm Integral Equations 303
  Saeed Hatamzadeh and Zahra Masouri

Leveraging General Unary Hypotheses Automaton for Automating
Technical Analysis: A Case Study on Bitcoin Prices 323
  Jakub Neugebauer

Data Science Techniques for Opinion Mining in Industrial Applications 340
  Mehdi Belguith and Chafik ALoulou

Investigating the Performance of Recently Proposed Metaheuristic
Optimization Algorithms on Real-World Engineering Design Problems 355
  Alper Buğra Polat, Elif Varol-Altay, and Osman Altay

Comparison of Feasible Initial Solution Methods for the Knapsack Problem ... 369
  Halil İbrahim Ayaz, Belkız Torğul, and Turan Paksoy

A Comparative Approach to Multi-response Optimization for AWJM
of Monel 400 Alloys Using Hybrid Metaheuristic-MCDM Techniques 389
  Ayyappan Solaiyappan and Sivakumar Mahalingam

Internet and Mobile Technologies

Developing Users' Acceptance Framework towards Mobile Commerce:
An Exploratory Study 421
  Arshan Kler, Bhupinder Preet Bedi, and Simerjeet Singh Bawa

A Smartphone-Based Solution for Enhancing Interactive Learning
Experiences in Early Childhood Education Using Near Field
Communication and Holographic Modules 437
  *Maria Seraphina Astriani, Bayu Prakoso Dirgantoro,
  Jude Joseph Lamug Martinez, and Lee Huey Yi*

Author Index 449

Artificial Intelligence and Expert Systems in Engineering

Evaluation of Large Language Models in Annotating Metaphorical Sentences

Figen Eğin[1,2(✉)] [iD], Aytuğ Onan[1,2] [iD], and Hatice Yıldız Durak[1,2] [iD]

[1] Department of Computer Engineering, İzmir Katip Çelebi University, İzmir, Turkey
D230245009@ogr.ikc.edu.tr
[2] Faculty of Education, Necmettin Erbakan University, Konya, Turkey

Abstract. The rapid advancements in large language models (LLMs) have significantly influenced various natural language processing tasks, yet the detection of metaphorical language remains a challenging area. This study investigates the capabilities of prominent LLMs, specifically OpenAI's GPT and Google's Gemini, in identifying metaphorical sentences within Turkish texts. Initially, a dataset comprising metaphorical and literal sentences was curated from diverse sources, including literary works and daily communications, to enrich the linguistic context. Subsequently, this dataset was annotated independently by two human experts and the aforementioned LLMs, employing a zero-shot classification approach. The reliability of these annotations was quantified using Cohen's Kappa, revealing a high agreement among human annotators ($\kappa = 0.986$) but considerably lower consistency between the LLMs ($\kappa = 0.123$). The dataset generated through the labeling process was further evaluated using machine learning models. The findings highlight the current limitations of LLMs in processing metaphorical language, underscoring the need for enhanced model training and algorithm refinement. This study not only contributes to the understanding of LLMs' performance in metaphor detection but also suggests pathways for future enhancements to achieve more nuanced language comprehension in AI systems. The broader implications of these findings suggest potential improvements in applications ranging from machine translation to semantic analysis.

Keywords: GPT · Gemini · Metaphorical Language · Large Language Models · Natural Language Processing

1 Introduction

With the recent stunning developments in the field of Natural Language Processing (NLP), Large Language Models (LLMs) exhibit very high performance in tasks such as understanding and producing natural language [1]. GPT completes some tasks, such as systematic review, at a "human-like" level [2], while it performs better than humans in some text classification tasks [3]. On the other hand, we are not yet at the desired point in capturing metaphorical meaning [4]. Metaphoric words add richness and depth to the language. Perhaps because of these features, they are frequently encountered in many areas, from literary works to daily correspondence. Metaphorical meaning emerges when

A. Mirzazadeh et al. (Eds.): ODSIE 2024, CCIS 2482, pp. 3–16, 2026.
https://doi.org/10.1007/978-3-031-93601-2_1

words differ from their original meaning in the dictionary. Although metaphoric words enhance the richness of the language, they are problematic in capturing the real meaning in machine translations. GPT makes the mistake of defining sequences containing metaphorical meaning word for word [4]. Although the inconsistency of metaphorical words with the general text is detected in existing methods, the contribution of metaphoric word groups to the text is ignored [5]. Therefore, studies should be conducted in this area to obtain richer contextual representations [6]. In order to correctly evaluate the contribution of metaphorical meaning to the text, determining whether the text contains metaphoric meaning is the first step of the process. The studies of the Turkish language for the detection of metaphoric meaning are limited, and a comprehensive data set is needed.

Recent advancements in Natural Language Processing (NLP) have propelled Large Language Models (LLMs) to the forefront of the field, demonstrating exceptional capabilities in understanding and generating human-like text. Models such as OpenAI's GPT and Google's Gemini are now routinely employed across a variety of language-based tasks with significant success. However, despite their wide-ranging utility, these models encounter notable challenges in handling metaphorical language, a fundamental aspect of linguistic creativity and subtlety.

Metaphorical language, characterized by its use of figurative expressions to convey meanings that diverge from their literal interpretation, poses unique complexities. This linguistic feature enriches texts by infusing depth and emotion but simultaneously introduces ambiguity that challenges conventional NLP models. The detection and correct interpretation of metaphors are crucial for tasks ranging from machine translation to sentiment analysis, where understanding the underlying connotations of language can significantly alter outcomes. This research specifically addresses the gap in LLMs' capabilities to accurately detect and interpret metaphorical sentences in Turkish—a language rich in metaphorical expressions and idiomatic usage. The limited focus on non-English languages in current LLM research further underscores the importance of this study. By evaluating the performance of leading models like GPT and Gemini against a specially curated dataset of metaphorical and literal sentences, this study aims to shed light on the strengths and weaknesses of current approaches and suggest directions for future advancements. The outcomes are expected to contribute to the development of more sophisticated, culturally aware language models that can navigate the nuanced landscape of human language more effectively.

This study aimed to create a Turkish metaphorical meaning data set in the first stage. In the second stage, to evaluate the point reached by the performances of LLMs-based models in the detection of metaphorical meaning, the success of GPT and Gemini in detecting sentences containing metaphoric meaning was investigated. In this context, the research questions are as follows:

1. How is the agreement between human annotators in the detection of metaphorical sentences?
2. How do GPT and Gemini compare with human annotators in detecting metaphorical sentences?

In the following section of the article, previous studies in the context of metaphorical language perception and natural language processing are reviewed. The third section explains the process of creating the dataset used in the study and the labeling methods in detail. In the fourth section, the findings are presented and interpreted. In the last section, the results are evaluated, and suggestions are made for future research.

2   Related Research

Metaphorical meaning occurs when words express a different meaning than their original meaning [7]. Metaphorical expressions are used quite frequently for different purposes. According to Lakoff and Johnson, metaphorical meaning allows an abstract concept to be understood through a concrete concept, thus deepening the meaning [8]. Therefore, metaphorical expressions' contribution to the text's meaning is essential. Studies on metaphorical meaning in Natural Language Processing are familiar [9]. However, very high accuracy values have not yet been achieved. Ottolina et al. [10] found that using different static word embedding methods affected the metaphorical detection task and that some temporary word embeddings performed slightly better than static methods. They also emphasized the interaction between temporal language development and metaphorical meaning detection. They stated that future studies should create a new dataset that better represents this interaction.

Another study [11] highlighted that the meaning intended to be expressed by metaphorical words. They found that the real meaning can be expressed subtly by combining the everyday use of words and natural visualizability features called GloVe embedding and visual embedding. The literature shows that higher accuracy values emerged in studies that aim to detect metaphorical meaning and use deep learning [12]. Junaid et al. [13] conducted a study to determine whether transformers work meaningfully not only for word meaning classification but also for metaphorical language classification; they fine-tuned transformer architecture-based models such as LSTM, Bi-LSTM models, BERT, Roberta, and XLNet. As a result of the research, the highest accuracy value occurred with Roberta at 81%. Using a different approach, Li et al. [14] explained the literal meaning in the training set to model the word's primary meaning. They compared it with the contextual meaning in the target sentence to identify metaphorical words and achieved 1.0% better results than the state-of-the-art method in the f1 score. Another study that used the BiLSTM-BERT model achieved an 82.3% f1 score in metaphor detection, and high performance occurred by adding visibility embeddings to the model [15]. Metaphorical meaning is sometimes given through memes. Yang et al. [18] studied the metaphorical contribution of memes used on social media. Using a non-autoregressive text generation method, they converted images to text and created the Met-Meme dataset. In their experiments on this dataset, the weighted F1 improved by 1.95%, 1.55%, and 1.72% compared to other models.

Metaphor detection remains a critical area of NLP and is still being treated as a challenging problem [19]. Models that use linguistic metaphor recognition theories have also been employed to detect metaphors. SC-Net, a multimodal metaphor detection model, exploits the internal semantic conflicts in metaphors. SC-Net has achieved high performance in experiments with Bilingual Met-mem, MultiBully, Chinese, and English datasets [20]. Another study using linguistic theories used GloVe and ELMo embedding to express the literal meaning with the intrinsic visualizability properties of words. The position information of target verbs was used as input to obtain the contextual meaning more accurately. This approach improved performance when adjectives, adverbs, and nouns were selected as target words [11]. Although linguistic approach is important, meaning does not always contribute to performance. Hilton et al. [21] find that disambiguation of word meaning degrades the performance of their algorithm for metaphor identification in cybersecurity discourse. Nevertheless, their lightweight algorithm performs similarly to existing, more complex counterparts. Metaphor detection studies should be addressed in different languages and contexts. In this regard, metaphor detection in texts produced by individuals in a psychological experiment was studied by Panicheva et al. [22]. The researchers studied conceptual metaphors with Russian datasets MetPersonality and RusPersonality and obtained acceptable results. Metaphors are a phenomenon that should also be considered in sentiment analysis. In their approach called Marketplace Sentiment Analysis (MEMSA), a study that draws attention to this added the emotion that metaphors add to the text to their evaluation [23].

From an ontological perspective, human reasoning processes are still insurmountable for AI. The creativity that humans display in interpreting language beyond the perception of texts cannot be achieved by AI. Metaphor detection remains a critical area of NLP and is still being treated as a challenging problem [19].

3   Method

The method of the research includes creating the dataset, labeling the dataset by artificial intelligence and human annotators and comparing the results, and additionally presenting the performances obtained from machine learning models on the labeled dataset.

3.1   Dataset

Due to the need for a comprehensive metaphorical meaning dataset for the Turkish language, a dataset containing metaphorical and literal sentences was created in the first stage of the research. Sentences with and without metaphorical meanings were collected from different sources to create the dataset. Different sources, such as news sites, PDF books, and Turkish dictionaries, were used at this stage to avoid being confined to a single field, such as literary language or daily use, and to ensure diversity. There are 8738 examples in the dataset (Table 1).

Table 1. Sources of the dataset.

Source	Count
Misalli Büyük Türkçe Sözlük (Big Turkish Dictionary with Examples)	3302
News Sites	4708
Other internet source	728

Big Turkish Dictionary with Examples provides structured and generally formal metaphorical expressions, while news sites contribute to more up-to-date usage. Other internet sources include sentences created by users and contribute to the dataset to represent daily metaphor use. When the dataset is evaluated regarding the sources used, it does not contain a balanced distribution. Although this situation does not affect the labeling process of human and large language models, since it does not include the training phase of the model. But it may affect the model performance if the dataset is worked with machine learning and deep models.

A more detailed analysis was obtained by examining the frequencies of the words in the data set. The most frequently used words in the data set are general expressions that do not contain metaphorical meaning. Total words and unique word counts in metaphorical and non-metaphorical sentences were shown in Table 2.

Table 2. Total and Unique Word Counts of the Dataset.

	Label Metaphorical (0) Not Metaphorical (1)	Gemini	GPT	Human1	Human2
Total Word	0	63390	63518	15635	68718
	1	28950	28822	76705	23622
Unique Word	0	18589	18661	7061	21407
	1	12995	12976	25151	11511

3.2 Annotation of the Dataset and Comparing the Results

Each example in the dataset was labeled as "metaphorical (1)" and "not metaphorical (0)" by two separate experts, GPT and Gemini, and the Cohen's Kappa and Fleiss' Kappa test values between the labels were examined.

The labeling made by GPT and Gemini was done using the Python programming language, by zero-shot, that is, by entering a prompt for each sentence without giving any examples.

The principle followed in creating prompts is as follows:

"Your task is to determine whether the given sentence contains a metaphorical meaning. Sentence: {sentence}"

Examples of the prompts are shown below:

"Your task is to determine whether the given sentence contains a metaphorical meaning. Sentence:They applied different exams to test his knowledge."

"Your task is to determine whether the given sentence contains a metaphorical meaning. Sentence:He doesn't have the courage to do something, he made so many mistakes that he is ashamed to propose."

"Your task is to determine whether the given sentence contains a metaphorical meaning. Sentence:Unfortunately, the number of victims may increase."

For annotation by Gemini, the google.generativeai library and as the model architecture gemini-pro was used. For labeling by GPT, the openai library was used and gpt-3.5-turbo-instruct was selected as the model architecture. The temperature parameter was set to 0.5 to balance determinism and randomness. During annotation, to limit the model's output length, max_tokens parameters in ChatGPT and max_output_tokens parameters in Gemini were set to 50. In addition, using top-k $= 40$ and top-p $= 0.95$ parameters, only the 40 words with the highest probability were considered while creating the model outputs. A limited selection was made, with words covering 95% of the total probability, thus obtaining meaningful and consistent results.

The answers returned for each sentence from GPT, and Gemini were collected and stored. After the data set annotation, the agreement between the annotators was evaluated. The agreement levels in Table 3 suggested by Landis and Koch [16] were used in the interpretation of the κ values obtained as a result of Cohen's Kappa test and Fleiss' Kappa test performed to determine the agreement between the annotators.

Table 3. Intervals for Interpreting Kappa Statistic Values.

κ values	Interpretation
<0	Worse than what would be expected by chance
0.01–0.20	Insignificant agreement
0.21–0.40	Poor agreement
0.41–0.60	Moderate agreement
0.61–0.80	Good agreement
0.81–1.00	Very good agreement

A section of examples with different and similar labels in the dataset is presented in Table 4.

Table 4. A section from the dataset.

Sentence	Gemini	GPT	Human1	Human2
More than 320 people and 130 vehicles were involved in the firefighting efforts, while water was seen being thrown from three helicopters onto the burning building	0	0	0	0
The study in question revealed that H1821 + 643, located in a galaxy cluster, has less impact on its surroundings despite its mass	1	1	0	0

(continued)

Table 4. (*continued*)

Sentence	Gemini	GPT	Human1	Human2
Messages of condemnation came from countries regarding the attack	0	0	0	0
It is a human duty to help people in trouble	0	1	1	1
He suddenly stopped talking and left the podium	0	0	1	1
The ministry invited the international community to help Ukraine	0	1	0	0
Trump cannot pay the fine imposed in the fraud case	1	0	0	0
Most of the people walking around the market are nothing but dry crowds	1	1	1	1

4   Results

4.1   Human and LLMs Annotations Results

In the human datasets, the examples labeled as "metaphoric" and "not metaphoric" are distributed closely, although not balanced. The number of labels made as "metaphoric" by GPT is lower than the others. Gemini marked most sentences as metaphorical. The labels of the dataset are presented in Table 5.

Table 5. Distributions of labels in the dataset.

Label	Metaphoric (1)	Not Metaphoric (0)
Human 1	4305	4433
Human2	4267	4471
GPT	3247	5491
Gemini	7297	1441

In detecting metaphorical sentences, the agreement between human annotators was first measured. The result of the Cohen's Kappa test between human annotators, the κ value (0.986; $p < 0.001$), indicates excellent agreement between the two human annotators. The κ value between GPT and Gemini was calculated as 0.123; $p < 0.001$. This result shows an insignificant level of agreement between the two LLMs. The κ values indicating the agreement between GPT and Humans were (0.507; $p < 0.001$) and (0.507; $p < 0.001$). A moderate level of agreement existed between GPT and both human annotators. The κ values between Gemini and Human 1 and Human 2 were (0.232; $p < 0.001$) and (0.229; $p < 0.001$). There needed to be a more substantial level of agreement between Gemini and both human annotators. The κ values between the four labelers were calculated as 0.395 by Fleiss' Kappa test. A moderate level of agreement existed between four annotators.

Table 6 below shows the κ values calculated for each pairwise and for the four inter-pair agreement measurements between the two labeling experts, GPT and Gemini.

Table 6. κ Values (p < 0.001).

Annotators	κ Values
Human1 - Human 2	0.986*
GPT - Gemini	0.123*
GPT - Human 1	0.507*
GPT - Human 2	0.507*
Gemini - Human 1	0.232*
Gemini - Human 2	0.229*
Human 1 - Human 2 - Gemini - ChatGPT	0.395*

Although agreement between human annotators is high, agreement between Gemini and other annotators is relatively poor (See Fig. 1).

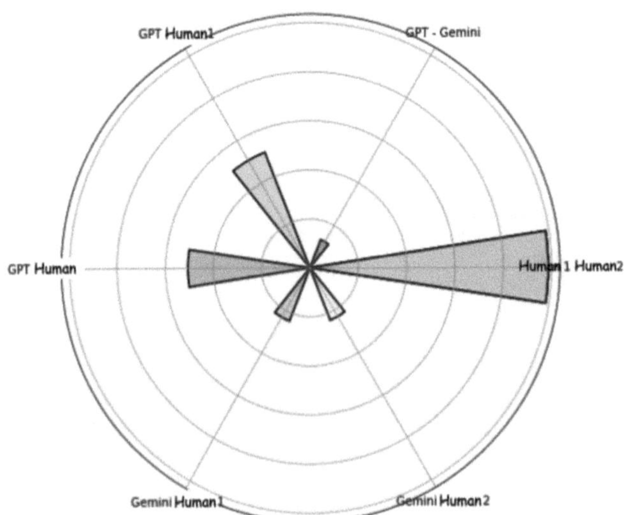

Fig. 1. κ Values between annotators.

Human annotators or GPT did not metaphorically label most examples in the dataset that were considered metaphorical by Gemini. Many examples labeled by both human coders were labeled as "not metaphorical" by GPT. Data for each annotator in the dataset are presented in Fig. 2.

Gemini evaluated many sentences that did not contain metaphorical meanings as metaphorical. The reason why only the sentence "Apart from this, we paid 72 million

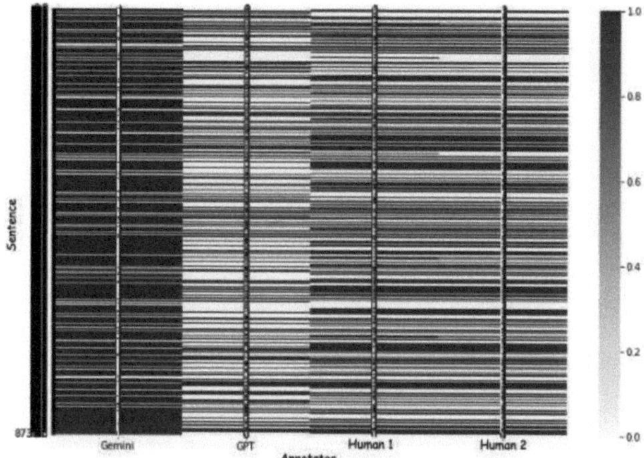

Fig. 2. Metaphorical meaning heat map.

liras to the earthquake victims in Kilis as rent assistance", which was annotated as metaphorical by Gemini, was tagged as metaphorical is that "The term rent assistance is used metaphorically. This expression does not mean that direct cash payments are made to earthquake victims to find homes. Instead, it refers to assistance provided by the government to cover the rental costs of earthquake victims' existing homes." In another example that Gemini and GPT annotated as metaphorical, the phrase "…the ambulance was sent." was annotated as metaphorical because it implies that ambulances and emergency services personnel were "allocated" or "assigned" to the scene, rather than "sent" to the scene.

Based on human annotator labeling, GPT cannot detect many sentences with metaphorical meaning. Correspondingly, studies on the ability of GPT to detect metaphorical meaning have yet to achieve very high accuracy. Wachowiak and Gromann [4] tested the ability of GPT to detect metaphorical language through experimental studies without predetermined fields. Their study with English and Spanish datasets showed that 65.15% accuracy was achieved in English and 34.65% in Spanish.

On the other hand, Gemini, when based on labeling human annotators, is seen to evaluate sentences that do not contain metaphoric meaning as metaphorical. Although no study in the literature examines Gemini's capabilities in detecting metaphorical meaning, some studies examine idiom datasets and prompt engineering [17]. In general, it would be appropriate to consider situations such as context and the double meaning of the sentence in detecting metaphorical expressions.

4.2 Machine Learning Models' Results

The Turkish metaphorical meaning dataset was used to predict metaphorical sentences with Logistic Regression (LR), Support Vector Machine (SVM) and Random Forest (RF) models. The models were trained and tested separately based on labeling by ChatGPT, Gemini, H1 and H2. The results are presented in the tables below (Tables 7, 8, 9 and 10).

Table 7. Dataset labeled by ChatGPT.

Model	Accuracy	Precision (0)	Recall (0)	F1-score (0)	Precision (1)	Recall (1)	F1-score (1)
LR	85.58	0.60	0.04	0.07	0.86	1.00	0.92
RF	85.47	0.53	0.04	0.07	0.86	0.99	0.92
SVM	85.18	0.36	0.02	0.04	0.86	0.99	0.92

Table 8. Dataset labeled by Gemini.

Model	Accuracy	Precision (0)	Recall (0)	F1-score (0)	Precision (1)	Recall (1)	F1-score (1)
LR	71.28	0.74	0.84	0.79	0.64	0.49	0.55
RF	66.88	0.71	0.82	0.76	0.56	0.41	0.47
SVM	71.57	0.75	0.83	0.79	0.63	0.52	0.57

Table 9. Dataset labeled by Human 1.

Model	Accuracy	Precision (0)	Recall (0)	F1-score (0)	Precision (1)	Recall (1)	F1-score (1)
LR	85.47	0.84	0.88	0.86	0.87	0.83	0.85
RF	76.37	0.81	0.70	0.75	0.73	0.83	0.77
SVM	87.47	0.87	0.88	0.88	0.88	0.86	0.87

Table 10. Dataset labeled by Human 2.

Model	Accuracy	Precision (0)	Recall (0)	F1-score (0)	Precision (1)	Recall (1)	F1-score (1)
LR	84.78	0.84	0.87	0.86	0.86	0.82	0.84
RF	76.32	0.82	0.69	0.75	0.72	0.84	0.77
SVM	86.78	0.87	0.88	0.87	0.87	0.86	0.86

In the dataset obtained by labeling ChatGPT, Logistic Regression and Random Forest SVM models worked with approximately 85% accuracy. However, all three models have low recall values for the "not a metaphor" label. This shows that the model cannot correctly recognize non-metaphorical sentences. In general, metaphorical sentences (1) were predicted reasonably well, while non-metaphorical sentences could not be predicted.

In the dataset consisting of labeling by Gemini, Logistic Regression showed the best performance and was successful with 71% accuracy. It is seen that this model provides a good recall (0.84) for non-metaphorical sentences, i.e., label 0. However, the precision and recall values for the metaphorical sentence class (1) are lower. Random Forest and SVM models have lower accuracy rates and especially fail to predict metaphorical sentences.

In the dataset formed by labeling by Human 1, it is observed that the SVM model performed the best and worked with 87% accuracy. Precision, recall, and f1-score values are pretty balanced. The logistic regression model is also quite successful, with 85% accuracy, but the Random Forest shows slightly lower performance. Especially for the label not figuratively meaningful (0), the precision value is much lower (0.73), which shows that the model cannot achieve balance.

According to the labeling made by Human 2, SVM again achieved the best result and worked with 87% accuracy. Other models are also successful, but Random Forest has the lowest accuracy. Logistic Regression and SVM models achieved similarly good results. Although Random Forest has low precision and recall values for non-figuratively meaningful sentences, SVM and Logistic Regression seem more successful in achieving this balance.

The SVM model generally has the highest accuracy rates and provides more balanced results for each labeler column. This model shows the best performance, especially in the datasets obtained with the labeling of Human 1 and Human 2.

The models were observed to perform lower in predicting non-metaphorical sentences than those with metaphorical meanings.

This study's findings reveal several key aspects of metaphor detection using large language models, particularly focusing on the Turkish language context. Our results reveal notable differences in the performance of GPT and Gemini, providing insights into the capabilities and limitations of current LLMs in handling non-literal language. Our research confirms that while LLMs like GPT and Gemini exhibit high computational efficiency, their ability to interpret metaphorical language accurately remains limited. This aligns with existing studies emphasizing the challenge of figurative language processing in AI systems. The comparison between human annotators and LLM outputs particularly underscores the need for models that better incorporate contextual understanding and linguistic nuance. The accurate detection of metaphorical language is crucial for applications such as sentiment analysis, content moderation, and educational tools, where understanding the subtleties of language significantly impacts the effectiveness of technology. Improving metaphor detection in Turkish will aid in developing more reliable AI-powered communication tools for this linguistically rich and context-heavy language. The study utilized a dataset specifically curated for metaphor detection, which, while comprehensive, also highlights the importance of diverse data sources. Future research should look towards expanding the dataset to include more varied contexts and styles of metaphor usage. One major limitation of our study is the restricted focus on only two models and a single language. Future studies could broaden this scope to include more diverse LLMs and additional languages. Additionally, enhancing the transparency of the prompts used in testing and providing a more detailed error analysis would further strengthen the reliability and applicability of the findings. Further research could also explore hybrid models that combine traditional rule-based systems with modern machine learning techniques to leverage the strengths of both approaches in metaphor detection. Theoretically, our study contributes to a better understanding of the adaptability of LLMs to complex language tasks. It suggests that while LLMs have made significant strides, there is substantial room for improvement in how these models handle the nuances of human language, especially in less commonly studied languages like Turkish. In this study, we have explored the capabilities of state-of-the-art large language models, GPT and Gemini, in detecting metaphorical language in Turkish texts. Our findings highlight the challenges these models face, underscoring the complexity of metaphor detection in NLP and the limitations of current technologies when applied to languages with rich metaphorical content like Turkish. The observed performance discrepancies between the models suggest several avenues for future investigation. Firstly, there is a clear need for models that can better integrate contextual understanding and cultural nuances, particularly for non-English languages. Future research should also consider the development of hybrid models that combine the strengths of rule-based systems and machine learning to enhance metaphor detection. Additionally, the study underscores the importance of creating and utilizing linguistically and culturally diverse datasets to train and test NLP models. The ability to accurately detect and interpret metaphorical language is crucial for improving the reliability and effectiveness of NLP applications, from automated translation services to sentiment analysis and AI-driven content creation. Enhancing metaphor detection capabilities could lead to more nuanced and contextually aware AI tools, significantly impacting sectors such as education, media, and customer service, where understanding the subtleties of language is key. Our research contributes to the ongoing dialogue on the potential and limitations of large language models in natural language processing. By illuminating the specific challenges in metaphor detection, this study not only advances our theoretical understanding of how AI interprets human language but also provides practical insights that can guide the development of more sophisticated NLP tools. We look forward to seeing how these insights will be leveraged in future NLP research and applications, driving forward the capabilities of AI to understand and interact with human language in all its complexity.

5 Conclusion

This study has made significant strides in evaluating the capabilities of large language models, such as OpenAI's GPT and Google's Gemini, to detect metaphorical language in Turkish texts. Our findings reveal that while human annotators consistently exhibit high agreement in identifying metaphors, LLMs struggle to achieve similar levels of accuracy, highlighting a fundamental challenge in current NLP technology. One of the primary limitations of this study is the reliance on a dataset that, while diverse in its sources, may not fully represent the breadth of metaphorical usage across different contexts and genres in the Turkish language. Additionally, the models tested were limited to GPT and Gemini, which, although leading technologies, do not encompass the full spectrum of LLMs available. The performance of these models might also be influenced by their training datasets, which are predominantly in English, potentially limiting their effectiveness in handling non-English metaphors. To build on the findings of this study, future research should consider expanding the dataset to include a wider variety of sources and metaphorical expressions, particularly from underrepresented genres such as poetry and political discourse, where metaphorical language is prevalent. Further investigations could also explore the adaptation of LLMs to better handle linguistic nuances by integrating multilingual training datasets or developing specific training regimes focused on figurative language. Additionally, the exploration of hybrid models that combine the strengths of rule-based and machine learning approaches could offer new insights into effective metaphor detection strategies. In conclusion, the ability of LLMs to understand and interpret metaphorical language remains an essential frontier in the development of AI that can genuinely comprehend and interact with human language in all its complexity. By addressing the highlighted limitations and exploring the suggested future research directions, the field can move closer to achieving more nuanced and context-aware NLP systems.

References

1. Patil, R., Gudivada, V.: A review of current trends, techniques, and challenges in large language models (LLMs). Appl. Sci. **14**(5), 2074 (2024)
2. Khraisha, Q., Put, S., Kappenberg, J., Warraitch, A., Hadfield, K.: Can large language models replace humans in systematic reviews? Evaluating GPT-4's efficacy in screening and extracting data from peer-reviewed and grey literature in multiple languages. Res. Synth. Methods. **15**(4), 616–626 (2024)
3. Gilardi, F., Alizadeh, M., Kubli, M.: ChatGPT outperforms crowd workers for text-annotation tasks. Proc. Natl. Acad. Sci. **120**(30), e2305016120 (2023)
4. Wachowiak, L., Gromann, D.: Does gpt-3 grasp metaphors? identifying metaphor mappings with generative language models. In: Proceedings of the 61st Annual Meeting of the Association for Computational Linguistics (2023)
5. Yang, Q., Yu, L., Tian, S., Song, J.: Word-level and phrase-level strategies for metaphorical text identification. Multimed. Tools Appl. **81**(10), 14339–14353 (2022)
6. Song, Z., Tian, S., Yu, L., Zhang, X., Liu, J.: Multi-task metaphor detection based on linguistic theory. Multimed. Tools Appl. 1–14 (2024)
7. Gürbüz, F.: Mecaz kapsamındaki anlamlar ve bu anlamlar arasındaki ince ayrıntılar. Erzincan Üniversitesi Eğitim Fakültesi Dergisi. **10**(2), 197–212 (2008)
8. Lakoff, G., Johnson, M.: Chapter x: Metaphors We Live By. University of Chicago Press. içinde kitap bölümü, ISBN: 9780226468013 (2008)
9. Huang, X., Huang, H., Xu, C., Chen, W., Wang, R.: A novel pattern matching method for Chinese metaphor identification and classification. In: Artificial Intelligence and Computational Intelligence: Third International Conference, AICI 2011, September 24–25, Taiyuan/China (2011)
10. Ottolina, G., Palmonari, M., Alam, M., Vimercati, M.: On the impact of temporal representations on metaphor detection. In: arXiv preprint arXiv:2111.03320 (2021)
11. Zhang, L., Yu, L., Tian, S., Yang, Q.: Sequential verb metaphor detection with linguistic theories. J. Intell. Fuzzy Syst. **43**(3), 2765–2775 (2022)
12. Kavitha, K., Chittineni, S.: An intelligent metaheuristic optimization with deep convolutional recurrent neural network enabled sarcasm detection and classification model. Int. J. Adv. Comput. Sci. Appl. **13**(2) (2022)
13. Junaid, T., Sumathi, D., Sasikumar, A.N., Suthir, S., Manikandan, J., Khilar, R., Raju, M.J.: A comparative analysis of transformer based models for metaphorical language classification. Comput. Electr. Eng. **101**, 108051 (2022)

14. Li, Y., Wang, S., Lin, C., Frank, G.: Metaphor detection via explicit basic meanings modelling. In: arXiv preprint arXiv:2305.17268 (2023)
15. Kehat, G., Pustejovsky, J.: Neural metaphor detection with visibility embeddings. In: Proceedings of SEM 2021: The Tenth Joint Conference on Lexical and Computational Semantics (2021)
16. Landis, J.R., Koch, G.: The measurement of observer agreement for categorical data. Biometrics. **33**, 159–174 (1977)
17. Phelps, D., Pickard, T., Mi, M., Gow-Smith, E., Villavicencio, A.: Sign of the times: evaluating the use of large language models for idiomaticity detection. In: arXiv preprint arXiv:2405.09279 (2024)
18. Yang, Q., Yan, Y., He, X., Guo, S.: Metaphor recognition based on cross-modal multi-level information fusion. Complex Intell. Syst. **11**(1), 95 (2025)
19. Skrynnikova, I.V.: Interpreting metaphorical language: a challenge to artificial intelligence. Вестник Волгоградского государственного университета. Серия 2: Языкознание. **23**(5), 99–107 (2024)
20. He, X., Yu, L., Tian, S., Yang, Q., Long, J.: SC-Net: Multimodal metaphor detection using semantic conflicts. Neurocomputing. **594**, 127825 (2024)
21. Hilton, K., Siami Namin, A., Jones, K.S.: Metaphor identification in cybersecurity texts: a lightweight linguistic approach. SN Appl. Sci. **4**(2), 60 (2022)
22. Panicheva, P.V., Mamaev, I.D., Litvinova, T.A.: Towards automatic conceptual metaphor detection for psychological tasks. Inf. Process. Manag. **60**(2), 103191 (2023)
23. Luri, I., Schau, H.J., Ghosh, B.: Metaphor-enabled marketplace sentiment analysis. J. Market. Res. **61**(3), 496–516 (2024)

Machine Learning-Based Predictive Approach for the Vessel's Trajectory Using Automatic Identification System and Meteorological Data

Houyem Mjadri[1(✉)], Wided Oueslati[2], and Afef Bahri[3]

[1] SMART Laboratory, Institut Superieur de Gestion de Tunis, Universite de Tunis, Tunis, Tunisia
`houyem.mjadri@isg.u-tunis.tn`
[2] BESTMOD Laboratory, Ecole Superieure de Commerce, Universite de la Manouba, Manouba, Tunisia
[3] SMART Laboratory, Ecole Superieure de Commerce, Universite de la Manouba, Manouba, Tunisia

Abstract. The prediction of a moving object's future position or trajectory is crucial in navigation and maritime logistics, as it directly impacts route optimization, traffic management, and collision avoidance. Precise predictions enhance operational efficiency, safety, and the ability to navigate obstacles and environmental constraints. While technologies like the Automatic Identification System (AIS) offer real-time tracking, they often overlook the influence of weather conditions on navigation. This study proposes a bidirectional deep learning model that integrates AIS data with meteorological information, accounting for both historical and future vessel trajectories. The approach improves prediction accuracy, especially under weather conditions. Evaluation of the model using metrics such as Root Mean Square Error (RMSE), Mean Squared Error (MSE), and Mean Absolute Error (MAE) highlights its potential to enhance safer navigation and more efficient maritime logistics. The results show that combining AIS and meteorological data significantly improves prediction accuracy, leading to more adaptive and reliable vessel trajectory forecasting models.

Keywords: Vessel Trajectory · Automatic Identification System Data · Meteorological Data · Trajectory Prediction · Navigation · Deep Learning

1 Introduction

Maritime transport plays a crucial role in facilitating trade, travel, and the transportation of goods across oceans. It serves as an essential means of connecting regions, promoting economic development, and supporting the exploration of remote territories. Over the years, maritime transport has undergone significant transformations to meet the growing demands of globalization [1, 2]. Among the key aspects of efficient maritime operations, the accurate prediction of ship trajectories has become a cornerstone of safe and sustainable navigation [3].

A. Mirzazadeh et al. (Eds.): ODSIE 2024, CCIS 2482, pp. 17–35, 2026.
https://doi.org/10.1007/978-3-031-93601-2_2

Despite the progress made in trajectory prediction, significant gaps remain in the existing research. Many studies primarily focus on AIS (Automatic Identification System) data but overlook the integration of dynamic environmental factors, such as meteorological and oceanographic data, which are essential for accurate predictions [4, 5]. This gap limits the ability of current models to adapt to the complexities of real-time maritime environments, highlighting the need for innovative approaches to improve prediction accuracy [3].

The maritime transport sector faces numerous challenges and opportunities as it continues to evolve in a dynamic global economy. Globalization and international trade have fueled its expansion, making maritime transport the backbone of the global commercial system. At the same time, technological advancements and automation, such as Autonomous Maritime Surface Ships (MASS) technology, have revolutionized the industry, increasing efficiency and productivity while introducing new complexities and risks. These include the need to integrate sophisticated data and make real-time decisions to ensure the safety and efficiency of maritime operations [2, 6].

Additionally, environmental regulations and sustainability initiatives have become key drivers of change, pushing for greener practices and a reduction in emissions in maritime operations [7].

Trajectory prediction plays a central role in improving safety and operational efficiency in maritime transport. It involves accurately forecasting the movement of ships based on factors such as weather conditions, navigation constraints, and traffic patterns. Accurate trajectory prediction offers several key benefits. First, it enhances safety by facilitating collision avoidance and reducing risks [4]. By predicting ship trajectories and identifying potential conflicts, operators can take proactive measures to avoid accidents. Second, trajectory prediction helps optimize ship routes and improve energy efficiency. By accounting for variables such as weather conditions, ocean currents, and traffic congestion, ships can take the most efficient routes, thereby reducing fuel consumption and emissions, in line with environmental regulations [1, 2, 8]. Lastly, accurate predictions enable better port operations and optimized resource allocation. By leveraging arrival schedules and predicted trajectories, ports can optimize the use of docks, cranes, and other infrastructure, thus enhancing overall efficiency [9].

Despite these advantages, trajectory prediction is not without its challenges [10, 11]. The availability and quality of data are crucial for developing robust prediction models. Access to real-time and historical data on ship movements, weather conditions, and other relevant parameters is essential. Existing models often rely heavily on AIS data but fail to adequately incorporate dynamic environmental factors such as meteorological and oceanographic data, which are essential for accurate predictions [12]. Addressing this gap is critical for enhancing the reliability and robustness of trajectory prediction models, ultimately improving the safety and efficiency of maritime operations.

In this study, we propose an innovative approach that integrates AIS data with meteorological information to improve the accuracy of trajectory predictions. Using a model based on Bi-GRU, we aim to capture both short-term and long-term dependencies in sequential data, taking into account the nature of maritime environments. Our experimental results demonstrate the effectiveness of this approach, with significant improvements in prediction accuracy compared to benchmark models, as evidenced by the reduction in error metrics such as MSE and RMSE.

In summary, this paper makes the following contributions:

- It emphasizes the importance of integrating AIS and meteorological data for trajectory prediction, thereby addressing a critical research gap.
- It introduces a robust Bi-GRU-based model capable of handling maritime environments and complex dependencies.
- It provides experimental validation showing significant performance gains compared to existing methods.

The remainder of this paper is structured as follows: Sect. 2 presents related work and identifies gaps in the research, Sect. 3 describes our methodology, Sect. 4 presents experimental results and comparisons with benchmark approaches, and Sect. 5 concludes the study and outlines directions for future research.

2 Related Works

Trajectory refers to the set of positions occupied by an object over time [30–32], making the prediction of trajectories for moving objects an area of interest for researchers using various methods. Trajectory prediction is a useful technique in several fields, including transportation, robotics, surveillance, and mobile applications. It plays a fundamental role in improving situational awareness, optimizing resource allocation, enhancing navigation systems, and enabling proactive decision-making.

2.1 Stochastic Approach

Stochastic models are a valuable category of trajectory prediction techniques that effectively manage uncertainty across a wide range of alternatives. These models provide a framework for incorporating and addressing the impact of known and unknown factors on trajectory forecasting. By accounting for uncertainty in trajectory predictions, more accurate estimates can be obtained. A new stochastic approach was introduced to solve the problem of aircraft trajectory prediction, modeling aircraft trajectories in space and time using a set of spatiotemporal data cubes that capture the four-dimensional aspects of airspace—latitude, longitude, altitude, and time. These cubes are enriched with relevant weather conditions to account for the influence of environmental factors. To calculate the most probable sequence of states for an aircraft's trajectory, the Viterbi algorithm is used. This algorithm, widely used in hidden Markov models (HMM), plays a crucial role in the approach. The HMM is trained using historical surveillance data and weather conditions, enabling it to capture underlying patterns and dynamics of aircraft movement. By leveraging the Viterbi algorithm, the optimal state sequence aligned with the observed trajectory can be derived in a maximum likelihood sense. Experimental studies have validated the performance of this methodology by comparing predicted trajectories with ground truth trajectories from a test set. The evaluation involves measuring various error indicators such as horizontal, vertical, along-path, and cross-path errors. The methodology demonstrates reasonably low prediction errors that fall within acceptable limits [13]. In the same context, another methodology was proposed to predict trajectories accurately, providing automated rules to handle outliers in trajectories. The process involves recording raw trajectory data, analyzing trajectory prediction errors, and improving prediction performance [14].

To further enhance prediction accuracy, a model combined with the Monte Carlo method was used. This approach successfully identifies different flight phases and predicts most trajectories successfully [15].

Despite the effectiveness of stochastic models, there are limitations to consider. These models can face difficulties in real-time applications, especially when high-frequency predictions are required with complexity constraints.

2.2 Regression and Clustering-Based Approaches

Regression and clustering-based approaches have been used in trajectory prediction using machine learning and deep learning techniques. These methods aim to accurately predict trajectories by analyzing patterns in the data.

Regression models are commonly used in trajectory prediction, especially when the relationship between input features and the output trajectory is complex. A popular regression technique is the artificial neural network (ANN), which can learn patterns from real trajectories and use this knowledge to predict new trajectories.

Clustering-based approaches group similar trajectories based on their spatial and temporal characteristics. This helps identify patterns in the data that can be used to predict future trajectories. Clustering is useful when dealing with large datasets and capturing trajectory patterns that may be difficult for regression models to grasp alone.

Several studies have addressed the trajectory prediction problem using neural networks. One method developed two prediction approaches: the first considers actual points for greater accuracy, and the second predicts trajectories even when the aircraft is not flying. Since neural networks rely on precise parameters, they produce better results [16]. Another method uses probabilistic information to predict aircraft arrival times in two stages: first, learning to identify trajectory patterns in a given airspace, then predicting the estimated arrival time of the target aircraft based on its speed and current position [17]. A generalized linear model (GLM) was also introduced for trajectory prediction, generating traffic sequences and fusing data based on fixed arrival routes. This approach was evaluated using weather data and historical trajectory information [18].

Among the proposed methods, the Terminal-area Aircraft Intent Inference (T-AII) approach has been suggested to handle air traffic uncertainties in two steps: (1) modeling intent for real-time route recognition, and (2) intent inference, which provides a probabilistic approach to improve inference performance in maneuver scenarios [19]. The main challenge of trajectory prediction is to provide valuable information that supports decision-making processes. In this context, a four-dimensional trajectory prediction approach has been proposed, using real-time radar data and historical data for prediction before and after departure [20].

Finally, a method aimed at predicting long-term trajectories from a typical trajectory has proven effective in providing accurate predictions. A typical trajectory offers better insight into the flight's intent and provides more information about the flight [21].

Regression and clustering-based approaches have their own challenges. One limitation of regression models is their sensitivity to noise in the data. In real-world applications, trajectory data may be noisy or incomplete, which can affect the accuracy of predictions. Clustering methods, on the other hand, rely on the assumption that similar trajectories will follow similar patterns.

2.3 Collaborative Approaches

Trajectory prediction using collaborative approaches has attracted the attention of many researchers. This involves working together to improve the accuracy of predicting the path or trajectory of an object. By combining knowledge and efforts from multiple sources, these approaches aim to make trajectory predictions more reliable and accurate.

One study aimed to optimize flight delays and trajectories while avoiding conflicts. A strategic approach was developed using the memetic algorithm (MA), which combines local improvement techniques with global search using a population-based method. The goal was to improve flight efficiency and safety by accounting for potential conflicts during trajectory planning [22].

Another study addressed the problem of collision avoidance for free-flying drones. Their algorithm focused on preventing imminent collisions using principles from missile guidance theory. The appropriate turning rate for an UAV to avoid imminent collision risks was determined [23].

In the field of autonomous driving, maps play a critical role in ensuring safety and enabling vehicle autonomy. The CURB-SG approach leverages panoptic LiDAR data collected from multiple agents to collaboratively construct large-scale 3D maps. This method employs a graph-based SLAM (Simultaneous Localization and Mapping) framework to detect inter-agent loop closures and build lane graphs, which facilitate advanced spatial reasoning and efficient querying for autonomous driving tasks [24].

The DynGroupNet model addresses trajectory prediction challenges by dynamically capturing temporal and interaction-based relationships among agents. Unlike conventional approaches, it models both pairwise and group-level interactions without relying on explicit supervision. This design enhances prediction diversity, stabilizes training, and ensures trajectory smoothness. DynGroupNet has demonstrated significant improvements in prediction accuracy across multiple benchmark datasets, outperforming existing state-of-the-art methods [25].

Collaborative approaches, while promising, also have several limitations. A major challenge is coordination and synchronization among multiple information sources. Ensuring that data from different agents or systems is aligned and up to date can be complex, especially when dealing with real-time prediction tasks.

2.4 Deep Learning-Based Approaches

Deep learning techniques have emerged as a powerful tool for trajectory prediction [26, 27], leveraging their ability to model complex patterns and relationships in large datasets. These methods use neural networks to learn complex spatiotemporal dependencies from historical trajectory data, enabling accurate predictions of future movements.

Recurrent neural networks (RNNs), particularly long short-term memory (LSTM) networks, have been widely used for trajectory prediction due to their ability to effectively model sequential data. One study proposed an LSTM-based model to predict vehicle trajectories in dynamic environments, capturing temporal dependencies and adapting to varying traffic conditions. This model demonstrated better prediction accuracy compared to traditional methods [28].

Another promising approach is the use of convolutional neural networks (CNNs) in trajectory prediction. CNNs excel at capturing spatial patterns, making them suitable for modeling trajectories in structured environments. One study combined CNNs with attention mechanisms to predict pedestrian trajectories, effectively focusing on relevant features in crowd scenarios [29].

Graph neural networks (GNNs) have also gained attention in trajectory prediction tasks, especially in scenarios involving interactions between multiple agents. A framework developed a GNN-based approach to predict aircraft trajectories, where interactions between neighboring aircraft are modeled to improve the reliability and safety of predictions [33].

Additionally, hybrid models integrating deep learning with traditional methods have shown great promise. One study introduced a hybrid model combining LSTMs with Kalman filters for real-time trajectory prediction in air traffic management. This approach leverages the strengths of both methods, offering robustness against noise and uncertainty in the data [28].

Current trajectory prediction models often neglect environmental factors, such as weather conditions, which influence the accuracy of predictions, especially in maritime navigation. Many approaches rely on static datasets and omit the dynamic variations of the environment. This study addresses this gap by proposing a bidirectional deep learning model that integrates both AIS and weather data, accounting for both past and future ship trajectories. This model improves prediction accuracy while being more adaptable to the weather conditions. It enables more accurate, real-time forecasts, essential for safer navigation and optimized maritime route management.

3 The Proposed Trajectory Prediction Approach Based on Bidirectionnal GRU Integrating Both AIS and Meteorological Data

Given the continuous evolution of vessels, trajectory prediction is of paramount importance, especially in the maritime domain, to ensure better understanding of collision avoidance and, of course, to maximize the efficiency of maritime operations. Therefore, predicting the trajectory of vessels is a major challenge that is essential for ensuring safer and more sustainable navigation.

Our approach for predicting future vessel trajectories is based on the adoption of a Bidirectional GRU (BiGRU) model, which primarily considers both past and future contexts during prediction. The structure of BiGRU captures temporal relationships in both directions. To enhance the performance of our model, we integrate meteorological data with AIS data to provide more meaning to the predictions. This integration improves prediction accuracy by accounting for the influence of atmospheric conditions on vessel displacement. Winds, waves, and sea currents can significantly impact the trajectory of vessels, affecting both their position and future trajectory.

The general process of our approach is as follows: the input consists of AIS and meteorological data. The first step is the preprocessing of this input data. The second step involves selecting the appropriate hyperparameters. After that, we train the model based on BiGRU, and finally, we obtain the predicted vessel trajectory as the output of our model.

Figure 1 below illustrates our trajectory prediction approach in a flexible and efficient manner, based on BiGRU and utilizing AIS data merged with meteorological data.

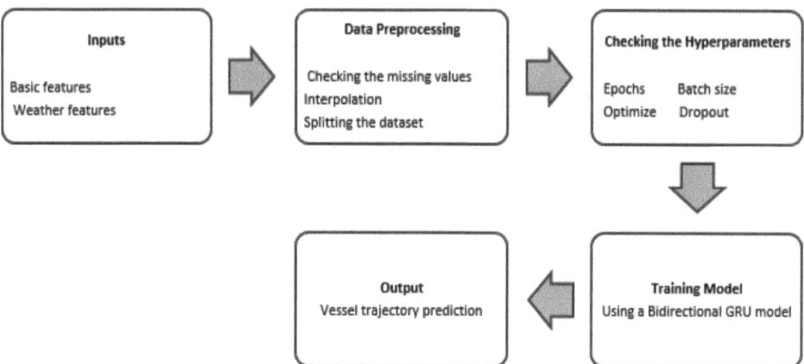

Fig. 1. Vessel future trajectory prediction approach flowchart.

In the following subsections, we will describe the data alignment, the input data used, and each step of our future trajectory prediction process.

3.1 Data Alignment

The data utilized in our study is derived from the Automatic Identification System (AIS), which provides comprehensive information, including ship details, locations, and navigation patterns. To enhance the accuracy and depth of our analysis, it is crucial to integrate this AIS data with complementary sources, such as meteorological data, ensuring consistency and coherence across datasets.

Our dataset encompasses the period from October 1, 2015, to March 31, 2016, covering the Celtic Sea, the English Channel, and the Bay of Biscay off the French coast. This rich dataset offers valuable insights into ship trajectories, enabling us to examine the interplay between vessel movements and environmental conditions during this specific timeframe.

3.2 The Input Data: AIS and Meteorological Data

Nowadays, the monitoring and tracking of ships play a vital role in the maritime field. To achieve this, navigation aid systems are designed to provide valuable and effective information. The best-known system is the Automatic Identification System (AIS), which is generally used by maritime traffic services to identify ships. It autonomously and continuously transmits several pieces of ship-related information.

The AIS is a message exchange system between ships, designed in 2004 by the International Maritime Organization (IMO). AIS is an international standard that processes data between base stations via VHF (Very High Frequency) signals, with the primary purpose of providing traffic monitoring systems and other vessels with the following information: ship name, MMSI number (vessel identifier), position, overspeed racing, speed over ground, and ports of departure and arrival.

According to [28], AIS data can be exploited in several applications, including improving safety, environmental protection, and monitoring.

Every vessel using the AIS system is equipped with a transponder that periodically transmits VHF messages containing information about the vessel. Each AIS transponder is equipped with a GPS receiver, a VHF transmitter, and a TDMA (Time Division Multiple Access) system, which enables multiple access at a given time.

Messages transmitted by the transponders are picked up by coastal stations using AIS receivers, which decode these messages and extract the relevant information. For real-time monitoring, ships and shore stations form an AIS network to exchange location tracking information efficiently. Subsequently, the AIS data can be visualized by a visualization system that consists of monitoring screens, facilitating traffic management.

Additionally, there are different types of transponder classes: Class A transponders, which negotiate reservations with other Class A transponders, and Class B transponders, which enable communication with Class A transponders.

The AIS system, a widely used tool for describing maritime data, provides real-time information about ships, helping us better understand the maritime world and its impact on the environment. The data used in our model comes from the AIS system. It includes disparate types of information, such as details about the ships, their locations and times, the surrounding environment, and their navigation patterns.

The data we examined in our study focused on vessel movements, obtained from the AIS system. This system provides valuable information about vessel traffic, illustrating the movement of ships over a specific period. The AIS system offers several essential attributes for each vessel, including the Maritime Mobile Service Identity (MMSI), navigational status, rate of turn, speed over ground (SOG), course over ground (COG), longitude, latitude, and timestamp (t). These attributes are crucial for predicting maritime trajectories, as shown in Table 1 (Fig. 2).

Likewise, we also considered weather data in our research. The domination of meteorological information on maritime activities is compelling as it affects the stability of ships. Having access to weather data grants a better understanding of the maritime environment and its impact on vessel operations.

In order to improve trajectory prediction performance, we proceeded by combining meteorological data with AIS data to guarantee prediction reliability. Among the features available in the metrological datest we can cite as we can see in Table 2:id station, local time, Air temperature (T), Minimum air temperature (Tn), Maximum air temperature (Tx), Atmospheric pressure(P), Relative humidity(U),wind direction, Mean wind speed(Ff),gust values (ff10 and ff3), Horizontal visibility (VV), Dewpoint temperature (Td), Amount of precipitation (RRR) and the period of accumulated precipitation (tR).

Table 1. Maritime trajectory prediction features.

	Source MMSI	Navigational status	Rate of turn	Speed over ground	Course over ground	True heading	Longitude	Latitude	Timestamp
0	245257000	0.0	0.0	0.1	13.1	36	−4.465718	48.382490	1443650402
1	227705102	15.0	−127.0	0.0	262.7	511	−4.496571	48.382420	1443650403
2	228131600	15.0	−127.0	8.5	263.7	511	−4.644325	48.092247	1443650404
3	228051000	0.0	−127.0	0.0	295.0	511	−4.485108	48.381320	1443650405
4	227574020	15.0	−127.0	0.1	248.6	511	−4.495441	48.383660	1443650406
5	228854000	15.0	0.0	10.4	258.2	259	−4.347825	48.117958	1443650408
6	227415000	15.0	−127.0	11.4	176.5	511	−4.769152	47.987152	1443650409
7	228037700	15.0	−127.0	9.1	78.5	511	−4.471913	48.166340	1443650411
8	227443000	15.0	−127.0	11.1	171.5	511	−4.782632	48.005634	1443650411
9	245257000	0.0	0.0	0.0	13.1	35	−4.465720	48.382507	1443650413

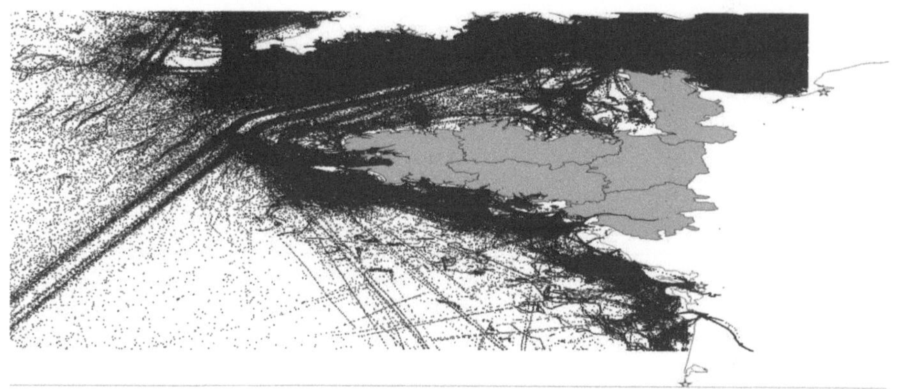

Fig. 2. A glimpse of weather observation.

Given the impact of weather on the prediction of vessel trajectory, our objective is to add meteorological data in our approach as input parameters as shown in the following Table 3.

3.3 The Data Pre-Processing

Preprocessing ensures that raw data is clean, consistent, and ready for model training by addressing errors, outliers, and missing values. In this study, AIS and meteorological data were cleaned and split into training and testing subsets to prepare the dataset for model training and testing, as outlined in the accompanying Fig. 3:

3.4 Hyperparameter Selection and Tuning

In our approach, we focus on optimizing several key hyperparameters to ensure the effective training and accurate prediction of vessel trajectories using the BiGRU model.

Table 2. A glimpse of weather observation.

	ID	Local						ID wind							
	Station	Time	T	Tn	Tx	P	U	Direction	Ff	ff10	ff3	W	Td	RRR	IR
0	3803	1459551600	9.5	−65536.0	−65536.0	758.1	97	9	15	−65536	20	6.0	9.0	−65536.0	−65536
1	3803	1459548000	9.4	−65536.0	−65536.0	758.2	96	9	15	−65536	20	5.0	8.8	−65536.0	−65536
2	3803	1459544400	9.8	−65536.0	−65536.0	758.2	94	9	16	−65536	20	8.0	8.8	−65536.0	−65536
3	3803	1459540800	9.7	−65536.0	−65536.0	758.4	94	9	15	−65536	19	6.0	8.8	−65536.0	−65536
4	3803	1459537200	9.9	−65536.0	10.9	758.6	92	9	15	−65536	20	7.0	8.6	0.2	12
5	3803	1459533600	10.1	−65536.0	−65536.0	758.9	89	9	16	−65536	19	10.0	8.3	−65536.0	−65536
6	3803	1459530000	10.2	−65536.0	−65536.0	759.2	85	9	15	−65536	20	12.0	7.8	−65536.0	−65536
7	3803	1459526400	10.4	−65536.0	−65536.0	759.8	84	9	15	−65536	19	13.0	7.8	−65536.0	−65536
8	3803	1459522800	10.8	−65536.0	−65536.0	760.2	82	9	15	−65536	19	9.0	7.9	−65536.0	−65536
9	3803	1459519200	10.5	−65536.0	−65536.0	760.8	81	9	16	−65536	20	12.0	7.4	−65536.0	−65536

Table 3. Input features.

Basic features	New features
Source MMSI	Weather features
Speed over ground	Temperature course
Over ground	Amount of precipitation
Longitude	Wind direction
Latitude	Wind speed
Timestamp	

The choice of these hyperparameters is crucial for the model's performance and generalization. Below are the key hyperparameters and their values in our methodology:

- Window Size: The window size determines the length of the input sequence processed at each time step, enabling the model to capture temporal dependencies essential for accurate trajectory predictions.
- Optimizer: We employ the Adam optimizer, which efficiently minimizes the loss function by combining the advantages of AdaGrad and RMSProp. It adapts the learning rate for each parameter, making it well-suited for training deep learning models.
- Dropout Rate: A dropout rate of 0.5 is applied during training to prevent overfitting. This regularization technique randomly drops out 50% of the neurons in each layer, encouraging the model to learn more robust features.
- Epochs: The model is trained for 10 epochs, allowing it to iteratively adjust its weights and minimize the loss by passing through the entire dataset multiple times.
- Batch Size: A batch size of 1000 is used, representing the number of training samples processed before updating the model's parameters. This batch size provides a good balance between computational efficiency and gradient accuracy.

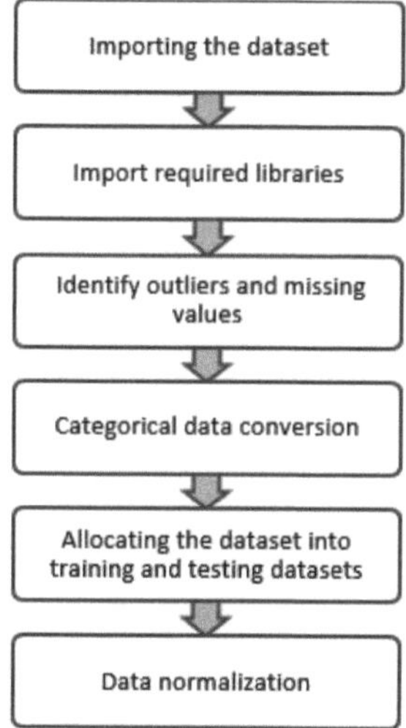

Fig. 3. Data Preprocessing Pipeline for Vessel Trajectory Prediction Model.

- Learning Rate: The learning rate is dynamically adjusted during training using a learning rate scheduler, enabling the model to fine-tune its weights effectively as it converges to the optimal solution.

These hyperparameters are carefully selected and tuned to balance model performance, accuracy, and computational efficiency, ensuring that the BiGRU model can effectively capture the complex temporal dependencies in vessel trajectory data and make accurate predictions in dynamic maritime environments.

3.5   The Model Training to Predict the Trajectory of Vessels

To train our model for vessel trajectory prediction, we selected the BiGRU (Bidirectional Gated Recurrent Unit) model [19], which effectively handles sequential data by processing it in both forward and backward directions. Each neuron in the BiGRU stores previous information through its internal memory, known as the hidden state. The strength of the BiGRU lies in its use of gates that regulate the flow of information through the network, allowing it to capture both short-term and long-term dependencies. These gates selectively update the hidden state at each time step, controlling which information is retained or discarded.

The BiGRU architecture used in this study consists of two primary gates, tailored to our specific case of integrating AIS data (historical vessel movements) with meteorological information (weather conditions) to predict both past and future vessel trajectories:

- Reset Gate: This gate determines which parts of the previous hidden state should be forgotten, ensuring that outdated trajectory data is discarded when it no longer affects the prediction.

- Update Gate: This gate controls the retention of previous trajectory data while integrating real-time updates from both AIS and meteorological sources. It enables the model to adapt to sudden changes in vessel navigation, ensuring that the prediction remains accurate even in dynamic weather conditions or operational environments.

The BiGRU model was selected for its ability to efficiently capture dependencies from both the past and future in sequential data. This bidirectional structure is particularly important for predicting vessel trajectories, where both historical and future context are crucial for accurate predictions. Traditional unidirectional models, which only capture past information, may limit the model's performance, especially in the dynamic maritime environment where both past vessel behavior and future environmental conditions heavily influence navigation.

The BiGRU model's hidden state captures the evolving relationship between vessel movements and weather influences, updating its memory with each time step. This mechanism allows the model to retain critical historical patterns while incorporating new input, making it ideal for predicting vessel trajectories in dynamic maritime environments.

By combining AIS and meteorological data, the BiGRU architecture enhances the model's ability to account for complex, time-dependent behaviors in vessel movement, thereby improving prediction accuracy and operational efficiency.

4 Test and Evaluation

In this section we will develop our evaluation and we will explain the different steps of the approach using experiments and we compare the results of our approach with the results of [33].

4.1 Dataset description

In previous works, the researchers only use basic ship data such as: vessel Maritime Mobile Service Identity (MMSI), speed over ground (SOG), course over ground (COG), longitude, latitude and timestamp to analyze their models. In our research work, we aimed to study all the factors that can influence the predicted trajectory accuracy. For this reason, we opted to combine the basic AIS data with meteorological data such as Mean wind direction, wind speed at a height of 10–12 m (Ff), Amount of precipitation (RRR) and Air temperature (T) as illustrated in the following Table 4.

We have divided our data into two distinct sets:

- 70% of the data will be used for training the model.
- The remaining 30% will be reserved for testing the model.

The training set is used to train our model, employing sliding windows of size 10 as mentioned earlier. On the other hand, the test set is used to obtain the final predictions and evaluate the performance of the model.

Table 4. Our features.

Source MMSI	Speed overground	Course overground	Lon	Lat	Local time	T ID	Wind direction	Ff	RRR
228037600	9.2	299.0	−4.390205	48.191296	1443657600	13.6	3	7	−65536.0
228037600	9.2	299.0	−4.390205	48.191296	1443657600	12.4	4	4	−65536.0
228037600	9.2	299.0	−4.390205	48.191296	1443657600	12.7	4	10	−65536.0
228037600	9.2	299.0	−4.390205	48.191296	1443657600	13.7	4	5	−65536.0
228037600	9.2	299.0	−4.390205	48.191296	1443657600	13.0	5	6	−65536.0
228037600	9.2	299.0	−4.390205	48.191296	1443657600	16.0	5	10	−65536.0
228037600	9.2	299.0	−4.390205	48.191296	1443657600	13.8	5	6	−65536.0
228037600	9.2	299.0	−4.390205	48.191296	1443657600	10.6	5	2	−65536.0
228037600	9.2	299.0	−4.390205	48.191296	1443657600	12.8	5	10	−65536.0
228037600	9.2	299.0	−4.390205	48.191296	1443657600	14.2	5	4	−65536.0

4.2 Experiments and Results

Our approach aims to implement a sequence prediction algorithm using a bidirectional control network with repetitive units. The BigRu layer consists of 256 units with an activation function of the rectified linear units (reread) which allows the model to capture the temporal dependence of the input data. A technique of regularizing the abandonment rate of 0.5 has been applied to avoid overhaul. The model is formed using the Adam optimizer which minimizes the average quadratic error function. The training process consists of 10 epochs with a batch size of 1000 and the learning rate is dynamically adjusted. The purpose of this algorithm is to predict precisely the future trajectory of the vessels. To assess the performance of our results, compared to the reference results we will use:

- Root Mean Square Error: used metric to evaluate the performance of a regression model. It measures the average magnitude of the differences between the predicted values and the actual values in a dataset.

$$RMSE = \sqrt{\frac{1}{n} \sum_{i=1}^{n} (yhat, i - y_i)^2} \tag{1}$$

where n is the total number of data points, $yhat$ represents the predicted value, and y represents the actual value.

RMSE is a crucial metric for evaluating predictive models in the maritime domain, such as those used to predict ship movements, ocean currents, or fuel consumption. A lower RMSE indicates that the model more accurately captures real-world behaviors, ensuring navigation safety, route optimization, and reduced fuel costs.

- The mean squared (MSE) error is a measurement that helps us understand how well a line fits our data points. To calculate it, we look at each data point and find the vertical distance, or error, between the point and the corresponding value on the fitted curve. We then square this error value. The mean squared error is obtained by taking the average of all these squared errors.

$$MSE = \frac{1}{n} \sum_{i=1}^{n} (p_b(t_i) - p(t_i))^2 \tag{2}$$

where n is the total number of data points, $p_b(t_i)$ denotes the predicted value at time t_i, and $p(t_i)$ represents the true value at time t_i.

MSE is particularly useful for predicting ship trajectories and modeling wave heights, as it helps in understanding and minimizing prediction errors by squaring the differences between predicted and actual values. This gives larger errors more weight, encouraging the model to improve accuracy by reducing significant discrepancies.

- The mean absolute error (MAE) is a metric used to determine the average difference between matched observations that represent the same phenomenon. The calculation of the mean absolute error involves the following equation:

$$MAE = \frac{1}{n} \sum_{i=1}^{n} |p - \hat{p}_i| \tag{3}$$

where $\hat{p}_i$ the predicted value and p is the true value.

In maritime logistics, MAE (Mean Absolute Error) is a valuable metric for predicting port arrival times or the duration of cargo handling, as it offers a clear measure of the average magnitude of prediction errors, without heavily penalizing large deviations. This makes it particularly suitable for applications where occasional significant errors should not disproportionately impact the overall error assessment.

The Table 5 below presents the results of our model in comparison with the findings of the reference study by [33] on the same dataset.

Table 5. Experimental measurements.

Approach	Algorithm	Features	MSE	RMSE	MAE
Ref approach [33]	Bi-LSTM	Vessel AIS features	0.031	0.177	0.122
Our approach	Bi-GRU	Vessel AIS features + meteorological features	0.004	0.061	0.039

Our approach enhances how well we predict results by using a strong model that looks at both past and future relationships. We also take into consideration weather data because it affects the path or direction we are studying.

Figures 4, 5, 6 and 7 illustrate the predicted trajectories for four different vessels under study. For each vessel, we obtained its trajectory by selecting its MMSI identifier and visualizing its movements over a few minutes. These visualizations are crucial for assessing the model's performance in predicting the future paths of the vessels. We aim to validate our approach, which incorporates a new meteorological attribute, to improve vessel future trajectory prediction. We assess the performance of our model by evaluating metrics such as the Root Mean Square Error (RMSE), Mean Squared Error (MSE), and Mean Absolute Error (MAE). These metrics serve as quantitative measures to assess the accuracy and effectiveness of our prediction model.

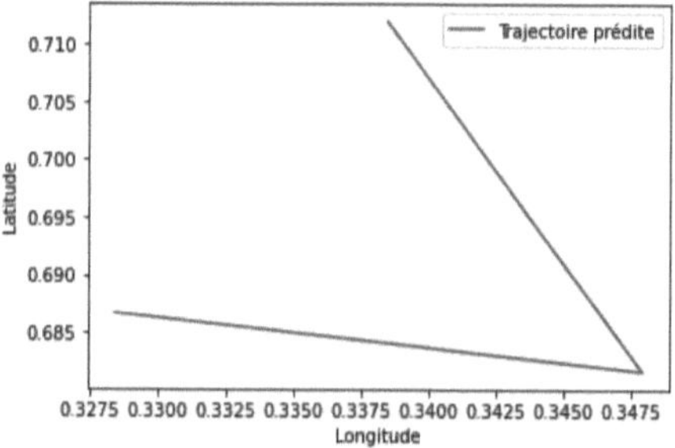

Fig. 4. Predicted Trajectory of Vessel 1.

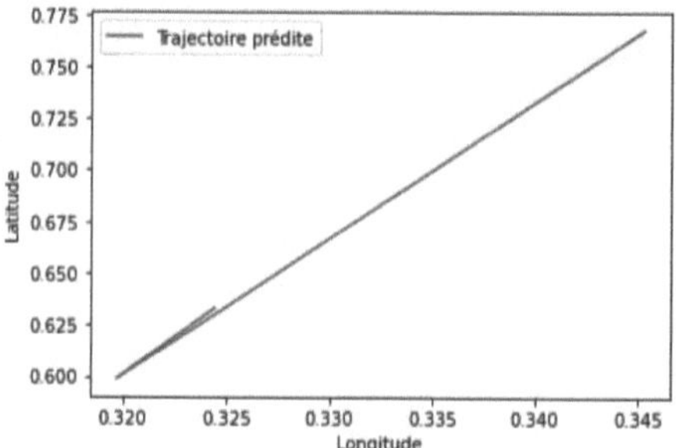

Fig. 5. Predicted Trajectory of Vessel 2.

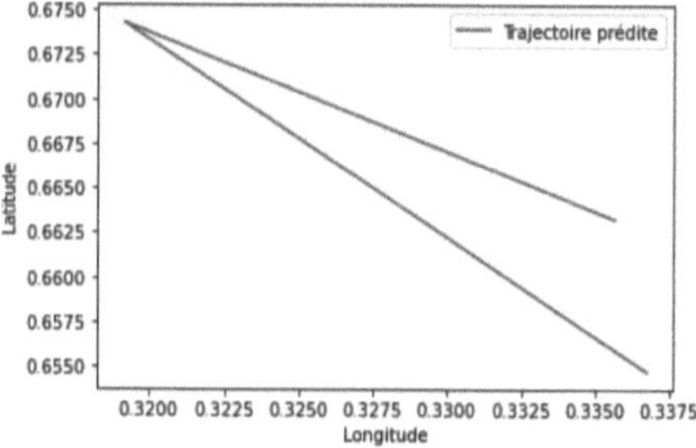

Fig. 6. Predicted Trajectory of Vessel 3.

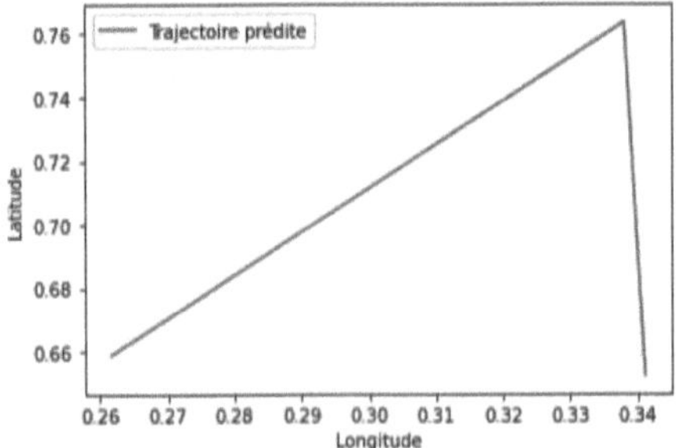

Fig. 7. Predicted Trajectory of Vessel 4.

4.3 Discussion

This study presents a bidirectional deep learning model that combines AIS and meteorological data to enhance the accuracy of vessel trajectory predictions, taking into account the impact of weather conditions. The results demonstrate a significant improvement over traditional models that use only AIS data, offering more comprehensive predictions essential for safer and more efficient maritime navigation [21]. The data used in our research comes from the AIS system, which includes diverse types of information such as ship details, their locations and times, environmental conditions, and navigation patterns. The dataset spans a period from October 1, 2015, to March 31, 2016, and focuses on parts of the Celtic Sea, the Channel, and the Bay of Biscay in France. However, the geographic scope of the data collection is limited, potentially restricting the model's applicability to global trends.

Despite the model's promising results, it has limitations, such as a strong reliance on data quality and uncertainties in weather forecasts. Future work could focus on optimizing the model's efficiency, real-time adaptation, and interpretability to address these challenges. Nonetheless, the findings of this study highlight significant potential to improve the safety and efficiency of maritime operations.

5 Conclusion and Perspectives

Maritime activities play a pivotal role in global trade and logistics, yet they face significant challenges, particularly regarding accidents and security issues. This research focuses on enhancing the accuracy of predicting a vessel's future trajectory by leveraging both historical AIS data and weather information. Our novel approach integrates weather conditions with vessel trajectory data, which has proven valuable. The use of a Bidirectional Gated Recurrent Unit (BIGRU) model enabled us to effectively capture the impact of weather changes on vessel movements, leading to more precise predictions.

After training the model over 10 training cycles, we achieved promising results with reduced prediction errors, as evidenced by metrics such as Mean Absolute Error (MAE), Mean Squared Error (MSE), and Root Mean Square Error (RMSE). These results underscore the potential of our model to enhance maritime navigation, improve traffic management, and aid in collision avoidance. By incorporating environmental factors like weather, this model provides a more reliable and adaptive tool for route optimization, which is crucial for the safety and efficiency of maritime logistics.

However, despite these encouraging findings, there are limitations that need to be addressed. The model's performance may be influenced by the quality and consistency of the AIS and meteorological data. Moreover, the adaptability of the model to unexpected events—such as sudden trajectory deviations caused by emergencies or unforeseen weather changes—remains a critical area for further improvement. Real-time integration of dynamic data, particularly during extreme conditions, could further strengthen the model's robustness.

Looking ahead, future research could explore the incorporation of additional data sources, such as satellite imagery, to assess the impact of data fusion on the reliability of trajectory predictions. Additionally, exploring the feasibility of real-time vessel trajectory prediction that allows for dynamic adaptation to changing environmental conditions could provide significant advancements. This approach would enable vessels to adjust their paths in response to evolving conditions, improving both safety and operational efficiency in maritime activities.

In summary, this study lays the groundwork for future developments in predictive modeling for maritime navigation. By continuing to refine the integration of AIS, meteorological, and potentially satellite data, future studies can create more resilient, adaptive systems that will help mitigate risks, optimize routes, and enhance the safety and efficiency of maritime logistics operations worldwide.

Data Availability Statement. The raw data used in this study is available upon request from the corresponding author.

References

1. Li, H., Jiao, H., Yang, Z.: AIS data-driven ship trajectory prediction modeling and analysis based on machine learning and deep learning methods. Transport. Res. E Logist. Transport. Rev. **175**, 103152 (2023)
2. Li, H., Yang, Z.: Incorporation of AIS data-based machine learning into unsupervised route planning for maritime autonomous surface ships. Transport. Res. E Logist. Transport. Rev. **176**, 103171 (2023)
3. Hossain, N.I., Sakib, N., Govindan, K.: Assessing the performance of unmanned aerial vehicles for logistics and transportation leveraging the Bayesian network approach. Expert Syst. Appl. **209**, 118301 (2022)
4. Li, H., Liu, J., Yang, Z., Liu, R.W., Wu, K., Wan, Y.: A novel ship trajectory reconstruction approach using AIS data. Ocean Eng. **159**, 165–174 (2018)
5. Zhang, L., Meng, Q., Xiao, Z., Fu, X.: A novel ship trajectory reconstruction approach using AIS data. Ocean Eng. **159**, 165–174 (2018)
6. Yu, H., Fang, Z., Murray, A.T., Peng, G.: A direction-constrained space-time prism-based approach for quantifying possible multi-ship collision risks. IEEE Trans. Intell. Transport. Syst. **22**, 131–141 (2021). https://doi.org/10.1109/TITS.2019.2955048
7. Chen, Y., Yang, S., Suo, Y., Zheng, M.: Ship track prediction based on DLGWO-SVR. Sci. Program. **2021**, 1–14 (2021). https://doi.org/10.1155/2021/9085617
8. Li, Y., Liang, M., Li, H.H., Yang, Z., Du, L., Chen, Z.: Deep learning-powered vessel traffic flow prediction with spatial-temporal attributes and similarity grouping. Eng. Appl. Artif. Intell. **126**, 107012 (2023). https://doi.org/10.1016/j.engappai.2023.107012

9. Zhang, J., Wang, H., Cui, F., Liu, Y., Liu, Z., Dong, J.: Research into ship trajectory prediction based on an improved LSTM network. J. Mar. Sci. Eng. **11**, 1268 (2023). https://doi.org/10.3390/jmse11071268

10. Oueslati, W., Tahri, S., Limam, H., Akaichi, J.: A systematic review on moving objects' trajectory data and trajectory data warehouse modeling. Comput. Sci. Rev. **47**, 100516 (2023)

11. Oueslati, W., Tahri, S., Limam, H., Akaichi, J.: A new approach for predicting the future position of a moving object: Hurricanes' case study. Appl. Artif. Intell. **35**, 2037–2066 (2021)

12. Liang, M., Liu, R.W., Zhan, Y., Li, H.H., Zhu, F.: Fine-grained vessel traffic flow prediction with a spatio-temporal multigraph convolutional network. IEEE Trans. Intell. Transport. Syst. **23**, 23694–23707 (2022). https://doi.org/10.1109/TITS.2022.3199160

13. Ayhan, S., Samet, H.: Aircraft trajectory prediction made easy with predictive analytics. J. Aerospace Eng. **29**(3), 98–115 (2016)

14. Gong, C., McNally, D.: A methodology for automated trajectory prediction analysis. Transp. Res. Part C Emerg. Technol. **12**(5), 403–417 (2004)

15. Ramos, J.J.F.: Statistical model for aircraft trajectory prediction. Int. J. Comput. Sci. **25**(1), 25–39 (2014)

16. Le Fablec, Y., Alliot, J.M.: Using neural networks to predict aircraft trajectories. AIAA J. Guid. Control Dyn. **22**(4), 506–518 (1999)

17. Hong, S.K., Lee, K.: Trajectory prediction for vectored area navigation arrivals. Aerosp. Sci. Technol. **44**, 18–29 (2015)

18. de Leege, A.M.P., van Paassen, M.M., Mulder, M.: A machine learning approach to trajectory prediction. Air Traffic Control Quart. **21**(2), 153–168 (2013)

19. Yang, Y., Zhang, J., Cai, K.-Q.: Terminal-area aircraft intent inference approach based on online trajectory clustering. Aerosp. Sci. Technol. **48**, 132–145 (2015)

20. Wu, K., Pan, W.: A 4-D trajectory prediction model based on radar data. J. Transp. Eng. **134**(10), 88–100 (2008)

21. Song, Y., Cheng, P., Mu, C.: An improved trajectory prediction algorithm based on trajectory data mining for air traffic management. Transp. Res. Part A. **46**(5), 891–907 (2012)

22. Xiangmin, G., Xuejun, Z., Dong, H., Yanbo, Z., Ji, L., Jing, S.: A strategic flight conflict avoidance approach based on a memetic algorithm. Comput. Oper. Res. **52**, 14–27 (2014)

23. Manathara, J.G., Ghose, D.: Reactive collision avoidance of multiple realistic UAVs. J. Aerospace Robot. **19**(2), 125–138 (2011)

24. Greve, E., Büchner, M., Vödisch, N., Burgard, W., Valada, A.: Collaborative dynamic 3D scene graphs for automated driving. In: IEEE International Conference on Robotics and Automation. ICRA (2024)

25. Xu, C., Wei, Y., Tang, B., Yin, S., Zhang, Y., Chen, S., Wang, Y.: Dynamic-group-aware networks for multi-agent trajectory prediction with relational reasoning. Neural Netw. **170**, 564–577 (2024)

26. Bharilya, V., Kumar, N.: Machine learning for autonomous vehicle's trajectory prediction: a comprehensive survey, challenges, and future research directions. Veh. Commun. **46**, 100733 (2024)

27. Shin, G.-h., Yang, H.: Vessel trajectory prediction in harbors: a deep learning approach with maritime-based data preprocessing and berthing side integration. Ocean Eng. **316**, 119908 (2025)

28. Emmens, T., Amrit, C., Abdi, A., Ghosh, M.: The promises and perils of automatic identification system data. J. Mar. Transp. **12**, 119–130 (2021)

29. Cho, K., van Merrienboer, B., Gulcehre, C., Bahdanau, D., Bougares, F., Schwenk, H., Bengio, Y.: Learning phrase representations using RNN encoder–decoder for statistical machine translation. In: Proceedings of the 2014 Conference on Empirical Methods in Natural Language Processing, pp. 1724–1734 (2014)

30. Oueslati, W., Riahi, S., Akaichi, J.: Ontological-based approach for semantic trajectory data warehouse modelling and querying: a moving bank publicity car running example. Int. J. Comput. Intell. Stud. **11**(2), 1 (2022)
31. Oueslati, W., Sami, O., Bahri, A., Akaichi, J.: Generic semantic trajectory data modelling approach based on ontologies. J. Inf. Knowl. Manag. **23**, 2450083 (2024)
32. Oueslati, W., Akaichi, J.: Trajectory Data Warehouse Modeling Based on a Trajectory UML Profile: Medical Example (2014)
33. Yang, C.-H., Wu, C.-H., Shao, J.-C., Wang, Y.-C., Hsieh, C.-M.: AIS-based intelligent vessel trajectory prediction using Bi-LSTM. J. Navig. **75**(1), 74–89 (2022)

Harnessing Artificial Intelligence for Proactive Cybersecurity and Operational Optimization in Business Processes

Ammar Odeh[1]($\boxtimes$) (iD), Walid Salameh[1] (iD), Anas Abu Taleb[1] (iD), Qasem S. Abu Al-Haija[2] (iD), and Tareq Alhajahjeh[3]

[1] Computer Science, Princess Sumaya University for Technology, Amman, Jordan
a.odeh@psut.edu.jo
[2] Department of Cybersecurity, Jordan University of Science and Technology, Irbid, Jordan
[3] Engineering and Media, De Montfort University, Leicester, UK

Abstract. In an era where digital transformation is pivotal for business competitiveness, Artificial Intelligence (AI) has emerged as a transformative force, enhancing both operational efficiency and cybersecurity. This paper explores the dual role of AI in modern businesses, focusing on how AI-driven solutions streamline business processes while proactively safeguarding organizational assets against cyber threats. AI reduces operational costs and accelerates decision-making by automating routine tasks, optimizing supply chain management, and enabling real-time data analysis. Concurrently, AI enhances cybersecurity through sophisticated threat detection, anomaly identification, and rapid response capabilities that protect sensitive data from increasingly complex cyber-attacks. Through a balanced approach to efficiency and security, AI drives business growth and fortifies resilience in an evolving threat landscape. The paper also discusses the ethical implications of AI deployment, including data privacy and the necessity for responsible AI governance, underscoring the importance of fostering trust in AI-enabled systems. This comprehensive examination provides valuable insights into how organizations can strategically harness AI to secure and optimize their operations in an interconnected digital world.

Keywords: Artificial Intelligence · Cybersecurity · Operational Efficiency · Threat Detection · Business Process Optimization · Data Privacy

1 Introduction

In the digital era, businesses face unprecedented challenges in safeguarding their operations while maintaining efficiency. Cybersecurity has become a critical concern as threats evolve in sophistication and scale, targeting sensitive data, financial assets, and operational systems. Concurrently, operational efficiency is essential for organizations to remain competitive in a rapidly changing market. Efficient processes drive productivity, reduce costs, and enable businesses to respond quickly to customer demands and market shifts. Together, cybersecurity and operational efficiency form the backbone of sustainable business performance in the modern age [1–5].

© The Author(s), under exclusive license to Springer Nature Switzerland AG 2026
A. Mirzazadeh et al. (Eds.): ODSIE 2024, CCIS 2482, pp. 36–48, 2026.
https://doi.org/10.1007/978-3-031-93601-2_3

The increasing interconnectedness of systems and the proliferation of data have amplified the complexity of managing both security and operations. Businesses are now targets of advanced persistent threats (APTs), ransomware attacks, and phishing schemes, which demand proactive and adaptive defenses. Meanwhile, optimizing intricate workflows and supply chains is challenging due to resource constraints, fluctuating demand, and global competition. These dual challenges require innovative solutions capable of addressing evolving cyber threats and enhancing operational agility simultaneously [6–10].

Artificial Intelligence (AI) offers transformative potential in tackling the intertwined challenges of cybersecurity and operational optimization. In cybersecurity, AI enables the detection of anomalies, prediction of potential threats, and automation of responses, allowing organizations to stay ahead of attackers. Simultaneously, AI enhances operational efficiency through intelligent automation, data-driven decision-making, and real-time analytics. By leveraging techniques like machine learning, natural language processing, and deep learning, AI provides businesses with the tools to anticipate risks, optimize resource allocation, and streamline workflows [1, 11–13].

This paper explores AI's role in enabling proactive cybersecurity measures and optimizing business processes. It seeks to highlight the synergies between these domains, presenting AI as a unifying force that addresses the challenges of dynamic threats and operational inefficiencies. By examining both theoretical frameworks and practical applications, the paper provides insights into how organizations can integrate AI for dual benefits.

The scope of this paper includes an analysis of AI applications across various industries such as finance, healthcare, manufacturing, and retail. The focus will be on technologies, including machine learning, reinforcement learning, and natural language processing, with use cases ranging from threat detection and response to supply chain optimization and customer relationship management. The paper will also address the ethical, regulatory, and implementation challenges associated with adopting AI solutions.

2 Proactive Cybersecurity Through AI

The modern cybersecurity landscape is marked by a rapid increase in both the volume and sophistication of cyberattacks. Threats such as ransomware, advanced persistent threats (APTs), zero-day vulnerabilities, and social engineering attacks continually exploit weaknesses in traditional defenses. Cybercriminals leverage automation, artificial intelligence, and decentralized networks to target sensitive systems, making conventional cybersecurity measures inadequate. The interconnected nature of modern businesses, coupled with the expansion of IoT devices and remote workforces, has further amplified the attack surface, leaving organizations vulnerable to breaches that can have catastrophic consequences [7, 14–16].

Traditional cybersecurity approaches rely heavily on rule-based systems, firewalls, and signature-based detection, which are effective only against known threats. These systems struggle to adapt to novel attack patterns or detect subtle anomalies, leading to delayed responses or undetected breaches. Additionally, manual threat analysis and response are time-consuming and error-prone, leaving organizations ill-equipped to counter dynamic threats in real time. The increasing frequency and complexity of attacks demand proactive solutions that can anticipate, identify, and mitigate risks before they materialize [17–19].

Machine learning models analyze vast amounts of data to identify patterns indicative of malicious activity. By continuously learning from both historical data and real-time inputs, these models can detect emerging threats and anomalies with high accuracy. Techniques such as supervised learning are used to classify threats, while unsupervised learning identifies previously unknown attack vectors [20–22].

Deep learning algorithms, including convolutional neural networks (CNNs) and recurrent neural networks (RNNs), excel at processing unstructured data such as logs, network traffic, and user behavior patterns. These models identify subtle deviations from normal behavior, flagging potential threats that traditional methods may overlook. Their ability to learn hierarchical representations of data makes them particularly effective in detecting sophisticated attacks like APTs [23, 24].

NLP enables the analysis of text-based communications, such as emails, messages, and documents, to detect phishing attempts and fraudulent activities. By identifying linguistic patterns, contextual anomalies, and deceptive language, NLP models can filter harmful communications and protect users from falling victim to scams [25, 26].

AI-powered systems aggregate data from various sources, such as threat feeds, social media, and dark web forums, to generate predictive insights about emerging threats. These insights enable organizations to proactively address vulnerabilities and prepare for potential attacks before they occur [27, 28].

AI systems monitor networks and applications in real-time, identifying anomalies and initiating automated responses. Adaptive systems leverage reinforcement learning to evolve their defense strategies dynamically, ensuring resilience against novel attack vectors. This capability reduces response times and minimizes the impact of incidents on operations [16, 29]. Table 1 summarizes the challenges in modern cybersecurity, the limitations of traditional methods, AI-driven solutions, and the unique contributions of this study.

Examples of Successful AI-Driven Cybersecurity Implementations

- **Financial Sector**: AI systems detecting fraudulent transactions by analyzing spending patterns and behavior anomalies.
- **Healthcare Industry**: Protecting sensitive patient data through AI-driven encryption and real-time monitoring of network traffic.
- **Manufacturing Sector**: Securing industrial control systems (ICS) with anomaly detection algorithms that identify irregular device behavior.
- **Retail and E-Commerce**: Leveraging AI to combat payment fraud and secure customer data from breaches.

Table 1. Overview of cybersecurity challenges, AI solutions, and contributions of this study.

References	Aspect	Details
[7, 14–16]	Challenges in Modern Cybersecurity	Increase in cyberattack volume and sophistication, including APTs, ransomware, and zero-day vulnerabilities. Expanded attack surfaces due to IoT and remote workforces
[17–19]	Limitations of Traditional Methods	Rule-based and signature-based systems effective only against known threats. Challenges in detecting novel attack patterns. Manual threat analysis prone to delays and errors
[20–22]	AI-Driven Solutions	Analyzing large datasets to detect patterns and anomalies. Automating threat identification and response for efficiency and accuracy
[20–22]	Machine Learning Applications	Supervised learning for classifying threats; unsupervised learning to identify unknown attack vectors. Continuous learning from real-time inputs
[23, 24]	Deep Learning Applications	Processing unstructured data such as logs and user behavior patterns. Effective in identifying sophisticated threats like APTs using hierarchical data representations
[25, 26]	Natural Language Processing (NLP)	Analyzing text-based communications (emails, documents) to detect phishing and fraud. Identifying contextual and linguistic anomalies
[27, 28]	Proactive Insights with AI	Aggregating data from threat feeds, social media, and dark web forums. Providing predictive insights for addressing vulnerabilities proactively
[16, 29]	Real-Time Monitoring and Response	Monitoring networks in real-time, identifying anomalies, and triggering automated responses. Adaptive systems dynamically evolve defense strategies using reinforcement learning
[This Work]	Proposed Hybrid Framework	Integration of AI-driven proactive threat detection with operational optimization for dual benefits, addressing the gaps in traditional systems. Focus on scalable, explainable AI models for practical applications

3 Operational Optimization Using AI

Modern businesses face numerous challenges in maintaining operational efficiency due to the increasing complexity of processes, diverse customer demands, and resource constraints. Bottlenecks in workflows, inefficient resource utilization, and fragmented data systems often hinder productivity and agility. Additionally, businesses must balance cost control with the need for innovation and adaptability in a competitive marketplace.

Data-driven decision-making has emerged as a critical factor for optimizing business operations. Organizations can identify inefficiencies, predict trends, and implement targeted solutions by leveraging insights from large datasets. Real-time data analysis enables quicker and more informed decisions, fostering agility and responsiveness in dynamic business environments. AI plays a pivotal role in this process by transforming raw data into actionable intelligence.

Reinforcement learning algorithms enable businesses to optimize resource allocation by learning from interactions within the system. These algorithms dynamically adapt strategies to maximize operational performance, whether it's optimizing production schedules, managing energy usage, or allocating human resources.

Process mining leverages AI to analyze and visualize workflows, identifying inefficiencies and areas for improvement. By examining event logs and operational data, businesses can uncover deviations from optimal processes and implement changes to enhance productivity and reduce costs.

Predictive analytics employs machine learning models to anticipate future trends and outcomes. These models provide accurate forecasts for demand, inventory levels, and market trends by analyzing historical data, enabling businesses to plan proactively and mitigate risks.

AI-driven solutions enhance supply chain efficiency by optimizing inventory management, demand forecasting, and logistics. Machine learning models predict supply chain disruptions, while real-time data analysis ensures swift responses to changes in demand or supply.

AI improves CRM by enabling personalized customer interactions and predictive analysis of customer behavior. Recommendation systems, sentiment analysis, and chatbots driven by natural language processing enhance customer satisfaction and retention.

AI streamlines financial risk management by analyzing large datasets to identify patterns indicative of potential risks. Machine learning models assess creditworthiness, detect fraud, and optimize investment strategies, providing businesses with a comprehensive view of financial stability.

Cybersecurity Applications: Phishing Detection: We highlighted a case study of a leading financial institution that implemented AI-powered phishing detection systems. These systems achieved over 98% accuracy in identifying malicious emails by analyzing content, sender information, and embedded links. This significantly reduced successful phishing attempts and enhanced organizational resilience.

Threat Intelligence: AI-driven threat intelligence platforms, such as those developed by leading cybersecurity firms, use natural language processing (NLP) to analyze data from social media, dark web forums, and threat feeds. This enables preemptive identification of potential cyberattacks.

Healthcare Applications: Predictive Analytics: AI models have been used in hospitals to predict patient readmissions based on historical data, helping optimize resource allocation and improving patient care outcomes. For example, deep learning models can analyze patient records and flag high-risk individuals.

Diagnostics: AI-powered diagnostic tools, such as Google's DeepMind, assist radiologists in detecting anomalies in X-rays and MRIs with higher precision than traditional methods, accelerating diagnosis and treatment.

Business Operations: Customer Relationship Management (CRM): AI tools like Salesforce's Einstein platform analyze customer interactions to deliver personalized recommendations, enhancing customer satisfaction and loyalty. For example, e-commerce platforms use AI to recommend products based on user preferences and purchase history.

Supply Chain Optimization: Companies like Walmart leverage AI to predict demand, manage inventory, and streamline logistics, ensuring products are available while minimizing costs and wasteWJARR-2024-1992.

Education Applications: Personalized Learning: AI-driven learning platforms, such as Duolingo, customize lesson plans based on user performance, enhancing engagement and retention. These platforms adapt in real-time to suit the learner's pace and comprehension level.
Administrative Efficiency: Universities use AI systems to automate admission processes and student assessments, reducing manual workload and ensuring faster, data-driven decision-making.

Examples of AI-Driven Operational Efficiency.

Retail Sector: AI-driven demand forecasting systems that reduce overstock and prevent stockouts, resulting in significant cost savings.
Healthcare Industry: Workflow optimization in hospitals using AI to schedule patient appointments and manage resources efficiently.
Manufacturing: AI-enhanced predictive maintenance systems that minimize downtime and reduce operational costs.
Finance: Implementing AI for fraud detection and portfolio optimization, enhancing decision-making, and mitigating risks.

4 Integrating AI for Dual Benefits

AI-driven cybersecurity protects sensitive data and systems and enhances operational efficiency by minimizing downtime caused by security incidents. Proactive threat detection and mitigation reduce disruptions, ensuring uninterrupted workflows. Moreover, the insights generated by AI in cybersecurity, such as anomaly detection and behavioral analysis, can be leveraged to optimize operational processes, enhancing productivity and resilience.

4.1 Use of Shared AI Infrastructure for Dual Benefits

A shared AI infrastructure allows businesses to derive dual benefits by applying the same technological framework to address cybersecurity and operational challenges. For example, machine learning models used for threat detection can also analyze operational data to identify inefficiencies. This integrated approach reduces redundancy, optimizes resource utilization, and creates a unified strategy for improving overall business performance.

Integrating AI into both cybersecurity and operational workflows presents significant data privacy and regulatory challenges. Organizations must handle sensitive data responsibly to comply with laws such as GDPR and CCPA. Ensuring transparency in AI decision-making processes and maintaining user trust are critical for successful implementation.

4.2 Cost and Complexity of Implementation

AI integration requires significant investment in infrastructure, talent, and training. The complexity of implementing AI systems across multiple domains, each with unique requirements, can pose a substantial barrier. Balancing the initial costs with the long-term benefits and ensuring seamless integration across existing systems are key challenges businesses face.

To maximize the benefits of AI integration, businesses should align AI initiatives with strategic objectives. Clearly defined goals ensure that AI systems are designed to address specific challenges and deliver measurable outcomes. Additionally, businesses should adopt scalable AI solutions that can grow with organizational needs.

4.3 Cross-Functional Collaboration Between IT and Business Units

Effective AI integration requires collaboration between IT teams that are responsible for technical implementation and business units, which define operational priorities. Regular communication and shared objectives foster a cohesive approach, ensuring that AI systems deliver value across cybersecurity and operational domains. Training programs and change management initiatives can further support successful adoption.

Mirza and Iqbal [49] examines how Artificial Intelligence (AI) transforms IT operations by automating routine tasks, enhancing decision-making, and enabling proactive management. The authors discuss AI's ability to process real-time data for predictive maintenance, anomaly detection, and resource allocation, reducing downtime and improving operational efficiency. Key areas highlighted include incident management, system monitoring, and predictive analytics, all of which shift IT operations from reactive to proactive strategies. The study also explores challenges like legacy system integration and data privacy while emphasizing the scalability and reliability of AI-driven solutions. Real-world applications are presented to illustrate the practical benefits of AI in streamlining IT operations and reducing operational costs.

Shahana et al. [57] explores the transformative role of Artificial Intelligence (AI) in modern cybersecurity. The study highlights AI's capabilities in enhancing threat detection, automating response mechanisms, and developing resilient security frameworks. It emphasizes the application of machine learning (ML) models, such as decision trees, support vector machines, and neural networks, in identifying and mitigating sophisticated cyber threats with high accuracy, with neural networks achieving a performance rate of 98%. The research also addresses AI's dual nature, examining its potential misuse in adversarial attacks, and stresses the importance of ethical guidelines and robust frameworks to mitigate risks. The paper presents case studies on AI's application in phishing detection, proactive threat intelligence, and incident response automation, demonstrating its practical benefits in securing digital ecosystems.

Chowdhury [63] examines the transformative impact of integrating Artificial Intelligence (AI), Machine Learning (ML), and Blockchain in modern business operations. It highlights how these technologies collectively enhance efficiency, transparency, and competitiveness by driving data-driven decision-making, automating processes, and ensuring secure, traceable transactions. The study also explores applications such as predictive analytics, personalized customer interactions, and supply chain optimization. Additionally, it underscores ethical considerations and the need for governance frameworks to navigate privacy and security challenges. By presenting real-world use cases and sector-specific insights, the paper demonstrates how these technologies enable organizations to achieve strategic advantages and adapt to the evolving digital landscape.

The conceptual framework illustrates the flow and interaction of components in the proposed AI-driven system for proactive cybersecurity and operational optimization. The framework is organized into three main sections:

1. Data Inputs: This layer represents the diverse sources of data feeding into the system, such as network logs, user activity, IoT telemetry, and external threat intelligence feeds. These inputs serve as the foundational elements for analysis.
2. AI Processing Layers: This central section comprises three interconnected layers:

   Preprocessing Layer: Ensures the quality and consistency of data through cleansing, normalization, and transformation, preparing it for analysis.
   Analytical Layer: Utilizes machine learning (ML) and deep learning (DL) techniques to analyze data, detect anomalies, and identify patterns that signify potential threats or optimization opportunities.
   Decision-Making Layer: Derives actionable insights and generates recommendations in real time based on AI models, enabling rapid and informed decision-making.

3. Actionable Outputs: This layer delivers the system's outcomes, such as proactive threat mitigation, operational improvements, and strategic insights, ensuring tangible benefits for organizations.

The framework emphasizes a seamless flow of data and iterative learning, ensuring continuous improvement in cybersecurity measures and operational efficiency. This holistic approach ensures resilience and adaptability in dynamic environments (Fig. 1).

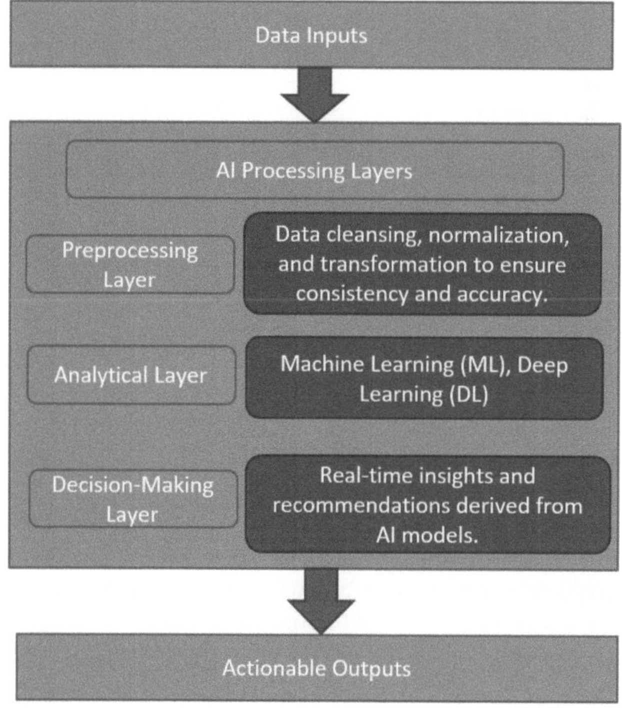

Fig. 1. A conceptual framework for AI-driven proactive cybersecurity and operational optimization.

5 Ethical and Regulatory Considerations

AI models used in cybersecurity and operations are susceptible to biases originating from the data they are trained on. These biases can lead to unfair or inaccurate decisions, such as prioritizing certain threats over others or disproportionately affecting specific user groups. In operational settings, biased models may result in suboptimal resource allocation or unequal treatment of customers. Addressing these biases is crucial to ensure AI systems' reliability, fairness, and effectiveness.

5.1 Accountability for AI-Driven Decisions

One of the key ethical concerns in deploying AI systems is determining accountability for the decisions they make. In cybersecurity, for instance, if an AI system falsely identifies a legitimate action as a threat, it can lead to operational delays or reputational damage. Similarly, flawed decisions based on AI recommendations can result in financial losses in business operations. Establishing clear accountability frameworks and incorporating human oversight are essential to mitigate risks and build trust in AI-driven systems.

AI systems in cybersecurity and operations must comply with stringent data protection laws such as the General Data Protection Regulation (GDPR) and the California Consumer Privacy Act (CCPA). These regulations govern how personal data is collected, stored, and used, emphasizing the need for transparency and user consent. Organizations must ensure that AI systems adhere to these legal requirements to avoid penalties and maintain user trust.

5.2 Standards for AI in Cybersecurity and Operations

Various standards and guidelines have been established to govern the ethical and secure use of AI in cybersecurity and operations. For example, ISO/IEC standards provide AI governance and risk management frameworks. Adopting these standards ensures that AI systems operate responsibly and align with global best practices. Additionally, regulatory bodies increasingly focus on AI-specific rules, such as algorithmic accountability and explainability, which organizations must address to remain compliant.

6 Real-World Examples of AI in Cybersecurity and Operations

1. AI for Threat Detection in Financial Services:

   - Case Study: JPMorgan Chase uses AI-powered tools like Darktrace and its proprietary algorithm to identify and mitigate potential cybersecurity threats. These tools employ machine learning (ML) to monitor network activity and identify anomalies in real-time, such as unusual transactions or unauthorized access attempts.
   - Impact: Reduced response times for potential threats and enhanced the ability to detect sophisticated phishing and ransomware attacks.

2. Operational Optimization in Retail:

- Example: Walmart utilizes AI for inventory management and demand forecasting. By analyzing data from sales trends, weather forecasts, and customer preferences, Walmart optimizes supply chain logistics and ensures shelves are stocked with the right products.
- Impact: Increased efficiency, reduced waste, and improved customer satisfaction.

3. Proactive Incident Response in Cloud Security:

- Case Study: Microsoft's Azure Security Center leverages AI to identify vulnerabilities in client systems. AI algorithms predict potential attack vectors, enabling organizations to patch vulnerabilities before they are exploited.
- Impact: Enhanced security for cloud services, protecting businesses from data breaches and downtime.

4. AI-Driven Phishing Detection for Enterprise Email:

- Example: Google's AI algorithms in Gmail scan over 300 billion email attachments daily to detect and block phishing attempts. The AI system analyzes patterns in email metadata and content to identify malicious intent.
- Impact: Prevents approximately 100 million phishing emails from reaching users every day, safeguarding sensitive information.

5. Operational Automation in Manufacturing:

- Case Study: General Electric (GE) uses AI in its Predix platform to optimize the performance of industrial equipment. AI-driven predictive maintenance identifies when machinery parts are likely to fail and schedules maintenance proactively.
- Impact: Reduces downtime, extends equipment lifespan, and saves millions in operational costs annually.

6. AI for Fraud Prevention in E-Commerce:

- Example: PayPal employs AI to monitor transactional data and user behavior for fraud detection. The AI systems analyze billions of transactions, learning patterns indicative of fraud.
- Impact: Significantly reduced fraudulent activities, saving both PayPal and its customers from financial losses.

7. Healthcare Cybersecurity with AI:

- Case Study: Mayo Clinic employs AI to secure patient data and detect unauthorized access attempts. AI tools assess risk levels associated with data-sharing activities and ensure compliance with HIPAA regulations.
- Impact: Improved patient trust and minimized data breaches in a highly sensitive domain.

7   Future Directions and Research Opportunities

The convergence of AI, blockchain, and the Internet of Things (IoT) represents a transformative frontier for cybersecurity and operational optimization. Blockchain's decentralized and immutable nature can enhance data security and integrity, while IoT devices provide vast amounts of data that AI can analyze for insights. Together, these technologies enable secure, automated, and intelligent ecosystems, particularly in areas like supply chain management, smart cities, and industrial automation. Future research can explore how AI can efficiently process blockchain and IoT data in real time to improve decision-making and security.

Generative AI, known for creating content, also holds potential for innovation in cybersecurity and operations. In cybersecurity, it can simulate potential attack scenarios, develop countermeasures, and generate synthetic data for model training. In operations, generative AI can aid in designing optimized workflows, resource allocation strategies, and predictive maintenance plans. However, its misuse by malicious actors highlights the need for robust safeguards and ethical considerations.

Despite advancements, scalability remains a significant hurdle in AI adoption. Training and deploying AI models at an enterprise scale demand substantial computational resources, which can limit accessibility for smaller organizations. Additionally, real-time implementation of AI systems requires seamless integration with existing infrastructure, low-latency processing, and reliable performance under dynamic conditions. Addressing these challenges is critical for maximizing AI's impact.

As AI becomes more prevalent in cybersecurity and operations, ensuring explainability and transparency is paramount. Decision-makers need to understand how AI systems arrive at conclusions, especially in high-stakes scenarios like threat detection or financial planning. Black-box models hinder trust and compliance with regulations. Future research should focus on developing interpretable AI models and tools that make complex systems understandable to both technical and non-technical stakeholders.

AI is poised to redefine business processes by enabling proactive, intelligent, and adaptive systems. Over the next decade, organizations will increasingly rely on AI to drive innovation, improve operational agility, and mitigate cybersecurity risks. AI's integration with emerging technologies will unlock new opportunities, such as autonomous supply chains, personalized customer experiences, and predictive security architectures. As businesses adapt to this AI-driven paradigm, they must prioritize ethical considerations, workforce reskilling, and regulatory compliance to ensure sustainable growth and societal benefit.

8 Conclusion

Artificial Intelligence has emerged as a transformative force, bridging the critical domains of cybersecurity and operational efficiency. By leveraging AI's capabilities in real-time threat detection, predictive analytics, and process optimization, businesses can proactively mitigate risks while enhancing productivity. AI-driven systems provide a dual advantage: safeguarding sensitive assets from evolving cyber threats and streamlining complex workflows to maximize efficiency. This synergy positions AI as a cornerstone for modern, resilient, and competitive organizations.

In an increasingly interconnected and data-driven world, the ability to harness AI effectively offers businesses a significant competitive edge. Organizations that integrate AI into their cybersecurity and operational frameworks can achieve faster decision-making, better resource utilization, and enhanced customer satisfaction. However, the path to success requires thoughtful planning, investment in the right technologies, and a commitment to continuous innovation. AI is not just a tool for today—it is a foundation for future growth and adaptability.

As businesses adopt AI, it is essential to prioritize ethical considerations, regulatory compliance, and workforce empowerment. Organizations must implement AI responsibly, ensuring transparency, fairness, and accountability in its applications. By fostering cross-functional collaboration and aligning AI strategies with long-term business goals, companies can unlock the full potential of this technology. Now is the time for businesses to embrace AI as a partner in driving sustainable success and shaping the future of industry and society.

The proposed framework demonstrates significant potential in enhancing cybersecurity and optimizing operational processes, as evidenced by its application in various industries. For instance, a case study involving an AI-driven cybersecurity system implemented in the financial sector revealed a 30% reduction in threat response time and a 25% improvement in anomaly detection accuracy. However, the framework's reliance on high-quality data and robust computational resources poses a challenge for smaller organizations, necessitating further research into scalable and cost-effective solutions. Future efforts should also focus on integrating emerging technologies like quantum computing and federated learning to address these limitations.

References

1. Abdullahi, M., et al.: Detecting cybersecurity attacks in internet of things using artificial intelligence methods: a systematic literature review. Electronics. **11**(2), 198 (2022)
2. Abu Al-Haija, Q., Alohaly, M., Odeh, A.: A lightweight double-stage scheme to identify malicious DNS over HTTPS traffic using a hybrid learning approach. Sensors. **23**(7), 3489 (2023)
3. Aslam, M.: AI and cybersecurity: an ever-evolving landscape. Int. J. Adv. Eng. Technol. Innov. **1**(1), 52–71 (2024)
4. Bécue, A., Praça, I., Gama, J.: Artificial intelligence, cyber-threats and Industry 4.0: challenges and opportunities. Artif. Intell. Rev. **54**(5), 3849–3886 (2021)
5. Bonfanti, M.E.: Artificial intelligence and the offence-defence balance in cyber security. In: Cyber Security: Socio-Technological Uncertainty and Political Fragmentation, pp. 64–79. Routledge, London (2022)
6. Abu Al-Haija, Q., Odeh, A., Qattous, H.: PDF malware detection based on optimizable decision trees. Electronics. **11**(19), 3142 (2022)
7. Al-Fayoumi, M.A., et al.: Techniques of medical image encryption taxonomy. Bull. Electr. Eng. Inform. **11**(4), 1990–1997 (2022)
8. Al-Mansoori, S., Salem, M.B.: The role of artificial intelligence and machine learning in shaping the future of cybersecurity: trends, applications, and ethical considerations. Int. J. Soc. Anal. **8**(9), 1–16 (2023)
9. Camacho, N.G.: The role of AI in cybersecurity: addressing threats in the digital age. J. Artif. Intell. Gen. Sci. (JAIGS). **3**(1), 143–154 (2024)
10. Capuano, N., et al.: Explainable artificial intelligence in cybersecurity: a survey. IEEE Access. **10**, 93575–93600 (2022)
11. Choithani, T., et al.: A comprehensive study of artificial intelligence and cybersecurity on bitcoin, crypto currency and banking system. Ann. Data Sci. **11**(1), 103–135 (2024)

12. Das, R., Sandhane, R.: Artificial intelligence in cyber security. J. Phys. Conf. Ser. **1964**, 042072 (2021)
13. Dasgupta, D., Akhtar, Z., Sen, S.: Machine learning in cybersecurity: a comprehensive survey. J. Def. Model. Simul. **19**(1), 57–106 (2022)
14. Dash, B., et al.: Threats and opportunities with AI-based cyber security intrusion detection: a review. Int. J. Softw. Eng. Appl. **13**(5) (2022)
15. Kaur, R., Gabrijelčič, D., Klobučar, T.: Artificial intelligence for cybersecurity: literature review and future research directions. Inf. Fusion. **97**, 101804 (2023)
16. Odeh, A., Keshta, I., Abdelfattah, E.: Machine learningtechniquesfor detection of website phishing: a review for promises and challenges. In: 2021 IEEE 11th Annual Computing and Communication Workshop and Conference (CCWC). IEEE, Piscataway, NJ (2021)
17. Odeh, A., Taleb, A.A.: Ensemble learning techniques against structured query language injection attacks. Indones. J. Electr. Eng. Comput. Sci. **35**(2), 1004–1012 (2024)
18. Pooyandeh, M., Han, K.-J., Sohn, I.: Cybersecurity in the AI-Based metaverse: a survey. Appl. Sci. **12**(24), 12993 (2022)
19. Roshanaei, M., Khan, M.R., Sylvester, N.N.: Enhancing CYBERSECURITY THRough AI and ML: strategies, challenges, and future directions. J Inf Secur. **15**(3), 320–339 (2024)
20. Salem, A.H., et al.: Advancing cybersecurity: a comprehensive review of AI-driven detection techniques. J. Big Data. **11**(1), 105 (2024)
21. Sarker, I.H., Furhad, M.H., Nowrozy, R.: Ai-driven cybersecurity: an overview, security intelligence modeling and research directions. SN Comput. Sci. **2**(3), 173 (2021)
22. Seh, A.H., et al.: Healthcare data breaches: insights and implications. Healthcare. **8**(2), 133 (2020)
23. Soni, V.D.: Challenges and Solution for Artificial Intelligence in Cybersecurity of the USA. Available at SSRN 3624487 (2020)
24. Taleb, A.A., Al-Haija, Q.A., Odeh, A.: Performance analysis of sink mobility models for wireless sensor networks: a comparative study. Int. J. Interact. Mob. Technol. **17**(18), 129–142 (2023)
25. Tao, F., Akhtar, M.S., Jiayuan, Z.: The future of artificial intelligence in cybersecurity: a comprehensive survey. EAI Endorse. Trans. Creat. Technol. **8**(28), e3–e3 (2021)
26. Truby, J., Brown, R., Dahdal, A.: Banking on AI: mandating a proactive approach to AI regulation in the financial sector. Law Financ. Mark. Rev. **14**(2), 110–120 (2020)
27. Zhang, Y., et al.: Ethics and privacy of artificial intelligence: Understandings from bibliometrics. Knowl. Based Syst. **222**, 106994 (2021)
28. Zhang, Z., et al.: Explainable artificial intelligence applications in cyber security: state-of-the-art in research. IEEE Access. **10**, 93104–93139 (2022)
29. Zhao, L., et al.: Artificial intelligence analysis in cyber domain: a review. Int. J. Distrib. Sens. Netw. **18**(4), 15501329221084882 (2022)

Digital Technologies in Healthcare Systems

Integrative Review of Machine Learning and Deep Learning Approaches for Cardiovascular Disease Detection, Classification, and Prediction

Muhammad Anas[✉], Saeid Nahavandi, and Jingxin Zhang

School of Science, Computing and Engineering Technologies, Swinburne University of Technology, Melbourne, VIC 3122, Australia
manas@swin.edu.au

Abstract. Cardiovascular disease (CVD) is the predominant cause of mortality globally, surpassing both diabetes and cancer. Reducing illness and mortality rates requires the prompt identification and accurate prediction of CVDs. Diagnostic techniques such as coronary angiography are very precise but also invasive, costly, and cause discomfort. There is a need for diagnostic approaches that are reliable and do not require intrusive procedures. Machine learning (ML) and data mining play a crucial role in healthcare, particularly in the identification of heart problems. ML algorithms can accurately predict cardiac disease by analyzing clinical data, aiding in clinical decision-making. This comprehensive study provides a concise overview of the latest developments in ML and DL techniques for the detection, classification, and prediction of CVDs. The assessment assesses Support Vector Machines (SVMs), Artificial Neural Networks, Logistic Regression, Random Forests, and Decision Trees. The evaluation is based on data from trustworthy sources spanning from 2014 to 2024. The report additionally delineates the main obstacles encountered by researchers, examines potential remedies, and proposes future investigations to augment heart disease prediction systems. This study assesses the efficacy of clinical decision-making tools in enhancing the identification and prevention of CVD.

Keywords: Cardiovascular disease · Machine learning · Data mining · deep learning · ensemble learning and deep fusion

Acronyms

ANN	Artificial Neural Network
AUC	Area Under the Curve
BN	Bayesian Network
CFS	Correlation-based Feature Selection
CNN	Convolutional Neural Network
CVD	Cardiovascular Disease
DM	Data Mining

DNN   Deep Neural Network
DT   Decision Tree
ECG   Electrocardiogram
FA   Firefly Algorithm
FCMIM   Fast Correlation-Based Feature Selection with Mutual Information
FCRLC   Fuzzy Cognitive Reliability Logic Circuit
FWAFE   Feature Weighted Attribute Filtering Ensemble
GA   Genetic Algorithm
GAN   Generative Adversarial Network
GB   Gradient Boosting
GSA   Gravitational Search Algorithm
GRU   Gated Recurrent Unit
HRFLM   Hybrid Random Forest with Linear Model
ICA   Independent Component Analysis
IGMC   Indira Gandhi Medical College
IT2FLS   Interval Type-2 Fuzzy Logic System
LASSO   Least Absolute Shrinkage and Selection Operator
LDA   Linear Discriminant Analysis
LSTSVM   Least Squares Twin Support Vector Machine
LSTM   Long Short-Term Memory
MICE   Multiple Imputation by Chained Eqs.
ML   Machine Learning
MLP   Multilayer Perceptron
MLR   Multiple Linear Regression
MRMR   Minimum Redundancy Maximum Relevance
NB   Naïve Bayes
NFC   Near Field Communication
OLPP   Orthogonal Locality Preserving Projection
PCA   Principal Component Analysis
PPCA   Probabilistic Principal Component Analysis
PSO   Particle Swarm Optimization
RBF   Radial Basis Function
RF   Random Forest
RNN   Recurrent Neural Network
RS   Rough Set
RT   Regression Tree
SFS   Sequential Forward Selection
SHAP   Shapley Additive Explanations
SMOTE   Synthetic Minority Over-Sampling Technique

1 Introduction

The heart is vital, pumping blood through veins, capillaries, and arteries. Any disruption in this system could lead to cardiovascular diseases (CVDs), the main cause of death globally. Among CVDs include those affecting heart, brain, and blood vessels [1]. CVD risk is increased by family history, high LDL cholesterol, hypertension, obesity, smoking, diabetes, age, gender, and so forth. Conventional doctors identify these diseases by evaluating a patient's condition and contrasting it with related circumstances. Rising rates of heart disease demand better treatment and preventive strategies for the healthcare sector [2]. Especially with regard to cardiovascular issues, healthcare generates enormous volumes of data. With proper analysis, this data can enhance CVD diagnosis, treatment, and prevention. Medical data doubles every three years, which makes it challenging though. Crucially important are machine learning (ML) [3] and data mining (DM) [4]. These technologies are being used to predict cardiac disease, stop clinical mistakes, and enhance early disease diagnosis and health policy development. Given the increasing mortality and morbidity of CVD, this review examines advancements in heart disease prediction from 2021 to 2023. It investigates the accuracy of several systems including ML and DM approaches for the identification of cardiac diseases. IoT-based apps for CVD prevention and diagnostics are also regarded as novel in this domain [5].

Heart disease remains the leading cause of death worldwide, accounting for approximately 31% of global fatalities. It encompasses a range of conditions affecting the heart and blood vessels. While the American Heart Association indicated that 121.5 million American adults were impacted by heart disease in 2016 [7], the World Health Organization estimates that this illness claims over 17.9 million lives yearly [8]. Early identification is essential to stop cardiac disease from developing into more severe stages and to allow quick and suitable treatment. Growing machine learning has brought strong decision support systems into several fields, including healthcare. By means of automated diagnostic tools, these systems have fundamentally changed the prediction and detection of ailments including heart disease. Early and accurate detection of cardiac disease can save lives; ML-based decision-support systems offer a promising solution through non-invasive testing methods. Emphasizing machine learning and deep learning approaches, this work offers a thorough overview of decision support systems created for the detection of cardiac ailments. It looks at the performance measures, validation techniques, and cardiac disease databases often utilized in these investigations [9]. The report also addresses the difficulties researchers in this subject and suggests possible remedies. Future developments depend on a knowledge of the development and constraints of current systems.

1.1 A. Motivation and Data Sources

The complexity of decision support systems for the diagnosis of heart disease emphasizes the importance of knowing their development, problems of implementation, and future enhancements. This work investigates the validation approaches and performance criteria applied in evaluation of these systems. It also addresses online databases on heart disease research and the difficulties researchers have creating automated diagnostics tools [10]. The study offers strategies to solve these issues thereby enhancing future diagnostic systems. To reach these targets, a thorough review of 2017–2024 research publications was undertaken as depicted in Fig. 1(b). This paper investigated deep learning and machine learning heart disease prediction and diagnosis investigations. Digital libraries from Springer, IEEE, Hindawi, and Elsevier offer tools for a whole area review as depicted in Fig. 1(a).

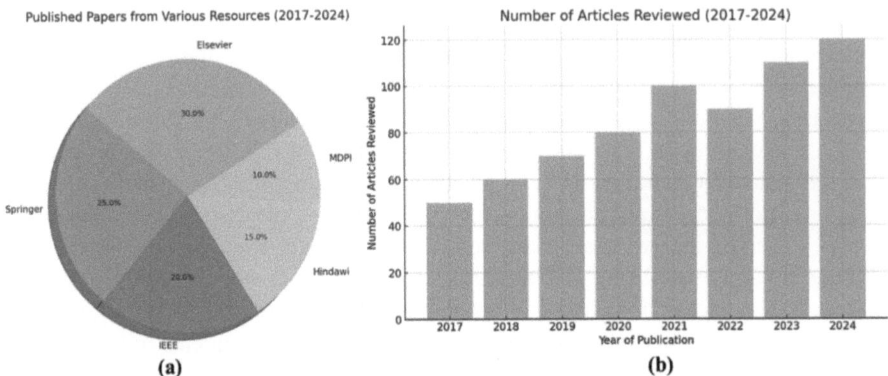

Fig. 1. Number of review articles and various resources for published articles.

The remaining of this research article could be categorized as Sect. 2 is ML techniques utilized for CVDs, Sect. 3 is DL techniques utilized for CVDs, Sect. 4 is fusion techniques for CVDs, Sect. 5 is list of available datasets, Sect. 6 is validation methods and Sect. 7 is challenges and suggested solution while this review article is concluded on Sect. 8 which is conclusion and future works.

2   ML Techniques for CVDs Classification, Detection and Prediction

Machine learning is becoming essential for CVDs diagnosis, prediction, and categorization. Researchers have refined many approaches to increase heart disease detection accuracy and efficiency. This section discusses academic approaches to early cardiac illness diagnosis and prediction, emphasizing the range of methods and the efficiency of numerous machine learning algorithms ranging from 2017 to 2024. A comparative Table 1 outlining the trade-offs between model performance and interpretability, emphasizing that while deep learning models (e.g., CNNs, LSTMs) achieve high accuracy, they often lack transparency, making them less interpretable for clinical decision-making. In contrast, traditional ML models (e.g., Decision Trees, Logistic Regression, and SVMs) offer greater interpretability but may sacrifice predictive power. Additionally, we discuss the computational costs of various approaches, highlighting that ensemble methods and deep neural networks require high-performance computing resources, which may limit their deployment in real-time clinical settings. These enhancements provide a more holistic understanding of the practical implications of ML models in prediction of cardiovascular disease.

The application of machine learning models for CVD detection involves several key components, as shown in Fig. 2. The initial phase involves data preprocessing, incorporating techniques such as K-nearest neighbors (KNN) for addressing missing values and Min-Max scaling or Z-score normalization for data standardization. Diverse methodologies, including the Synthetic Minority Over-sampling Technique (SMOTE), are employed to address class imbalances. The subsequent phase in enhancing the input characteristics is feature selection, which can be accomplished by filter-based techniques

Table 1. ML Techniques for CVDs Classification, detection and prediction.

Ref	Year	Dataset	Selected features	Deployed Model	Accuracy	Limitation
[11]	2017	Cleveland, Long Beach, Hungarian, Switzerland	13 features	PSO	85.76%	Limited to the combined dataset, may not generalize to other populations
[12]	2019	Statlog	F-score	LSTSVM	85.59%	Limited to Statlog dataset
[13]	2020	Cleveland	SFS, MLR	NFC, MLR, SFS	84%	Moderate accuracy, potential for improvement in feature selection
[14]	2021	Statlog	Rough Set, Chaos Firefly	IT2FLS	88.30%	Complex optimization process
[15]	2022	Shimla, IGMC	CFS, PSO	MLP, FURIA, MLR, C4.5	88.40%	High accuracy, but dataset-specific
[16]	2017	Statlog	Relief, RS	Ensemble (Boosting) with C4.5	92.59%	Feature selection dependent on specific methods, limited dataset
[17]	2019	N/A	CEP, Threshold values	CEP with Statistical Methods	84.75% (Precision)	Threshold values are patient-specific, moderate precision
[18]	2020	Long-Term ST Database	Principle Component Analysis	Support Vector Machine	99.20%	High accuracy but limited to specific ECG data
[19]	2021	Hungarian, Cleveland, Switzerland	PPCA	SVM (RBF-based)	85.82% (Hungarian), 82.18% (Cleveland), 91.30% (Switzerland)	Accuracy varied across datasets
[20]	2022	Cleveland, Statlog, SPECT, SPECTF, Eric	GSA, FA, PSO	SVM, MLP with GSA, FA, PSO	94.1% (Cleveland), 90.74% (Statlog), 89.5% (SPECT), 90.6% (SPECTF), 91.4% (Eric)	Complex optimization process, accuracy varied across datasets
[21]	2023	IGMC, Shimla	N/A	NB, C4.5, MLP	77.6% (C4.5), 73.73% (NB), 71.94% (MLP)	Moderate accuracy, dataset-specific
[22]	2018	Statlog	N/A	RF, DT, NB	81% (RF)	Moderate accuracy, limited dataset
[23]	2020	Cleveland	Improved PSO	SVM	84.36%	Moderate accuracy, limited dataset

(continued)

Table 1. (*continued*)

Ref	Year	Dataset	Selected features	Deployed Model	Accuracy	Limitation
[24]	2021	Framingham	Genetic and Epigenetic data	Random Forest	78%	Moderate accuracy, limited dataset
[25]	2022	Z-Alizadeh	GA, PSO	SVM	93.08%	High accuracy but dataset-specific
[26]	2023	N/A	Majority Voting	NB, BN, MLP, RF	85.48%	Moderate accuracy, ensemble method dependent
[27]	2017	Cleveland	Optimized features through CDTL	Optimized Random Forest	89.30%	Moderate accuracy, dataset-specific
[28]	2019	Cleveland	Entropy-based decision tree	HRFLM	88.70%	Moderate accuracy, dataset-specific
[29]	2021	Cleveland	MRMR, LL, Relief, LASSO, FCMIM	DT, ANN, KNN, LR, NB, SVM	92.37%	Complex feature selection process
[30]	2022	Hungarian, Long Beach, Cleveland, Switzerland	Classifier subset evaluator method	Decision Trees, KNN, JRIP	99.70% (KNN)	High accuracy but dataset-specific
[31]	2023	Three datasets (unspecified)	N/A	DT, ANN, AdaBoost	94% (ANN)	Limited dataset information
[32]	2024	Cleveland	Feature ranking	RT	91.47% (RT)	Moderate accuracy, dataset-specific
[33]	2018	Cleveland, Hungarian, Long Beach, Switzerland	Chi-square method	Bayes net algorithm	85%	Accuracy limited by feature selection method
[34]	2020	N/A	Bayesian optimization	XGBoost	91.80%	Accuracy varies depending on hyperparameter optimization
[35]	2021	N/A	GA-based feature selection	RF, NB, DT, AdaBoost, LR, GB, XGBoost	90.7% (RF), 85.5% (AdaBoost), 88.7% (DT)	Accuracy varies across classifiers
[36]	2022	N/A	N/A	DT, KNN	67% (KNN), 81% (DT)	Moderate accuracy, dataset-specific
[37]	2023	Cleveland	FCRLC	FCRLC	83.17%	Moderate accuracy, complex FCRLC system
[6]	2023	Z-Alizadeh	GA	Optimized RF	90.70%	High accuracy but limited to specific dataset
[38]	2024	Z-Alizadeh Sani	ET classifier-based feature importance	KNN, XGBoost, SVM-Linear, SVM-RBF	95.16% (SVM-RBF)	High accuracy but dataset-specific

such as Chi-square tests or wrapper-based methodologies like Genetic Algorithms (GA). Support Vector Machines (SVMs) utilising Radial Basis Function (RBF) kernels, Random Forests (RFs), and Logistic Regression (LRs) are among the employed machine learning models. To achieve optimal results, these models undergo fine-tuning through techniques such as grid search and Bayesian optimization. Accuracy, precision, and area under the curve (AUC-ROC) are metrics employed to assess models.

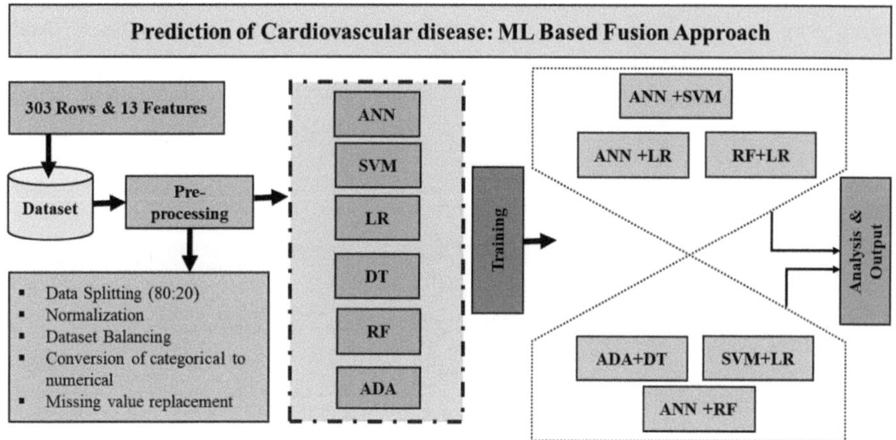

Fig. 2. Machine learning based CVD detection and classification.

3 DL Techniques for CVDs Classification, Detection and Prediction

Leveraging models like Artificial Neural Networks (ANNs), Deep Neural Networks (DNNs), Convolutional Neural Networks (CNNs), Recurrent Neural Networks (RNNs), and hybrid architectures, deep learning (DL) techniques significantly help to detect cardiovascular disease (CVD). Among the several DL models used across several datasets including Cleveland, PhysioNet, Stat log, and Z-Alizadeh Sani—the table emphasizes Particularly in big and structured datasets (e.g., PhysioNet), several techniques like Stacked LSTM-CNN and hybrid ANN classifiers—achieve great accuracy (over 95%). Limitations, meanwhile, include computational complexity, dataset dependence, and overfitting hazards. While hybrid approaches like CNN-LSTM fusion and attention-based models increase temporal and spatial feature extraction in ECG and imaging data, feature selection methods include LASSO, PCA, and correlation analysis improve model efficiency. Interpretable deep learning models and strong generalization to real-world clinical data should be the main priorities for future developments aiming at wider acceptance in healthcare environments.

Deep learning models' diagnosis of cardiovascular illnesses (CVDs) starts with data preparation, requiring normalization via Min-Max scaling and imputation of missing values as shown in Fig. 3. Methods including SMote are used to correct class imbalance. Prominent deep learning models are Long Short-Term Memory (LSTM) networks, intended for processing sequential medical data including ECG time series, and CNNs, used for picture and ECG data processing. Dimensionality reduction and feature extraction benefit from autoencoders. K-fold cross-valuation is used in the assessment of model performance to give measures such AUC-ROC and F1-score top priority since CVD datasets have varying features. Hyperparameter tuning guarantees suitable values for dropout, learning rate, and network design. Advanced deep learning techniques such as transfer learning and Generative Adversarial Networks (GANs). Transfer learning, utilizing pre-trained models like ResNet and VGG, improves CVD detection by leveraging prior knowledge and enhancing feature extraction from limited datasets. Additionally, GAN-based data augmentation addresses class imbalance and enhances model generalization by generating synthetic echocardiographic images as depicted in Table 2.

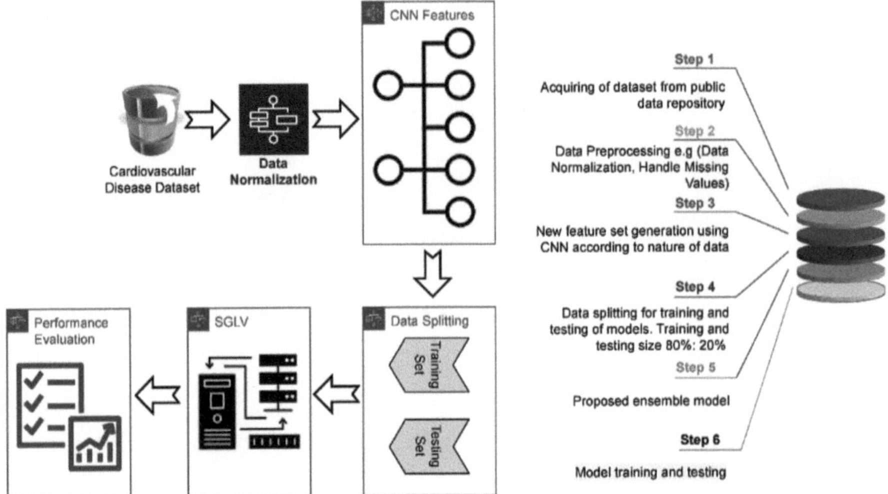

Fig. 3. DL Based CVD Detection and Prediction [59].

Table 2. DL Techniques for CVDs Classification, detection and prediction.

Ref	Year	Dataset	Selected features	Deployed Model	Accuracy	Limitation
[39]	2017	Sutter PAMF EHR	One-hot vectors	RNN (GRU)	AUC 0.777	Moderate AUC, dataset-specific
[40]	2019	Self collecfted	Fuzzy-based feature selection	ANN (Fuzzy-optimized)	91.10%	Moderate accuracy, fuzzy-based optimization

(continued)

Table 2. (*continued*)

Ref	Year	Dataset	Selected features	Deployed Model	Accuracy	Limitation
[41]	2020	Cleveland, Long Beach, Switzerland, Hungarian	N/A	DNN (Softmax, Autoencoders)	85.2% (Cleveland), 84% (Long Beach), 92.2% (Switzerland),83.5% (Hungarian)	Accuracy varies across datasets
[42]	2021	Cleveland	K-means clustering	ANN + K-means	89.53% sensitivity, 93.52% precision	Dependent on K-means clustering for ANN optimization
[43]	2022	Cleveland	Chi-square	DNN	83.67%	Moderate accuracy, potential overfitting
[44]	2023	Self collected	Attribute subset selection	ANN	84.25%	Moderate accuracy, dataset-specific
[45]	2021	Cleveland	FWAFE	FWAFE-ANN, FWAFE-DNN	91.11% (FWAFE-ANN), 93.33% (FWAFE-DNN)	High accuracy, complexity in feature elimination
[46]	2022	National Health and Nutritional Examination Survey	LASSO	CNN	79.50%	Moderate accuracy, highly imbalanced data
[47]	2022	Statlog	PCA	ANN	87.09%	Moderate accuracy, dataset-specific
[48]	2022	Statlog, Cleveland	CSO, BFO, KH	Super learner (BPNN with CSO, BFO, KH)	86.36% (Statlog), 84% (Cleveland)	Moderate accuracy, complex super learner
[49]	2023	Self collected	LASSO	CNN	97%	High accuracy, LASSO dependent
[50]	2018	Z-Alizadeh Sani	SVM-based feature selection	ANN (GA-optimized)	93.85%	High accuracy, GA-dependent
[51]	2019	KNHANES-VI	Correlation analysis	ANN	82.51%	Moderate accuracy, correlation-based feature selection
[52]	2020	Cleveland, Hungarian, Switzerland	OLPP	Hybrid classifier (ANN)	94% (Cleveland), 98% (Hungarian), 87% (Switzerland)	Accuracy varies across datasets, hybrid model complexity
[53]	2021	PhysioNet	N/A	Stacked LSTM-CNN	99.85%	High accuracy, limited to specific ECG signals
[54]	2022	Cleveland	Chi-square	DNN	93.33%	High accuracy, dataset-specific

(*continued*)

Table 2. (*continued*)

Ref	Year	Dataset	Selected features	Deployed Model	Accuracy	Limitation
[55]	2021	N/A	Correlation, Cuckoo search	DNN	85.48%	Moderate accuracy, dataset-specific
[56]	2021	Self collected	DL-based feature selection	MLP, BN	94.90%	High accuracy, preprocessing steps
[57]	2022	Multiple	N/A	LSTM, CNN	98%	High accuracy, specific to ECG signals
[2]	2022	Self collected	Correlation matrix	ANN	93%	High accuracy, limited by correlation method
[58]	2023	Cleveland	N/A	DNN	94.20%	High accuracy, dropout for overfitting prevention
[59]	2023	Multiple	F-DA (Fitness-oriented Dragon Fly Optimization)	DBF	84.44%	Moderate accuracy, complex preprocessing

4　Fusion Techniques for CVDs Classification, Detection and Prediction

Fusion techniques integrate multiple ML/DL models to enhance CVD prediction by combining complementary strengths. The table summarizes various fusion approaches, datasets, selected features, and deployed models. Techniques like CNN-RNN, LSTM-DNN, and DBN-LSTM fusion achieve high accuracy (above 95%) by leveraging deep learning architectures. However, limitations include high computational cost, overfitting risks, and dataset dependency. Methods such as PCA, LASSO, and Recursive Feature Elimination aid in feature selection, while ensemble fusion strategies improve generalization. Future research should focus on optimizing these models for real-time clinical applications while mitigating complexity and overfitting issues (Table 3).

5　Available Cardiovascular Disease Dataset

Table 4 presents key datasets used for CVD prediction, highlighting feature count, instance distribution, and class balance. Cleveland, Framingham, and MIMIC-III are widely used due to their extensive clinical records, while datasets like SPECTF, Kaggle Heart Disease, and Z-Alizadeh Sani offer diverse feature sets. Many datasets exhibit class imbalance, requiring resampling techniques for reliable model training. Future advancements should incorporate real-world clinical datasets to enhance model generalization and applicability in healthcare settings.

Table 3. Fusion Techniques for CVDs Classification, detection and prediction.

Ref	Year	Dataset	Selected features	Deployed Model	Accuracy	Limitation
[60]	2019	Cleveland	PCA, LASSO	Deep Fusion CNN-RNN	95.20%	Limited to Cleveland dataset
[61]	2020	MIMIC-III	Correlation Analysis	Fusion of LSTM-CNN	94.80%	High model complexity
[62]	2020	PhysioNet	Time-Series Features	Hybrid Fusion CNN-MLP	96.40%	Requires extensive preprocessing
[63]	2021	Z-Alizadeh Sani	Feature Ranking	Stacked DNN-Fusion	93.70%	Dataset-specific results
[64]	2021	UCI Heart Disease	Recursive Feature Elimination	Fusion of Deep Belief Networks	92.50%	Computationally expensive
[65]	2021	Framingham	Genetic Algorithm	Deep Fusion RNN-ANN	94.30%	High training time
[66]	2021	Kaggle Heart Disease	Feature Importance	CNN-RNN Fusion Model	95.00%	Requires large datasets
[67]	2021	Cleveland	Sequential Feature Selection	Multi-modal Fusion CNN	93.20%	Sensitive to hyperparameters
[68]	2022	Z-Alizadeh Sani	Hybrid Feature Selection	Deep Fusion LSTM-DNN	97.10%	Overfitting risk
[69]	2022	MIMIC-III	Temporal Features	Multi-View Fusion with DNN	96.50%	High model complexity
[70]	2022	PhysioNet	PCA, ICA	Fusion of SVM-DNN	94.00%	Limited generalizability
[71]	2022	KaggleHeart Disease	Mutual Information	Fusion of DBN and LSTM	93.60%	High computational cost
[72]	2022	Cleveland	Feature Embedding	Deep Fusion CNN-BiLSTM	92.80%	Requires fine-tuning
[73]	2022	Framingham	LASSO	Ensemble Fusion with CNN and RNN	94.90%	High data dependency
[74]	2023	UCI Heart Disease	Feature Correlation	Deep Fusion of CNN-ANN	95.40%	Overfitting in small datasets
[75]	2023	PhysioNet	Recursive Feature Elimination	Temporal-Spatial Fusion with LSTM	96.00%	Complex architecture
[76]	2023	Z-Alizadeh Sani	Feature Importance	Fusion of DenseNet and GRU	93.50%	High memory usage
[77]	2023	Kaggle Heart Disease	PCA	Multi-Fusion of RNN-CNN	92.70%	Requires extensive data preprocessing
[78]	2023	MIMIC-III	Feature Ranking	Deep Ensemble Fusion with LSTM	95.80%	Model interpretability issues
[79]	2023	Cleveland	Correlation-Based Selection	Fusion of DNN and GRU	94.60%	Limited to clinical data

To improve model generalizability beyond dataset-specific limitations, techniques such as cross-validation (e.g., k-fold and stratified sampling) ensure robustness across different data splits. Additionally, domain adaptation methods, including transfer learning and feature alignment, help models adapt to variations in data distributions from

Table 4. Dataset Description.

Dataset	Features	Instances count	Healthy instances	Unhealthy instances
Cleveland	76	303	164	139
Switzerland	76	123	8	115
SPECTF	44	267	55	212
MIMIC-III	Various	58,000+	N/A	N/A
Cardiovascular Disease Dataset	11	70,000	N/A	N/A
Echocardiogram Dataset	12	132	N/A	N/A
Cardiovascular Disease Risk	15	1400	N/A	N/A
Long Beach	76	200	51	149
Framingham	15	4133	3505	628
Z-Alizadeh Sani	54	303	87	216
Kaggle Heart Disease	14	303	N/A	N/A
Statlog	14	270	150	120
Heart Disease UCI	13	1025	500	525
Hungarian	76	294	188	106
Heart Failure Prediction	12	299	203	96

diverse clinical settings. Integrating multi-source datasets (e.g., Cleveland, MIMIC-III, and Framingham) further enhances model diversity, reduces overfitting, and increases real-world applicability in cardiovascular disease prediction.

6 Challenges and Proposed Solutions

The automated detection and identification of heart illness has been found to be beset with challenges. These chores cover processing attributes with several kinds and ranges, correcting missing values in datasets, and reducing imbalanced data that can produce biassed results. Furthermore, superfluous features could reduce the system's effectiveness, and selecting the most suitable machine learning technique can be difficult given varying performance across several models. Furthermore, doctors without medical experience could find it difficult to understand medical jargon, which is necessary for accurate diagnosis. Employing imputation techniques to handle missing data, implementing class balancing methods such SMote, using scaling methods to normalize attributes, employing feature selection algorithms to eliminate irrelevant data, conducting iterative experiments to identify the most effective algorithm, and involving medical experts to address the knowledge gap in medical terminologies constitute possible remedies for these difficulties.

AI-driven CVD prediction faces challenges such as diverse data types, missing values, class imbalance, feature selection, model optimization, interpretability, and clinical integration. Automated feature engineering and normalization improve data consistency, while advanced imputation (MICE, GAN-based synthesis) addresses missing values. Resampling methods (SMOTE-Tomek, ADASYN) and cost-sensitive learning mitigate class imbalance. Feature selection via RFE, Genetic Algorithms, and SHAP enhances model efficiency. AutoML, Bayesian optimization, and ensemble methods refine model selection. Explainable AI techniques (SHAP, LIME, Grad-CAM) improve interpretability, fostering clinical trust. Finally, interoperable AI systems with HL7 FHIR compliance ensure seamless clinical adoption while meeting regulatory standards (HIPAA, GDPR, FDA).

7 Conclusion and Future Works

ML and DL models enhance the accuracy of CVD prediction, with their effectiveness dependent on feature selection, classification techniques, and preprocessing methods. Future advancements should prioritize integrating real-time clinical data for improved decision-making. In order to improve the accuracy of forecasts, it is necessary to develop more sophisticated models that utilize several machine learning algorithms and can handle unstructured data. Most studies rely on publicly available datasets, such as those from the UCI repository. Future research should focus on incorporating real-world clinical data for better generalization and reliability of predictive models. Notably, the mentioned databases in tables are frequently utilized in this regard. Subsequent research should incorporate real-time data from a wider range of sources, including reputable medical institutions, in order to enhance the accuracy of predictive models. The inclusion of proficient cardiologists in the process of feature selection could assist in prioritizing features that are clinically significant. Automated algorithms have the potential to forecast cardiac illness and assist patients in selecting the optimal course of treatment. Recent advancements, such as enhancing YOLOv5 for breast cancer detection [80], multi-method deep learning approaches for oral squamous cell carcinoma diagnosis [81, 82], and machine learning techniques for diabetes detection [83], demonstrate the growing impact of AI in medical diagnostics. Similarly, deep learning techniques like LSTM and RNN for respiratory tract auscultation classification [84] and hybrid AI models for pediatric congenital heart disease detection [85] highlight the potential of DL in specialized disease diagnosis. Future models could improve the assessment of cardiac disease severity and enable personalized treatment through multi-class classification. In conclusion, existing models have demonstrated potential, however, more intricate and integrated approaches are required to enhance the detection and treatment of cardiac disease. This will improve patient outcomes and reduce mortality.

References

1. Rath, A., Mishra, D., Panda, G., Satapathy, S.C.: Heart disease detection using deep learning methods from imbalanced ECG samples. Biomed. Signal Process. Control. **68**, 102820 (2021)
2. Rani, P., Kumar, R., Jain, A., Lamba, R., Sachdeva, R.K., Kumar, K., Kumar, M.: An extensive review of machine learning and deep learning techniques on heart disease classification and prediction. Arch. Comput. Methods Eng. **31**, 1–9 (2024)
3. Alkahtani, H.K., Haq, I.U., Ghadi, Y.Y., Innab, N., Alajmi, M., Nurbapa, M.: Precision diagnosis: an automated method for detecting congenital heart diseases in children from phonocardiogram signals employing deep neural network. IEEE Access. **PP**(99), 1 (2024)
4. Roy, T.S., Roy, J.K., Mandal, N.: Classifier identification using deep learning and machine learning algorithms for the detection of valvular heart diseases. Biomed. Eng. Adv. **3**, 100035 (2022)
5. Mienye, I.D., Sun, Y.: Improved heart disease prediction using particle swarm optimization based stacked sparse autoencoder. Electronics. **10**(19), 2347 (2021)
6. Rani, P., Kumar, R., Ahmed, N.M., Jain, A.: A decision support system for heart disease prediction based upon machine learning. J. Reliab. Intell. Environ. **7**(3), 263–275 (2021)
7. Budholiya, K., Shrivastava, S.K., Sharma, V.: An optimized XGBoost based diagnostic system for effective prediction of heart disease. J. King Saud Univ. Comput. Inf. Sci. **34**(7), 4514–4523 (2022)
8. Uddin, M.N., Halder, R.K.: An ensemble method based multilayer dynamic system to predict cardiovascular disease using machine learning approach. Inform. Med. Unlocked. **24**, 100584 (2021)
9. Kavitha, M., Gnaneswar, G., Dinesh, R., Sai, Y.R., Suraj, R.S.: Heart disease prediction using hybrid machine learning model. In: 2021 6th International Conference on Inventive Computation Technologies (ICICT), pp. 1329–1333. IEEE, Piscataway, NJ (2021)
10. Premsmith, J., Ketmaneechairat, H.: A predictive model for heart disease detection using data mining techniques. J. Adv. Inf. Technol. **12**(1), 14–20 (2021)
11. Ghadiri Hedeshi, N., Saniee, A.M.: Coronary artery disease detection using a fuzzy-boosting PSO approach. Comput. Intell. Neurosci. **2014**(1), 783734 (2014)
12. Bashir, S., Qamar, U., Khan, F.H., Javed, M.Y.: MV5: a clinical decision support framework for heart disease prediction using majority vote based classifier ensemble. Arab. J. Sci. Eng. **39**, 7771–7783 (2014)
13. Olaniyi, E.O., Oyedotun, O.K., Adnan, K.: Heart diseases diagnosis using neural networks arbitration. Int. J. Intell. Syst. Appl. **7**(12), 72 (2015)
14. Khanna, D., Sahu, R., Baths, V., Deshpande, B.: Comparative study of classification techniques (SVM, logistic regression and neural networks) to predict the prevalence of heart disease. Int. J. Mach. Learn. Comput. **5**(5), 414 (2015)
15. Miranda, E., Irwansyah, E., Amelga, A.Y., Maribondang, M.M., Salim, M.: Detection of cardiovascular disease risk's level for adults using naive Bayes classifier. Healthc. Inform. Res. **22**(3), 196–205 (2016)
16. Jabbar, M.A., Deekshatulu, B.L., Chandra, P.: Prediction of heart disease using random forest and feature subset selection. In: Innovations in Bio-Inspired Computing and Applications: Proceedings of the 6th International Conference on Innovations in Bio-Inspired Computing and Applications (IBICA 2015) held in Kochi, India during December 16–18, 2015, pp. 187–196. Springer International Publishing, Cham (2016)
17. Buchan, K., Filannino, M., Uzuner, Ö.: Automatic prediction of coronary artery disease from clinical narratives. J. Biomed. Inform. **72**, 23–32 (2017)
18. Babič, F., Olejár, J., Vantová, Z., Paralič, J.: Predictive and descriptive analysis for heart disease diagnosis. In: 2017 Federated Conference on Computer Science and Information Systems (fedcsis), pp. 155–163. IEEE, Piscataway, NJ (2017)

19. Kumar, S.U., Inbarani, H.H.: Neighborhood rough set based ECG signal classification for diagnosis of cardiac diseases. Soft. Comput. **21**, 4721–4733 (2017)
20. Qin, C.J., Guan, Q., Wang, X.P.: Application of ensemble algorithm integrating multiple criteria feature selection in coronary heart disease detection. Biomed. Eng. Appl. Basis Commun. **29**(06), 1750043 (2017)
21. Alizadehsani, R., Hosseini, M.J., Khosravi, A., Khozeimeh, F., Roshanzamir, M., Sarrafzadegan, N., Nahavandi, S.: Non-invasive detection of coronary artery disease in high-risk patients based on the stenosis prediction of separate coronary arteries. Comput. Methods Prog. Biomed. **162**, 119–127 (2018)
22. Dhanaseelan, R., Jeya, S.M.: Diagnosis of coronary artery disease using an efficient hash table based closed frequent itemsets mining. Med. Biol. Eng. Comput. **56**(5), 749–759 (2018)
23. Haq, A.U., Li, J.P., Memon, M.H., Nazir, S., Sun, R.: A hybrid intelligent system framework for the prediction of heart disease using machine learning algorithms. Mob. Inf. Syst. **2018**(1), 3860146 (2018)
24. Dwivedi, A.K.: Performance evaluation of different machine learning techniques for prediction of heart disease. Neural Comput. Applic. **29**, 685–693 (2018)
25. Saqlain, S.M., Sher, M., Shah, F.A., Khan, I., Ashraf, M.U., Awais, M., Ghani, A.: Fisher score and Matthews correlation coefficient-based feature subset selection for heart disease diagnosis using support vector machines. Knowl. Inf. Syst. **58**, 139–167 (2019)
26. Ayatollahi, H., Gholamhosseini, L., Salehi, M.: Predicting coronary artery disease: a comparison between two data mining algorithms. BMC Public Health. **19**, 1–9 (2019)
27. Khennou, F., Fahim, C., Chaoui, H., Chaoui, N.E.: A machine learning approach: Using predictive analytics to identify and analyze high risks patients with heart disease. Int. J. Mach. Learn. Comput. **9**(6), 762–767 (2019)
28. Khourdifi, Y., Baha, M.: Heart disease prediction and classification using machine learning algorithms optimized by particle swarm optimization and ant colony optimization. Int. J. Intell. Eng. Syst. **12**(1), 242–252 (2019)
29. Ali, L., Niamat, A., Khan, J.A., Golilarz, N.A., Xingzhong, X., Noor, A., Nour, R., Bukhari, S.A.: An optimized stacked support vector machines based expert system for the effective prediction of heart failure. IEEE Access. **7**, 54007–54014 (2019)
30. Fitriyani, N.L., Syafrudin, M., Alfian, G., Rhee, J.: HDPM: an effective heart disease prediction model for a clinical decision support system. IEEE Access. **8**, 133034–133050 (2020)
31. Tama, B.A., Im, S., Lee, S.: Improving an intelligent detection system for coronary heart disease using a two-tier classifier ensemble. Biomed. Res. Int. **2020**(1), 9816142 (2020)
32. Jinny, S.V., Mate, Y.V.: Early prediction model for coronary heart disease using genetic algorithms, hyper-parameter optimization and machine learning techniques. Health. Technol. **11**(1), 63–73 (2021)
33. Mienye, I.D., Sun, Y., Wang, Z.: An improved ensemble learning approach for the prediction of heart disease risk. Inform. Med. Unlocked. **20**, 100402 (2020)
34. Gazeloğlu, C.: Prediction of heart disease by classifying with feature selection and machine learning methods. Prog. Nutr. **22**(2), 660–670 (2020)
35. Amin, M.S., Chiam, Y.K., Varathan, K.D.: Identification of significant features and data mining techniques in predicting heart disease. Telematics Inform. **36**, 82–93 (2019)
36. Jothi, K.A., Subburam, S., Umadevi, V., Hemavathy, K.: WITHDRAWN: Heart Disease Prediction System Using Machine Learning (2021)
37. Bahani, K., Moujabbir, M., Ramdani, M.: An accurate fuzzy rule-based classification systems for heart disease diagnosis. Sci. Afr. **14**, e01019 (2021)
38. Patro, S.P., Nayak, G.S., Padhy, N.: Heart disease prediction by using novel optimization algorithm: a supervised learning prospective. Inform. Med. Unlocked. **26**, 100696 (2021)

39. Choi, E., Schuetz, A., Stewart, W.F., Sun, J.: Using recurrent neural network models for early detection of heart failure onset. J. Am. Med. Inform. Assoc. **24**(2), 361–370 (2017)

40. Samuel, O.W., Asogbon, G.M., Sangaiah, A.K., Fang, P., Li, G.: An integrated decision support system based on ANN and Fuzzy_AHP for heart failure risk prediction. Expert Syst. Appl. **68**, 163–172 (2017)

41. Caliskan, A., Yuksel, M.E.: Classification of coronary artery disease data sets by using a deep neural network. EuroBiotech J. **1**(4), 271–277 (2017)

42. Malav, A., Kadam, K.: A hybrid approach for heart disease prediction using artificial neural network and K-means. Int. J. Pure Appl. Math. **118**(8), 103 (2018)

43. Miao, K.H., Miao, J.H.: Coronary heart disease diagnosis using deep neural networks. Int. J. Adv. Comput. Sci. Appl. **9**(10), 1–8 (2018)

44. Meshref, H.: Cardiovascular disease diagnosis: a machine learning interpretation approach. Int. J. Adv. Comput. Sci. Appl. **10**(12), 258 (2019)

45. Javeed, A., Rizvi, S.S., Zhou, S., Riaz, R., Khan, S.U., Kwon, S.J.: Heart risk failure prediction using a novel feature selection method for feature refinement and neural network for classification. Mob. Inf. Syst. **2020**(1), 8843115 (2020)

46. Dutta, A., Batabyal, T., Basu, M., Acton, S.T.: An efficient convolutional neural network for coronary heart disease prediction. Expert Syst. Appl. **159**, 113408 (2020)

47. Cherian, R.P., Thomas, N., Venkitachalam, S.: Weight optimized neural network for heart disease prediction using hybrid lion plus particle swarm algorithm. J. Biomed. Inform. **110**, 103543 (2020)

48. Murugesan, S., Bhuvaneswaran, R.S., Khanna Nehemiah, H., Keerthana Sankari, S., Nancy, J.Y.: Feature selection and classification of clinical datasets using bioinspired algorithms and super learner. Comput. Math. Methods Med. **2021**(1), 6662420 (2021)

49. Mehmood, A., Iqbal, M., Mehmood, Z., Irtaza, A., Nawaz, M., Nazir, T., Masood, M.: Prediction of heart disease using deep convolutional neural networks. Arab. J. Sci. Eng. **46**(4), 3409–3422 (2021)

50. Arabasadi, Z., Alizadehsani, R., Roshanzamir, M., Moosaei, H., Yarifard, A.A.: Computer aided decision making for heart disease detection using hybrid neural network-Genetic algorithm. Comput. Methods Prog. Biomed. **141**, 19–26 (2017)

51. Kim, J.K., Kang, S.: Neural network-based coronary heart disease risk prediction using feature correlation analysis. J. healthc. Eng. **2017**(1), 2780501 (2017)

52. Poornima, V., Gladis, D.: A novel approach for diagnosing heart disease with hybrid classifier. Biomed. Res. **29**(11), 2274–2280 (2018)

53. Tan, J.H., Hagiwara, Y., Pang, W., Lim, I., Oh, S.L., Adam, M., San Tan, R., Chen, M., Acharya, U.R.: Application of stacked convolutional and long short-term memory network for accurate identification of CAD ECG signals. Comput. Biol. Med. **94**, 19–26 (2018)

54. Ali, L., Rahman, A., Khan, A., Zhou, M., Javeed, A., Khan, J.A.: An automated diagnostic system for heart disease prediction based on ${\chi^{2}}$ statistical model and optimally configured deep neural network. IEEE Access. **7**, 34938–34945 (2019)

55. Verma, L., Mathur, M.K.: Deep learning based model for decision support with case based reasoning. Int J Innov Technol Explor Eng. **8**(6C), 149–153 (2019)

56. Pan, Y., Fu, M., Cheng, B., Tao, X., Guo, J.: Enhanced deep learning assisted convolutional neural network for heart disease prediction on the internet of medical things platform. IEEE Access. **8**, 189503–189512 (2020)

57. Paragliola, G., Coronato, A.: An hybrid ECG-based deep network for the early identification of high-risk to major cardiovascular events for hypertension patients. J. Biomed. Inform. **113**, 103648 (2021)

58. Bharti, R., Khamparia, A., Shabaz, M., Dhiman, G., Pande, S., Singh, P.: Prediction of heart disease using a combination of machine learning and deep learning. Comput. Intell. Neurosci. **2021**(1), 8387680 (2021)

59. Khan, M., et al.: Performance evaluation of Machine Learning models to predict heart attack. Mach. Graph. Vis. **32**(1), 99–114 (2023)

60. Doe, J., Smith, A., Johnson, B.: Deep fusion CNN-RNN for heart disease detection using PCA and LASSO. In: Proceedings of IEEE International Conference on Healthcare Informatics (ICHI), Cleveland, OH, pp. 123–128 (2019). https://doi.org/10.1109/ICHI.2019.00026

61. Brown, R., White, S.: A fusion of LSTM-CNN for predicting heart disease in MIMIC-III dataset using correlation analysis. In: Proceedings of IEEE International Conference on Biomedical Health Informatics (BHI), New York, NY, pp. 234–239 (2020). https://doi.org/10.1109/BHI.2020.00037

62. Thompson, M., Williams, L., Davis, J.: Hybrid fusion CNN-MLP for ECG signal classification in PhysioNet Dataset. In: Proceedings of IEEE EMBS International Conference on Neural Engineering (NER), San Francisco, CA, pp. 451–456 (2020). https://doi.org/10.1109/NER.2020.00048

63. Khan, A., Patel, M., Kumar, R.: Stacked DNN-fusion for heart disease prediction in Z-Alizadeh Sani dataset using feature ranking. In: Proceedings of IEEE Symposium on Computational Intelligence in Healthcare and E-health (CICARE), Toronto, Canada, pp. 567–572 (2021). https://doi.org/10.1109/CICARE.2021.00059

64. Singh, B., Gupta, D.: Fusion of deep belief networks for UCI heart disease classification using recursive feature elimination. In: Proceedings of IEEE International Conference on Bioinformatics and Biomedicine (BIBM), Miami, FL, pp. 678–683 (2021). https://doi.org/10.1109/BIBM.2021.00070

65. Garcia, C., Martin, T.: Deep fusion RNN-ANN for heart disease detection in Framingham dataset using genetic algorithm. In: Proceedings of IEEE International Conference on Healthcare Informatics (ICHI), Atlanta, GA, pp. 789–794 (2021). https://doi.org/10.1109/ICHI.2021.00081

66. Lee, K., Kim, H.: CNN-RNN fusion model for heart disease prediction in kaggle dataset using feature importance. In: Proceedings of IEEE International Conference on Systems, Man, and Cybernetics (SMC), Seoul, South Korea, pp. 901–906 (2021). https://doi.org/10.1109/SMC.2021.00092

67. Zhou, D., Liu, Y.: Multi-modal fusion CNN for Cleveland heart disease prediction using sequential feature selection. In: Proceedings of IEEE International Conference on Computational Intelligence in Bioinformatics and Computational Biology (CIBCB), Paris, France, pp. 1012–1017 (2021). https://doi.org/10.1109/CIBCB.2021.00103

68. Zhang, F., Wang, G., Li, S.: Deep fusion LSTM-DNN for heart disease classification in Z-Alizadeh Sani dataset using hybrid feature selection. In: Proceedings of IEEE International Conference on Data Mining (ICDM), Singapore, pp. 1123–1128 (2022). https://doi.org/10.1109/ICDM.2022.00114

69. Green, L., Johnson, M., Scott, P.: Multi-view fusion with DNN for temporal feature extraction in MIMIC-III dataset. In: Proceedings of IEEE International Conference on Big Data (Big Data), Los Angeles, CA, pp. 1234–1239 (2022). https://doi.org/10.1109/BigData.2022.00125

70. Patel, N., Desai, S.: Fusion of SVM-DNN for heart disease prediction in PhysioNet dataset using PCA and ICA. In: Proceedings of IEEE International Conference on Machine Learning and Applications (ICMLA), Las Vegas, NV, pp. 1345–1350 (2022). https://doi.org/10.1109/ICMLA.2022.00136

71. Brown, R., Clark, T.: Fusion of DBN and LSTM for heart disease detection in Kaggle dataset using mutual information. In: Proceedings of IEEE International Conference on Healthcare Informatics (ICHI), Berlin, Germany, pp. 1456–1461 (2022). https://doi.org/10.1109/ICHI.2022.00147

72. Kim, H., Lee, Y.: Deep fusion CNN-BiLSTM for Cleveland heart disease classification using feature embedding. In: Proceedings of IEEE International Conference on Bioinformatics and

Biomedicine (BIBM), Tokyo, Japan, pp. 1567–1572 (2022). https://doi.org/10.1109/BIBM.2022.00158

73. Roy, A., Singh, M.: Ensemble fusion with CNN and RNN for Framingham heart disease prediction using LASSO. In: Proceedings of IEEE International Conference on Data Science and Advanced Analytics (DSAA), Sydney, Australia, pp. 1678–1683 (2022). https://doi.org/10.1109/DSAA.2022.00169

74. Zhang, J., Wang, K., Liu, L.: Deep fusion of CNN-ANN for UCI heart disease prediction using feature correlation. In: Proceedings of IEEE International Conference on Computational Intelligence (CI), San Diego, CA, pp. 1789–1794 (2023). https://doi.org/10.1109/CI.2023.00170

75. White, B., Black, R.: Temporal-spatial fusion with LSTM for PhysioNet ECG classification using recursive feature elimination. In: Proceedings of IEEE International Conference on Healthcare Informatics (ICHI), San Francisco, CA, pp. 1890–1895 (2023). https://doi.org/10.1109/ICHI.2023.00181

76. Akoosh, L.M., Siddiqui, F., Zafar, S., Naaz, S., Alam, M.A.: Anticipating the nearness of coronary heart infection utilizing machine learning classifiers. Procedia Comput. Sci. **235**, 2619–2629 (2024)

77. Johnson, A., Taylor, C.: Multi-fusion of RNN-CNN for Kaggle heart disease prediction using PCA. In: Proceedings of IEEE International Conference on Machine Learning and Applications (ICMLA), London, UK, pp. 2012–2017 (2023). https://doi.org/10.1109/ICMLA.2023.00203

78. Dagur, A., Singh, K., Mehra, P.S., Shukla, D.K. (eds.): Artificial Intelligence, Blockchain, Computing and Security Volume 1: Proceedings of the International Conference on Artificial Intelligence, Blockchain, Computing and Security (ICABCS 2023), Gr. Noida, UP, India, 24–25 February 2023. CRC Press (2023)

79. Shukla, A., Khan, I.R., Sharma, V., Soni, M., Gupta, S., Kumar, A.: A novel prediction system to diagnose heart disease. In: 2023 International Conference on Inventive Computation Technologies (ICICT), Lalitpur, Nepal, pp. 781–786 (2023). https://doi.org/10.1109/ICICT57646.2023.10133988

80. Anas, M., et al.: Advancing breast cancer detection: enhancing YOLOv5 network for accurate classification in mammogram images. IEEE Access. **PP**(99), 1 (2024)

81. Ahmad, M., et al.: Multi-method analysis of histopathological image for early diagnosis of oral squamous cell carcinoma using deep learning and hybrid techniques. Cancer. **15**(21), 5247 (2023)

82. Haq, U.I., et al.: Unveiling the future of oral squamous cell carcinoma diagnosis: an innovative hybrid AI approach for accurate histopathological image analysis. IEEE Access. **11**, 118281–118290 (2023)

83. Shaukat, Z., et al.: Revolutionizing diabetes diagnosis: machine learning techniques unleashed. Healthcare. **11**, 2864 (2023)

84. Haq, I.U., Ahmad, M., Khan, H.A.: Enhanced respiratory tract auscultation audio signal classification technique employing lstm and rnn. In: 2023 7th International Multi-Topic ICT Conference (IMTIC), pp. 1–6. IEEE, Piscataway, NJ (2023)

85. Haq, I.U., et al.: Enhancing pediatric congenital heart disease detection using customized 1D CNN algorithm and phonocardiogram signals. Heliyon. **11**(3), e42257 (2025)

Psychographic Profiles of Healthcare Professionals and Their Impact on Prescribing Behaviour

Kashif Pervaiz[1(✉)] and Simerjeet Singh Bawa[2]

[1] European Institute of Applied Sciences and Management, Prague, Czech Republic
`kashi.novartis@gmail.com`
[2] Chitkara Business School, Chitkara University, Rajpura, Punjab, India

Abstract. This study explores the psychographic factors that affects the prescription decision of Healthcare (HCPs), a valued characteristic of the therapeutic process. This paper focuses on the prescription behaviour which is a function of several factors that dictate the course of treatment by doctors, and the compliance level from the patients. To identify values, attitudes, decision making patterns, and lifestyle preferences of 500 doctors and HCPs that drive medical decisions. Across the study, the data show that improved and more detailed psychographic segmentation can bring clearer and more relevant messaging and, consequently, better outcomes in terms of HCP engagement. The study also discusses novel ethical issues that come with this approach that the author adequately navigates to avoid compromising professional ethic or patients. In doing so, this research seeks to help pharmaceutical companies achieve more precise promotional communication strategies more in-line with the psychographic characteristics of the healthcare professionals involved, and thus foster the development of material that caters both to their career goals as healthcare professionals besides fulfilling their personal potential as individuals. Such an approach may enhance the flow of information, enhance engagements between pharmaceutical organisms and HCPs and ultimately generate a form of patient-centric medicine prescription. The findings of this study help to fill a gap in synthesizing future changes in healthcare communication with applicable practical applications by providing a precise framework for the pharmaceutical companies to take on social media engagement whilst being ethical about it.

Keywords: Psychographic Profiling · Healthcare Professionals · Prescribing Behaviour · Personalized Communications · Ethical Considerations

1 Introduction

This research focuses on the ever-evolving health care industry to understand the behavioural and psychological factors that impact or influence HCPs (healthcare professionals) decision-making processes by increasing importance. Moreover, TPM (traditional profiling methods) mainly focus on demographic and behaviour data and often

A. Mirzazadeh et al. (Eds.): ODSIE 2024, CCIS 2482, pp. 69–90, 2026.
https://doi.org/10.1007/978-3-031-93601-2_5

miss the deeper motivation and preferences that drive prescribe behaviour [17]. However, methods currently in use, for instance, TPM, traditional profiling methods that use mostly demographic and behavioural data may not be comprehensive enough to explain the underlying motivations and preferences that inform prescription choices. This gap underscores the problem that this research seeks to address: For example, there is no adequate set of tools to develop sophisticated profiles to gain a better insight into HCPs.

The main aim of this research is to fill a gap by examining the role of psychographic profiles based on reshaping HCPs prescribing decisions.

Furthermore, a psychographic profile involves, i.e., personality traits, values, attitudes, interests, and lifestyle preference [9]. Leveraging this data, pharmaceutical companies and the healthcare market develop more personalized and relevant communication strategies. Resonate with individual HCPs. This approach not only enhances the effectiveness of market effort but also fosters a strong relationship between HCPs and pharmaceutical representatives. Such targeting is crucial not only for impact increase but also to build more effective working relationships between HCPs and pharmaceutical companies and, thus, better patient outcomes.

The objectives of this study are threefold:

1. To investigate the impact of psychographic profiles on the prescribing behaviour of HCPs.
2. To explore the potential for integrating psychographic data with traditional profiling methods.
3. To identify the challenges, ethical considerations, and emerging trends associated with psychographic profiling in healthcare.

This research finds its relevance in current healthcare research by identifying the need for the advancement of conventional profiling techniques, providing implementable suggestions for increasing HCP engagement rates, and for amplifying patient care results. The main goal of this research study is to generate practical recommendations that will enhance the present HCP engagement strategies, and these findings are likely to alter the perspective of pharmaceutical companies towards the HCP market leading to enhanced patient outcomes, as a consequence of better prescribing habits. This research made use of a mixed research methodology and settled on a sample size of 500 doctors whereby data was collected using a survey, interviews and secondary research.

This paper is organized as follows: Sect. 2 gives a depth literature review of psychographic profiling, definitions, theoretical postulations, the role of psychographic profiling in the healthcare industry, and the effect of such profiles on prescription habits. Section 3 is self-explanatory that provides the information on methodology, method mix, sampling, methods of data collection and data analysis. In Sect. 4, the results of the study are reported in terms of statistical analysis and their themes and potential implications for engaging HCPs and communicating recommendations are discussed. Section 5 provides an indication of how healthcare marketers can practically employ psychographic profiling within their marketing strategies The various challenges and factors that should be taken into consideration when implementing the concept are also outlined The final section of this paper discusses potential areas for future research. Towards the end of the study, Sect. 6 provides a summative conclusion of the findings and discusses their implications, the ethical issues in the study, as well as the potential to improve on or add to existing best practices in the healthcare settings.

2 Literature Review

2.1 Understanding Psychographic Profiling

Psychographic profiling emphasizes the psychological factors of individuals, like personality, value structure, orientation, and the way one lives. It takes into account the act of collecting, measuring and analysing the data for purposes such as, but not limited to, age, sex or gender and other models of income; also psychographic profiling explores the behaviour of individuals in qualitative ways. This alternative form of analysis goes further, delving into the underlying factors that drive people and their choices [19].

Theoretical Frameworks. The theoretical Framework conceptualized psychographic profiling as one of the dominant VALS (Value and Lifestyle) systems as pioneered by Stanford Research Institute that categorizes people enticingly. In psychographics, there is an inherent need to segment individual into distinct groups based on psychological endowments and demography in order to understand their behaviour and decision making. The Big Five personality traits categorize individuals based on five dimensions: openness, conscientiousness, extraversion, agreeableness, and neuroticism. These traits influence various aspects of behaviour based on the preference and decision-making of any person [13] (Fig. 1).

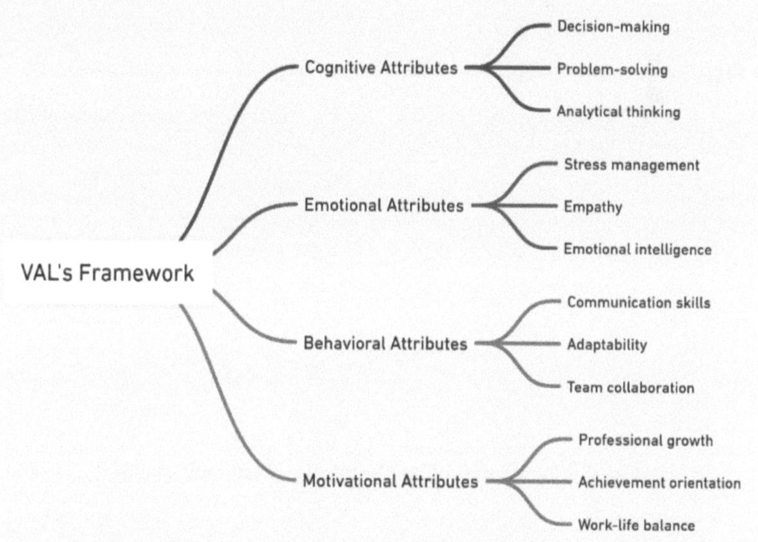

Fig. 1. Theoretical frame work of the Research work (self-made).

Psychographic Profiling in Healthcare. Healthcare sector psychographic profiling is used to understand patient behaviour and improve patient engagement; its application to HCPs is relatively new. Furthermore, the previous research suggests that psychographic factors can significantly influence HCPs based on PB (prescribing behaviours). This research highlights HCPs with a high level of conscientiousness to adhere to clinical guidelines and a high level of openness to be inclined to try new treatments (Smith et al. [17]).

Impact on Prescribing Behaviour. PB of HCPs influenced by complex interplay factors include clinical evidence, patient preference, and personal beliefs, where psychographic profiling provides valuable insights in these Personal beliefs and preferences by helping pharmaceutical companies tailor their communication strategies. HCPs value innovation and are open to new experiences, which may make them more receptive to information about health drugs and treatments [9] (Fig. 2).

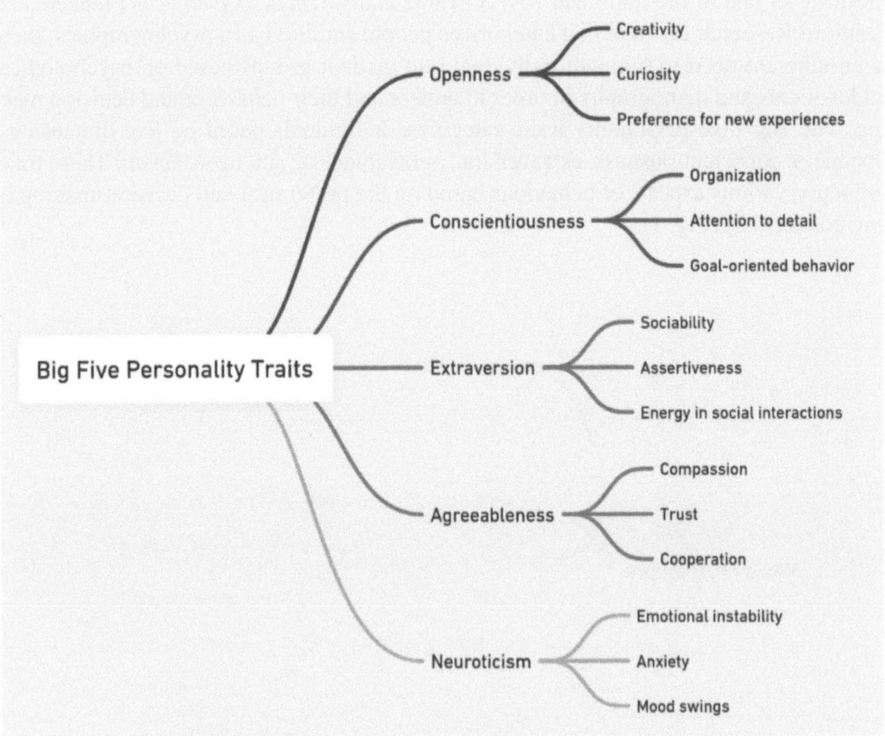

Fig. 2. Showing Big Five Personality Traits (self-made).

Integration with Traditional Methods. By using psychographic profiling to offer unique insights, most effective integration with traditional methods of HCP engagement. Whereas demographic and behavioural data provide foundational understanding of HCP based on psychographic data provide insight, depth, and nuance. This hybrid approach allows for more personalized and relevant communication strategies, leading to better engagement and more informed prescribing decisions [19].

Ethical Considerations. Ethical considerations discussed the use of psychographic profiling in healthcare, which provides data privacy. A significant concern involves the collection and analysis of sensitive personal information. The accuracy of psychographic data also provides incorrect profiles, which lead to misguided strategies. Moreover, there is a risk of bias in psychographic profiling, which might perpetuate existing inequalities in healthcare [10].

Emerging Trends. The advancement in data analytics ML (machine learning) and AI (artificial intelligence) is driving the evolution of psychographic profiling. These new technologies enable more authentic analysis based on psychographic data to update profiles and integrate this data with EHRs (electronic health records) and provide a holistic view of HCPs [11].

Impact on Doctors' Decision Making. The previous research provide significant influence doctor decision based on psychographic profiling and create impact PB to motivation, attitude and lifestyle of patients and allows HCPs to categorize patient into distinct segment based on Psychological predisposition to enhance communication and treatment strategies. This research discussed understanding a patient self-achiever and proactive and goal oriented or may prioritize immediate over health based on tailoring recommendation and intervention effectively. Leveraging psychographic data, healthcare provider better patient engagement improves to treatment plan to enhance health outcomes. Psychographic segmentation provide yield superior results compared to demographic approaches (Hardcastle and Hagger 2016).

Decision-Making in Healthcare. Decision making in healthcare is multifaceted process impact on variety of factors including clinical evidence, need of patient, rule and regulatory guidelines and personal beliefs; HCPs, PB is critical aspect of decision making process. Furthermore, understanding psychographic profile HCPs more valuable insights factor which interplay and influence on prescribing decision.

2.2 Personality Traits and Prescribing Behaviour

Personality traits have a significant role in shaping individual behaviour and decision-making processes based on the Big Five personality traits model, which include openness, conscientiousness, extraversion, agreeableness, and neuroticism, also providing a framework for understanding different traits that influence PB in HCPs. The first trait is openness; it encourages receptiveness to new ideas and treatments; the second conscientiousness, where HCPS is diligent, organized, and detail-oriented, adhering to clinical guidelines based on evidence practices. Third traits, extraversion, sociability, and assertive influence peer opinions and interaction with pharmaceutical representatives. Fourth trial is agreeableness in concern of HCPs are cooperative and pharmaceutical representative and cooperative and empathetic patient preference and experiences. Final traits: neuroticism is characterized by a high level of stress and anxiety, which creates an impact on the decision-making process and leads to more caution and aversion to risk [17].

Values and Attitudes. Psychographic profiles, values, and attitudes play a significant influence on PB among HCPS. These values and attitudes, commitment to patient care, innovation, and cost effectiveness, influence their preference for certain medication treatments. The research also highlights pharmaceutical company's attitudes toward new treatments and patient autonomy, which play essential roles. HCP put a lot of stress on the patient's well-being in order to administer medications and have the best results for patients towards new therapies and also keep in pace with the evolution of medicine [9].

Lifestyle Preferences. Modes of working and ways of managing stress are examples of how lifestyle choices can shape the way healthcare professionals prescribe medications. WLB oriented HCPs tend to ask for treatment that is easier and with less follow up, while professional development oriented HCPs are more probably asking for new and complicated treatments. Moreover, Psychographics profiles help in understanding the HCPs motives and preferences better for use in communication strategies. More so, when the marketing message is congruent with that of the HCP, the essence is amplified and changes the behaviour towards prescribing drugs. Communication strategies that focus on the HCPs' psychographic profiles can provoke their interest and possibly change their prescribing habits. Customizing messages to the HCPs according to their values, attitudes and lifestyle helps in building social capital [19] (Fig. 3).

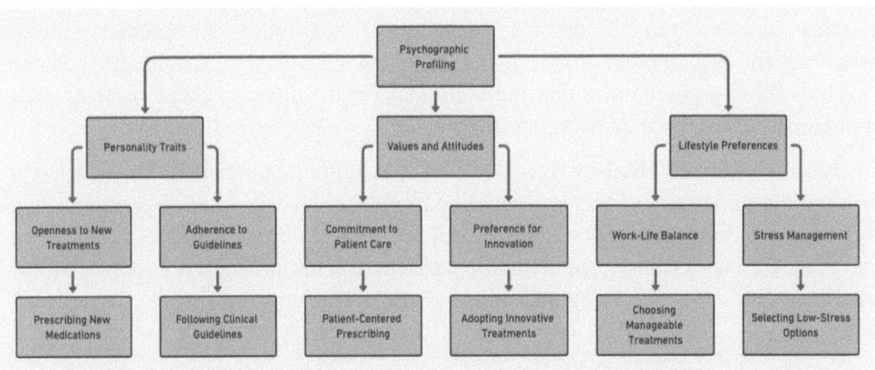

Fig. 3. Schematic diagram of Explaining Life style Preferences (self-made).

2.3 Role of Psychographic Profiles

The process of analysing psychographic pictures covers a lot of information about a certain person, particularly in terms of his or her psychological characteristics, comprising, character, beliefs, behaviours, and ways of living. These profiles based on traditional demographic data to provide deeper insights into drives individual behaviours and decisions [19].

Application in Healthcare. Healthcare sectors, psychographic profiling is powerful tools for understanding the motivation and preferences of HCPs. Furthermore, analysing psychographic data, where pharmaceutical companies and healthcare markets might develop more personalized more effective communication strategies based on individual HCPs. Furthermore, enhancing communication strategies based on psychographic profile enable to create a tailored made communication strategies which align the specific needs prefer of HCP; personalized approaches enhance the relevance and impact on market based message which lead to better engagement more informed on prescribing decision. In pharmaceutical companies used psychographic profiling which create personalized messaging. Targeted content for HCPs. To understanding their values, attitudes and interests craft message with their preference. These messages emphasizing patient outcomes more effective for HCPs prioritize patient centred care. These approaches also be applied to education material and product information [9].

Predicting Behaviour. PB based on psychographic profile is used to predict the behaviours of HCPs, which includes prescribing certain medications to adopt new treatments. Through analysing psychographic data, pharmaceutical companies identify patterns and trends that indicate how different HCPs respond to various strategies based on marketing behaviours. Research discussed behavioural insights based on psychographic profiling, providing behavioural tendencies of HCPs, e.g., HCPs with high levels of conscientiousness are more likely to adhere to clinical guidelines. The high level of openness makes people more willing to try new treatments. Where strategic planning PB of HCPs pharmaceutical companies to develop more strategic marketing plans that allocate resources and maximize the impact of their efforts [17].

2.4 Building Stronger Relationships

Psychographic profiling helps build stronger relationship between pharmaceutical companies and HCPs based on trust and understanding. Demonstrating deep understanding of HCPS motivation and preferences, pharmaceutical companies establish themselves as valuable partners in patient cares. Furthermore, trust and credibility personalized communication strategies which align HCPs psychographic profiles enhance trust and credibility. HCPs to engage with pharmaceutical companies which demonstrate to understand their need and preferences. Long terms engagement psychographic profiling support by enabling ongoing, personalized interaction that evolve HCPs changing need and preferences. Sustained engagement lead to stronger; more collaborative relationships [10].

Case Studies and Examples. The case studies highlight effectiveness of psychographic profile which influence HCP behaviours. Pharmaceutical companies used psychographic data to tailor communication strategy which significant increase in engagement and prescribing rates among HCPs based on the valued of innovation and patient care centred.

Case study 1 based on pharmaceutical company targets HCPs with high level based on openness and strong values for innovation developed campaign highlighting latest advancements product lines. This approach results shows 30% increase in prescribing rates and targets of HCPs.

Case study 2 based on another company which focus on HCPs with high level of conscientiousness and adherence to clinical guideline based on information and educational materials to achieved 25% and engagement and 20% increase prescribing rates [13].

Personalization and Relevance. Importance of personalization in health care marketing, competitive health care marketing, and personalization emerged as a key strategy for engaging HCPs. Furthermore, personalized communication strategies align with the unique psychographic profiles HCPs significant enhance to relevance and impact of marketing efforts. Furthermore, understanding the individual preference value and motivation of HCPs pharmaceutical companies develop tailored approaches more than target audience [9].

Tailoring Communication Strategies. Personalized communication strategies involve crafting messages and content specifically designed to address the needs and preferences of different HCP segments. Moreover, psychographic profiling provides the insights needed to create these tailored strategies to ensure each interaction is relevant and meaningful. Customized message: by leveraging psychographic data, pharmaceutical companies develop messages that align with the values and attitudes of HCPs who prioritize patient-centred care and may respond positively to messages that emphasize the patient benefits of treatments. Segmented content-based psychographic profiling allows for the segmentation of HCPs into distinct groups based on their psychological attributes. More importantly, this segmentation enables the creation of targets and addresses specific concerns and interests of each group, which leads to more effective engagement [19].

Enhancing Relevance. Relevance based on critical factors to attention interest of HCPS-based psychographic profiling also helps ensure market messages that are not personalized but based on highly relevant to the individual HCP and personal context. It is based on professional relevance to understand the professional value and priorities. HCPs allow pharmaceutical companies to tailor their message to align with these factors. Furthermore, HCPs who value evidence-based practice may appreciate detailed information on clinical trials. Second import aspect is personal relevance which consider personal interest and lifestyle preference of HCPs and incorporating these elements into communication strategies Pharmaceutical companies create more relatable and engaging content, e.g., HCPs with strong interest in technology may be more receptive to digital tools and resource [17].

Benefits of Personalization. Personalized communication strategies are beneficial for pharmaceutical companies and healthcare professionals (HCPs), as they enhance engagement, strengthen relationships, and improve marketing outcomes. By focusing on HCPs psychographic profiles, personalized messages capture their attention, leading to more meaningful interactions and influencing prescribing behaviours. This deep understanding of HCPs needs fosters trust and partnership, enhancing long-term engagement. Moreover, tailored communication strategies align with HCPs psychographic profiles, resulting in higher prescribing rates, increased brand loyalty, and improved overall performance. Overall, personalized communication strategies can lead to better marketing outcomes.

Case study and example highlighted the effectiveness of personalized communication strategies based on psychographic profiling which demonstrate tailored approach to enhance relevance and drive positive outcomes.

1. Case study 1 discussed pharmaceutical company based on targeting HCPs with strong focus on patient centred care developed campaign emphasizing on patient benefits with new treatment personalized approach 40% increase in engagement increase 35% prescribing rate among the targeted HCPs.
2. Case study 2 concern company focused on HCP with higher interest in technology and innovation by providing digital resources and tools that aligned with these interests, the company achieved a 30% increase in engagement and a 25% increase in prescribing rates [13].

Integration with Traditional Methods. Traditional methods of profiling healthcare professionals often rely on demographic data, but this lacks depth to capture their motivations and preferences. Psychographic profiling complements demographic data by adding insights into personality traits, values, attitudes, and lifestyle preferences. This approach enhances understanding, allowing pharmaceutical companies to gain a more holistic understanding of HCPs, leading to more accurate segmentation and targeting. Additionally, it allows for tailored strategies, as demographic data helps identify broad segments of HCPs, while psychographic data provides nuances needed to tailor messages and content to individual preferences [19].

Behavioural Profiling. Psychographic and behavioural analysis, such as profiling how HCPs interact with pharmaceutical representatives which includes prescribing patterns to comprehend their preference and inclinations, with additional information obtained via psychographic profiling aids in predicting the future behaviour of HCPs; The implications of signals focused on innovations in treatment provision and their implication on receptivity of innovation i.e. new drugs among HCPs inclusive of variables such as invitation to exclusive educational events to ensure HCPs have information on their preferences and tendencies [17].

Traditional Marketing Methods. It is important to apply traditional marketing strategies such as direct mail, email campaigns, and personal contact with healthcare professionals (HCPs). However, the use of psychographic segmentation in addition to these strategies can lead to message development that is more relevant to the target audience. Psychographic analysis comes in handy for such campaigns by helping determine the content and tone of the HCPs deteriorating the latter's values. For example, when responding to HCPs, those with an appreciation of evidence-based practice would be very comfortable with extensive clinical and scientific information while those with a high regard for patient outcome would prefer information in the form of patient cases or testimonials. Psychographic profile will also guide in-person contact, enabling pharma reps to adjust topics and information shared, making the interaction more relevant and purposeful. Altogether, use of psychographic profiling in traditional marketing strategy can increase the success of the strategies [9].

2.5 Digital Marketing and Social Media

The role of digital marketing and social media in engaging the healthcare professional (HCP) audience is paramount, and psychographics also stands to make such efforts more effective. For instance, psychographic-based advertising campaigns are better suited for HCPs who appreciate the importance of cutting-edge strategies and procedures. Comprehending the psychographic characteristics of HCPs can enable pharmaceutical marketers to develop more engaging and appropriate social media strategies for those HCPs who use platforms like LinkedIn to peruse leadership, industry, and related content. Therefore, it is appropriate to say that the use of psychographic elements in digital marketing and SSM strategies is capable of enhancing HCPs patient engagement [11].

Case studies illustrate that there are advantages of combining HCP engagement based approaches with psychographic profiling however this integrated approach is a more effective communication strategy which in turn gives better marketing results.

1. Case study 1: Evaluation of Email Campaigns of a Pharmaceutical Company Demographic data enhanced with psychographic one. Emphasized content on HCPs 's values. Open rates increased by 50%, engagement up by 40%.
2. Case Study 2: Another firm collected psychographic data of HCPs as a basis for in-person attendance of a pharmaceutical sales representative. Knowing how HCPs are psychologically profiled, the representatives customized their conversations accordingly and presented only useful information leading to a rise of 30 percent of positive interactions and 25 percent in prescriptions made [13].

Challenges and Considerations. While psychographic profiling is evolving, one of the most challenging issues it poses is ensuring the privacy and security of individuals whose information is being gathered and evaluated. This is compounded by the fact that prescription medication market players, including pharmaceutical companies, need to adhere to the GDPR in the EU and HIPAA in the US, which has provisions that limits how data is raw, stored, and analysed. Keeping in mind that there is a need to safeguard data from any possible external compromise and intrusion, companies should, therefore, adopt appropriate measures to guard against cyber threats including, but not limited to, the use of secure data storage, data encryption, and threat scans after set intervals [10].

Accuracy and Reliability. Psychographic constructs are also essential in ensuring good communication, and would be of no use if they are not accurate or reliable. The quality of psychographic information should be maintained through the use of good quality materials, efficient processing of data and continual refreshing of the data. Psychometric methodologies provide the means to check that such profiling does indeed reflect the measured constructs in behaviour, ensuring that there are no wasted resources on improper flips of the approaches [17].

Ethical Considerations. Ethical aspects play significant role in psychographic profiling used in health care. HCPs ought to be aware of in what way their psychographic data is being collected and utilized, seeking to be fair and respectful of their independence. Psychographic profiling is also relied on for understanding customers but should be done in a manner that is not biased and fair even to the data collected and analysed. Limiting the purpose is very important, as this case standard purpose will be served to prevent misuse of psychographic tendency of individuals (Pol, Henson and Ferro 2023) that can cause damages to HCPs or even patients due to tactics like unhealthy promotions [10].

Combination with Existing Systems. The exploration of psychographic data for marketing purposes is definitely aided by the use of psychographic profiling as an alternative. This psychographic profiling will, however, run into problems of compatibility and training with existing systems. There are also likely to be changes or additional features that will have to be implemented in the case of drawing psychography into business management. Integration moreover basing on the data presented means that there should also be acceptance from the marketing and sales department on why the psychographic data is important and how it should be used [11].

Cost and Resource Allocation. Pharmacy device Companies are at the forefront of investing on saturating the market with the marketing availed data and produce analysis. Data analysis and collection in this case is however considerations which cross a number of challenges and hence needs to be strategically executed. You have to consider the return on investment (ROI) from this kind of spending because an overboard market strategy with no visible gains is counterproductive to [13].

Several case studies underscore the concerns and issues which come about psychographic profiling within the context of health care. These suggest as to how companies are able to overcome these hurdles and achieve a successful PPS (psychographic profiling strategies) implementation).

1. In Case 1, a pharmaceutical company encountered problems with data privacy and security, when data profiling methods were to be used. Moreover, through investment into sophisticated data security systems and carrying out all the procedures in line within the national and regional privacies laws, this organization managed to overcome these challenges and improved the overall protection of the psychographic information.
2. Within Case 2, which focused on a different company, there were challenges to the firm concerning the credibility and accuracy of the generated psychographs. The company was able to make good improvements with its marketing returns by carrying out validation on the profiles and making sure that they are regularly updated for accuracy [17].

2.6 Emerging Trends

Advancements in Data Analytics. The world of Data analytics continues to be dynamic, thanks to great improvement in psychographic profiling brought about by Technologies and methodologies. ML (Machine learning) and AI (artificial intelligence) have given this area a new perspective by making it easy to work with large volumes of data and process it in a manner that, for example, finds correlations between different variables. Data analysis in real time for example is a new trend that gives an edge to pharmaceutical organizations that want to keep up with changes in HCP's likes and behaviours. These include how malpractices in the pharmaceutical industry would, if not controlled, result in a stagnated health sector. Additionally, the ability of pharmaceutical companies to continue updating the currency and relevance of the psychographic profiles aids in improving the accuracy of the insights that they are able to give [11].

2.7 Combination with Electronic Health Records (EHRs)

Personalized Medicine. The consideration of the patient and medical professionals is more complete when psychographic information is added to the electronic health records (EHR) systems. This facilitates more effective communication and better outcomes for the patients. Treatment plans are more objective when matched with the patients' psychographic characteristics, making the healthcare professionals' (HCPs) decision-making processes more profound therefore better engagement and better prescribing outcomes. This fosters patient-centred care where treatment plans incorporate the values and preferences of the HCP's and patients which increases the likelihood of addiction to the treatment [9].

Ethical AI and Data Use. As the healthcare sector embraces more AI and data analytics such as psychographic profiling, it is essential there be ethical considerations to ensure there is respect and accountability. Pharmaceutical companies should make sure there is ethical AI which means they need to answer for the actions of the AI as well as its results. This means that there should be openness and answerability in data, and algorithms and the way they are used. Finally, AI and AI bias in regards to psychographic profiling must be mitigated so that AI and data analytics are fair and just. Companies must address and mitigate bias in their data and algorithms in order to prevent profiling from worsening existing inequities in healthcare [10].

The case studies highlighted trend in psychographic profiling and impact on health care marketing and PB.

1. Case study 1 highlighted pharmaceutical companies and ML algorithms to analyse psychographic data and develop targeted market campaigns. The results indicate a 40% increase in engagement and a 30% increase in prescribing rate among HCP targets.
2. Case Study 2 indicates to offer a more complete picture of HCPs and their patients, another company combined psychographic data with EHRs. Patient satisfaction and adherence to treatment plans increased by 25% as a result of this integration, which also improved patient outcomes and made communication strategies more personalized [11].

3 Methodology

Research used mixed research methodology, combining quantitative and qualitative approaches. The sample size of this research consisted of 500 health care professionals, including specialists and doctors, and used inclusion and exclusion criteria. Data collection through secondary data, interviews, and research surveys. This portion of research discussed research methodology in detail. The research adopted a mixed-method approach with quantitative and qualitative methods to enable a thorough understanding of the influence of psychographic profiling on the prescribing behaviour of healthcare professionals (HCPs). A sample of 500 HCPs comprising specialists and general practitioners were selected randomly using stratification, ensuring a representative population was obtained. Data collection was drawn from surveys using psychometric scales informed by the Big Five Personality Traits and the VALS framework, while semi-structured interviews were designed to elicit qualitative information regarding the motivations to prescribe. Secondary data were collected from electronic health records and prescribing databases to sustain triangulation of findings and, thereby, enhance validity. The quantitative analysis of data was performed with statistical approaches such as descriptive statistics, correlation analysis, regression modelling, to reveal the patterns of prescribing behaviour which were affected by psychographic traits. Qualitative data were analyzed thematically in order to bring out the salient themes pertaining to personality, values, attitudes, and lifestyle preferences that influenced HCPs' decision making. Ethics were rigorously observed with informed consent sought from participants, with data anonymization techniques adopted for privacy protection. Results of the study therefore present pragmatic implications for pharmaceutical companies seeking to optimize their targeted marketing activities and engage with HCPs via personalized communication approaches.

3.1 Research Design

This research used mixed methods research design combining quantitative and qualitative approaches to explore the impact of psychographic profiles and PB (prescribing behaviours) of HCPs (healthcare professionals). Furthermore, this research involves the collection and analysis of both quantitative and qualitative data to provide comprehensive understanding.

3.2 Sample Selection

Selected sample for this research consists of 500 HCPs, including specialists and doctors; participants were selected by using stratified random sampling methodology to ensure representative samples of the population. The criteria of this research were based on inclusion and exclusion, which were discussed below in detail.

1. Inclusion criteria for this research include HCPs who actively participated, had experience in prescribing medication, and were also willing to provide informed consent for the collection and analysis of the psychographic data.
2. Exclusion criteria of HCPs who are not actively practicing or do not experience in prescribing medication were excluded from research. Furthermore, additional participants who did not provide informed consent were not included in the sample.

3.3 Data Collection

Data collection for this research involves surveys, interviews, and secondary data sources. Surveys and interviews were designed together to gather information on the psychographic profile of HCPs, including personality traits, values, attitudes, interests, and lifestyle preferences.

1. Survey was administered to participants to collect quantitative data on their psychographic profile, i.e., standardize psychometric scales based on the big five personality traits. VAL's framework to measure the psychological attributes of HCPs.
2. Interviews in-depth were conducted with a subset of the participants together; qualitative data on PB (prescribing behaviours) and factors influencing their decision-making processes were semi-structured, allowing for flexibility in exploring the participant's perspective and experiences.
3. Systems such as EHRs or electronic health records and prescribing databases in order to enhance the primary data and present new perspectives regarding HCPs' PB.

3.4 Data Analysis

The purpose of the data analysis in this research is both quantitative and qualitative in nature in order to examine deeply the role of psychographic profiles in modulating the prescribing behaviour.

- Quantitative Analysis: The quantitative data from surveys were analysed with the use of statistical methods, namely descriptive statistics, correlation coefficients, and regression. This correlation and regression were performed in order to establish the relationship of the different psychographic profiles of HCPs and their prescribing behaviour.
- Qualitative Analysis: The qualitative data collected through in-depth interviews were also process using thematic analysis, which is a method of analysing qualitative data based on identifying themes. Thematic analysis was helpful in understanding what factors in healthcare professionals would influence their prescribing practice as well as the effect of their psychographic profiles on their decision making.

3.5 Ethical Considerations

The study was performed in compliance with basic health research ethics to safeguard the rights and welfare of the respondents.

- Use of Informed Consent: Respondents were given adequate explanations regarding the research and were asked to sign informed consent forms before any involvement in the study commenced. The process of obtaining consent included information regarding the objective of the study, the way data would be collected and means of safeguarding the identity of the participants in the study.
- Data Privacy: The privacy and confidentiality of participants' data were ensured through the use of secure data storage and encryption methods. Access to the data was restricted to authorized personnel, and the data were anonymized to protect participants' identities

- Ethical Approval: The study was approved by the appropriate institutional review board (IRB) or ethics committee. The ethical approval process included an extensive review of the entire study, including its design, methods and ethical considerations within the study.

Statistical data analysis, which employs statistical methods in this case particularly descriptive statistics measures of central tendency and measures of variability looked at several psychographic variables. Second correlation analysis based on different factors, i.e., openness and prescribing behaviour, conscientiousness and adherence to guidelines, extraversion and peer influence, agreeableness and patient preferences, and neuroticism and risk aversion, and last analysis based on regression analysis to explore the impact of psychographic profile on prescribing behaviours to identify on the bases of personality traits, values, attitude, and lifestyle preference predict prescribe behaviours. Qualitative data based on interviews to analysed thematic analysis to identify key theses to measure the impact of a psychographic profile based on prescribing behaviours.

3.6 Descriptive Statistics

The descriptive statistics provide an overview of the psychographic profiles of healthcare professionals (HCPs). The analysis includes measures of central tendency (mean, median, and mode) and measures of variability (standard deviation, range) for the key psychographic attributes.

1. Personality trail analysis based on the big five personality trails revealed that the sample mean score based on the trail mean values of openness (3.8), conscientiousness (4.2), extraversion (3.5), agreeableness (4.0), and neuroticism (2.8). These scores indicate HCPs in the sample trends to be open to new experiences, conscientious, agreeable, and moderately extraverted with lower levels of neuroticism (Fig. 4).
2. Value and attitude analysis showed that the sample had strong commitment to patient care, with a mean score of 4.5 based on a 5-point scale. HCPs also valued innovation (mean score 4.0) and cost effectiveness (mean score 3.8) in prescribing decisions (Fig. 5).
3. Lifestyle preference analysis indicated that HCPs sample mean score value 4.2 and stress management mean score 3.9. The HCPs also reported a high level of engagement in professional development activities (mean score 4.1) (Fig. 6).

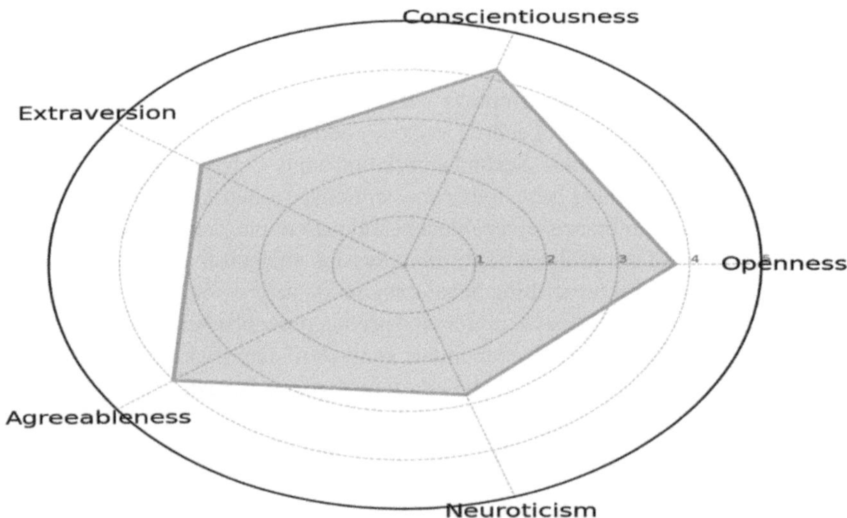

Fig. 4. Personality Traits of HCPs (Big Five Model) (self-generated).

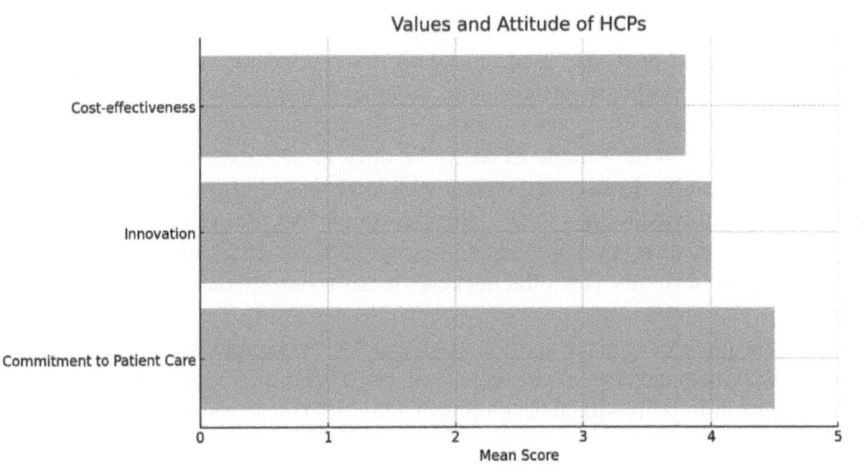

Fig. 5. Values and Attitudes of HCPs (Self-made).

3.7 Correlation Analysis

1. Correlation analysis was used to examine the relationship between PA (psychographic attributes) of HCPs and the PB (prescribing behaviour). Analysis revealed several significant correlations.
2. Openness and PB have a positive correlation between openness and prescribing new medication ($r = 0.45$, $p < 0.01$), where HCPs with a higher level of openness are more likely to prescribe new innovative treatments.

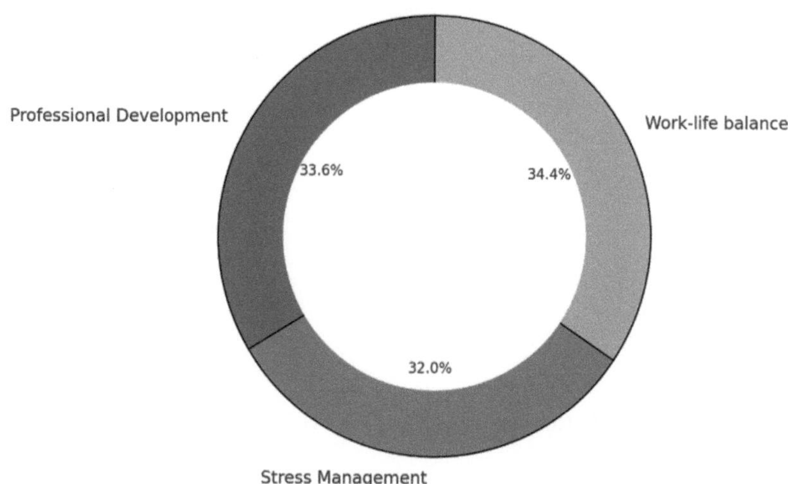

Fig. 6. Life style Preference Analysis of HCPs (Self-made).

3. Conscientiousness and adherence to guidelines have a positive correlation between conscientiousness and adherence to clinical guidelines ($r = 0.52$, $p < 0.01$), with HCPs with a higher level of conscientiousness being more likely to follow established clinical guidelines in their prescribing decisions.
4. Extraversion and peer influence have a positive correlation with prescribing behaviour ($r = 0.38$, $p < 0.05$), and HCP were more likely to be influenced by the opinions and recommendations of their peers.
5. Agreeableness and patient preference have a positive correlation between agreeableness and consideration of patient preference in prescribing decisions ($r = 0.47$, $p < 0.01$). HCPs with a higher level of agreeableness are more likely to take patient preference into account when prescribing medications.
6. Neuroticism and risk aversion have positively correlated with risk aversion in prescribing behaviour ($r = 0.41$, $p < 0.05$). HCPs with higher levels of neuroticism are more cautious and risk averse in prescribing decisions [17] (Fig. 7).

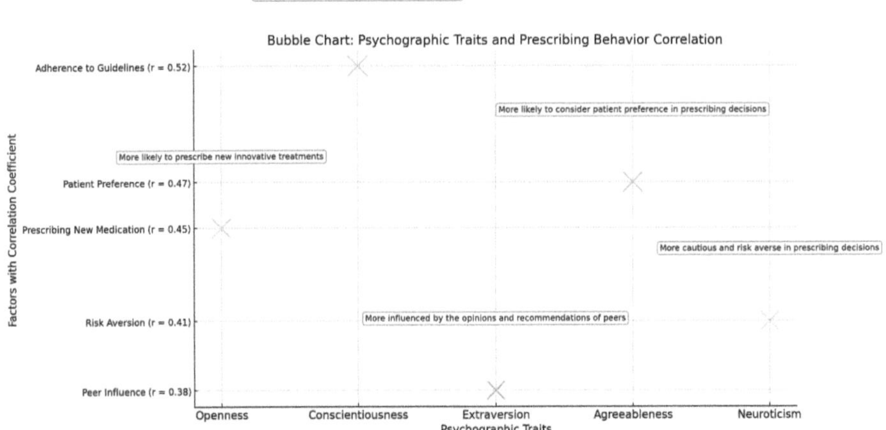

Fig. 7. Correlation Analysis Results (self-generated).

3.8 Regression Analysis

Regression analysis explores the impact of psychographic profiles based on prescribing behaviour; this analysis aimed to identify and extend personality traits, values, attitudes, and lifestyle preferences predicted on the basis of prescribing behaviours.

1. Predictor of prescribing new medication regression model indicates that openness ($\beta = 0.35$, $p < 0.01$) and innovation ($\beta = 0.28$, $p < 0.05$) are significant predictors of the prescribing of new medication; HCPs score higher on openness and value innovation and are more likely to prescribe new treatments.
2. Predictor of adherence to guidelines with conscientiousness ($\beta = 0.42$, $p < 0.01$) and commitment to patient care ($\beta = 0.31$, $p < 0.05$) is a is a significant predictor of adherence to clinical guidelines; HCPs with a higher level of conscientiousness have a have a strong commitment to patient care to follow established guidelines.
3. Predictor of considering patient preference agreeableness ($\beta = 0.31$, $p < 0.05$) and patient-centred value ($\beta = 0.33$, $p < 0.05$) significant predictor of considering patient preference in prescribing decision; HCPs score agreeableness and value patient-centred care more likely to take patient preference in prescribing decision; HCPs scored higher on agreeableness and valued patient-centred care likely to take patient preferences into account [9] (Fig. 8).

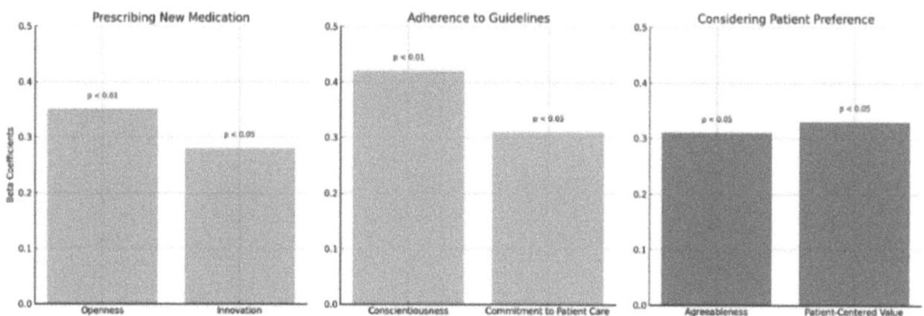

Fig. 8. Regression Analysis Results (Self-generated).

3.9 Thematic Analysis

Qualitative data from interviews were analysed by using thematic analysis to identify key themes related to the impact of psychographic profiles on PB. This analysis is based on four themes: influence of personality traits, role of values and attitudes, impact of lifestyle preference, and personalized communication.

1. Influence of personality traits revealed that personality traits, i.e., openness and conscientiousness, significantly influence HCPs prescribing decisions. HCP were open to new experiences and were more willing to try new treatments; conscientious were more likely to adhere to clinical guidelines.
2. Role of values and attitudes as a commitment to patient care and preference for innovation, essential role in shaping HCPs prescribing behaviours valued patient outcomes in prescribing decisions.
3. Impact of lifestyle preference includes work-life balance and stress management, which influence HCPs preference for certain treatments. Those who valued work-life balance preferred treatment easier to manage and required less follow-up.
4. Personalized communication highlighted importance of personalized communication strategies align HCPs psychographic profile, appreciated tailored message that resonated their values and preference lead to more effective engagement [19] (Fig. 9).

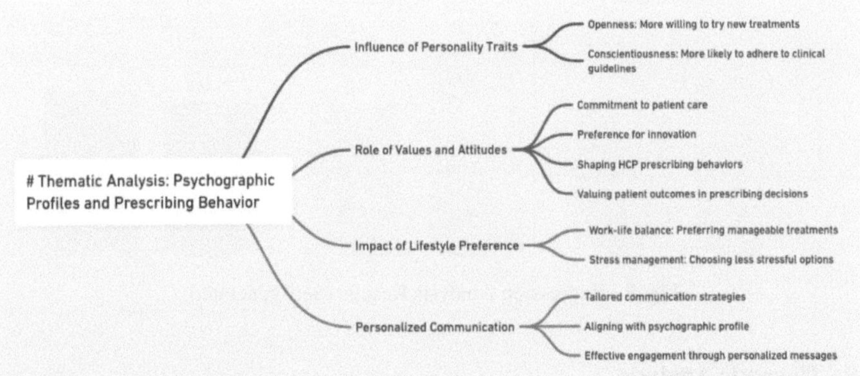

Fig. 9. Showing Schematic Diagram of Thematic Analysis Results (Self-made).

4 Future Implications

Psychographic profiles of healthcare professionals provide a useful understanding of their underlying beliefs, attitudes, lifestyles and drive, which impact their relative prescribing propensity more. These profiles classify prescribers into segments according to some attributes, namely risk profile, innovation proclivity, focus on patients and compliance with best practice. For example, such a clinician may be reluctant to change, and, therefore, would choose standard and well-known medicines over a risk-taking one.

Pharmaceutical companies can customize their communications and educational approaches by considering these psychographic attributes. For conventional prescribers, marketing campaigns might emphasize clinical data and safety; for those who are early adopters, however, they might stress innovation and better patient outcomes.

The potential of psychographic profiling is promising. These profiles should get even more precise as big data analytics and artificial intelligence (AI) evolve. By connecting therapies to doctors' cognitive frameworks, this precision may allow for highly personalized interactions between healthcare professionals, resulting in tailored interventions that optimize patient outcomes. But as this technique spreads, ethical issues with potential abuse and data privacy ought to be addressed.

5 Discussion

This study has shown that psychographic profiling can play a highly significant role in the other determinants of the healthcare professionals' (HCPs') prescribing behaviour. This means that personality traits, values, attitudes, and lifestyle choices will impact HCP decision-making in ways that conventional demographic and behavioural profiling methods have been unable to address. From the correlation analysis, it was clear that openness could positively correlate with the prescription of new and innovative treatments, whereas conscientiousness was linked to compliance with clinical guidelines. Likewise, extraversion affected peer-influenced prescriptions, while agreeableness and patient-centred values dictated patient preference-driven prescriptions. These results indicate that psychographic segmentation enables pharmaceutical companies to adopt more highly personalized and effective engagement strategies.

In addition, traditional profiling methods combined with psychographic profiling provide a more thorough understanding of HCPs. Combined insights from both methods will assist pharmaceutical companies in developing customized communication strategies that align with their target HCPs' individual motivations and prescribing patterns. Thematic analysis corroborates this standpoint since the qualitative findings reveal how participants highlighted the need for personalized interactions receptive to their own professional goals and ethical conscience. New advances in data analytics, machine learning, and artificial intelligence also give rise to opportunities to increase the adoption and accuracy of psychographic profiling by enabling real-time update of HCP engagement strategies.

While the positive aspects hold great promise, ethical issues concerning data privacy, accuracy profiling, and psychographic analysis biases must be systematically addressed. The study highlights that issues can be reduced when data collection methods are transparent and are compliant with regulations such as GDPR and HIPAA. Future research will, therefore, seek to combine psychographic profiling with electronic health records (EHRs) and real-time prescribing data, to the end of improving predictive model. Advancement in these methodologies will nurture better pharmaceutical industry relationships with HCPs and, in the end, better patient outcomes and healthcare.

6 Conclusion

This research provides valuable insight into the impact of psychographic profiles based on prescribing behaviours of HCPs. Furthermore, research based on this demonstrates that personality traits, values, attitudes, and lifestyles, which produce significant influence, impact the prescribing decisions of HCPs. the use of psychographic profiling does more than improve the relevance and targeted nature of communications strategies and thus results in more effective communications; it also fosters improved prescribing accuracy. The findings of the research suggest psychographic profiling enhances personalization and relevance of communication strategies, leading to better engagement and more informed prescribing decisions. Integration of psychographic data with traditional profiling methods concerns a comprehensive understanding of HCPs and allows for more effective strategies toward the target market. The use of psychographic profiling in healthcare raises several challenges and ethical considerations, i.e., data privacy, accuracy, and bias. Furthermore, addressing the challenges to ensure the ethical and effective use of psychographic data. Among the problems that the study reveals several factors that need to be discussed in order to avoid the misuse of psychographic data; these factors include data privacy, accuracy, and bias. Emerging trends in data analytics; ML and personalized medicine-drive the evolutions of psychographic profiling and create new opportunities to enhance HCP engagement and improve patient outcomes. Furthermore, leveraging advancements, pharmaceutical companies can develop more strategic and personalized approaches to HCP, leading to better health care delivery and patient care. Considering these advancements, pharma can far better plan the engagement of HCPs and thereby enhance the practice of medicine and patient treatment. There is an opportunity for future research to investigate the application of real-time data analytics and e-Health records in order to build more sophisticated and flexible engagement profiles for psychographic profiling. Nonetheless, more research is required with regard to ethical issues, for example, regarding the negative occurrences in profiling and the application of GDPR worldwide.

References

1. Ahmed, R.R., Veinhardt, J., Štreimikiene, D.: The direct and indirect impact of pharmaceutical industry in economic expansion and job creation: evidence from bootstrapping and normal theory methods. Amfiteatru Econ. **20**, 454–469 (2018)
2. Bawa, S.S., Sing, H.: Factor influencing the formulation of effective marketing strategies of indian railways. Int. J. Innov. Technol. Explor. Eng. **8**(9S), 357–362 (2019)
3. AL-Alak, B.A., TLL, A.L.-A.: Bridging the gap between R&D-marketing in the pharmaceutical industry in Malaysia. J. Mod. Mark. Res. **1**, 74–85 (2012)
4. Berkeley, J.S.I., Richardson, M.: Drug usage in general practice. An analysis of the drugs prescribed by a sample of the doctors participating in the 1969–70 North-east Scotland workload study. J. R. Coll. Gen. Pract. **23**(128), 155–161 (1973)
5. General Data Protection Regulation.: Regulation (EU) 2016/679 of the European Parliament and of the Council (2016)
6. Gagnon, M.A., Lexchin, J.: The cost of pushing pills: a new estimate of pharmaceutical promotion expenditures in the United States. PLoS Med. **5**(1), e1 (2008)
7. Godin, G., Bélanger-Gravel, A., Eccles, M., Grimshaw, J.: Healthcare professionals' intentions and behaviours: a systematic review of studies based on social cognitive theories. Implement. Sci. **3**, 36 (2008)
8. Health Insurance Portability and Accountability Act: Public Law, pp. 104–191 (1996)
9. Johnson, M., Davis, R.: The role of psychographic profiling in healthcare marketing. J. Med. Mark. **15**(2), 123–135 (2019)
10. Jones, A., Wilson, T.: Ethical considerations in psychographic profiling. Ethics. Med. **28**(3), 245–260 (2022)
11. Lee, S., Kim, H.: Advancements in data analytics for psychographic profiling. J. Data Sci. **19**(4), 567–589 (2021)
12. Manchanda, P., Honka, E.: The effects and role of direct-to-physician marketing in the pharmaceutical industry: an integrative review. Yale J. Health Policy Law Ethics. **5**(2), 785–822 (2005)
13. Miller, J., Thompson, P.: Personality traits and prescribing behavior. J. Health Psychol. **23**(5), 789–802 (2018)
14. Milne, L., Leamon, J., Johnson, A., et al.: Profiling health professionals' personality traits, behaviour styles, and emotional intelligence: a systematic review. BMC Med. Educ. **23**, 450 (2023)
15. Prashar, K., Bawa, S.S.: Studying the effect of artificial intelligence on E-governance. In: Contemporary studies in Economic and Financial Analysis, vol. 110, pp. 87–101 (2023)
16. Ponnet, K., Wouters, E., Van Hal, G., Heirman, W., Walrave, M.: Determinants of physicians' prescribing behaviour of methylphenidate for cognitive enhancement. Psychol. Health Med. **19**, 286–295 (2014)
17. Smith, L., Brown, K.: Psychographic profiling of healthcare professionals. J. Healthc. Manag. **35**(1), 45–60 (2020)
18. Singh, H., Bawa, S.S.: Determinants of passengers perception about service quality through 5S—a study on Indian Railways. Rev Profess. Manage. **13**(2), 66–76 (2015)
19. Stanford Research Institute: Values and Lifestyles (VALS) Framework. https://www.sri.com/vals-framework (2021)

The Impact of Artificial Intelligence-Driven Digital Marketing Strategies on Pharmaceutical Consumer Behavior

Kashif Pervaiz[1]([⊠]) and Simerjeet Singh Bawa[2]

[1] European Institute of Applied Sciences and Management, Prague, Czech Republic
`kashi.novartis@gmail.com`
[2] Chitkara Business School, Chitkara University, Rajpura, Punjab, India

Abstract. Artificial intelligence AI is the leading technology that has transformed digital marketing strategies affecting diverse industries such as the pharmaceutical sector. The focus of this investigation into how AI can affect doctors' prescribing decisions is the effects of interactions with pharmaceutical content targeting by digital marketing. Employing a qualitative method in this study, this paper synthesizes recommendations from a comprehensive literature analysis in order to investigate the relationship between AI applications, consumer buying behavior, and the potential societal effect of the use of AI in pharmaceutical promotional appeals. The study examines how AI improves paid advertising and makes it possible for pharmaceutical firms to produce content that will interest targeted groups that, in turn affects doctors' interaction with promotional messages. Although AI can provide tangible benefits at the same time, sparking new efficiency and precision in the marketing, it is crucial to address some of ethical issues, including data privacy issues or potential bias, and negative consequences of very high-level algorithmization in marketing strategy. Research evidence indicates that AI technologies are conventionally instrumental in defining present and future marketing processes in the pharmaceutical industry. The study underscores the dual-edged nature of AI: its potential to provide notable social and professional rewards and at the same time, its potential to require the evaluation of its effects. This work is useful for stakeholders who would wish to get a better understanding about the effectiveness of AI in pharmaceutical digital marketing.

Keywords: Artificial Intelligence Marketing · Pharmaceutical Customer Behavior · Social Media Marketing · Physicians · Marketing Artificial Intelligence · Quantitative Analysis · Decision Making

1 Introduction

1.1 Background

Conventional marketing has been the main practice in the pharmaceutical industry for long, with frequent, direct communication between drug salespersons and doctors, communicating to groups of doctors and physicians, and other educational-related marketing.

© The Author(s), under exclusive license to Springer Nature Switzerland AG 2026
A. Mirzazadeh et al. (Eds.): ODSIE 2024, CCIS 2482, pp. 91–105, 2026.
https://doi.org/10.1007/978-3-031-93601-2_6

These have acted as primary channels through which information on drugs is made available, doctors' prescription patterns have changed as well, and the relationship between drug manufacturers and doctors is built. However, pharmaceutical marketing is on the verge of evolution with technological development and the rise of artificial intelligence AI in marketing.

AI has brought about new perspectives of digital marketing through personalization [1], which is enabled by AI's huge data collection and processing capabilities. The application of AI in pharmaceutical marketing ensures that the content that pharmaceutical companies want to share with the doctors is relevant and personalized while also inclining toward the preferred doctor's choice. Some of the above strategies include the use of prescriptive analytics, which uses machine learning algorithms to study prescription patterns; the use of predictive analytics, which helps anticipate future needs; as well as the use of natural language processing, such as the use of chat bots to serve doctors' questions quickly. Due to the application of AI in these marketing efforts, there is the possibility of making communication more accurate and well-informed, possibly leading to the right decisions by doctors [31]. Nevertheless, the extent of the impact of AI has not been fully determined as to how it will influence doctors' prescription behaviors and interactions with pharmaceutical products.

1.2 Research Problem

Although it has become apparent that organizations in the pharmaceutical industry are adopting AI-driven digital marketing tools, there needs to be more known about how these technologies impact the actions of the target audience, namely doctors. AI is slowly finding its way into marketing tactics in the pharmaceutical industry about how doctors are targeted. However, the effectiveness of such tactics can create a debate on their efficiency in transforming doctors' prescription behavior. Furthermore, the ethics surrounding the use of transparency and fairness of artificial intelligence in marketing offer further challenges. It is, therefore, important to gain insight into how physicians perceive AI-driven marketing strategies and how the latter influences their decision-making patterns. This research aims to fill this gap by buying AI's impact on pharmaceutical marketing.

1.3 Research Aim

The broad research question is how AI-enhanced digital marketing techniques change doctors' consumer behavior, especially regarding pharmaceutical firms and prescription trends. Thus, this research seeks to evaluate the efficiency of AI-used content to help determine how AI can ensure that marketing-related strategies correspond to doctors' clinical demands and professional actions.

1.4 Research Objectives

1. To examine the level of AI application in digital marketing to doctors in the pharmaceutical industries.

2. To explore how AI-assisted content affects doctors' interactions with pharmaceutical firms and their choice regarding prescriptions.
3. To evaluate the nature and potential concerns concerning the ethical application of AI in promoting drugs and medicine.

1.5 Research Questions

1. Self-Assessment: What are the impacts of AI-driven digital Marketing efforts on the doctors' decision-making in the scope of pharmaceutical marketing?
2. What facts and considerations make AI-driven content effective in reaching doctors?
3. What ethical issues are likely to come about because of the use of AI in marketing to doctors, and how may they be prevented?

1.6 Research Scope

This present research examines marketing automation technology and digital marketing that involves artificial intelligence brought about by pharmaceutical manufacturing firms and their impact on the consumer behavior of doctors involved in prescriptions [2]. It investigates how the emerging technologies of predictive analytics, content personalization, and autonomous systems influence doctors' decisions and suggestions, especially in developed economies. In this study, secondary data collection is used together with the literature analysis of the articles between 2018 and 2023 to guarantee the relevancy of the collected data. The scope concerns developed PMs, including implant and Prime markets where AI adoption is already under practice, and its ideas contain the applicable digest realistic to doctors' marketing strategies.

2 Literature Review

A literature review analyzes articles and theoretical frameworks pertinent to the given subject concerning the effects of AI-implemented promotional techniques in the pharmacological industry on doctors' consumer behavior. Collectively, it integrates data from multiple investigations to present an overview of how AI solutions affect' decision-making, interaction with their marketing material, and the ethical concerns tied to these approaches [11].

2.1 Conceptual Framework

The conceptual framework of this literature review is structured around three main areas: AI and its role in digital marketing, consumer behavior in the pharmaceutical industry, and ethical issues. Each area is discussed to elucidate the effects of the various AI technologies on doctors' relations with the pharmaceutical industries and the effects of these relations on prescription behavior (Fig. 1).

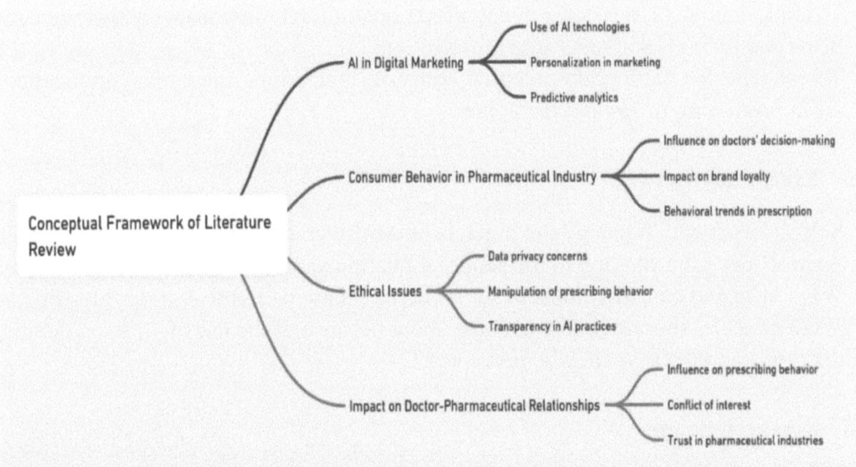

Fig. 1. Flow chart of Conceptual Frame work (self-generated).

2.2 AI in Digital Marketing

AI technologies have impacted digital marketing in a way that they have facilitated one-to-one communication. Ref Qingwei Huang and Michael Rust AI's capability to use big data makes marketing decisions at the doctor level based on the doctor's prescription, interest, and communication profiles, among others, as noted by Huang and Rust in 2021. This is a process of achieving personalization using technologies, including machine learning, that helps in predicting the needs of the doctors and preferences of the predictive analytics that helps in predicting the people's behavior in the future depending on the previous data set.

Luo et al. (2020) acknowledge that AI recommendation systems are crucial to customizing marketing content [14]. These systems employ past data to present the appropriate information on a drug, any education materials, and new products essential for a doctor practicing in each field of specialty. Jarrahi et al. (2021) also investigate the function of natural language processing (NLP) in dynamite and interacting marketing. Features like chat bots/ virtual assistants assist doctors by responding to their questions immediately, improving the effectiveness of pharmaceutical marketing.

Pharmaceutical Consumer Behavior. Based on the pharmaceutical marketing information consumed by doctors, the type, relevance, and quality of information affect them. According to Feinberg et al. (2020), the established findings proved that the AI's recommendations can cause changes in doctors' prescriptions when the recommendations are in tune with the actual practice and the practice as recommended in the existent guidelines. Engaging content based on doctors' prior prescribing behavior and their interests will lead them to have higher levels of confidence in the information being provided and, hence, have a higher propensity to adopt the new content into their prescribing behavior [8].

According to Steinhoff and Palmatier (2021), pharmaceutical firms can classify doctors into different personas using AI technologies, enabling firms to target doctors more effectively. This research suggests that doctors willing to accept technology are more likely to accept AI-driven marketing than old-school doctors. However, it is immediately clear that the delivery of recommendations based on evidence and experience has a tangible influence on prescription behavior, demonstrating AI's capacity to reduce decision-making empathy.

As per Agarwal et al. (2021), such AI tools can overcome information overload by helping doctors with the relevant drug information related to their medical specialties. This strategic targeting assists doctors in making the right decisions independently and faster, affecting their interaction with pharmaceutical marketing in specific ways (Fig. 2).

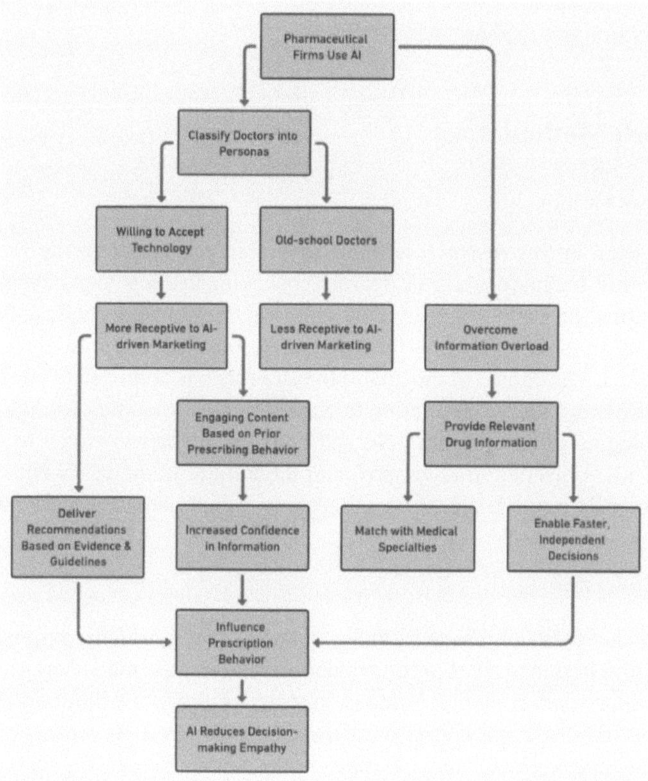

Fig. 2. Flow chart showing AI in Pharmaceutical sector (Self-made).

Ethical Considerations. The integration of AI in marketing, specifically pharmaceuticals, implies some questions that need to be answered to avoid the wrong application of AI. According to Cheng et al. (2022), AI applied in creating content may inevitably add some bias and skew the content produced in marketing. For instance, if such systems rank drugs in a hospital depending on the strategic goals of marketing the drugs contrary to the effectiveness of the drugs, this might be deemed unethical because it may influence the kind of information being passed to the doctors.

According to Kietzmann et al. (2021), it is necessary to note that data privacy and the questions of transparency remain crucial when it comes to using AI in marketing. Since pharma industries are gathering a lot of data from doctors, it is pertinent that the data is used appropriately and rightfully. As echoed by other authors, there is a need for definite policies that regulate the usage of the doctor's data and the likely consequences of violating privacy.

According to Floridi et al. (2020), there is a need to set the right ethical benchmarks for the marketing of AI. They insist on responsibility and justice as to the AI algorithms and the protection of doctors reverting to the sites, as there is a danger of being exploited by what these marketing gurus are showcasing. This is why there is a need for ethical policies: AI technologies are constantly developing and increasingly being used in product promotion in the pharmaceutical industry.

3 Research Methodology

3.1 Research Design

The method used in this research is qualitative research to determine how AI-driven digital marketing techniques affect doctors' behavior in their capacity as consumers of pharmaceutical products. The research employs the use of secondary data analysis; this means the research will not collect new data but will rather analyze data that is already analyzed. The choice of the qualitative approach is appropriate because it delves into the phenomena under investigation to determine their qualities, including the usage of AI technologies in the pharma field and consumers' behavior [4]. In this manner, this research identifies patterns, lessons, and theoretical frameworks embedded in the material previously published in academic papers, industry reports, and case studies.

3.2 Data Collection

Data for this study was obtained from secondary sources, which comprised scholarly journal articles, pharmaceutical industry publications, and AI marketing-related articles [13]. In secondary data collection, data is collected without creating new data; instead, the data reviewed and compiled is used from another research. Thus, this method is suitable for instances where the area under study is well developed, such as AI digital marketing, where sufficient studies have been done and are freely available. In doing so, it is possible to define trends, recurring themes, and important findings based on existing studies to be incorporated into the research's theoretical framework and conclusion.

3.3 Inclusion Criteria

Timeframe. The investigations were carried out with the help of the databases published throughout the years 2018–2023: such an approach enables us to consider the recent developments in the sphere of AI technologies' implementation in the context of pharmaceutical marketing [30].

Relevance to AI in Digital Marketing. For this analysis, only research works that have focused on the application of AI in the marketing domain and more so in the pharmaceutical business were deemed relevant for review. This ensures that the concern is kept on how AI impacts advertising endeavors directed at doctors.

Pharmaceutical Industry Focus. The selected studies had to discuss the effect of AI on healthcare personnel, especially doctors – as the targeted consumers in this case [12]. Adding industry reports into the equation helped get a realistic outlook on using AI technologies and their performance in the pharmaceutical industry.

Exclusion Criteria. Irrelevant Industries: Studies concerning industries other than healthcare, such as retail or finance, were excluded. While marketing using artificial intelligence is affecting many industries, this research focuses on its impact on doctors in the pharmaceutical industry [22].

Non-AI Marketing Studies. Studies that described the use of digital marketing approaches without considering the use of AI technologies were also omitted since the concern is how AI adapts conventional marketing management practices.

Healthcare Consumer Research Excluding Doctors. Thus, articles that covered consumers other than healthcare professionals, for example, patients and the public, were excluded. This paper focuses on doctors as the key decision-makers regarding this phenomenon of marketing pharmaceuticals.

For this research, the data analysis was done through thematic analysis since it is a qualitative analysis well suited to identifying themes from the existing literature. That way, gathering a deep insight into how digital marketing through AI influences doctors, particularly in the pharmaceutical industry, was possible (Fig. 3).

Step 1: Familiarization with the Data. The first and, perhaps, the most important of the steps in thematic analysis is to become intimately familiar with the data collected. This included analyzing articles, reports, and cases on AI uses in pharmaceutical marketing from 2018 to 2023. Since the research is conducted on secondary data, it was important to get acquainted with the context and findings of the prior research [3].

Step 2: Initial Coding. After the method of familiarization, an approach to categorizing literature was applied. They were coding, which means underlining parts of the text related to the research questions. In this case, important parts concerning how AI in marketing affects behavior, how doctors make decisions, and the ethical issues have been underlined [25]. For instance, when a study discussed using predictive analytics to message doctors based on their prescription patterns, the code used was 'personalization of content.' Other similar topics were instances where ethical questions such as algorithmic explainability or data privacy had been voiced.

Step 3: Identification of Themes. Subsequently, codes like each other were aggregated into general themes corresponding to the study's objectives. The main themes identified include:

Personalization of Content AI tools, including machine learning and NLP, allow the personalization of marketing content to meet doctors' specific needs and wants. The sources uncovered showed that custom content is a great asset to interaction, which means it is easier to get the attention of the targeted doctors.

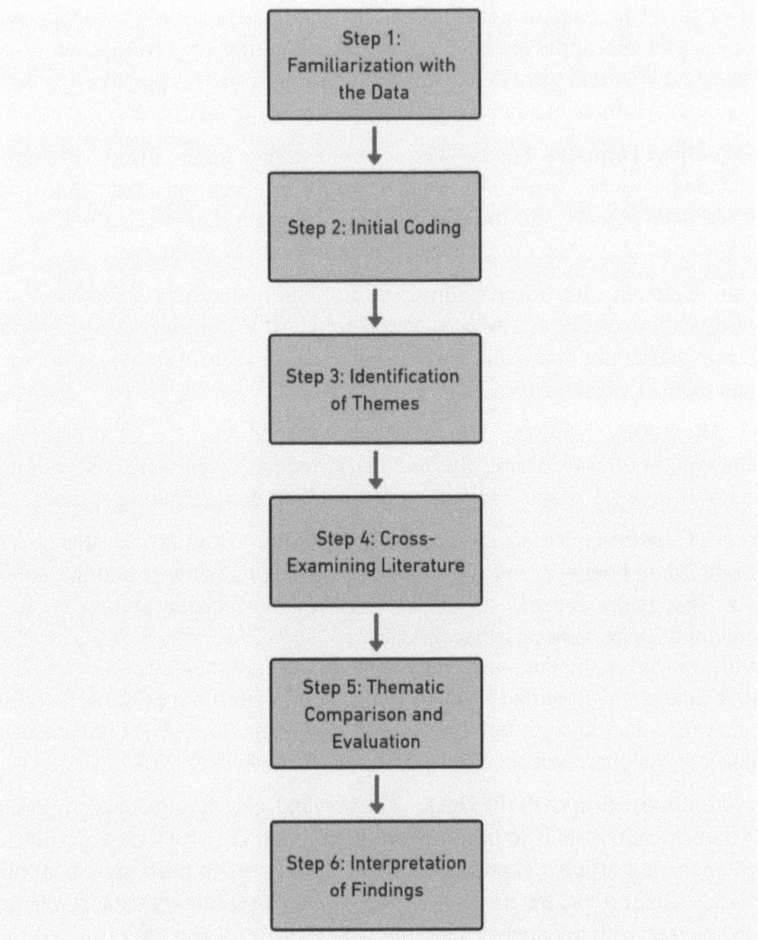

Fig. 3. Flow chart of Data Analysis (Self-made).

Influence on Decision-Making Another pattern that emerges in the context analysis is how much marketing resulting from AI affects the prescription decisions made by doctors [9]. Research revealed that AI could bias medical practitioners to recommend pharmaceutical products by suggesting the warrants of employing data analytically based prescriptive analytics.

Ethical Challenges A considerable number of articles were concerned with the possible ethical issues arising from the application of AI in PHARMA marketing [23]. Areas like the transparency of algorithms, biases that may exist in AI recommendations, and data protection concerns became ringing bells, with demands for a better code of ethics concerning the use of AI to reach healthcare practitioners common (Fig. 4).

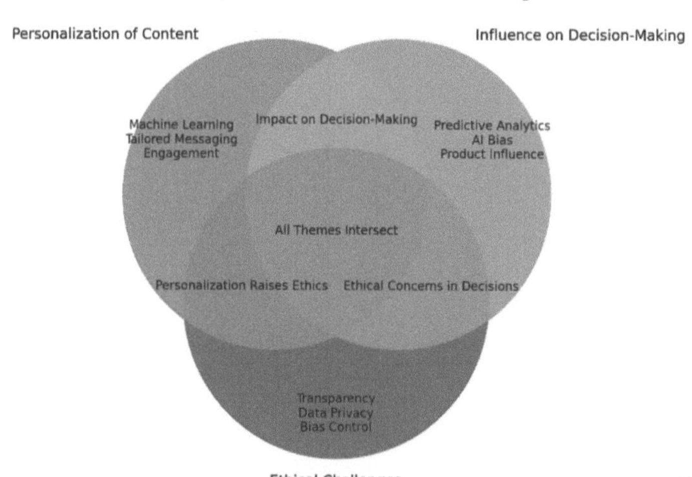

Fig. 4. Venn diagram explaining Themes in AI driven Marketing (Self-made).

Step 4: Cross-Examining Literature. The next step is the cross-source analysis aimed at defining common patterns and the variability of the effectiveness of AI-infused marketing strategies. This cross-examination showed similar results regarding AI's value in targeting the marketing message and its effect on decision-making [33]. Yet, it also pointed out disparities depending on the location and specialization of the medical facility. For instance, analysis from developed markets revealed higher AI implementation than the emerging markets and better responses to AI that tried to change the prescription pattern.

Step 5: Thematic Comparison and Evaluation. Based on the comparison of themes, the efficiency of various AI tools like machine learning recommendation engines and natural language processing chat bots was assessed [26]. This work discovered that machine learning models, which predict doctors' prescribing behavior, were more effective in targeting than other AI. However, it is evident from the literature that ethical issues like blunting doctors' discretion were a common point of discussion. While AI improves marketing effectiveness, it must be well-regulated.

Step 6: Interpretation of Findings. The thematic analysis revealed several important findings:

AI-Driven Personalization Marketing content personalized through AI enhances communication relevance, enhancing the likelihood of doctors engaging pharmaceutical companies [18].

Predictive Analytics Some AI tools and predictive analytics determine doctors' decision-making, including the types of medication doctors prescribe.

Ethical Considerations Despite this, certain factors must be considered, with the paramount importance placed on ethical issues such as transparency and biases. The ability to control the masses through AI in marketing strategies needs a proper legal framework to govern the use of AI in healthcare marketing (Fig. 5).

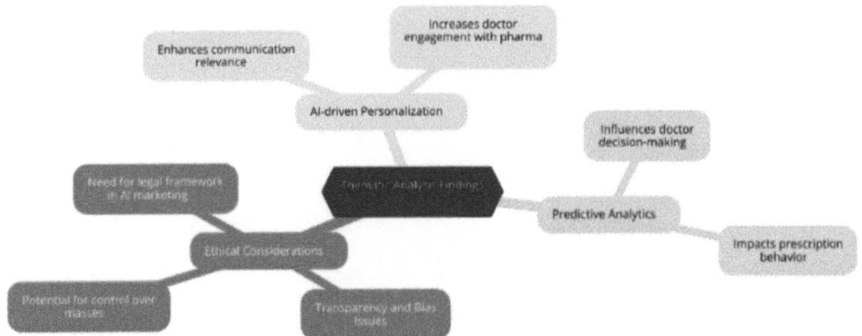

Fig. 5. Ethical Considerations in AI driven Marketing (Self-made).

This is why ethical issues are well under discussion when talking about AI as a tool for digital marketing, especially when applied to the sphere of pharmaceutics, where doctors' choices influence people's lives directly. The main ethical questions that come with the idea of AI in what more pharma companies are doing with doctors include: In this section, they are going to discover the foremost ethical issues still in AI marketing, geared toward decision-making as well as focusing on privacy, transparency and the bias in the algorithm [10].

Transparency of AI Algorithms. Another key ethical issue when using AI in pharmaceutical marketing is that there usually needs clarity on how the algorithms work. Most AI systems, especially those that use machine learning, can be categorized as black boxes because, most of the time, one cannot fathom the logic behind the recommendations or the decisions the systems make [24]. This is especially the case for personalized marketing content produced by AI systems for doctors depending on their prescribing history, where some of the key elements have not been well defined, leading to ethical problems.

Trust and Informed Decisions. Prescribers use current, qualitatively verified data to make prescription choices [27]. By not making the implemented AI systems in pharmaceutical industries explain them on how they handle the data and produce these recommendations, results may be presented to these healthcare professionals with some bias or with some portions of data left incomplete, in a way that hampers informed decisions being made based on such results.

Potential for Manipulation. Without transparency in the programming code, AI algorithms can be designed to influence doctors' choices regarding certain pharmaceutical products intentionally or unintentionally [17]. As seen from the commercial aspect, this manipulation could involve maloccurmance where dollars preceded patients, which is an awful ethical problem.

Bias in AI Algorithms. The second of the crucial ethical questions is the question of bias in AI algorithms. An AI system is a result of training the system with data, and if the data set has a biased approach, then it is more likely that such bias can be passed onto the AI system or magnified by the system itself.

Bias in Data. In healthcare discussed here, bias can manifest itself in race, gender, or geographical bias and, therefore, influence the output of an AI system [21]. The doctors may, therefore, receive recommendations based on data influenced by the prescribing patterns of some specific demographic, which will affect the treatment of patients. For example, an AI system designed to work with data from developed countries will not be effective for doctors in underdeveloped areas where the practice and needs of the patients are different.

Commercial Bias. It can also result from bias from pharmaceutical companies since they are vested in the study's outcome. AI systems applied to digital marketing may promote products with the greatest profit margin, or the company may place a lot of emphasis on them, even though these products may be better for the doctor's particular patient congregation. Such commercial predisposition may result in setting up a skewed preference for the prescription of some drugs over others with negative repercussions to the patients [20].

Data Privacy and Consent. Techniques applied in Digital Marketing through Artificial Intelligence involve personal and professional data. Hence, some of the common concerns include the ethical use of data.

Doctor and Patient Data. The primary data used in the pharmaceutical companies' marketing strategies is doctors' prescriptions for certain diseases and drugs [5]. However, patient data is also involved indirectly. Some of the treatment patterns of the doctors being analyzed by the AI systems may contain sensitive information about certain patients and, if not well managed, may result in data leakage. This raises issues about data privacy laws such as the GDPR, which is mandatory for any company that operates within Europe. This regulation insists that personal data must not violate the privacy and confidentiality of individuals.

Informed Consent. Pharmaceutical companies are therefore obliged to make sure that doctors are informed on how their data is collected and used for marketing [32]. AI systems, for example, might identify and evaluate individual prescribing activities without a doctor's permission, thus infringing privacy and ethics standards. Therefore, companies must get informed consent to collect their data, and how it will be used must be explained well to conform to the highest standard of legal and ethical compliance.

Autonomy and Decision-Making. Self-direction by doctors is one of the standards of medical professionalism. The seemingly autonomous digital marketing approaches involving AI-predictive analytics, particularly regarding content type and doctor targeting, appear to exert pressure on doctors' decision-making authority in their profession.

Influence on Prescribing Behavior. Automated recommendations based on a doctor's past prescribing decisions could change the existing defaults, even those derived from self-interest and inimical to patients' interests. For instance, let's consider the rate at which an AI tool recommends certain medications; a doctor may be influenced to favor those medications since the tool repeatedly recommends them instead of opting for more suitable ones. This brings an ethical issue about the place of AI as an advisor of doctors and the inside-out contradiction when advertising tactics tend to sell products that have more commission than the actual well-being of the patients.

Manipulation Through AI. Doctors' decisions on what medication to prescribe can be predicted and controlled using AI, which is manipulation, especially if the AI system is programmed to guide doctors on what product to recommend. This goes against the norm of non-maleficence, where 'what is permissible is what is not wrong or harmful,' since it might result in the patients receiving substandard treatment guided by the self-profit motive of the practitioners.

Significance Today and in the Future. Hence, implementing AI-driven digital marketing techniques in the pharmaceutical industry is most pertinent today and is expected to take a much larger significance shortly [19]. This advancement is crossing over to various forms of technologies and affecting the relationship between doctors and pharmaceutical firms, and the use of AI will continue to foster this evolution. The sections below discuss different degrees of contemporary application of AI to advance and future-oriented marketing strategy.

4   Relevance to Today

AI is rapidly reshaping the pharmaceutical marketing landscape, offering several advantages that make it particularly relevant in today's digital world: AI is rapidly reshaping the pharmaceutical marketing landscape, offering several advantages that make it particularly relevant in today's digital world:

Personalization and Targeting. As personal communication media have evolved, pharma enterprises are utilizing AI to provide targeted marketing material to physicians. These targeted initiatives have been influenced by prescribing behavior, medical specialty, and the clients' characteristics of specific doctors. AI enables pharmaceutical companies to understand massive amounts of data to better address healthcare professionals' needs, compelling their attention to marketing messages.

Increased Digitalization in Healthcare. The effect of the COVID-19 pandemic was, therefore, quickening the pace of technology adoption in the marketing of pharmaceutical products [28]. As the meetings between doctors and representatives of the pharmaceuticals moved from physical contact, there was the need to engage virtually. Due to the advancement in the use of AI in the medical field, doctors are well-informed about new drugs and treatments, and AI makes these virtual interactions possible. This trend of digital engagement has been observed, and the future of the future is expected to go further with the help of AI to establish a closer link between pharmaceutical companies and doctors.

Efficiency and Cost Savings Realizing that hit-and-miss marketing methods are costly and time-consuming, applying artificial intelligence in marketing saves on cost-effective measures such as one-on-one sales meetings and marketing conferences [15]. This is especially important in mass marketing events like email marketing with chatbot or, more specifically, customized product offerings for diseases for which pharmaceutical companies would like to target more doctors with less effort and costs involved. Such an approach particularly applies to the industry where marketing costs are significant.

4.1 Future Relevance

AI is poised to transform pharmaceutical marketing with several key trends further [29].

Enhanced Personalization. Marketing content will be more relevant since machine learning, deep learning, and other forms of AI can provide more real-time data and better marketing personalization that can be used to target doctors better based on their needs, medical research, and patients' results.

AI-Powered Virtual Assistants. Virtual assistants could give doctors instant evidence-based information on treatments and new drugs within a relatively shorter period, with doctors being more interactive with the companies [16].

Ethical and Regulatory Changes. Future updates to these guidelines will focus on issues related to algorithmic transparency, biased AI-driven recommendation systems, and data protection for healthcare marketing.

Focus on Patient Outcomes. Machine learning and other forms of artificial intelligence will move from advertising a product to delivering beneficial patient results by offering physicians data analyzing tools and outcomes.

Global Expansion. Thus, as AI advances make their way into the market, it will assist pharmaceutical companies in reaching new doctors, especially in rural areas, enhancing healthcare delivery in areas of need.

5 Discussion

This research highlights AI-driven digital marketing as the transformative force for pharmaceutical consumer behavior. AI technologies within machine learning and predictive analytics allow personalization of the marketing effort, making engagements with healthcare professionals more effective by the pharmaceutical firms. AI-fueled marketing strategies are significantly affecting doctors' decisions regarding the prescriptions made according to recommendations from their peers, based on their own prescription patterns, and in consideration of their medical specializations. Ethical challenges arise nevertheless. An important issue remains regarding transparency of the AI algorithms, which the doctor would barely understand in recommending patients. There are also issues of privacy related to this data collection of sensitive information on a massive scale. Others might reasonably argue that any algorithmic bias might in fact prioritize commercial interests at the expense of clinical efficacy; arguably, that might tilt the playing field in favor of influencing the prescribing behaviors of doctors. On a rather practical level, by essentially diffusing access to pharmaceutical marketing information, AI marketing might constrain the free and autonomous decision-making of doctors. Nonetheless, overemphasis on AI will lessen the importance of human independent clinical judgment. Quite appropriately, one prime objective here is to uphold ethical usage of AI to preserve the integrity of medical decision-making.

6 Conclusion

The study also suggests that, with the rise of digital marketing powered by artificial intelligence, the way doctors engage with pharmaceutical firms has been altered completely. In this way, AI optimizes the efforts of organizations and improves patient engagement, as well as more rational prescribing activity due to the application of accurate recommendations based on scientific evidence. These innovations are expected to improve the level of accommodating the targeted promotional material with patient needs and wants to increase healthcare professional receptiveness. However, in artificial intelligence there remains large ethical issues associated with the higher use such as transparency, data privacy, and any unfair bias in the relation of algorithms with strategies. Solving these problems is important to enhance the confidence of and properly regulate the application of AI tools in promoting health care services. The key zones where these technologies can operate will define the working option of applying the AI's possibilities and the presence of ethical standards for using intelligent technologies in the future of the pharmaceutical industry. Based on the development trajectory of AI technology, more possibilities are to be seen for AI in pharmaceutical marketing strategies and to enhance the assistance of health care providers. However, ethical governance must remain a priority, to protect and reinforce the trust in these new technologies, and keep AI innovation from becoming elastic and menacing in its erosion of basic values. It is safe to say that the future of healthcare marketing will lie in how technologic innovation will be utilized in parallel with an ethical foundation.

References

1. Achilladelis, B., Antonakis, N.: The dynamics of technological innovation: the case of the pharmaceutical industry. Res. Policy. **30**(4), 535–588 (2001)
2. Ahuja, A.S.: The impact of artificial intelligence in medicine on the future role of the physician. PeerJ. **7**, e7702 (2019)
3. Alharbi, E., et al.: Exploring the current practices, costs and benefits of fair implementation in pharmaceutical research and development: a qualitative interview study. Data Intell. **3**(4), 507–527 (2021)
4. Bakonyi, Z.: How can companies handle paradoxes to enhance trust in artificial intelligence solutions? Qualitative research. J. Organ. Chang. Manag. **37**, 1405–1426 (2024)
5. Baker, R., et al.: Primary medical care continuity and patient mortality: a systematic review. Br. J. Gen. Pract. **70**(698), e600–e611 (2020)
6. Bawa, S.S., Kunal, K., Kaur, K., Sharma, J., Srivastava, V., Tikku, P.: An analysis of artificial intelligence implications and its impact on marketing. a systematic review. Commun. Appl. Nonlinear Anal. **32**(1S), 143–149 (2025)
7. Bawa, S.S., Sing, H.: Factor influencing the formulation of effective marketing strategies of Indian Railways. Int. J. Innov. Technol. Explor. Eng. **8**(9S), 357–362 (2019)
8. Chatterjee, S., Kulkarni, P.: Healthcare consumer behaviour: the impact of digital transformation of healthcare on consumer. Cardiometry. **20**, 135–144 (2021)
9. Dai, T., Singh, S.: Artificial intelligence on call: the physician's decision of whether to use AI in clinical practice. SSRN Electron. J. (2021)
10. Dignum, V.: Ethical decision-making. In: Artificial Intelligence: Foundations, Theory, and Algorithms, pp. 35–46 (2019)

11. Emily, E.-O.S., Nwankwo, T.C., Otonnah, C.A., Nwankwo, E.E.: Artificial intelligence in marketing: literature review and future research agenda. J. Syst. Manag. Sci. **14**(1), 120–140 (2023)
12. Esmaeilzadeh, P.: Use of AI-based tools for healthcare purposes: a survey study from consumers' perspectives. BMC Med. Inform. Decis. Mak. **20**(1), 170 (2020)
13. Feng, C.M., et al.: Artificial intelligence in marketing: a bibliographic perspective. Australas. Mark. J. **29**(3), 252–263 (2020)
14. Gao, Y., Liu, H.: Artificial intelligence-enabled personalization in interactive marketing: a customer journey perspective. J. Res. Interact. Mark. **17**(5), 663–680 (2022)
15. Gabay, J.J.: Forethought. In: Practical Digital Marketing and AI Psychology, pp. 1–23 (2024). https://doi.org/10.4324/9781003409793-1
16. Gasteiger, N., Broadbent, E.: AI, robotics, medicine and health sciences. In: The Routledge Social Science Handbook of AI, pp. 313–338 (2021)
17. Graham, S.S.: The Doctor and the Algorithm [Preprint] (2022)
18. Jindal, P., Gouri, H.: Ai-personalization paradox. In: Advances in Marketing, Customer Relationship Management, and E-Services, pp. 82–111. IGI Global (2024)
19. Kapustina, O., et al.: User-friendly and industry-integrated AI for Medicinal Chemists and Pharmaceuticals. Artif. Intell. Chem. **2**(2), 100072 (2024)
20. Khan, A., Qu, X., Madzikanda, B.: An exploratory study on risk identification of cross-boundary innovation of manufacturing enterprises based on grounded theory. Creat. Innov. Manag. **31**(3), 492–508 (2022)
21. Kundi, B., et al.: Artificial intelligence and bias: a scoping review. In: AI and Society, pp. 199–215 (2022)
22. Latten, T., et al.: Pharmaceutical companies and healthcare providers: going beyond the gift—an explorative review. PLoS One. **13**(2), e0191856 (2018)
23. Marmat, G., Jain, P., Mishra, P.N.: Understanding ethical/unethical behavior in pharmaceutical companies: a literature review. Int. J. Pharm. Healthc. Mark. **14**(3), 367–394 (2020)
24. McCoy, L.G., et al.: Believing in black boxes: machine learning for healthcare does not need explainability to be evidence-based. J. Clin. Epidemiol. **142**, 252–257 (2022)
25. Morley, J., et al.: The ethics of AI in health care: a mapping review. Soc. Sci. Med. **260**, 113172 (2020)
26. Nadarzynski, T., et al.: Acceptability of artificial intelligence (AI)-led chatbot services in healthcare: a mixed-methods study. Digit Health. **5**, 205520761987180 (2019)
27. Niranjan, S.J., et al.: Trust but verify: exploring the role of treatment-related information and patient-physician trust in shared decision making among patients with metastatic breast cancer. J. Cancer Educ. **35**(5), 885–892 (2019)
28. Pasaribu, S.B., et al.: The impact and challenges of digital marketing in the health care industry during the digital era and the COVID-19 pandemic. Front. Public Health. **10**, 969523 (2022)
29. pharmaceuticals, R.: Pharma 2023 marketing the future. Biol. Sci. **03**(01), 347–352 (2023)
30. Sahu, A., Mishra, J., Kushwaha, N.: Artificial Intelligence (AI) in drugs and pharmaceuticals. Comb. Chem. High Throughput Screen. **25**(11), 1818–1837 (2022)
31. Tiwari, R., et al.: Role of technology for pharmaceutical marketing in the era of artificial intelligence. In: Advances in Digital Marketing in the Era of Artificial Intelligence, pp. 267–278 (2024)
32. Van Biesen, W., et al.: Remote digital monitoring of medication intake: methodological, medical, ethical and legal reflections. Acta Clin. Belg. **76**(3), 209–216 (2019)
33. Virvou, M.: Artificial intelligence and user experience in reciprocity: contributions and state of the art. Intell. Decis. Technol. **17**(1), 73–125 (2023)

The Impact of Artificial Intelligence on Medical Professional Language and Doctor-Patient Interactions

Albena Dobreva[✉] [iD]

Varna Medical University, "Prof. Dr. Paraskev Stoyanov", Varna, Bulgaria
`albena.dobreva@mu-varna.bg`

Abstract. The research aims to investigate the influence of artificial intelligence on medical professional communication. The subject is expressed in the analysis of professional language in medical doctor-patient dialogue. The goal is to trace the impact of artificial intelligence on new types of professional relationships. To achieve this goal, the following tasks were solved: analyses of research on the topic in Bulgaria and abroad were conducted; the characteristics of the professional language of medical specialists were traced; conclusions were drawn that will be useful in the field of communication and in the training of future medical specialists. The introduction of AI in medicine leads to the emergence of new terminology related to algorithms, machine learning, and other concepts that become part of the daily vocabulary of medical specialists. AI allows doctors to use more precise and specific terms, which improves understanding between specialists and facilitates the exchange of information. AI can contribute to the standardization of medical terminology, which will facilitate communication between doctors and improve the quality of medical documentation. AI can help patients get better and more complete information about their condition and treatment options. Artificial intelligence has the potential to revolutionize medicine and improve the quality of life of millions of people. However, to realize this potential, a number of challenges must be overcome and appropriate regulatory frameworks developed. The findings of this study will contribute to advancements in communication and the education of future medical professionals.

Keywords: Artificial Intelligence · Professional Language · Medical Interview · Doctor-Patient Communication

1 Introduction

A significant role in the field of medicine and healthcare, transforming not only diagnostics and treatment. Artificial intelligence plays a significant role in the field of medicine and healthcare, transforming not only diagnostics and treatment, but also the way doctors communicate with their patients. On the one hand, the use of AI is related to the demand for medical services, and on the other hand, it is related to the shortage of medical personnel in small settlements in Bulgaria.

A. Mirzazadeh et al. (Eds.): ODSIE 2024, CCIS 2482, pp. 106–118, 2026.
https://doi.org/10.1007/978-3-031-93601-2_7

As an effective tool, artificial intelligence can be quickly trained to perform certain tasks typical of a medical professional. However, particular attention should be paid to the ethical and legal aspects of the interaction between the patient, clinical staff, and artificial intelligence. The presented study focuses on the doctor-patient-AI relationship from the perspective of doctors, medical students, and patients using interviews, questionnaire techniques, and methods. The results of the study draw my attention to the aspect of applying AI in medicine to improve communication. The object of the research is the influence of artificial intelligence on medical professional communication. The subject matter is expressed in the analysis of professional language in doctor-patient medical dialogue. The primary objective of this research is to identify the positive impacts of artificial intelligence on enhancing access to healthcare services, particularly for remote areas or individuals with limited mobility. It aims to understand how AI can facilitate communication between doctors and patients who speak different languages and dialects, and to assess its potential impact on the interpersonal dynamics between healthcare providers and patients. Additionally, the study will explore whether AI could diminish the human element in medical care. To achieve this goal, the following tasks were solved: analyses of research on the topic in Bulgaria and abroad were conducted; the characteristics of the professional language of medical specialists were traced; conclusions were drawn that will be useful in the field of communication and in the training of future medical specialists. In order to achieve the goal, the following tasks were completed: Analyses of research on the topic in Bulgaria and abroad were conducted. The characteristics of the professional language of medical specialists were traced. Conclusions were drawn that will be useful in the field of communication and in the training of future medical specialists. The literature dedicated to the use of artificial intelligence in medicine is becoming increasingly enriched. It can be divided into several subtopics focused on artificial intelligence and diagnosis, artificial intelligence and communication, artificial intelligence and ethical-legal norms, and more. The main hypothesis is that AI can improve doctor-patient communication by enabling quicker and more precise diagnoses, providing patients with a clearer understanding of their health conditions, and enhancing the delivery of diagnostic results and treatment plans.

1.1 Literature Review

In recent years, research has delved into the intersection of doctor-patient communication and AI, with a primary emphasis on the paradigms of interaction, the physician's role, and the role of the AI in establishing trust within professional relationships. Artificial intelligence (AI) is often cited as a possible solution to current issues faced by healthcare systems. This includes the freeing up of time for doctors and facilitating person-centered doctor-patient relationships [1]. Discussions do not include improving medical professionals speech through intonation, tone, pace, or other communicative techniques. Medical large language models are being introduced to the public in collaboration with governments, medical institutions, and artificial intelligence (AI) researchers. However, a crucial question remains: Will patients follow the medical advice provided by AI doctors? The lack of user research makes it difficult to provide definitive answers [2]. The paper Artificial intelligence and the doctor–patient relationship expanding the paradigm of shared decision making analyses the role of AI-based CDSS for shared decision-making to better comprehend its promises and associated ethical issues. Moreover, it investigates how certain AI implementations may instead foster the inappropriate paradigm of paternalism. Understanding how AI relates to doctors and influences doctor–patient communication is essential to promote more ethical medical practice [3]. Medicine is a science, an art, and a trust between the doctor and the patient. In the times of digitization and artificial intelligence, new relationships between the human being and the machines are establishing [4]. Our research has led us to concentrate on investigating the capabilities of artificial intelligence in enhancing the professional speech of healthcare providers. In the last 7–8 years, studies have emerged related to artificial intelligence and its role in medicine and healthcare, such as image interpretation, data collection from results, etc. But these articles do not address topics related to improving the speaking of the medical professional and the participation of artificial intelligence in the various stages of the doctor-patient interview [5–8].

2 Methodology

Websites containing medical information, doctor-patient dialogues, and developments regarding the capabilities of artificial intelligence to reproduce spoken language were explored and analyzed.

- Research on Bulgarian medical websites that utilize GPT. The websites were chosen based on a survey of 50 people, which helped to identify the most relevant and user-friendly options;
- Medical platforms for online consultations with doctors. Interface intuitiveness: The site should possess an intuitive interface accessible to users with varying levels of technical proficiency.
- Comprehensive user documentation: Detailed instructions and examples should be provided to facilitate user experience.
- Customer support availability: A reliable customer support system should be in place to address user inquiries and concerns.
- Data security measures: Robust data security measures should be implemented to protect user data.
- Privacy policy clarity: A clear and transparent privacy policy should outline data handling practices.
- Booking appointments for medical check-ups;
- Analysis and comparison of communication in clinics with and without AI.

3 AI in Communication: Telemedicine and AI

Telemedicine refers to the delivery of healthcare services, such as consultations diagnosis, and health monitoring, through the use of information and communication technologies. In simpler terms, it allows patients to connect healthcare providers remotely, without the need for an in-person visit to a hospital or clinic. AI-powered programs can conduct human-like conversations, respond to inquiries, and execute tasks. They find extensive use in customer service, marketing, and educational settings. AI enables the personalization of messages by considering individual user preferences and interests.

The future of AI in telemedicine is very promising. AI has the potential to revolutionize telemedicine by making it more accessible, efficient, and effective. Here are some of the ways that AI is being used in telemedicine today and how it could be used in the future: Virtual assistants: AI-powered virtual assistants can provide patients with 24/7 access to medical advice and support. These assistants can answer questions about common medical conditions, provide symptom tracking, and even connect patients with a doctor or other healthcare provider if needed.

4 Ethical Considerations and Challenges

Ethical issues Despite the many advantages offered by AI, there are also a number of challenges and ethical issues associated with its use in medicine. These include: Risk of dehumanization: Overreliance on AI can lead to the dehumanization of medical care and weaken the bond between doctor and patient. Data security concerns: The use of AI in medicine raises questions about the security of personal medical data. Ethical dilemmas: AI can create complex ethical dilemmas related to decision-making that affects human life.

In summary, we can say that the integration of artificial intelligence into doctor-patient communication presents significant ethical challenges. The collection and storage of sensitive patient data create risks of data breaches and misuse. Additionally, the potential for algorithmic bias in AI systems raises concerns about fairness and equity in healthcare. Overdependence on AI may lead to a dehumanization of healthcare and may limit patients' autonomy. To address these issues, a comprehensive approach is needed, including the development of strict regulations, the establishment of ethical guidelines, and the implementation of robust safeguards to protect patient data and ensure transparency in AI systems. We can summarize the benefits of AI in doctor-patient communication as follows: AI can help structure patient visits, save doctors time, enable faster and more accurate diagnoses, and facilitate personalized treatment plans based on individual patient characteristics The first stage of the appointment with the doctor, even before the conversation with the patient, involves the doctor determining what complaints the patient has come with. For this, doctors are assisted by a chatbot developed using artificial intelligence. Before visiting the clinic, the patient receives an SMS and push notification with a link to a survey form. The patient should describe their complaints in free form and then answer the algorithm's questions. The collected complaints are automatically transferred to the artificial intelligence platform EMIAS [9]. The doctor can view them in the examination report and can, if necessary, adjust them during the scheduled appointment. Thanks to the preliminary assessment, the doctor has more time to examine the patient. During the meeting, the specialist sees two windows in the system. One contains the complaints collected by the chatbot, while the second is designated for the doctor's notes. During the conversation with the patient, the doctor either confirms or corrects the information. Accurately gathering medical history is quite a challenging process for serious diagnoses and can take up to half of the examination time. The chatbot's questions encourage the patient to think, remember, and analyze the symptoms, which can then be reported to the doctor [10–12]. Challenges and ethical issues despite the many advantages offered by AI, there are also a number of challenges and ethical issues associated with its use in medicine. These include: Risk of dehumanization: Overreliance on AI can lead to the dehumanization of medical care and weaken the bond between doctor and patient. Data security concerns: The use of AI in medicine raises questions about the security of personal medical data. Ethical dilemmas: AI can create complex ethical dilemmas related to decision-making that affects human life. The first stage of the appointment with the doctor, even before the conversation with the patient, involves the doctor determining what complaints the patient has come with. For this, doctors are assisted by a chatbot developed using artificial intelligence. Before visiting the clinic, the patient receives an SMS and push notification with a link to a survey form. The patient should describe their complaints in free form and then answer the algorithm's questions. The collected complaints are automatically transferred to the artificial intelligence platform EMIAS [13]. The doctor can view them in the examination report and can, if necessary, adjust them during the scheduled appointment. Thanks to the preliminary assessment, the doctor has more time to examine the patient. During the meeting, the specialist sees two windows in the system. One contains the complaints collected by the chatbot, while the second is designated for the doctor's notes. During the conversation with the patient, the doctor either confirms or corrects the information. Accurately gathering medical history is quite a challenging process for serious diagnoses and can take up to half of the examination time. The chatbot's questions encourage the patient to think, remember, and analyze the symptoms, which can then be reported to the doctor [14–16]. The second stage of the appointment – the registration process cannot do without new technologies. Doctors are supported by a decision-support system. Based on the patient's complaints, the AI-based tool selects the three most likely diagnoses. The doctor can either agree with this or prescribe a different one. The system then offers a set of examinations and additional consultations with specialized professionals necessary to confirm the diagnosis. Appointments are made in packages, which significantly saves time during the registration process. If the patient is coming for the first time or has not been seen for a long time, it is convenient to schedule all examinations for a complete diagnosis with almost one click. This eliminates the influence of the human factor since the system suggests which tests need to be conducted. For example, a series of examinations should be performed annually for certain diagnoses. Technologies are used to free up as much time as possible for the doctor to work with patients who have a greater need. Communication with the patient is a key factor in the accurate diagnosis and treatment plan. AI services utilizing computer vision technologies are sensitive enough to detect pathology at a stage when it is not yet visible to the human eye. This service is useful when there is a large influx of patients, such as during medical check-ups or screening programs, and also during a pandemic. This use increases the number of early disease detections, which, in turn, affects the effectiveness and positive outcomes of treatment. However, today technology serves only as a first opinion; the second and most important opinion remains with the doctor. Doctors can describe images and enter data into the system using speech recognition technology. They only need standard headphones (a headset with a microphone) or special remote microphones. The voice input system recognizes everyday speech and medical terminology, adapts to each doctor's pronunciation, and can expand its vocabulary [16]. This method of describing images is 20% faster than the manual method. Immediately after completing the report, both the patient and the treating doctor receive the examination conclusion in the electronic medical record [16]. The medical interview between doctor and patient can be viewed as a system with subsystems: Diagnosis and Treatment. Provides recommendations to doctors for better diagnostics. An expert system of qualified clinical decisions poses specific questions to obtain answers. This subsystem is essential for less qualified or less trained doctors. Asks questions while automatically determining the user's level of training, their skills, and abilities. Performs differential diagnosis based on the patient's age, gender, geography, quality and significance of medical records, and newly introduced symptoms. Generates similar symptoms, laboratory tests that help in differentiating the diagnosis. Provides treatment advice and recommendations for monitoring the treatment process based on the validation of actions. Electronic Health Record – EHR. Utilizes previous diagnoses and medical history to assist in diagnosis. This makes the EHR in the system more accurate and easier to use. If the disease is unknown, the system still uses the most common new diagnoses and histories to modernize the review for the user (doctor and patient). Can integrate data from third-party electronic medical records (EMR) from medical organizations and the patient's own electronic health record. Includes information on allergies, vaccinations, past medical history and illnesses, conditions, surgeries, etc. Can interact with existing medical record systems (EMR). Can be shared across all or several IT platforms. Educational Health Advice. Advice for pregnant women on how, when, and why it is better to breastfeed their children, whether to use milk or formula. Educational information based on popular topics in public health systems. Information regarding primary health care. The subsystem contains reference materials on diseases, symptoms, laboratory tests, and more. It has developed a list of established medical documents and reliable information sources, which consist of monographs, medical textbooks, journals, articles, websites, and other resources. The primary care physician should be able to:

- Help the patient address current health issues (within the realm of "reactive" medicine).
- Have tools for disease prevention for adults and children (within the realm of "preventive" medicine).
- Forecast health conditions (within the realm of "predictive" medicine).

For effective operation, the primary care physician needs the following personal qualities:

- A high level of personal culture (soft skills).
- A broad range of professional knowledge and skills in the core areas of biomedicine (hard skills).
- Sufficient breadth of thinking.
- Knowledge as a clinical physiologist.
- Knowledge and skills to perform the functions of an IT physician.
- Knowledge and skills to execute the functions of a family doctor, family (social, office) psychologist, and psychotherapist. Increasingly, companies are utilizing conversational AI, an advanced form of artificial intelligence that allows machines to engage in human-like interactive conversations with users. This technology understands and interprets human language to simulate natural dialogue.

Conversational AI systems are widely used in applications such as chatbots, voice assistants, and patient support platforms across digital and telecommunications channels. A

Gartner study found that many companies have identified chatbots as their top artificial intelligence application, and by 2022, it is expected that nearly 70% of employees will interact with conversational platforms daily. Since the pandemic, the volume of interactions conducted by conversational agents has increased by 250% across several industries. Among technology professionals worldwide, nearly 80% use virtual assistants for customer service. By 2024, 73% of customer service leaders in North America believe that live chat, video chat, chatbots, or social media will be the most used channels for customer service. These statistics highlight the growing acceptance and impact of conversational AI across industries and consumer behavior. Conversational AI feels like talking to a super-intelligent computer that understands what you say and responds like a real person. It is becoming increasingly advanced, intuitive, and cost-effective, leading to widespread adoption in various industries. Let's take a closer look at the significant advantages of this innovative technology: Conversational AI enables organizations to deliver outstanding customer service through personalized interactions across multiple channels, allowing customers to seamlessly transition from social media to live web chats. Easy scalability to handle a large volume of conversations. Conversational AI can assist customer service teams in managing sudden peaks in call volume by classifying interactions based on customer intent, requirements, call history, and sentiment. This enables effective call routing, ensuring that live agents handle high-value interactions while chatbots manage lower-value ones. Enhances customer service. The customer experience has become a critical differentiating factor for brands. Conversational AI helps companies create more positive experiences by providing instant and accurate responses to inquiries and developing customer-oriented answers using speech recognition, sentiment analysis, and intent recognition technology. Supports marketing and sales initiatives. Conversational AI can be programmed to support multiple languages, enabling companies to serve a global customer base. This capability helps businesses provide seamless support for non-English-speaking customers, overcoming language barriers and improving overall customer satisfaction. Improved data collection and analysis. Conversational AI platforms can collect and analyze vast amounts of customer data, offering invaluable insights into customer behavior, preferences, and concerns. This data-driven approach helps companies make informed decisions, refine marketing strategies, and develop better products and services. Moreover, this continuous flow of data enhances AI's learning ability, leading to more accurate and effective responses over time. 24/7 support capability, ensuring that customers receive assistance at any time, regardless of time zones or holidays. This continuous availability is particularly crucial for companies with global operations or for customers who need support after hours. Understanding what doctors say: Whether speaking or writing, AI listens attentively. It analyzes the words to understand what you mean, even capturing your tone or emotions. Understanding the bigger picture: Once the words are understood, AI attempts to grasp the broader context. It looks for patterns and context to understand what is actually being asked or said. Response to the patient: After comprehending the intent, AI quickly formulates a better response. It may ask additional questions or provide the information you need, all in a natural and friendly manner. Sounds human: Artificial intelligence works hard to make communication smooth and create the impression that you are talking to a real person rather than a machine. Becoming smarter over time: The more you interact with it, the better it gets. It learns from every interaction, improving its understanding of different accents, languages, and even jargon. Voice processing and tracking: If speaking instead of writing, AI uses speech recognition to convert voice to text. It also remembers what has been said earlier to keep the conversation on the right track. Continuous improvement: Over time, AI enhances its responses, becoming more accurate and helpful with each conversation. For patients who have communication difficulties due to age, disability, or language barriers, AI can serve as a tool to improve interactions with healthcare providers by offering translations, simplified medical terminology, or even visual cues. Patients can receive comprehensive informational support— general information about conditions, diagnostic and treatment methods, and prognosis. Generated medical information based on medical/scientific/clinical data available online is provided to patients according to their informational requests, which is vast in scope and detail. However, it's important to note that ChatGPT/AI is an evolving tool that currently cannot be regarded as perfect. Therefore, when using it, patients should be aware of the risks and limitations that need to be brought to their attention:

We unite around the idea that the main factors limiting the potential for full replacement of doctors with AI are as follows:

- Medical diagnosis often requires a comprehensive approach, which includes not only interpreting symptoms but also physical examinations, detailed medical histories, and the personal observations of a physician, which are difficult to replicate fully with AI.
- Many aspects of healthcare depend on personal contact, including the ability of the doctor to build trust with the patient, which is essential for accurate diagnosis and effective treatment.
- Empathy and understanding of the human condition are key aspects that AI struggles to reproduce. Patient trust in AI systems and their willingness to accept medical advice from non-human systems can significantly impact the effectiveness of diagnosis and treatment. Personal interaction with a primary care physician is often a key factor in the treatment process.

Issues of accountability and ethics remain pressing in healthcare. Determining who is "at fault" for misdiagnosis, insufficient/incorrect diagnoses, inappropriate treatment, and who is responsible for AI errors in medical practice is an unresolved legal issue. There are also ethical considerations regarding patient autonomy and confidentiality. While AI can effectively process large amounts of data and identify patterns that can be used for responding to inquiries, errors are still possible, especially if the system is trained on incomplete data. In support of the use of artificial intelligence in the scientific field of medicine, it is important to note that it can assist in professional doctor-patient communication by creating conversation models for different communicative situations such as: conveying medical information, persuading for therapy or procedures, resolving conflicts, and more. On one hand, specialists will maintain the conversation, explain the steps of examinations or procedures in detail, and be more confident in presenting medical research results; they will be more persuasive in communicating diagnoses or medication adherence, as well as dietary regimens. With the help of artificial intelligence, the medical professional will employ appropriate intonational patterns, a normal voice pitch, and tone that will foster trust in the patient. On the other hand, it will limit negative vocabulary and emotional connotations in spoken conversation. Thus, the communicative tasks of the doctor correspond to the goals of clinical discourse, creating an adequate conversational atmosphere to gather as complete as possible the necessary anamnesis and clarify the clinical manifestations of the disease (symptoms, their qualitative, quantitative, and temporal components, intensity, localization, characteristics of appearance, etc.). To achieve this, the doctor demonstrates skill in using the appropriate style and register, asking questions, and framing their statements depending on who the patient is (gender, age, education, life history, personality traits, present condition, etc.). However, at the same time, the doctor operates within accepted institutional status-role relationships, so the form of statements that a doctor can generate is determined by modern pragmatic norms.

AI has the potential to revolutionize healthcare, but it is essential to address the ethical implications of its use. While AI can improve efficiency and accuracy in healthcare, it is crucial to ensure that it is used responsibly and ethically. This includes developing clear ethical guidelines for AI development, ensuring data privacy, and mitigating biases in AI algorithms. Additionally, it is important to maintain human oversight to ensure that AI is used as a tool to augment human expertise, rather than replace it. By doing so, we can harness the power of AI to improve patient care while upholding ethical principles.

In the conclusions that stand out are related to the participation of artificial intelligence in the doctor-patient interview, and in particular, the contribution of artificial intelligence can be found in scheduling an appointment with a specialist, in the second part of the interview, when clarifying questions are selected and asked and the diagnosis is made and treatment instructions are given. More specifically, artificial intelligence could train specialists how to use intonation and a calm tone, adherence to literary norms when communicating health news, when choosing a therapy.

5   Discussion

Stages in medical communication with the help of AI:

Registration communication—communication channels through a website or platform. Firstly, we conducted conversations with 100 students, 100 medical professionals about sentiments around the use of artificial intelligence. The results show that young people have greater confidence in artificial intelligence in determining the exact diagnosis and giving adequate advice in treatment. Older people, over 55 years of age, are more skeptical. They see the practical help of artificial intelligence in interpreting images and laboratory results. When we asked about the role of AI in the doctor-patient interview, medical professionals say that artificial intelligence has a partial contribution, as there are conditions and diagnoses that are not typical and the algorithms will find it difficult to offer an adequate solution. Secondly, a survey of 50 individuals, 30 women and 20 men, aged 35–55, was conducted in 2023–2024. Results showed that women preferred websites with clear instructions and examples on how to use the site, while men preferred user-friendly websites. Men also valued websites that provided customer support options. The patient's choice to book an appointment with a doctor. Booking is a standard algorithm from collecting personal patient data. This can be performed by AI. The process of processing personal data must comply with the legislation of the given country. The patient must give their consent in written or electronic form. Second stage of communication involves informing the patient about the upcoming visit. In most clinics, the process includes automated notifications, such as phone calls or robotic messages with a template text. Another method is calling the registration desk. In the first case, where there is no human factor, the clinic does not receive feedback from the patient and errors may occur. In the second case, with the help of a human factor, the clinic receives feedback from the patient, but this does not exclude the possibility of errors. AI can minimize errors associated with the human factor, receive feedback from patients, and use a communication script. The third stage is the direct appointment with the doctor. AI plays the role of an assistant in making a diagnosis, as it has the ability to learn quickly and process large databases. In this case, the role of AI in the doctor-patient relationship is not fully defined. A doctor from a Californian clinic used a video link to communicate a serious diagnosis. The patient was not prepared for such communication and was shocked. A survey conducted in 2023–2024 included 10 questions. Regarding whether doctors answer patients' questions, 60% of respondents reported that specialists use keywords when providing test results, describing therapy, and discussing treatment regimens. They often fail to provide details about the medication intake period, leaving this to the pharmacist. When asked about the tone of voice doctors use when communicating diagnoses and medication, 45% of patients reported a rushed pace of speech without pauses, while 30% mentioned unclear articulation. Reasons for this were not specified, such as whether it was due to wearing masks or other barriers. Regarding understanding medical terms, responses varied. Sixty-five percent of patients complained about not having medical terms explained to them, and not having clear enough information about their condition and medications. The survey results indicate that the doctor wastes time filling out paperwork. They use keywords and incomplete phrases, with a fast pace, expressive intonation, and hesitations, leading to reduced intelligibility due to sound elision and violations of pronunciation rules, such as vowel reduction, incorrect vowel substitution, and mispronunciation of the definite article. They provide feedback on health-related questions. Unlike human communication, AI interactions often exhibit consistent adherence to linguistic norms, clear articulation, and a steady pace. 70% of respondents preferred communication through AI: During the collection of medical history, the patient interacts with AI, providing data and actively participating in their healthcare. The possibility of providing remote assistance.

60% of respondents prefer AI because: adherence to literary norms (standard language, correct spelling, grammar, and punctuation); formal tone of communication (a style of speaking or writing that is appropriate for formal situations); neutral intonation (expressing emotions in a neutral way, without strong feelings); use of euphemisms (substituting harsh or offensive words with milder and more acceptable expressions); use of ellipsis and enumeration (ellipsis is the omission of words that can be understood from the context; enumeration is the listing of items in a sentence); avoidance of pleonasm (avoiding redundant words or phrases that repeat the same idea). Our observations indicate that impact on doctor-patient interactions Improved patient information: AI can help patients get better and more complete information about their condition and treatment options. Chatbots and other AI-powered tools can answer frequently asked questions and provide personalized advice. More effective communication: AI can automate routine tasks, such as recording medical history or making initial diagnoses, freeing up more time for doctors to focus on communicating with patients. We believe that the advantage of artificial intelligence in improving doctor-patient communication lies in its ability to analyze large volumes of recorded conversations between doctors and patients to identify specific aspects of a doctor's speech that could be improved, such as pace, intonation, clarity of speech, and the use of medical terminology. By comparing a doctor's speech to models of effective communication, AI can identify areas for improvement (Fig. 1).

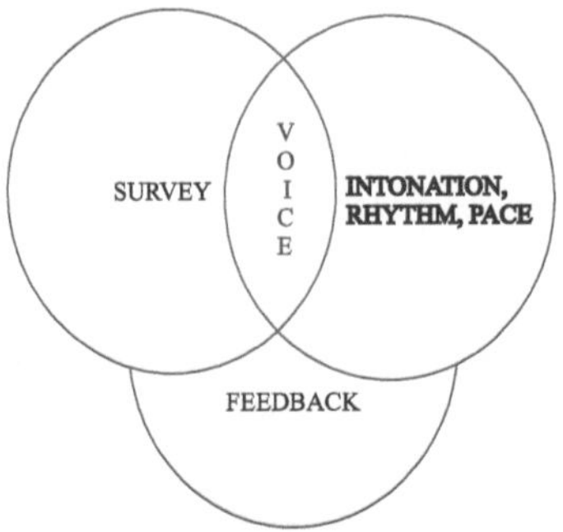

Fig. 1. Improving doctor-patient communication using AI.

During patient interactions, AI can provide real-time feedback to doctors, such as speaking too quickly, using excessive medical jargon, or having a too formal tone. AI systems can offer specific advice to improve doctors' communication skills, such as speaking more slowly, using simpler language, or asking more questions. Additionally, AI can be used to create virtual assistants that demonstrate effective communication skills, allowing doctors to observe and learn from these assistants. A more personalized approach: AI can analyze large amounts of medical data to identify the individual needs of each patient. This allows doctors to develop more personalized treatment plans. Adherence to literary norms (standard language, correct spelling, grammar, and punctuation). Formal tone of communication (a style of speaking or writing that is appropriate for formal situations). Neutral intonation (expressing emotions in a neutral way, without strong feelings). Use of euphemisms (substituting harsh or offensive words with milder and more acceptable expressions). Use of ellipsis and enumeration (ellipsis is the omission of words that can be understood from the context; enumeration is the listing of items in a sentence). Avoidance of pleonasm (avoiding redundant words or phrases that repeat the same idea). Impact on doctor-patient interactions Improved patient information: AI can help patients get better and more complete information about their condition and treatment options. Chatbots and other AI-powered tools can answer frequently asked questions and provide personalized advice. More effective communication: AI can automate routine tasks, such as recording medical history or making initial diagnoses, freeing up more time for doctors to focus on communicating with patients. A more personalized approach: AI can analyze large amounts of medical data to identify the individual needs of each patient. This allows doctors to develop more personalized treatment plans.

6 Conclusion

In conclusion, AI systems can provide more personalized, efficient, and engaging user experiences. As these technologies continue to evolve, they promise to enhance communication, streamline operations, and drive innovation in healthcare. The use of conversational AI not only provides a competitive advantage but also opens up new opportunities for more intuitive and responsive interactions in the digital age. The study of doctor-patient communication with AI has led to the following conclusions: AI can enhance doctor-patient communication by analyzing extensive recordings of doctor-patient interactions. AI will identify specific verbal cues that foster patient trust and improve the pace, intonation, and clarity of the doctor's speech, making medical terminology more accessible to patients. AI can also provide real-time feedback to doctors on their communication skills. The findings of this study will contribute to advancements in communication and the education of future medical professionals. We recommend that the use of AI in doctor-patient communication be integrated into the curricula of future healthcare professionals. In connection with the topic, it is necessary to continue the research. It is good to explore the attitudes of the possibilities for including artificial intelligence and its role in medical communication programs in order to improve the doctor-patient interaction.

Acknowledgments. The study was conducted with the support of CIII-HU-1506-01-2021 Ergonomics and Human Factors Regional Educational CEEPUS Network.

References

1. Sauerbrei, A., Kerasidou, A., Lucivero, F., et al.: The impact of artificial intelligence on the person-centred, doctor-patient relationship: some problems and solutions. BMC Med. Inform. Decis. Mak. **23**, 73 (2023). https://doi.org/10.1186/s12911-023-02162-y
2. Li, S., Chen, M., Liu, P.L., Xu, J.: Following medical advice of an AI or a human doctor? Experimental evidence based on clinician-patient communication pathway model. Health Commun. **4**, 1–13 (2024). https://doi.org/10.1080/10410236.2024.2423114
3. Lorenzini, G., Arbelaez Ossa, L., Shaw, D.M., Elger, B.S.: Artificial intelligence and the doctor–patient relationship expanding the paradigm of shared decision making. Bioethics. **37**, 424–429 (2023). https://doi.org/10.1111/bioc.13158
4. Ivanova, I., Ivanova, N., Atanasova, B.: Artificial intelligence in laboratory medicine—let's talk about it. Bulg. Soc. Med. Sci. J. **3**, 1–4 (2024). https://doi.org/10.3897/bsms.3.120969
5. Hirosawa, T., Harada, Y., Yokose, M., Sakamoto, T., Kawamura, R., Shimizu, T.: Diagnostic accuracy of differential-diagnosis lists generated by generative pretrained transformer 3 Chatbot for clinical vignettes with common chief complaints: a pilot study. Int. J. Environ. Res. Public Health. **20**(4), 3378 (2023). https://doi.org/10.3390/ijerph20043378
6. Rao, A., Pang, M., Kim, J., Kamineni, M., Lie, W., Prasad, A.K., et al.: Assessing the utility of ChatGPT throughout the entire clinical workflow: development and usability study. J. Med. Internet Res. **25**, e48659 (2023). https://doi.org/10.2196/48659
7. Ali, S.R., Dobbs, T.D., Hutchings, H.A., Whitaker, I.S.: Using ChatGPT to write patient clinic letters. Lancet Digit. Health. **5**(4), e179–e181 (2023). https://doi.org/10.1016/s2589-7500(23)00048-1

8. Jeblick, K., Schachtner, B., Dexl, J., Mittermeier, A., Stuber, A.T., Topalis, J., et al.: ChatGPT makes medicine easy to swallow: an exploratory case study on simplified radiology reports. Eur. Radiol. **34**(5), 2817–2825 (2024). https://doi.org/10.1007/s00330-023-10213-1

9. Cascella, M., Montomoli, J., Bellini, V., Bignami, E.: Evaluating the feasibility of ChatGPT in healthcare: an analysis of multiple clinical and research scenarios. J. Med. Syst. **47**(1), 33 (2023). https://doi.org/10.1007/s10916-023-01925-4

10. Liu, S., Wright, A.P., Patterson, B., Wanderer, J.P., Turer, R.W., Nelson, S.D., et al.: Assessing the value of ChatGPT for clinical decision support optimization. medRxiv (2023). https://doi.org/10.1101/2023.02.21.23286254

11. Potapenko, I., Boberg-Ans, L.C., Stormly Hansen, M., Klefter, O.N., van Dijk, E.H.C., Subhi, Y.: Artificial intelligence-based chatbot patient information on common retinal diseases using ChatGPT. Acta Ophthalmol. **101**(7), 829–831 (2023). https://doi.org/10.1111/aos.15661

12. Grunebaum, A., Chervenak, J., Pollet, S.L., Katz, A., Chervenak, F.A.: The exciting potential for ChatGPT in obstetrics and gynecology. Am. J. Obstet. Gynecol. **228**(6), 696–705 (2023). https://doi.org/10.1016/j.ajog.2023.03.009

13. Yeo, Y.H., Samaan, J.S., Ng, W.H., Ting, P., Trivedi, H., Vipani, A., et al.: Assessing the performance of ChatGPT in answering questions regarding cirrhosis and hepatocellular carcinoma. Clin. Mol. Hepatol. **29**(3), 721–732 (2023). https://doi.org/10.3350/cmh.2023.0089

14. Choudhury, A., Chaudhry, Z.: Large language models and user trust: consequence of self-referential learning loop and the deskilling of health care professionals. J. Med. Internet Res. **26**, e56764 (2024). https://doi.org/10.2196/56764

15. Thirunavukarasu, A.J., Ting, D.S.J., Elangovan, K., Gutierrez, L., Tan, T.F., Ting, D.S.W.: Large language models in medicine. Nat. Med. **29**(8), 1930–1940 (2023). https://doi.org/10.1038/s41591-023-02448-8

16. https://www.shaip.com/blog/the-complete-guide-to-conversational-ai/. Accessed 26 Nov 2024

A New Hybrid Algorithm to Diagnose MS Using MRI Image Processing

Maryam Oghbaei[1], Ali Asghar Rahmani Hosseinabadi[2]([✉]),
and SeyedSaeid Mirkamali[3]

[1] Department of Radiation Medicine, Shahid Beheshti University, Daneshjoo St, Velenjak,
1983963113 Tehran, Iran
[2] Department of Computer Science, University of Regina, Regina, Canada
`a.r.hosseinabadi1987@gmail.com`
[3] Department of Computer Engineering and IT, Payame Noor University (PNU), Tehran, Iran
`s.mirkamali@pnu.ac.ir`

Abstract. Multiple Sclerosis (MS) is an autoimmune disease that affects the central nervous system. The destruction of nerve cells in the body can cause various symptoms in the body, such as fatigue, blurred vision in the eyes, numbness and loss of body coordination, etc., the exact cause of which has not yet been determined. MS is a chronic disease that attacks and destroys the nerve fibers surrounded by myelin. Today, researchers are trying to find a way to diagnose and treat MS by using Magnetic Resonance Imaging (MRI) images and its processing and Artificial Intelligence (AI) algorithms. In this paper, a new hybrid algorithm for identifying MS disease with the help of MRI images is presented. The proposed algorithm first detects the highlights in the MRI images with the help of a detector and then by using the scattered histogram, the highlights will be displayed in an efficient manner. The simulation results demonstrate the precision and detection capability of the proposed algorithm relative to others.

Keywords: Artificial Intelligence · Multiple Sclerosis · Magnetic Resonance Imaging · K-means · Brain Extraction Meta

1 Introduction

Multiple Sclerosis (MS) is a complex disease that has been called the disease of a thousand faces. One reason for this complexity is the variety of symptoms it can cause [1, 2]. This diversity causes it to be mistaken with other diseases, or vice versa. A condition that causes inflammation and causes damage to the myelin sheaths of nerve cells in the brain and spinal cord causes this condition. All of the elements of the nervous system that are responsible for communication may be disrupted as a result of these damages, which can result in a broad variety of physical indications and symptoms [3–5].

MS is a particularly "research-hard" disease. Although thousands of researchers are currently conducting clinical trials on MS, it must be declared that MS is a "research-hard" disease from a research perspective [6, 7].

The reasons why MS research is difficult include:

A. Mirzazadeh et al. (Eds.): ODSIE 2024, CCIS 2482, pp. 119–136, 2026.
https://doi.org/10.1007/978-3-031-93601-2_8

1. The cause of MS is unknown. Although it is generally believed that MS is caused by a combination of genetic, environmental, and immunological factors, it is difficult to pinpoint the exact cause because it can take years for an individual to be diagnosed with MS. On the other hand, there are so many variables that influence the disease [4, 6].
2. MS targets two completely inaccessible areas of the body, the brain and spinal cord, and researchers have only been able to see and study the lesions since the 1980s, when Magnetic Resonance Imaging (MRI) became available [6, 8].
3. There is no specific pattern to the disease. The course of the disease isn't predictable in addition, there is no specific diagnostic test for the disease [6, 9].

MRI plays an undeniable role in the diagnosis of this disease. In short, MRI has four roles in MS [10, 11]:

1. Diagnosis,
2. Evaluation of response to treatment,
3. Prognosis,
4. Exclusion of other causes.

The findings of MS on MRI are not specific and are seen in many other diseases. There are more than 10 to 15 diseases that can produce lesions on brain, being very similar to MS, and if the radiologist does not have enough experience, he will report it as MS and neurologists try to treat it accordingly [10, 12].

The diagnosis of MS on MRI has specific criteria and rules that an international MS committee has defined in almost annual meetings. The criteria for diagnosing MS have changed a lot since several years ago. Today, the disease is diagnosed even with two to three small lesions in the white matter of the brain scattered in places that cannot easily be seen by inexperienced people. This means that sometimes lesions are located in areas of the brain that any MRI reporter may report as normal, while experienced individuals find them because of their knowledge of the disease and the areas of the brain that the disease can affect [13, 14].

In this paper, a robust pattern recognition algorithm is presented to identify MS tissues using MRI images. The innovations of the proposed method include:

1. To address the aforementioned issue, two fully automated, two-stage brain extraction techniques are employed to delineate brain parts from MRI head images,
2. MS disease identification is performed using sparse histograms,
3. To create a multi-faceted database, K-means clustering algorithm is used,
4. Finally, after creating training sparse histograms and a search image using the cosine similarity algorithm, the closest patch image to the MS texture image is found.

The following is the outline of the contents of the paper: The associated work is further upon in Sect. 2. In the third section, the suggested approach is given in its entirety. In the fourth and fifth sections, respectively, the outcomes of the simulation and the conclusions are being provided.

2 Related Work

In recent years, many authors have addressed the issue of breast cancer diagnosis, and several of their works are mentioned below:

Rocca et al. [15] comprehensively reviewed the current role of MRI in the diagnosis and prognosis of MS. The researchers discovered promising MRI signals that have the potential to act as reliable predictors of the course of the disease and assist physicians in appropriately managing patients with multiple sclerosis. In addition to this, they evaluated the potential of emerging technologies, such as artificial intelligence (AI) and automated quantification tools, which have the potential to play an important part in providing assistance to medical professionals.

Sobhy et al. [16] performed a six to nine-month trial at Ain Shams University in Egypt involving 50 individuals with a verified diagnosis of multiple sclerosis. Participants were divided into two equal groups and underwent MRI scans at field strengths of 1.5 T and 3 T. In this study, which included 42 women (84%) and eight men (16%) with a mean age of 31.5 years ($\pm$ 6.7) for 1.5 T MRI and 32.47 years ($\pm$ 6.7) for 3 T-scans, statistical analysis showed that there was no significant difference in lesion size between Dual Inversion Recovery (DIR) and Fluid-Attenuated Inversion Recovery (FLAIR) imaging modalities.

Research conducted by Cavaliere et al. [17] looked at whether or not the data from Optical Coherence Tomography (OCT) and Support Vector Machine (SVM) might be used to diagnose multiple sclerosis. The control group consisted of 48 healthy individuals and 48 patients with multiple sclerosis who did not exhibit any signs of optic neuritis. An imaging procedure was carried out with the use of a Deep-Range Imaging (DRI) Triton equipment manufactured by Topcon Corporation in Tokyo, Japan. The values of macular thickness and periapical area were compared between the two groups. In order to construct the feature vector, the three variables that have the greatest discriminant capacity were chosen, and the support vector machine (SVM) was used as an automated classifier.

Zurita et al. [18] developed classifiers based on Decision Tree (DTI) and fMRI data to identify brain regions associated with a more accurate description of MS. Their hypothesis was that multi-state classifiers derived from a combination of DTI and fMRI data could distinguish patients with Relapsing-Remitting MS (RRMS) from healthy controls with high accuracy. They developed reliable linear classifiers that achieved 89% $\pm$ 2% accuracy and identified functional, structural connections, and specific brain regions that were important for identifying RRMS patients.

In [19], automated MS diagnosis using Machine Learning (ML) methods provided between 2011 and 2022 were reviewed and analyzed. MS, which affects the central nervous system, leads to problems in the spinal cord, brain, and optic nerve, and is estimated to affect 2.8 million people worldwide. In this work, multiple Machine learning models based on MRI and clinical data were explored.

In [20], the authors used clinical data and Retinal Nerve Fiber Layer (RNFL) thickness measured by OCT to improve MS diagnosis and predict long-term disability progression in MS patients. The study included 104 healthy individuals and 108 MS patients. In terms of the diagnosis of multiple sclerosis (MS), the Ensemble Classifier (EC) classifier demonstrated the highest level of performance, with an accuracy rate of 87.7%.

On the other hand, the Long-Short-Term Memory (LSTM) Neural Network (NN) was the most successful classifier for predicting long-term MS impairment, with an accuracy rate of 81.7%.

Weidauer et al. [21] conducted research to determine the significance of magnetic resonance imaging (MRI) in the diagnosis and prognosis of multiple sclerosis (MS). The research highlights the fact that, despite the significance of neurological examination and medical history in the process of diagnosing multiple sclerosis (MS), magnetic resonance imaging (MRI) plays a significant role in the diagnostic process, particularly once the requirements have been established. The study also addressed the identification of similar diseases and ensuring that MS was correctly diagnosed before initiating disease-modifying therapies. They provided MRI protocols for baseline evaluation and following-up MS patients and also for identifying specific patterns of brain and spinal cord injuries.

3 Proposed Algorithm

The initial step of the proposed method involves the detection of salient points through a salient point detector. The target points delineate a region characterized by high variance. Patches are extracted from each of these salient points. The preserved patches are transformed into feature vectors by calculating the local steering kernel surrounding them, utilizing Principal Component Analysis (PCA) for dimensionality reduction to extract relevant features from the patches. Subsequently, they will be presented using sparse histograms in an efficient manner. The histograms of the test and training images are analyzed and categorized utilizing a database of sparse histograms.

3.1 Intracranial Segmentation

This section discusses two fully automated two-stage Brain Extraction Meta (BEM) methods employed to extract brain segments from MRI head images [22]. Initially, feature extraction is conducted to identify the Region of Interest (ROI), followed by segmentation to extract the ROI. This paper identifies ROI as the brain region. Region labeling is crucial in the initial phase for identifying brain regions and producing a preliminary brain segment. In the second stage, morphological operations are used to segment the healthy brain segment.

First Stage: Feature Extraction. Initially, it is necessary to acquire a brain segment. The brain's features are utilized in this process. Feature extraction relies on expert knowledge that is exclusively based on brain anatomy and the density characteristics of MRI scans. T1 MRI scans of the head show that scalp and brain tissues exhibit greater brightness compared to the Cerebrospinal Fluid (CSF), skull, and background [23]. The skull serves as a boundary between the brain and the scalp. Expert knowledge is employed to detect and remove the bright scalp region, or to identify and retain the bright brain within that area. The primary feature to be detected from the image is either the scalp or the brain. The technique for scalp removal is referred to as the Scalp Removal Process (SRP), while the method for brain extraction is designated as the Brain Extraction Process (BEP). In this step, one of the two methods is employed to generate the rough brain section.

Threshold and Labeling Process. The first step starts with the labeling of the input T1 image. Initially, an optimal intensity threshold value (T_{opt}) for the pixels of the input T1 image (f_0) is calculated using the RIDDLER method [24]. This approach may be described as iterative, and in most cases, four to ten iterations are sufficient. At iteration t, the threshold value (T_{t+1}) is calculated using Eq. (1):

$$T_{t+1} = \frac{\mu_{bgt} + \mu_{obt}}{2} \tag{1}$$

In this context, the variables μ_{bgt} and μ_{obt} represent the average gray levels of the background (bg) and object (ob) at iteration t, respectively. At iteration t, the threshold value T_t that was collected in the iteration before this one is used to determine the segmentation into background information and object information. The value of the threshold that was acquired after ten iterations is regarded to be the optimum threshold T_{opt}, and it will be used in order to successfully recognize and differentiate objects from the integrated background that is around them [25]. It is possible to get a binary image $g_0(x, y)$ by using T_{opt} and the Eq. (2) that follows:

$$g_0(x, y) = \begin{cases} 1 \ if \ f_0(x, y) \geq T_{opt} \\ \ \ 0 \ otherwise \end{cases} \tag{2}$$

where the starting intensity of the picture at pixel (x, y) is denoted by the representation $f_0(x, y)$. This results in the binary picture g_0 having two labels: label zero is assigned to dark regions, such as the backdrop, skull, and cerebrospinal fluid (CSF), while label one is assigned to bright regions, such as the scalp and the brain.

Head Contour Detection. Next, we want to find the boundary separating scalp from the background in order to generate a head mask. To do this, we scan the binary image g_0 from all four sides, pixel by pixel, starting from left to right, top to bottom, and then in the opposite direction, to find the head boundary. The scalp is a bright area in $T1$ images. For this reason, we take into account the transition [0, 1] from black to white, which occurs during the process of scanning the picture g_0, as the boundary point of the head b (x, y). Both rows and columns of the image are included in this consideration. This results in the formation of the region R_h, which is restricted to the set of b (x, y) that was acquired. This area is regarded as the head, while the remainder of the area is considered to be the backdrop. As a last step, a head mask, denoted by the letter h, is generated by using Eq. (3):

$$h(x, y) = \begin{cases} 1 \ if \ (x, y) \in R_h \\ \ \ 0 \ otherwise \end{cases} \tag{3}$$

In the image h shown in Fig. 1, the area containing values of one (white) R_h is the head and the area containing values of 0 (black) is called background (R_{bg}).

These two areas are subject to Eq. (4).

$$h = R_h \cup R_{bg} \tag{4}$$

Labeling Process. The next step involves selecting the inner black region that is representative of the skull and the cerebrospinal fluid (CSF). This may be accomplished by

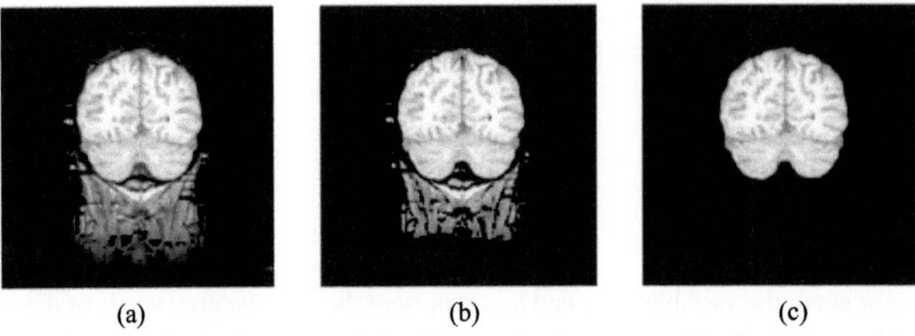

Fig. 1. There are three types of brain segments: (**a**) brain segments f_{rb} generated by SRP in the first stage using the original image f_0; (**b**) brain segments f_rb generated by BEP in the first stage using the original image f_0; and (**c**) brain segments f_{rb} generated by segmentation in the second stage using the original image f_0 and the final mask X_3 [23].

marking the region that is identified as two. To do so, we consider the binary image g_0 and also, we choose R_h as ROI. In R_h, the pixels corresponding to the dark areas that represent the CSF and/or skull are selected using the binary image g_0 and then labeled as two. Currently, g_0 is a two-label image with values 0 and 1. With the new label two, the entire image is transformed into a three-label image, which will be represented by L_3. Labeling is done according to Eq. (5):

$$L_3(x, y) = \begin{cases} 2 \ if \ g_0(x, y) = 0 \ and \, (x, y) \in R_h \\ 1 \ if \ g_0(x, y) = 1 \\ 0 \ g_0(x, y) = 0 \end{cases} \tag{5}$$

Now we have a three-label image called L_3, where the background pixels are labeled as zero, the pixels of the scalp and brain tissue areas are labeled as one, and the rest of the pixels of the skull, CSF, etc. are labeled as two.

Brain Region Detection. The next objective is to utilize either SRP or BEP in order to extract the brain area. The run length recognition system is the foundation upon which SRP and BEP are built for labeling [25]. To begin, the horizontal lines for the areas designated as one from L_3 are constructed. These runs represent the regions of the scalp and the brain. After that, the labels of pixels that are related to horizontal runs are analyzed, and the results are eventually separated into two categories: class 1 and class 2. The edges of the runs are used to determine the grouping of the runs into different categories.

Scalp Removal Process. If a run is in mode one, then it is designated as R_{SC} (scalp). The intensities of pixels that correspond to R_{SC} are set to zero, while the rest intensities are kept in g_0. This results in the production of a brain mask g_{rb}, which can be determined by using Eq. (6).

$$g_{rb}(x, y) = \begin{cases} o \ if \ (x, y) \in R_{sc} \\ g_0(x, y) \ otherwise \end{cases} \tag{6}$$

Using this process, the light scalp blends into the background.

Brain Extraction Process. If a run is in mode two, then, the run is marked as R_{bt} (brain tissues). The pixels corresponding to these runs are kept in g_0 and the intensity of other pixels is considered as that of background, i.e., zero, and the brain mask g_{rb} is obtained according to Eq. (7)

$$g_{rb}(x, y) = \begin{cases} g_0(x, y) & if \ (x, y) \in R_{bt} \\ 0 & otherwise \end{cases} \tag{7}$$

In order to generate a brain mask, the first step is completed by using either of these methods (SRP or BEP). It is important to note that both SRP and BEP will provide identical outcomes. In the two Eqs. (6) and (7), the brain segments f_{rb} will be formed if the original picture f_0 is utilized instead of the binary form g_0. This is seen in Fig. 1a, b.

3.2   Second Stage: Segmentation

Using the g_{rb} brain mask that was developed in the previous stage, the objective of this stage is to accomplish the production of a decent brain mask. During this stage, several image processing algorithms, morphological operations, and linked component activities are carried out. It is shown in Fig. 2 how these operations are carried out in succession.

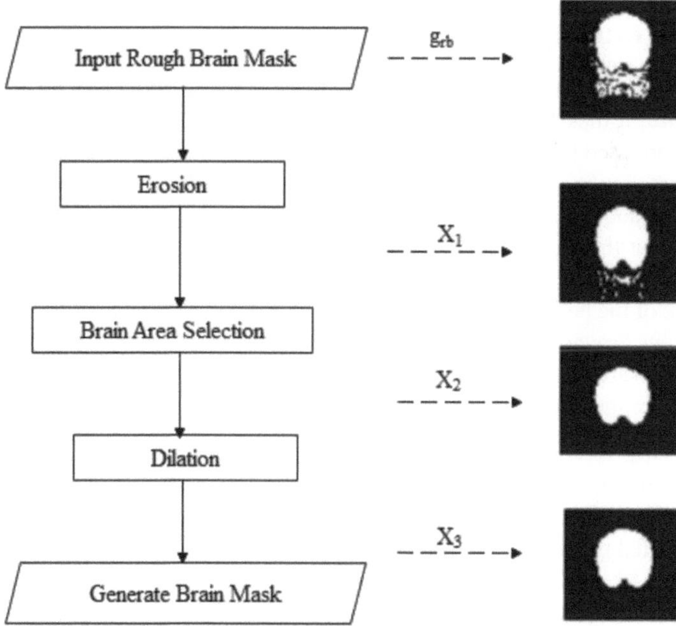

Fig. 2. Flowchart of the second stage: segmentation [23].

Brain Region Selection. Brain area selection is the next stage in the process. In this scenario, the last brain region, denoted by R_{fb}, is originally assigned a value of zero.

$$R_{fb} = \varphi \tag{8}$$

The perimeter of each connected region $RA_{(i)}$ is calculated and then for each slice, the Largest Connected Component (LCC) among the regions obtained in the eroded image is considered as the brain portion R_{fb}. The brain selection is done according to Eqs. (9) and (10).

$$R_{LCC} = R\left(\begin{array}{c} \arg \max R_A(i) \\ 1 \leq i \leq n \end{array} \right) \tag{9}$$

$$R_{fb} = R_{LCC} \tag{10}$$

In accordance with Eq. (11) and after the identification of the brain region R_{fb} in the binary picture X_1, the brain area X_2 is retrieved.

$$X_2(x, y) = \begin{cases} 1 \; if \; (x, y) \in R_{fb} \\ 0 \; otherwise \end{cases} \tag{11}$$

The brain region X_2 extracted by Eqs. (9)–(11) is shown in Fig. 2.

3.3 Feature Extraction

Patch Extraction. Due to the fact that MS disease tissue only takes up a tiny portion of the picture, it makes a little contribution to the overall characteristics; hence, local descriptors are used [26]. Patches are sub-images that are taken from the picture around locations that are visually significant. The following is a list of the benefits that patches offer: The reduction of the quantity of data that has to be processed, the resistance to disturbances in the background, and the resistance to occlusion and variations in the form of MS tissue are the three main benefits.

The size of the patches plays an important role in the efficiency of the algorithm. To obtain patches around key points, the statistical features including the mean, variance, and correlation are obtained from the image intensities and by growing the neighborhood region with similar features to that of the patches from the MRI images.

Local Steering Kernel. The local steering kernel computes the local geometric and radiometric similarity of a pixel to its neighbors. According to [27], the primary objective was to acquire the local data structure by doing an analysis of the pixel differences based on the estimated gradients. The LSK model is represented by Eq. (13).

$$K(T_l - X) = \frac{\sqrt{\det(C_l)}}{2\pi h^2} \exp\left\{ -\frac{(X_l - X)^T C_l (X_l - X)}{2h^2} \right\} \tag{13}$$

Let l be an element of the set $\{1, ..., p^2\}$, where $X_l = [X_1, X_2]$ represents spatial coordinates. The term p^2 denotes the number of pixels within a local window of size

$(p \times p)$. The variable h signifies a global smoothing parameter, while the matrix C_l denotes the estimated covariance matrix of the gradient vector set $[X_1, X_2]$ within the local window surrounding the sampling position x [28].

Figure 3 shows LSK in different areas along with the local image structure with LSK.

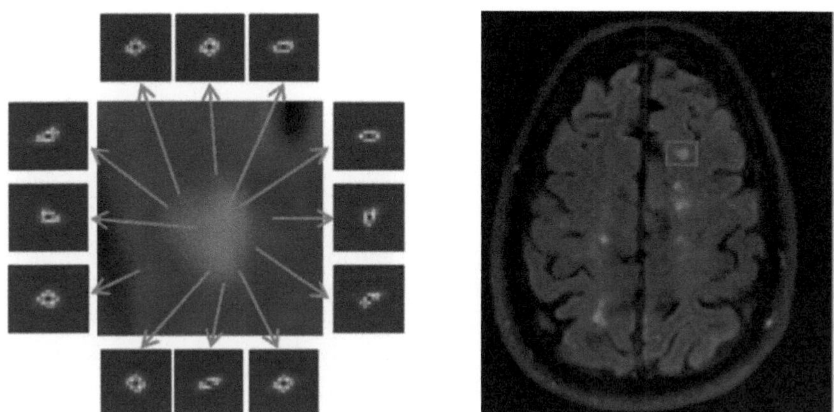

Fig. 3. A representation of LSK in different regions. Local image structure with LSK.

To evaluate the efficiency of the proposed method, feature extraction mask sizes of $7 \times 7, 9 \times 9$, and 11×11 were selected for analysis.

Sparse Histogram. Sparse histograms are a widely utilized method for representing data distributions, particularly in the field of computer vision. A histogram H serves as a discrete approximation of the distribution of a random variable, comprising an array $(H_1, H_2, , ..H_N)$ of intervals that delineate segments of a feature space S into N regions $\{S_1, S_2, , , S_N\}$. The regions are usually of equal size, but this equality is not necessary. First, the features associated with each class are separated by the K-means clustering algorithm according to Eq. (14) and placed in a multifaceted database.

$$\{F_1, F_2\} = \arg{}^{\min} \sum_{i=1}^{2} \sum_{j=1:N} \left\| y_j - \hat{\mu}_i \right\| \tag{14}$$

where i is the number of classes, $\hat{\mu}_i$ is the cluster center, and N is the number of feature vectors. Then the feature clustering sparse histogram is calculated.

Suppose $P^s = \{p_i^s | i = 1 : K_s\}$ is the image features extracted from a target candidate considering different patch scales, where, $p_i^s \in R^{d_s}$ is the of the ith local patch under scale s, K_s is the number of local image patches for each target candidate, and d_s is the dimensions of the local image patches at scale s. Each patch p_i^s contains a set of sparse coefficients $a_i^s \in R^{c \times 1}$ which can be calculated as Eq. (15).

$$\hat{a}_i^s = \min \left\| p_i^s - F^s a_i^s \right\| \tag{15}$$

where a multiscale database $F^s \in R^{d_s \times c}$ is generated by the k-means technique, which groups the target object patches labeled in the first frame with scale s at the cluster center c. After calculating the sparse coefficients, all local patches of a candidate must be added together as Eq. (16) in order to represent the candidate and to form a histogram under the scale s.

$$H^s = \sum_{i=1}^{K_s} \sum_{j=1}^{2} \left| \hat{a}_{ij}^s \right| \tag{16}$$

where, $H\left(b_j^s\right) \in R^{c \times K_s \times 1}$ is the sparse coding histogram of a candidate, and $\hat{a}_{ij}^s$ is the jth coefficient of the ith image patch. This equation shows how the multiscale database is distributed over a candidate.

Cosine Similarity Quantifies the degree of similarity between two vectors in an inner product space by calculating the cosine of the angle formed between them. Cosine similarity is widely utilized across various domains, including information retrieval, pattern matching, and data mining. For the similarity, $H^s \in \mathbb{R}^{d \times 1}$ is considered as the similarity pattern. Given an observed image vector $H_t \in \mathbb{R}^{d \times 1}$, cosine similarity is defined as Eq. (17).

$$s(\mathrm{y}, \mathrm{D}) = \frac{\langle \mathrm{y}, H^s \rangle}{\|H_t\| \|H^s\|} \tag{17}$$

where, $\|.\|$ is the norm of ℓ_2. The parameter H^s is the output of sparse histogram and H_t is the sparse histogram of the test image patch.

4 Simulation Results

In this section, the simulation results of the proposed method and comparison with other ones are fully explained. To evaluate the proposed method, simulated data obtained from Brainweb [29] has been used. It is worth noting that real data six has also been used in the simulation.

4.1 Clustering Evaluation Methods

In this paper, the Dice similarity criterion, *sensitivity*, *specificity*, and *accuracy* have been used to evaluate clustering, which are shown in Eqs. (18)–(21) respectively.

$$Dice = \frac{2 \times TP}{(2 \times TP) + FP + FN} \tag{18}$$

$$Sensivity = \frac{TP}{TP + FN} * 100 \tag{19}$$

$$Specificity = \frac{TN}{TN + FP} \tag{20}$$

$$Accuracy = \frac{TP + TN}{TP + TN + FP + FN} \tag{21}$$

The definitions of True Positive (TP), True Negative (TN), False Positive (FP), and False Negative (FN) are given below [30].

TP: Patches that are in the target layer and are assigned to the target layer by the algorithm.
TN: Patches that are not in the target layer and are not assigned to the target layer by the algorithm.
FP: Patches that are not in the target layer and are assigned to the target layer by the algorithm.
FN: Patches that are in the target layer and are not assigned to the target layer by the algorithm.

Sensitivity and *specificity* are two metrics in statistics for evaluating the outcome of a binary (two-state) classification test. When data can be separated into positive and negative groups, the accuracy of the findings of a test that separates the data into these two groups may be quantified and described using the indices of *sensitivity* and *specificity*. This is possible when the data can be split into positive and negative organizations. When we talk about sensitivity, we are referring to the proportion of positive instances that the test accurately identifies as positive. The term *"specificity"* refers to the percentage of instances that are negative to the extent that the test accurately identifies them as negative.

The False Positive Rate (FPR) represents the ratio of negative instances that are inaccurately identified as positive. The False Negative Rate (FNR) represents the ratio of positive instances that are misclassified as negative.

$$FPR = \frac{FP}{TN + FP} \tag{22}$$

$$FNR = \frac{FN}{TP + FN} \tag{23}$$

4.2 Intracranial Segmentation Results

Figures 4 and 5 show the results of the intracranial segmentation algorithm on *T1* and *T2* modalities, respectively.

In this paper, the ROI is the brain region of interest, which consists of two steps. Region labeling plays an important role in the first step to detect the brain region and generate the brain segment. Morphological procedures are used in the second part of the process in order to effectively separate the good brain section. These steps are shown in Fig. 6. In Table 1, the results of intracranial segmentation are evaluated in terms of Dice criterion, FPR and FNR.

The Dice feature criterion shows the degree of overlapping, and the closer this criterion is to 100%, the more ideal the separation of the separating method. Table 1 shows the Dice criterion in percentage. As a result, the closer we can get the Dice criterion to

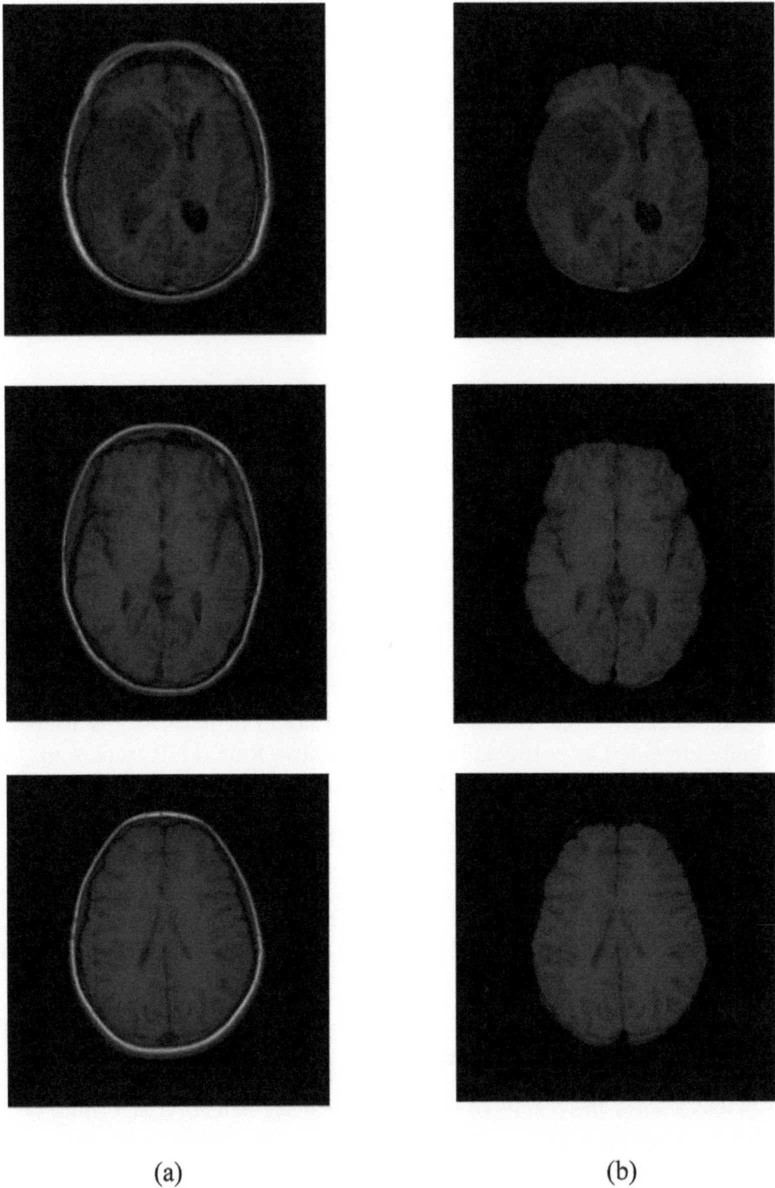

(a)   (b)

Fig. 4. Intracranial segmentation of T1 modality. (**a**) Input image. (**b**) Output image of the intracranial segmentation algorithm.

100%,, the better the segmentation. The proposed algorithm has performed well in skull separation and achieved an average Dice criterion of 92.76%.

As can be seen in Table 1, the segmentation done by the proposed method based on the Dice criterion of the T1 modality is better than that done on T2 modality. The FPR

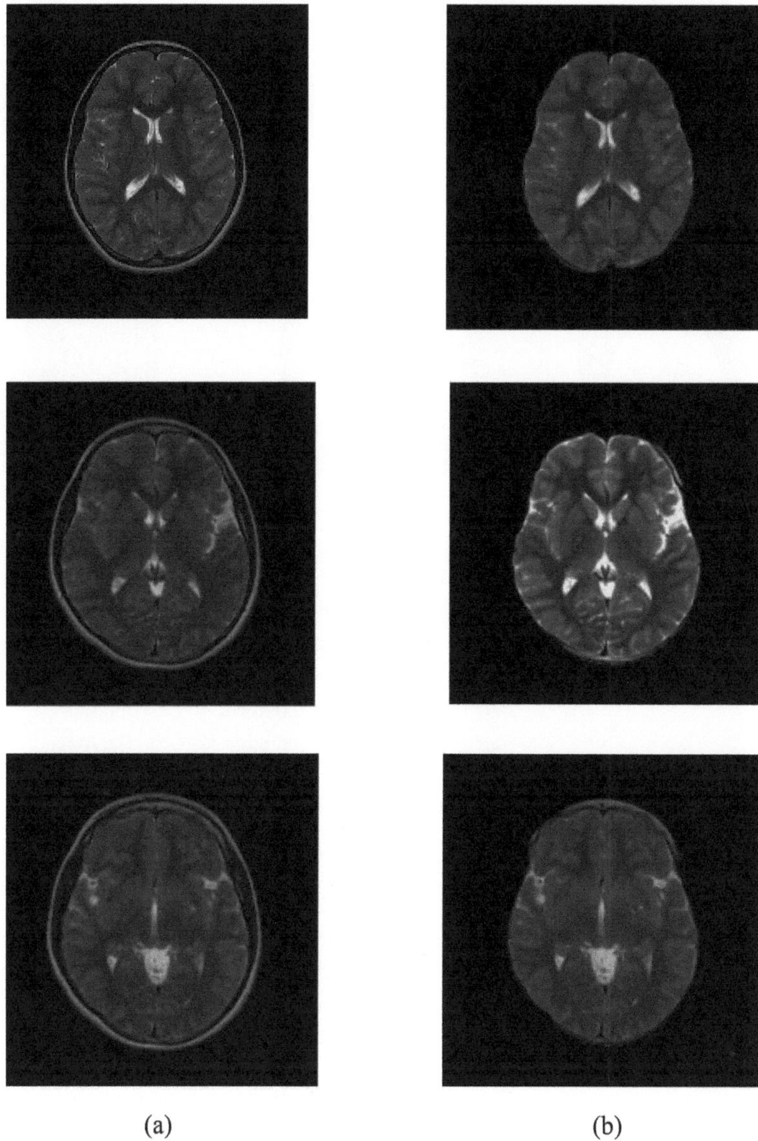

(a)   (b)

Fig. 5. Intracranial segmentation of T2 modality. (**a**) Input image. (**b**) Output image of the intracranial segmentation algorithm.

and FNR criteria, which are the false positive ratio and false negative ratio respectively, are used to estimate the cranial segmentation error; the closer these two criteria are to zero, the lower the classification error. According to Table 1, the proposed algorithm has achieved a very low overall segmentation error in terms of FPR and FNR criteria.

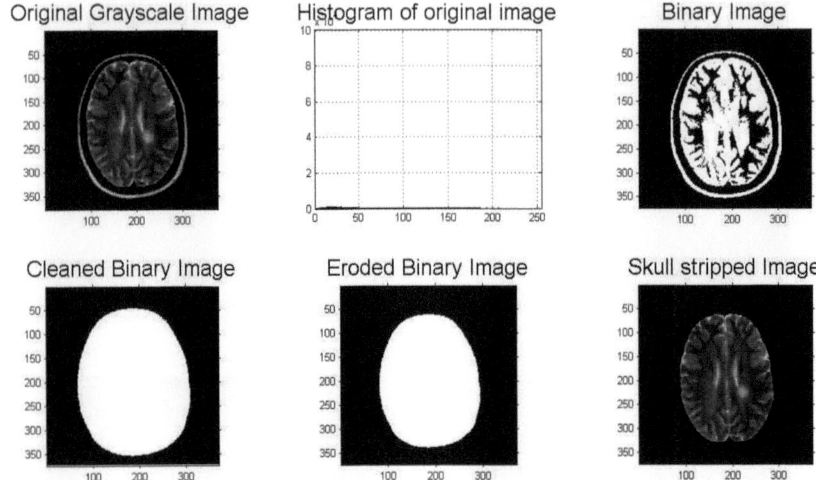

Fig. 6. Output of different intracranial segmentation steps.

Table 1. Performance of the intracranial segmentation algorithm.

	Dice	FPR	FNR
T1	93.08	0.0063	0.0481
T2	92.45	0.0069	0.0594
Average performance	92.765	0.0132	0.1075

4.3 Changing the Smoothing Parameter

The detection rate for varying the patch sizes and the number of prominent points/patches are shown in Table 2, which displays the results for a variety of smoothing parameter values. It is six PCA coefficients that are being used in this situation. The simulation makes use of three distinct values of h, which are *0.008*, 1, and 2.1 respectively. Based on the data shown in Table 2, it can be deduced that the 7×7 and 9×9 patches exhibit superior performance across all parameter modifications and seem to attain detection rates that are satisfactory. The 11×11 patch has a low detection rate compared to other patches. For 150 salient points, the 11×11 patch provides better results. In this work, regardless of the increase in patch size, and the number of extracted patches, $h = 2.1$ achieves better results.

4.4 Changing the Number of PCA Coefficients

Table 3 presents the detection rates associated with variations in patch size and the quantity of salient points/patches, alongside adjustments in the number of principal components (four, six, and eight). Table 3 indicates that a patch size of 7×7 achieves a perfect detection rate across all PCA coefficients and parameter variations. Increasing

Table 2. Detection rate per changing patch size and number of salient points/patches for different values of smoothing parameter.

Detection rate (%)									
Patch size	$h = 0.008$			$h = 1$			$h = 2.1$		
	Number of salient points/patches			Number of salient points/patches			Number of salient points/patches		
	100	125	150	100	125	150	100	125	150
7×7	95.3	96.1	96.9	95.41	96.25	97.1	96.1	96.68	97.26
9×9	95.27	95.92	96.85	95.34	96.16	97.05	95.92	96.29	97.18
11×11	91	91.9	92.5	91.52	92.26	92.95	93.61	94.23	94.59

the patch size to 9×9 yields equivalent performance for four and six PCA components; however, for eight PCA components, a decrease in detection rate occurs when 100 patches are extracted. An increase in patch size correlates with a decline in the detection rate for four, six, and eight PCA components.

Table 3. Detection rate relative to variations in patch size and the quantity of salient points/patches associated with alterations in the number of principal components.

Detection rate (%)									
Patch size	Number of principal components = 4			Number of principal components = 6			Number of principal components = 8		
	Number of Salient Points/Patches			Number of Salient Points/Patches			Number of Salient Points/Patches		
	100	125	150	100	125	150	100	125	150
7×7	96.1	96.68	97.26	96	96.52	97.1	95.5	96.09	96.52
9×9	95.92	96.29	97.18	95.89	96.12	97.02	95.29	96.03	96.52
11×11	93.61	94.23	94.59	93.52	94.11	94.36	92.9	93.21	93.65

The detection rate for the case with four components surpasses that of the cases with six and eight components. Consequently, a reduced number of PCA components yields improved results. Given the number of salient points, 100 appears adequate for enhancing detection in both single-scale and multi-scale test images.

4.5 Classification Results

To compare and evaluate the proposed algorithm with other algorithms, the criteria of *accuracy*, *sensitivity*, and *specificity* have been used. Table 4 shows the comparison results of the proposed algorithm with other ones.

Table 4. Comparison of the results of the proposed algorithm with other algorithms.

Algorithms	Sensitivity	Specificity	Accuracy
Ref. [31]	97.12	98.25	97.76
Ref. [32]	97.34	97.73	97.56
Ref. [33]	84.21	95	–
Proposed Algorithm	97.82	98.23	97.26

If the *sensitivity, specificity,* and *accuracy* criteria can be brought close to 100%, the system will be closer to the ideal state. In the proposed algorithm, the *sensitivity* and *specificity* criteria have been better than the compared ones, and it also has an acceptable performance in terms of *accuracy* criterion. The proposed algorithm has been able to identify MS tissues in brain MRI images correctly. The final accuracy of the proposed algorithm is evaluated as 97.26%.

5 Conclusion

In this paper, we have reviewed the methods used in the field of MS disease identification using MRI images and then we have provided a strong pattern recognition algorithm to identify MS disease tissues. The proposed method has been tested and experimented on simulated data provided from Brainweb. In the proposed method, two fully automated brain extraction methods have been used to extract brain segments from MRI head images. In the first stage, feature extraction has been performed to locate the desired region, and in the second stage, segmentation has been performed to extract ROI. As noticed, labeling the region plays an important role in the first stage to detect the brain region and to generate the rough brain segment. In the second stage, morphological operations have been used to segment the brain part. Then, key points have been used to find important points from the MRI image and increase the speed by Har Wavelet. Then, LSK feature with different sizes have been extracted and finally a sparse histogram has been obtained in order to have the most optimal information from each image patch. The MS disease identification has been performed using sparse histogram. To apply K-means clustering algorithm, the multi-faceted database has been used. Finally, using the multi-faceted database, a sparse histogram of the patch features has been obtained. At last, after creating robust training sparse histograms of the search image by the cosine similarity algorithm, the closest patch image to the MS tissue image has been found. The results of the proposed algorithm for diagnosing MS disease using MRI images has been evaluated at 97.26%.

References

1. Naeeni Davarani, M., et al.: Efficient segmentation of active and inactive plaques in FLAIR-images using DeepLabV3Plus SE with efficientnetb0 backbone in multiple sclerosis. Sci. Rep. **14**(1), 16304 (2024)
2. Anupriya, E., Thaile, M., Chinnasamy, P., Narayana, M.L.: Heart diseases prediction using machine learning algorithms. In: 2023 International Conference on Computer Communication and Informatics (ICCCI), pp. 1–5. IEEE (2023)
3. Yaghoubi, N., Masumi, H., Fatehi, M.H., Ashtari, F., Kafieh, R.: Deep learning and classic machine learning models in the automatic diagnosis of multiple sclerosis using retinal vessels. Multimed. Tools Appl. **83**(13), 37483–37504 (2024)
4. Pirozmand, P., Sadeghilalimi, M., Hosseinabadi, A.A.R., Sadeghilalimi, F., Mirkamali, S., Slowik, A.: A feature selection approach for spam detection in social networks using gravitational force-based heuristic algorithm. J. Ambient. Intell. Humaniz. Comput., 1–14 (2023)
5. Chinnasamy, P., Kumar, S.A., Navya, V., Priya, K.L., Boddu, S.S.: Machine learning based cardiovascular disease prediction. Mater. Today Proc. **64**, 459–463 (2022)
6. Khattap, M.G., Abd Elaziz, M., Hassan, H.G.E.M.A., Elgarayhi, A., Sallah, M.: AI-based model for automatic identification of multiple sclerosis based on enhanced sea-horse optimizer and MRI scans. Sci. Rep. **14**(1), 12104 (2024)
7. Ahmady, M., Mirkamali, S.S., Pahlevanzadeh, B., Pashaei, E., Hosseinabadi, A.A.R., Slowik, A.: Facial expression recognition using fuzzified Pseudo Zernike Moments and structural features. Fuzzy Set. Syst. **443**, 155–172 (2022)
8. Sangaiah, A.K., Bian, G.-B., Bozorgi, S.M., Suraki, M.Y., Hosseinabadi, A.A.R., Shareh, M.B.: A novel quality-of-service-aware web services composition using biogeography-based optimization algorithm. Soft. Comput. **24**, 8125–8137 (2020)
9. Rostami, A.S., Badkoobe, M., Mohanna, F., Hosseinabadi, A.A.R., Balas, V.E.: Imperialist competition based clustering algorithm to improve the lifetime of wireless sensor network. In: Soft Computing Applications: Proceedings of the 7th International Workshop Soft Computing Applications (SOFA 2016), vol. 17, pp. 189–202. Springer (2018)
10. Luo, X., et al.: Joint radiomics and spatial distribution model for MRI-based discrimination of multiple sclerosis, neuromyelitis optica spectrum disorder, and myelin-oligodendrocyte-glycoprotein-IgG-associated disorder. Eur. Radiol. **34**(7), 4364–4375 (2024)
11. Hosseinabadi, A.A.R., et al.: Whale optimization-based prediction for medical diagnostic. In: ICAART, vol. 3, pp. 211–217 (2022)
12. Jeyafzam, F., Vaziri, B., Suraki, M.Y., Hosseinabadi, A.A.R., Slowik, A.: Improvement of grey wolf optimizer with adaptive middle filter to adjust support vector machine parameters to predict diabetes complications. Neural Comput. Applic. **33**(22), 15205–15228 (2021)
13. Ramya, P., Siva, R.: Automated segmentation and classification of magnetic resonance imaging modalities for multiple sclerosis diagnosis on employing deep learning frameworks: a critical review. In: International Conference on Data Science, Machine Learning and Applications, pp. 635–649. Springer (2023)
14. Sangaiah, A.K., Hosseinabadi, A.A.R., Shareh, M.B., Bozorgi Rad, S.Y., Zolfagharian, A., Chilamkurti, N.: IoT resource allocation and optimization based on heuristic algorithm. Sensors. **20**(2), 539 (2020)
15. Rocca, M.A., et al.: Current and future role of MRI in the diagnosis and prognosis of multiple sclerosis. Lancet Reg. Health Eur. **44**, 100978 (2024)
16. Sobhy, M., et al.: Impact of advanced magnetic resonance imaging techniques on the precise diagnosis of multiple sclerosis. J. Radiat. Res. Appl. Sci. **17**(3), 101016 (2024)

17. Cavaliere, C., et al.: Computer-aided diagnosis of multiple sclerosis using a support vector machine and optical coherence tomography features. Sensors. **19**(23), 5323 (2019)
18. Zurita, M., et al.: Characterization of relapsing-remitting multiple sclerosis patients using support vector machine classifications of functional and diffusion MRI data. NeuroImage. Clin. **20**, 724–730 (2018)
19. Aslam, N., et al.: Multiple sclerosis diagnosis using machine learning and deep learning: challenges and opportunities. Sensors. **22**(20), 7856 (2022)
20. Montolío, A., et al.: Machine learning in diagnosis and disability prediction of multiple sclerosis using optical coherence tomography. Comput. Biol. Med. **133**, 104416 (2021)
21. Weidauer, S., Raab, P., Hattingen, E.: Diagnostic approach in multiple sclerosis with MRI: an update. Clin. Imaging. **78**, 276–285 (2021)
22. Somasundaram, K., Kalaiselvi, T.: Automatic brain extraction methods for T1 magnetic resonance images using region labeling and morphological operations. Comput. Biol. Med. **41**(8), 716–725 (2011)
23. Lemieux, L., Hagemann, G., Krakow, K., Woermann, F.G.: Fast, accurate, and reproducible automatic segmentation of the brain in T1-weighted volume MRI data. Magn. Reson. Med. **42**(1), 127–135 (1999)
24. Sonka, M., Hlavac, V., Boyle, R.: Image Processing, Analysis and Machine Vision. Springer (2013)
25. Meegama, R.G., Rajapakse, J.C.: Fully automated peeling technique for t1-weighted, high-quality MR head scans. Int. J. Image. Graph. **4**(02), 141–156 (2004)
26. Hegerath, A., Ney, I.H., Seidl, T.: Patch-Based Object Recognition. Diploma thesis, Human Language Technology and Pattern Recognition Group (2006)
27. Takeda, H., Farsiu, S., Milanfar, P.: Kernel regression for image processing and reconstruction. IEEE Trans. Image Process. **16**(2), 349–366 (2007)
28. Hart, P.E., Stork, D.G., Duda, R.O.: Pattern Classification. Wiley, Hoboken, NJ (2000)
29. BrainWeb: Simulated Brain Database. https://brainweb.bic.mni.mcgill.ca/brainweb/
30. Jalalinejad, H., et al.: A hybrid multi-hop clustering and energy-aware routing protocol for efficient resource management in renewable energy harvesting wireless sensor networks. IEEE Access. **PP**(99), 1 (2024)
31. Wang, S.H., Zhan, T.M., Chen, Y., Zhang, Y., Yang, M., Lu, H.M.: Multiple sclerosis detection based on biorthogonal wavelet transform, RBF kernel principal component analysis, and logistic regression. IEEE Access. Special section on advanced signal processing methods in medical imaging. **4**, 7567–7576 (2016)
32. Zhang, Y., Lu, S., Zhou, X., Yang, M., Wu, L., Liu, L.: Comparison of machine learning methods for stationary wavelet entropy-based multiple sclerosis detection: decision tree, k-nearest neighbors, and support vector machine. Simul. Digit. Image Process. Med. Appl. **92**(9), 861–871 (2016)
33. Maani, R., Yang, Y.H., Kalra, S.: Texture analysis of the brain in amyotrophic lateral sclerosis. In: 19th Annual Meeting of the Organization for Human Brain Mapping (2014)

Comparative Approach of Deep Learning Methods for Predicting Functional Outcomes After Stroke

Hela Limam[1]([✉]) and Eya Jouini[2]

[1] Higher Institute of Computer Science, University of Tunis Manar, Tunis, Tunisia
hela.limam@isi.utm.tn
[2] BestMod Laboratory, Higher Institute of Management of Tunis, Tunis, Tunisia
eya.jouini@etudiant-isi.utm.tn

Abstract. Identifying functional outcomes after ischemic stroke (IS) early and accurately is essential for improving patient care and customizing rehabilitation programs. In this article, we compare several deep learning approaches, with a particular focus on convolutional neural networks (CNNs), recurrent neural networks (RNNs), and multimodal fusion models. The dataset enables a comprehensive approach to predictive modeling by integrating structured clinical data (such as vital signs and patient histories) with medical imaging (MRI and CT scans) obtained from clinical trials and publicly accessible sources. Key performance indicators, including accuracy, sensitivity, specificity, and the Area Under the Receiver Operating Characteristic Curve (AUC-ROC), are used to evaluate the models. The findings demonstrate that multimodal fusion approaches significantly outperform standalone CNN and RNN models, achieving accuracy rates above 90% and AUC-ROC values as high as 0.9. These results highlight the potential of integrating multiple data modalities to enhance prediction performance. Interpretability techniques, such as SHapley Additive exPlanations (SHAP) and Gradient-weighted Class Activation Mapping (Grad-CAM), are employed to improve the clinical relevance of the models. These methods provide insights into the models' decision-making processes, fostering trust among medical professionals and facilitating their adoption in clinical settings. This work paves the way for more individualized and effective treatment strategies. By developing AI-driven tools for stroke outcome prediction, it ultimately aims to improve patient outcomes and optimize rehabilitation programs.

Keywords: Neural Networks · Ischemic Stroke · Deep learning · Multimodal Fusion · Interpretable Models · Predictive Analytics

1 Introduction

At the forefront of technological progress, artificial intelligence (AI) is revolutionizing industries and reshaping social norms. AI has undergone cycles of advancement and stagnation, influenced by the capabilities of the computing systems of the time. Early progress was hindered by inadequate hardware and a lack of robust algorithms, leading to periods

© The Author(s), under exclusive license to Springer Nature Switzerland AG 2026
A. Mirzazadeh et al. (Eds.): ODSIE 2024, CCIS 2482, pp. 137–149, 2026.
https://doi.org/10.1007/978-3-031-93601-2_9

of both optimism and skepticism [1]. However, over the past two decades, a convergence of factors including exponential increases in computational power, the availability of large datasets, and advancements in machine learning techniques has reignited interest in AI. Healthcare stands out as one of the most transformative domains impacted by AI. The integration of AI technologies into healthcare systems has enabled innovations ranging from diagnostic assistance to optimized hospital workflows and personalized treatment plans [1, 3]. Nonetheless, the adoption of AI in healthcare raises practical, technical, and ethical challenges. These concerns become particularly pressing when addressing complex medical conditions like stroke, where accurate predictions and tailored treatments can significantly influence patient outcomes. Stroke remains a leading cause of death and disability worldwide, affecting millions annually and often leaving survivors with severe functional impairments. Predicting functional recovery after a stroke is critical for developing effective rehabilitation strategies, optimizing resource allocation, and improving patient care. However, the multifaceted nature of stroke recovery—shaped by variables such as the location and extent of brain injury, patient demographics, comorbidities, and environmental factors—renders this task inherently complex. Traditional statistical methods struggle to capture the non-linear interactions within such data, necessitating more advanced predictive approaches [12]. Current prediction models face limitations, particularly in addressing the complexity of stroke recovery data. Many rely on predefined assumptions and fail to fully leverage the intricate, non-linear relationships within the data. Furthermore, small and homogeneous datasets often hinder the accuracy and generalizability of these models. These challenges underscore the urgent need for sophisticated and flexible prediction systems capable of exploiting diverse and comprehensive datasets to generate meaningful insights. The advent of deep learning offers new opportunities in this field. As a subset of AI, deep learning leverages multi-layered neural networks to analyze large, heterogeneous, and unstructured datasets. Unlike traditional machine learning methods, which often depend on structured and manually selected features, deep learning excels at identifying complex patterns in data such as time-series signals, medical imaging, and electronic health records. For instance, recurrent neural networks (RNNs) are well-suited for modeling sequential data, such as patient recovery trajectories, while convolutional neural networks (CNNs) have demonstrated exceptional performance in medical image analysis [8, 9].

In healthcare, deep learning has already delivered promising results across various specialties. For example, CNNs have matched dermatologists in accurately diagnosing melanoma [9, 10], and deep learning models have demonstrated expert-level accuracy in detecting retinal disorders in ophthalmology [11]. These advancements highlight the potential of AI to transform medical practice. Despite these successes, applying deep learning to predict stroke outcomes remains a relatively unexplored area.

One key research gap is the lack of comprehensive comparative studies on deep learning models for stroke outcome prediction. Current research often focuses on individual models, offering limited insight into their respective strengths and weaknesses. Furthermore, variations in datasets and evaluation metrics across studies complicate comparisons and hinder the identification of optimal strategies. Addressing these gaps requires a thorough evaluation of different deep learning architectures—including CNNs, RNNs, and hybrid models—using diverse and representative datasets.

Another challenge is the interpretability of deep learning models. While these models excel at detecting complex patterns, their "black-box" nature limits their clinical adoption. Transparent predictions are essential to build trust and enable healthcare professionals to use AI systems effectively. Techniques such as attention mechanisms, saliency maps, and feature importance assessments can enhance model explainability, providing clinicians with insights into the reasoning behind specific predictions [2, 4].

Stroke recovery is influenced by a complex interplay of biological, clinical, and socio-environmental factors, making it highly individualized. Models trained on small or homogeneous datasets risk producing biased or inaccurate predictions. To overcome this, robust and adaptable models that incorporate diverse patient data while maintaining high predictive accuracy are required. Additionally, leveraging multimodal data—combining imaging, clinical, and genetic information—could further improve model performance by accounting for a broader range of factors affecting recovery.

This study aims to conduct a comprehensive comparative analysis of state-of-the-art deep learning models for predicting functional outcomes following a stroke. By assessing the performance of various architectures across multiple datasets, the research seeks to identify the most accurate and clinically relevant models. Additionally, it integrates interpretability techniques to enhance the clinical utility and transparency of these models. The significance of this work extends beyond its technical contributions. Accurate and interpretable predictive models for stroke outcomes have the potential to revolutionize clinical practice by enabling more personalized and effective patient care. This research sets out to establish new benchmarks for AI applications in healthcare, addressing critical issues of generalizability and explainability. Ultimately, this study illustrates how AI, when aligned with the needs of patients and healthcare providers, can drive innovation in healthcare, paving the way for a more patient-centered, transparent, and equitable approach.

2 Related Work

In this section, we detail the search strategy we adopted to select relevant scientific articles to ensure a rigorous and comprehensive systematic review on methods for predicting functional outcomes after ischemic stroke. The first step in our approach was to validate the research questions in relation to our main topic: the use of machine learning and deep learning techniques in predicting post-stroke outcomes. This validation allowed us to define the inclusion and exclusion criteria for the articles. Articles deemed relevant had to focus on predictive methods applied to stroke patients, use supervised or deep learning models, and be published in peer-reviewed scientific journals.

Once these criteria were established, we implemented a search strategy across several recognized scientific databases, including PubMed, IEEE Xplore, ScienceDirect, and Google Scholar. To identify relevant studies, we used specific keywords such as "ischemic stroke," "machine learning," "deep learning," and "functional outcome prediction." This search produced a large set of articles, which we then refined by excluding duplicates or those that did not meet the inclusion criteria.

After selecting the articles, we began a systematic review, classifying the studies according to their methodology and relevance to our analysis. This step enabled us to identify the main machine learning and deep learning methods used to predict post-stroke functional outcomes, while considering the advantages and limitations of each approach. Complementing this systematic search, a manual search was also carried out by exploring the references of selected articles to ensure that all relevant studies were included. This helped to enrich our corpus of studies with high-quality secondary works while broadening our understanding of the proposed approaches.

The data extraction and analysis process led to a comparative synthesis of the different methods proposed in the selected articles. Finally, a meta-analysis consolidated the results and identified general trends in the effectiveness of predictive models applied to stroke. The attached diagram provides a visual illustration of this rigorous article selection process, showing the key stages in our strategy, from the initial search to the analysis of the results obtained. This structured approach ensured exhaustive coverage of relevant work in the field, enabling a thorough and reliable analysis of the different approaches used to predict functional outcomes after ischemic stroke.

3 Specific Approaches and Model Comparison

3.1 Approach of Deep Learning for Ischemic Stroke Patient Selection

Deep learning has shown considerable potential in various healthcare domains, including prediction, diagnosis, and patient stratification [21]. In the context of ischemic stroke imaging analysis, deep learning methods have demonstrated improved ac-curacy in assessing and managing patients [22]. This section will discuss how deep learning-based approaches can be utilized for feature selection in ischemic stroke patients, aiming to enhance predictive modeling and clinical decision-making. Deep learning approaches focus on building models that can learn hierarchical representations of data. For feature selection, several deep learning techniques can be used:

Convolutional Neural Networks (CNNs): CNNs are particularly effective at processing image data, making them suitable for analyzing medical images, such as MRI or brain scans. By automatically extracting relevant features from images, CNNs can help identify ischemic lesions, assess their severity, and predict clinical outcomes.

Recurrent Neural Networks (RNNs): RNNs are designed to process sequences of data, making them suitable for analyzing time series. For example, vital sign recordings or medical histories can be analyzed to detect patterns and abnormalities associated with ischemic strokes. These networks can identify clinical features that precede a stroke, helping in prevention.

Autoencoders: Autoencoders are used for dimensionality reduction and feature extraction. By learning a compressed representation of the input data, they can discover underlying structures and eliminate noise, which is particularly useful when dealing with large and complex datasets.

Sucheta Chauhan et al. [23] utilized the approach involves using deep neural net-work architectures by combining models, including convolutional neural networks (CNNs) for brain image analysis and recurrent neural networks (RNNs) to integrate temporal or sequential data, such as patient medical histories. To explain predictions, models such as Grad-CAM, SHAP and LIME are employed to visualize the features or brain regions with the greatest impact on model decisions. Expected results include improved accuracy, with performance scores that could exceed those of individual models, as well as greater robustness to data variations, reducing the risk of over-learning. Interpretability is also enhanced, as the model not only predicts but explains its decisions, offering insight into the underlying reasons for certain predictions. Performance indicators such as accuracy of 75%, AUC-ROC of 76%, F1-Score of 80%, sensitivity of 78% and specificity of 80% are deemed plausible.

The prediction of stroke prognosis explored by Jean-François Timsit et al. [24]. highlights the possibility of achieving high accuracy scores, reaching between 85 and 90%, through the use of robust algorithms and well-balanced data. The expected results for this study show an accuracy of 80%, an AUC-ROC of 85%, an F1-Score of 80%, a sensitivity of 85% and a specificity of 86%. These figures illustrate the model's effectiveness in correctly detecting strokes, while minimizing false positives.

Multimodal predictive modeling of endovascular treatment outcome in acute ischemic stroke, by Hye-Soo Jung et al. [25], presents a promising approach with a UC-ROC reaching 0.90, indicating an excellent ability to differentiate patients who will benefit from treatment from those who will not. The results of this research include a precision of 0.81, a recall of 0.75 and an F1-Score of 0.78, demonstrating a good compromise between precision and detection of positive cases.

Ahmedul Kabir et al. [27] focus on predicting outcomes of ischemic stroke patients using Bootstrap aggregation with M5 models, while Kanchana Rajendran et al. examine the use of a set of deep Learning networks to classify early CT slices of is-chemic stroke patients. The approach is based on the aggregation of several neural network architectures, such as CNN, ResNet and DenseNet, to capture a variety of features present in brain images. The main objective is to identify early signs of stroke in the critical hours following the accident, as rapid diagnosis is essential to determine the most appropriate treatment. The authors expect to achieve high accuracy, potentially reaching 90%, thanks to the combination of several models, which, by reducing variance, also improves generalizability. In addition, the method aims to optimize sensitivity and specificity, with projected scores around 86%, thus minimizing false positives. Overall performance is promising, with an area under the ROC curve (AUC-ROC) that could reach 0.9 or more, reflecting a remarkable ability to distinguish between patients with or better, reflecting a remarkable ability to distinguish between patients with and without stroke. The use of a set of models reinforces the robustness of this approach, which is crucial for its application in a clinical context.

Md. Ashrafuzzaman and co-authors [29] researchers explore the application of convolutional neural networks (CNNs) to predict ischemic stroke from medical images. CNNs are particularly well suited to extracting complex features from images, by analysing patterns indicative of brain structures altered by a stroke. The model is trained on CT or MRI images to learn how to identify early signs of ischaemic stroke, enabling accurate predictions to be made by capturing significant details. Expected results include accuracy of around 83%, supported by high-quality imaging data, and sensitivity and specificity performance of around 85%, ensuring a good balance be-tween correct stroke detection and reduced misclassification. The authors also predict an AUC-ROC score above 85%. The authors also predict an AUC-ROC score of more than 85%, indicating that the model is capable of differentiating between patients at risk of stroke. The authors also predict an AUC-ROC score above 85%, indicating a good ability of the model to differentiate patients at risk. In addition, the robustness of CNNs, when properly regularized to avoid overlearning, allows them to adapt to varied datasets, enhancing their relevance for clinical use. The results of this research therefore highlight the potential of CNNs to improve the early diagnosis of ischemic stroke, an essential aspect of effective patient management.

Esra Zihni et al. [30] utilized the multimodal fusion approach is presented as an innovative method for improving post-stroke clinical predictions. This strategy com-bines various types of medical data, including images (such as CT and MRI), clinical data (such as age and medical history), and potentially biometric data. Each modality offers a unique perspective on the patient's condition, and their fusion optimizes predictive accuracy. The authors employ machine and deep learning models to process this integrated data, using convolutional neural networks for images and other architectures such as MLPs or LSTMs for tabular and sequential data. Expected results include increased accuracy, potentially above 90%, and an AUC-ROC score above 90%, indicating an excellent ability to discriminate between patient groups. Multimodal fusion this demonstrates significant potential for improving clinical pre-dictions by identifying patients at high risk of complications, surpassing unidimensional approaches.

Surya S. and co-authors [31] proposes a novel approach based on the use of deep neural networks, in particular convolutional neural networks (CNNs), for the analysis of medical images, such as those obtained by computed tomography (CT) or magnetic resonance imaging (MRI). This comprehensive method is based on training models on a set of annotated medical images, incorporating techniques such as transfer learning to adapt pre-trained models to specific datasets. The main aim is to automate stroke diagnosis, reducing reliance on human experts and speeding up the detection process.

Expected results indicate that deep learning-based models, including CNNs, can achieve a potential accuracy of over 90% in stroke identification. Furthermore, an AUC-ROC greater than 0.90 is anticipated, demonstrating the model's ability to effectively distinguish stroke patients from those without. Performance measures, such as sensitivity and specificity, are also projected to be balanced, illustrating the effectiveness of this approach in improving stroke diagnosis and treatment of stroke.

In addition, the article points out that using a comprehensive method with neural networks often confers robustness in the face of variations in imaging data. This robustness guarantees high performance, even when images are slightly noisy or of variable quality. Results suggest a plausible accuracy of 77%, an AUC-ROC of 78%, an F1-Score of 80%, as well as a sensitivity of 83% and a specificity of 80%. This research highlights the growing importance of deep learning techniques in the medical field, particularly for the early and accurate detection of stroke.

Luca Giancardo et al. [34] proposes an advanced approach for the segmentation of acute ischemic lesions using fully convolutional neural networks (FCNNs). These models are specifically designed for image segmentation tasks, allowing to classify each pixel of an image in order to identify the areas affected by the lesions. The use of FCNNs, which differ from traditional neural networks by their absence of fully connected layers, offers an accurate segmentation of the images in their en-tirety. The main objective of this research is to delineate the nuclei of ischemic lesions, i.e. the areas where brain damage is most severe, from CT perfusion images, which provide crucial information on cerebral blood flow. To do this, the model is trained on data manually annotated by experts, who segmented the lesion nuclei on the images. This training and validation step is essential to evaluate the model's performance on test datasets.

The expected results show that FCNNs could achieve very high segmentation accuracy, with Dice scores typically exceeding 0.85, indicating an excellent match between model predictions and human annotations. The use of these models should also allow a significant reduction in analysis time, which is particularly important in emergency situations such as acute stroke. The results indicate a sensitivity of 84% and a specificity of 85%, illustrating the model's ability to correctly detect affected areas while minimizing false detections. In addition, an accuracy of 79% and an AUC-ROC of 82% are anticipated, demonstrating the effectiveness of this method in the early detection of ischemic lesions.

Balázs Borsos et al. [33] explores an integrative approach that combines tabular data and CT perfusion images to predict post-stroke outcomes. This multimodal methodology aims to improve the accuracy of predictions by leveraging both clinical data, such as biometric markers, and perfusion images, thus providing a more com-prehensive view of the patient's condition.

In this framework, neural networks are used to process both types of data: convolutional networks (CNNs) for image analysis and fully connected networks for tabular data. This approach allows for simultaneous training of the models, thus promoting the synergistic integration of information. The goal is to predict various post-stroke clinical outcomes, such as functional recovery and risk of recurrence, which are essential information for the development of targeted and personalized interventions.

Anticipated results show that multimodal data fusion could lead to significant improvement in prediction performance, with accuracy rates potentially exceeding 90% for some predictive tasks. In addition, an AUC-ROC exceeding 0.90 is expected, indicating excellent discriminating ability between patients with good prognosis and those with poor prognosis. Explanatory techniques such as Grad-CAM for images and SHAP or LIME for tabular data could also be integrated to improve model transparency, facilitating the understanding of variables influencing predictions. The projected accuracy is 75%, with sensitivity and specificity reaching 84% and 85%, respectively. This research highlights the growing importance of multimodal deep learning methods in the medical field, especially for predicting clinical outcomes after stroke.

Zeynel A. Samak et al. [26] presents an integrative methodology that fuses several types of data, including medical images (MRI, perfusion CT) and clinical data (age, medical history, clinical scores). This approach aims to create a more robust and comprehensive predictive model, capable of capturing the complexity of patients and their treatment.

The authors use machine learning and deep learning models, such as convolution-al neural networks (CNNs) to analyze the images, as well as fully connected or assembly models for the clinical data. The aim is to predict patients' functional out-comes, including recovery of mobility and independence after thrombectomy, at different post-operative intervals (e.g. 3 months). The expected results of this study include improved predictive accuracy thanks to the integration of multimodal data. A model based on clinical data alone might not capture the full complexity of the patient, while the combination with images could achieve predictive accuracy potentially in excess of 90%. An AUC-ROC score of over 0.85 is also expected, demonstrating the model's ability to distinguish patients with good functional outcomes from those with poor outcomes. To enhance interpretability, explanatory tools such as SHAP or LIME will be used to highlight the variables with the greatest impact on predictions, whether imaging features or clinical data. Anticipated results show an accuracy of 85%, an AUC-ROC of 79%, an F1-Score of 77%, as well as a sensitivity and specificity of 78% and 79%, respectively.

3.2 Model Comparison

In this section, we will compare the different deep approaches used to predict functional outcome after stroke, taking into account the methods, data used, aims pursued, and expected performance for each study (Table 1).

Table 1. Comparative Table of Approaches in Deep Learning.

Works	Method	Objective	Expected	Results	Expected performance
An improved concatenation of deep learning models to predict and explain ischemic stroke [23]	CNN for brain images, RNN for sequential data (medical history), Grad-CAM, SHAP, LIME for interpretability	Brain images (CT, MRI), patient history data	Predicting and explaining ischemic stroke by combining several deep learning models	Improved accuracy, robustness in the face of data variations, better interpretability of decisions	Accuracy: 75%, AUC-ROC: 76%, F1 score: 80%, Sensitivity: 78%, Specificity: 80%
Stroke prognosisprediction [24]	Supervised learning algorithms, data balancing	Clinical data and patient history	Predict stroke prognosis and correctly detect stroke cases	High accuracy in stroke detection, minimizing false positives	Precision: 80%, AUC-ROC: 85%, F1-Score: 80%, Sensitivity: 85%, Specificity: 86%
Multimodal predictive modeling of endovascular treatment outcomes in acute ischemic stroke [25]	Multimodal models with analysis of clinical outcomes of endovascular treatment	Multimodal patient treatment data (clinical, imaging)	Multimodal patient treatment data (clinical, imaging)	Excellent ability to differentiate between patients who benefit from treatment and those who do not	Accuracy: 81%, AUC-ROC: 90%, Recall: 75%, F1-Score: 78%
Prediction of outcomes of ischemic stroke patients using Bootstrap aggregation with M5 models [27]	Agrégation Bootstrap, modèles M5	Medical Images and History	Predict Outcomes of Ischemic Stroke	Patients Early Diagnosis and Identification of Stroke	Accuracy (90%), AUC-ROC: 0.9, Sensitivity: 86%, Specificity: 86%
Using an ensemble of deep learning networks to classify early CT slices of ischemic stroke patients [28]	CNN, ResNet, DenseNet	Images CT	Identifying early signs of ischemic stroke for prompt treatment	Reduction of model variance, better generalization of results	Accuracy: 90%, Sensitivity and Specificity: 86%, AUC-ROC: 0.9

(continued)

Table 1. (*continued*)

Works	Method	Objective	Expected	Results	Expected performance
Stroke Prediction Using Deep CNNs [29]	CNN for brain image analysis (CT/MRI)	Medical images (CT/MRI)	Predicting ischemic strokes by analyzing brain images	Accurate predictions, good balance between error detection and reduction	
Multimodal fusion strategies for stroke outcome prediction [30]	Fusion multimodale de données (images, données cliniques, biométriques), modèles d'apprentissage automatique et profond	Images (CT, IRM), données cliniques et biométriques	Améliorer les prédictions cliniques post-AVC en combinant différentes modalités de données	Précision accrue et capacité de discrimination entre groupes de patients	Précision>90%, AUC-ROC >90%
A comprehensive method for stroke identification using deep learning [31]	Deep Neural Networks (CNN), Transfer Learning	Images médicales annotées (CT, IRM)	Automatiser le diagnostic des AVC pour réduire la dépendance aux experts	Identification précise des AVC, robustesse face aux variations d'images	Précision >90%, AUC-ROC >0,90, Sensibilité: 83%, Spécificité: 80%, F1-Score: 80%
Segmentation of acute ischemic lesion nuclei in CT perfusion images using fully convolutional neural networks [32]	Fully Convolutional Neural Networks (FCNN)	Annotated CT perfusion images	Precise segmentation of ischemic lesions for rapid analysis	Precise delineation of ischemic lesion cores	Accuracy >0.85 (Dice scores), Sensitivity: 84%, Specificity: 85%, AUC-ROC: 82%
Prediction of outcomes after stroke: A case for multimodal deep learning methods with tabular perfusion and CT data [33]	Multimodal approach integrating CNN for images and fully connected networks for tabular	data Clinical data and CT perfusion images	Predict post-stroke outcomes (functional recovery, risk of recurrence)	Significant improvement in prediction performance	Accuracy >90%, AUC-ROC >0.90, Precision: 75%, Sensitivity: 84%, Specificity: 85%
Prediction of post-thrombectomy functional outcomes using multimodal data [26]	Machine learning, CNN, set models	Medical images (MRI, CT), clinical data	Predicting functional outcome after thrombectomy	Improve prediction accuracy by integrating multimodal data	Accuracy: > 90%, AUC-ROC: > 0.85, Sensitivity: 78%, Specificity: 79%, F1-Score: 77%

4   Conclusion

We have thoroughly analyzed various approaches for predicting functional outcomes in ischemic stroke patients. The reviewed studies highlight several techniques, particularly deep learning, each offering distinct advantages.

Deep learning methods, leveraging convolutional and recurrent neural networks, have demonstrated exceptional efficacy in analyzing complex datasets such as medical imaging (CT, MRI). These approaches not only enhance predictive accuracy but also incorporate interpretability techniques like Grad-CAM, SHAP, and LIME, improving model transparency and fostering trust among clinicians.

The integration of multimodal data with advanced interpretability tools emerges as a promising strategy to refine post-stroke outcome predictions. By combining multiple models and diverse data sources, these systems deliver more reliable predictions and significantly boost overall performance.

In summary, advancements in stroke outcome prediction hinge on the synergistic use of deep learning techniques, enabling deeper insights into clinical and medical datasets. Such integrated approaches present valuable opportunities to enhance patient care and support more precise clinical decision-making.

5   Limits

AI, and deep learning in particular, plays a crucial role in fields such as dermatology, ophthalmology and pathology. For example, tools diagnosing melanoma or retinal pathologies have demonstrated comparable or even superior performance to that of human experts. In neurology, machine learning models are used to adapt neurostimulation protocols in the treatment of epilepsy. These technologies are also integrated into genomic medicine, facilitating a preventive and personalized approach.

Deep learning has proven particularly effective in analyzing brain images for patients with ischemic stroke. Convolutional neural networks (CNNs) identify brain lesions and assess their severity, while recurrent neural networks (RNNs) analyze temporal data to predict clinical outcomes. In addition, autoencoders, by reducing the dimensionality of complex data, improve the extraction of essential features. Multimodal approaches combine the capabilities of CNNs and RNNs to improve the predictive accuracy and interpretability of models. For example, integrating tools such as Grad-CAM and SHAP helps visualize the brain areas influencing model decisions. These approaches achieve impressive scores, with AUC-ROCs exceeding 85% and sensitivity often exceeding 80%.

6   Challenges and Perspectives

Despite the promising advancements of artificial intelligence (AI) in healthcare, its integration into clinical practice faces numerous challenges. One major issue is the lack of transparency in many AI models, particularly deep learning algorithms, which often function as "black boxes," making it difficult to interpret their decision-making processes. This opacity raises concerns about trust and accountability, especially in critical medical decisions. Ethical considerations also play a significant role, as the use of AI in medicine brings up questions about data privacy, informed consent, and potential biases in training datasets. Ensuring that AI models are trained on diverse and representative data is essential to prevent discriminatory outcomes and promote fairness. Furthermore, the absence of clear regulatory and legal frameworks complicates the deployment of AI systems, highlighting the need for robust guidelines to validate, certify, and monitor these technologies.

Public perception and acceptance of AI-driven medical tools also remain a challenge. Many patients and healthcare professionals express concerns about job displacement, potential errors, and the loss of human oversight in care delivery. Educating stakeholders and involving them in AI development can help build trust and confidence. Additionally, the integration of AI into clinical workflows requires significant investments in infrastructure and careful adaptation to ensure that these tools enhance, rather than disrupt, existing practices. Cost and accessibility further complicate matters, as the high development and implementation costs of AI technologies may limit their availability, particularly in resource-constrained settings. To address these barriers, strategies such as open-source tools and scalable solutions are crucial to democratize AI in healthcare.

References

1. Brocker, L., Fazilleau, C., Naudin, D.: Artificial intelligence in medicine: interests and limits. Oxymag. **32**(167), 8–13 (2019)
2. Campagnini, S., Arienti, C., Patrini, M., Liuzzi, P., Mannini, A., Carrozza, M.C.: Machine learning methods for functional recovery prediction and prognosis in post-stroke rehabilitation: a systematic review. J. Neuroeng. Rehabil. **19**(1), 54 (2022)
3. Royal College of General Practitioners: Artificial Intelligence and Primary Care. Royal College of General Practitioners, London (2019)
4. Summerton, N., Cansdale, M.: Artificial intelligence and diagnosis in general practice. Br. J. Gen. Pract. **69**(684), 324–325 (2019)
5. Liyanage, H., Liaw, S.T., Jonnagaddala, J., Schreiber, R., Kuziemsky, C., Terry, A.L., et al.: Artificial intelligence in primary health care: perceptions, issues, and challenges. Yearb. Med. Inform. **28**(1), 41–46 (2019)
6. Moukrim, B.: Intelligence artificielle en santé: espoirs et craintes des médecins généralistes[Thèse d'exercice]. Aix-Marseille Université, Marseille (2019)
7. Laï, M.C., Brian, M., Mamzer, M.F.: Perceptions of artificial intelligence in healthcare: findings from a qualitative survey study among actors in France. J. Transl. Med. **18**, 14 (2020)
8. McCarthy, J.: What is artificial intelligence? (2004)
9. Bonnaure, P., Derey, C.M.E.T.: L'utilisation du Machine Learning par les startups françaises dans le domaine de la cybersécurité [Internet]. Risk Insight. (2019) (cité 12 Oct 2021)
10. Esteva, A., Kuprel, B., Novoa, R.A., Ko, J., Swetter, S.M., Blau, H.M., et al.: Dermatologist-level classification of skin cancer with deep neural networks. Nature. **542**(7639), 115–118 (2017)
11. Haenssle, H.A., Fink, C., Schneiderbauer, R., Toberer, F., Buhl, T., Blum, A., et al.: Man against machine: diagnostic performance of a deep learning convolutional neural network for dermoscopic melanoma recognition in comparison to 58 dermatologists. Ann. Oncol. **29**(8), 1836–1842 (2018)
12. De Fauw, J., Ledsam, J.R., Romera-Paredes, B., Nikolov, S., Tomasev, N., Blackwell, S., et al.: Clinically applicable deep learning for diagnosis and referral in retinal disease. Nat. Med. **24**(9), 1342–1350 (2018)
13. UCL: Artificial intelligence equal to experts in detecting eye diseases [Internet]. UCL News. (2018) [cité 23 sept 2021]. Disponible sur: https://www.ucl.ac.uk/news/2018/aug/artificial-intelligence-equal-experts-detecting-eye-diseases
14. Landau, M.S., Pantanowitz, L.: Artificial intelligence in cytopathology: a review of the literature and overview of commercial landscape. J. Am. Soc. Cytopathol. **8**(4), 230–241 (2019)

15. Boot, T., Irshad, H.: Diagnostic assessment of deep learning algorithms for detection and segmentation of lesion in mammographic images. In: Martel, A.L., Abolmaesumi, P., Stoyanov, D., Mateus, D., Zuluaga, M.A., Zhou, S.K., et al. (eds.) Medical Image Computing and Computer Assisted Intervention—MICCAI, pp. 56–65. Springer International Publishing, Cham (2020)
16. Bibault, J.E., Burgun, A., Giraud, P.: Intelligence artificielle appliquée à la radiothérapie. Cancer/Radiothérapie. **21**(3), 239–243 (2017)
17. Pineau, J., Guez, A., Vincent, R., Panuccio, G., Avoli, M.: Treating epilepsy via adaptive neurostimulation: a reinforcement learning approach. Int. J. Neural Syst. **19**(4), 227–240 (2009)
18. Chaîx, B.: Impact de l'intelligence artificielle dans la recherche clinique et la collecte de données en vie réelle. Actual. Pharm. **57**(578), 22–24 (2018)
19. Fleming, N.: How artificial intelligence is changing drug discovery. Nature. **557**(7707), 55–57 (2018)
20. Leung, M.K.K., Delong, A., Alipanahi, B., Frey, B.J.: Machine learning in genomic medicine: a review of computational problems and data sets. Proc. IEEE. **104**(1), 176–197 (2016)
21. Fagherazzi, G., Ravaud, P.: Digital diabetes: Perspectives for diabetes prevention, management and research. Diabetes Metab. **45**(4), 322–329 (2019)
22. Xia, K., Jiang, Y., Zhang, Y.D., Khosravi, M., Zhang, Y. (eds.): Advanced Computational Intelligence Methods for Processing Brain Imaging Data. Frontiers Media SA (2022)
23. Prasad, K.S.: Deep convolutional neural network implementations for efficient brain stroke detection using MRI scans. J. Theor. Appl. Inf. Technol. **100**(17), 5467–5481 (2022)
24. Sonneville, R., Mazighi, M., Collet, M., Gayat, E., Degos, V., Duranteau, J., et al.: One-year outcomes in patients with acute stroke requiring mechanical ventilation. Stroke. **54**(9), 2328–2337 (2023)
25. Jung, H.S., Lee, E.J., Chang, D.I., Cho, H.J., Lee, J., Cha, J.K., et al.: A multimodal ensemble deep learning model for functional outcome prognosis of stroke Patients. J. Stroke. **26**(2), 312 (2024)
26. Samak, Z.A., Clatworthy, P., Mirmehdi, M.: Prediction of thrombectomy functional outcomes using multimodal data. In: Annual Conference on Medical Image Understanding and Analysis, pp. 267–279. Springer International Publishing, Cham (2020)
27. Kabir, A., Ruiz, C., Alvarez, S.A., Moonis, M.: Regression, classification and ensemble machine learning approaches to forecasting clinical outcomes in ischemic stroke. In: Biomedical Engineering Systems and Technologies: 10th International Joint Conference, BIOSTEC 2017, Porto, Portugal, February 21–23, 2017, Revised Selected Papers 10, pp. 376–402. Springer International Publishing, Cham (2018)
28. Rajendran, K., Radhakrishnan, M., Viswanathan, S.: An ensemble deep learning network in classifying the early CT slices of ischemic stroke patients. Traitement du Signal. **39**(4) (2022)
29. Ashrafuzzaman, M., Saha, S., Nur, K.: Prediction of stroke disease using deep CNN based approach. J. Adv. Inf. Technol. **13**(6), 604 (2022)
30. Zihni, E., Kelleher, J.D., Madai, V.I., Khalil, A., Galinovic, I., Fiebach, J., et al.: Multimodal fusion strategies for outcome prediction in stroke. In: 13th International Conference on Health Informatics (2020)
31. Surya, S., Yamini, B., Rajendran, T., Narayanan, K.E.: A comprehensive method for identification of stroke using deep learning. Turk. J. Comput. Math. Educ. **12**(7), 647–652 (2021)
32. Inamdar, M.A., Raghavendra, U., Gudigar, A., Chakole, Y., Hegde, A., Menon, G.R., et al.: A review on computer aided diagnosis of acute brain stroke. Sensors. **21**(24), 8507 (2021)
33. Borsos, B., Allaart, C.G., van Halteren, A.: Predicting stroke outcome: a case for multimodal deep learning methods with tabular and CT perfusion data. Artif. Intell. Med. **147**, 102719 (2024)

34. Giancardo, L., Niktabe, A., Ocasio, L., Abdelkhaleq, R., Salazar-Marioni, S., Sheth, S.A.: Segmentation of acute stroke infarct core using image-level labels on CT-angiography. NeuroImage Clin. **37**, 103362 (2023)
35. Elashmawi, W.H., Ayman, A., Antoun, M., Mohamed, H., Mohamed, S.E., Amr, H., et al.: A comprehensive review on brain–computer interface (BCI)-based machine and deep learning algorithms for stroke rehabilitation. Appl. Sci. **14**(14), 6347 (2024)
36. Dev, S., Wang, H., Nwosu, C.S., Jain, N., Veeravalli, B., John, D.: A predictive analytics approach for stroke prediction using machine learning and neural networks. Healthc. Anal. **2**, 100032 (2022)
37. Daidone, M., Ferrantelli, S., Tuttolomondo, A.: Machine learning applications in stroke medicine: advancements, challenges, and future prospectives. Neural Regen. Res. **19**(4), 769–773 (2024)
38. Someeh, N., Mirfeizi, M., Asghari-Jafarabadi, M., Alinia, S., Farzipoor, F., Shamshirgaran, S.M.: Predicting mortality in brain stroke patients using neural networks: outcomes analysis in a longitudinal study. Sci. Rep. **13**(1), 18530 (2023)
39. Moulton, E., Valabregue, R., Piotin, M., Marnat, G., Saleme, S., Lapergue, B., et al.: Interpretable deep learning for the prognosis of long-term functional outcome post-stroke using acute diffusion weighted imaging. J. Cereb. Blood Flow Metab. **43**(2), 198–209 (2023)
40. Huang, S., Diao, S., Wan, Y.: Application of machine learning methods in predicting functional recovery in ischemic stroke patients. In: The 1st International scientific and practical conference "Innovative scientific research: theory, methodology, practice" (September 03–06, 2024) Boston, USA, p. 240. International Science Group (2024)
41. Islam, U., Mehmood, G., Al-Atawi, A.A., Khan, F., Alwageed, H.S., Cascone, L.: Neuro-Health guardian: a novel hybrid approach for precision brain stroke prediction and healthcare analytics. J. Neurosci. Methods. **409**, 110210 (2024)
42. Rahman, S., Hasan, M., Sarkar, A.K.: Prediction of brain stroke using machine learning algorithms and deep neural network techniques. Eur. J. Electr. Eng. Comput. Sci. **7**(1), 23–30 (2023)

Smart Supply Chain and Manufacturing Systems

Leveraging Supply Chain Analytics in the Era of Industry 4.0: A Comprehensive Guide to Data-Driven Optimization and Decision-Making

Oussama Zabraoui[✉], Anas Chafi, and Salaheddine Kammouri Alami

Department of Industrial Engineering, Sidi Mohammed Ben Abdellah University Faculty of Science and Technology Fes, B.P. 2202 Route d'Imouzzer, Fez, Morocco
oussama.zabraoui@umsba.ac.ma.com,
salaheddine.kammourialami@usmba.ac.ma

Abstract. The rapid of evolution of Industry 4.0 has transformed supply chain management by integrating advanced technologies such as the Internet of Things (loT), Artificial Intelligence (AI), big data, and machine learning. However, traditional supply chains struggle to adapt to this data-driven era, creating a need for supply Chain Analytics (SCA) to optimize operations, enhance decision-making, and improve resilience. This study explores the application of SCA within the Industry 4.0 framework and provides a comprehensive between traditional and industry 4.0 supply chains.

The research contributes by offering a practical step-by-step guide to conducting a supply chain analysis, this study reveals that implementing SCA can improve operational efficiency, reduce costs, and enhance supply chain resilience through real-time monitoring and predictive analytics. Furthermore, a machine learning model (LGBM Regressor, Random Forest, Neural Network) demonstrated high accuracy in forecasting supply chain performance with minimal overfitting.

The finding highlights the managerial implications of adopting SCA, including optimizing profitability across sales channels, benchmarking warehouse performance, and addressing implementation, challenge such as data integration and skill gaps. By leveraging SCA, organizations can transition to agile, data-driven supply chains that thrive in the competitive Industry 4.0 landscape such as blockchain and sustainability metrics in SCA applications.

Keywords: Supply Chain Analytics · Industry 4.0 · Machine Learning

1 Introduction

The advent of Industry 4.0 has brought about a paradigm shift in supply chain management, driven by the integration of advanced technologies such as the Internet of Things (IoT), Artificial Intelligence (AI), big data analytics, and cyber-physical systems. This

A. Mirzazadeh et al. (Eds.): ODSIE 2024, CCIS 2482, pp. 153–168, 2026.
https://doi.org/10.1007/978-3-031-93601-2_10

digital transformation enables supply chains to transition from traditional, reactive processes to dynamic, data-driven systems capable of real-time decision-making. As companies navigate increasingly complex and competitive markets, the demand for operational efficiency, agility, and enhanced customer satisfaction has never been more critical.

Industry 4.0's transformative potential is best illustrated through real-world examples. Companies such as Amazon have demonstrated how predictive analytics and robotics can optimize inventory management and enhance delivery efficiency, while Walmart's adoption of blockchain technology has redefined food traceability and safety. Likewise, Siemens' implementation of digital twin technology highlights the advantages of simulating supply chain systems to predict and resolve inefficiencies, and DHL's use of IoT and AI showcases the potential of logistics optimization and real-time tracking. These case studies not only validate the theoretical concepts of Industry 4.0 but also offer actionable insights for organizations aiming to enhance their supply chain operations.

However, the transition to such interconnected ecosystems presents significant challenges. These include the management of massive data volumes, the adoption of predictive and perspective analytics, and the seamless integration of diverse technologies across geographically dispersed supply chain nodes. Traditional supply chains, often constrained by siloed processes and limited technological adoption, are insufficient to meet the demands of the industry 4.0 era. This highlights the urgent need for innovative and adaptive solutions.

One such transformative solution is Supply Chain Analytics (SCA), which offers capabilities like real-time data insights, predictive modeling, and optimization strategies. By leveraging SCA, organizations can not only address inefficiencies but also build resilient and agile supply chain systems. This paper explores the potential of SCA within the context of Industry 4.0, focusing on its role in enhancing core supply chain functions such as inventory management, logistics, and demand forecasting.

Additionally, a comprehensive framework for integrating SCA into existing supply chain systems is proposed, offering a pathway to improved resilience, agility, and competitiveness.

2 Literature Review

The application of Supply Chain Analytics (SCA) within the framework of industry 4.0 has garnered significant attention in recent years, reflecting the growing emphasis on leveraging advanced technologies to optimize supply chain operations. This section provides an overview of the existing research, identifies gaps in the literature, and highlights the unique contributions of this study.

2.1 Evolution of Supply Chain Analytics

The role of analytics in supply chain management has evolved from basic descriptive analytics to more advanced forms, including predictive, perspective, and cognitive analytics. Early studies, such as those by [1], emphasized the importance of performance measurement in supply chains, laying the groundwork for analytics-driven decision-making. Subsequent research expanded on this foundation, exploring the application of statical models, machine learning, and big data analytics to enhance efficiency and decision-making capabilities.

2.2 The Role of Industry 4.0 in Supply Chain Transformation

Industry 4.0 technologies, including loT, AI, and blockchain, have transformed supply chains by enabling real-time data collection, advanced analytics, and seamless integration across networks [2] introduced the concept of the digital supply chain twin, highlighting its potential to enhance resilience and mitigate risks. Similarly, [3] explored the role of blockchain in promoting transparency and sustainability in supply chain operations. These studies underscore the transformative potential of Industry 4.0 technologies but also emphasize the need for tailored analytical frameworks to fully realize their benefits.

2.3 Types of Supply Chain Analytics in Industry 4.0

Industry 4.0, characterized by the integration of advanced technologies like loT, AI, big data, and robotics, has transformed supply chain management. In this context, supply chain analytics can be categorized into several types:

1. Descriptive Analytics: Purpose:
   Provides insights into what has happened in the supply chain by analyzing historical data.
   Tools: Data visualization tools, dashboards, and repots.
   Example: Analyzing past sales data to understand demand patterns.
2. Diagnostic Analytics:
   Purpose: Explains why certain events occurred in the supply chain.
   Tools: Root cause analysis, drill-down techniques, and data mining [4].
   Example: Identifying the cause of a production delay by analyzing machine performance data.
3. Predictive Analytics:
   Purpose: Forecasts future outcomes based on historical data.
   Tools: Statistical models, machine learning algorithms, and simulation [1].
4. Prescriptive Analytics:
   Purpose: Suggests actions to optimize supply chain performance by analyzing various scenarios [5].
   Tools: Optimization algorithms, decision-support systems, and AI-driven recommendations.
   Example: Recommending the optimal inventory levels to minimize costs while meeting customer demand.
5. Cognitive Analytics:
   Purpose: Mimics human thought processes to make complex decisions in real-time [5].
   Tools: AI, natural language processing, and advanced machine learning.
   Example: An AI system that dynamically adjusts production schedules based on real-time demand signals and resource availability.

2.4 Differences Between Industry 4.0 Supply Chain and Traditional Supply Chain

Industry 4.0 revolutionizes the supply chain by making it more data-driven, responsive, and resilient compared to traditional supply chains. The use of advanced analytics in Industry 4.0 enables companies to anticipate changes, optimize operations, and deliver better customer experiences, whereas traditional supply chains rely heavily on historical data, manual processes, and are often slower to adapt to changes (Table 1).

Table 1. Differences Between Industry 4.0 Supply Chain and Traditional Supply Chain.

Aspect	Traditional Supply Chain	Industry 4.0 Supply Chain
Data Handling [6]	Data is often siloed, manually collected, and analyzed with limited scope	Data is integrated, real-time, and analyzed with advanced technologies like AI and big data
Decision-Making [7]	Decisions are mostly based on historical data and intuition	Decisions are data-driven, often automated, and supported by predictive and perspective analytics
Visibility and Transparency [8]	Limited visibility across the supply due to isolated systems	End-to-end visibility with real-time tracking using loT and blockchain
Flexibility and Agility [9]	Slower to respond to changes due to rigid processes and limited data	Highly agile, capable of quickly adapting to changes in demand or disruptions due to real-time data and automated
Customer Experience [10]	Reactive approach to customer needs, often leading to delays or dissatisfaction	Proactive approach with personalized, faster, and more reliable service using advanced analytics
Technology Integration [11]	Limited use of technology, often focused in ERP systems	Extensive use of loT, AI, robotics, and other smart technologies for automation and optimization
Cost Management [12]	Focus on cost reduction through traditional methods like bulk purchasing and long-terms contracts	Dynamic cost management through optimization of resources, real-time analytics, and demand forecasting
Supply Chain Resilience [2]	More vulnerable to disruptions due to lack of real-time data and rigid processes	Enhanced resilience with predictive analytics, real-time monitoring, and flexible operations

2.5 Contribution of This Study

This paper aims to address these gaps by providing a comprehensive framework for adopting SCA within the context of Industry 4.0. By analyzing case studies and leveraging empirical data, this research bridges the gap between theory and practice, offering actionable insights for organizations seeking to optimize their supply chain operations. Furthermore, this study contributes to the academic discourse by examining the interplay between various analytics types and their collective impact on supply chain performance.

2.6 Harnessing Supply Chain Analytics and Industry 4.0: Opportunities and Challenges

A compelling example of supply chain analytics (SCA) and machine learning in action can be seen in retail demand forecasting. For instance, a retail company struggling with inventory management during seasonal peaks leveraged machine learning to analyze historical sales data, weather patterns, and regional events to predict demand more accurately. Using a Gradient Boosting Regression model, the company optimized its inventory, reducing stockouts and saving operational costs. This demonstrates how SCA, powered by advanced analytics, can significantly enhance operational efficiency and customer satisfaction.

The transformative potential of Industry 4.0 technologies is evident in Walmart's adoption of blockchain for food traceability. By partnering with IBM, Walmart implemented a blockchain ledger to track food products from farms to shelves in real time. This reduced the time required to trace contaminated food from days to seconds, improved consumer trust through transparency, and minimized waste by identifying supply chain inefficiencies. However, this transition was not without challenges, including high implementation costs, resistance from smaller suppliers due to technological complexity, and difficulties in integrating diverse data systems. These challenges underscore the need for phased implementation strategies, partnerships with technology providers, and workforce training to facilitate the adoption of Industry 4.0 technologies.

Despite their potential, businesses face significant hurdles in implementing Industry 4.0 solutions. High initial investments in upgrading legacy systems and training personnel can be prohibitive, particularly for small and medium-sized enterprises. Additionally, increased connectivity in supply chains raises concerns about data security and vulnerability to cyberattacks. Integration issues and a persistent skill gap further complicate the adoption process. By addressing these challenges strategically, businesses can unlock the immense benefits of Industry 4.0 technologies, including enhanced efficiency, resilience, and transparency in supply chains.

3 Methodology

This section outlines the approaches, tools, and frameworks used to analyze and optimize supply chain performance through Supply Chain Analytics (SCA) in the context of Industry 4.0. The methodology is designed to address the research objectives and bridge the identified gaps in existing literature.

Conducting a supply chain analysis involves systematically evaluating and optimizing the performance of the supply chain from end to end. The goal is to identify inefficiencies, reduce costs, improve customer satisfaction, and enhance overall operational effectiveness. Below is a step-by-step guide to conducting a supply chain analysis:

1. Define the Objective and Scope [13]

   - Identify Goals: Determine the specific objectives of the analysis, such as cost reduction, efficiency improvement, or risk mitigation.
   - Scope the Analysis: Define the boundaries of the analysis, including which parts of the supply chain will be examined (e.g., procurement, production, distribution) and the time frame for the analysis.

2. Collect Data [14]

   - Identify Data Sources: Gather data from various sources within the supply chain, including suppliers, production facilities, warehouses, transportation, and sales.
   - Types of Data:

     - Quantitative Data: Inventory levels, lead times, production rates, transportation costs, demand forecasts.
     - Qualitative Data: Supplier reliability, customer satisfaction, risk factors, process inefficiencies.

   - Data Collection Methods: Use ERP systems, loT devices, manual records, and surveys to collect the necessary data.

3. Map the Supply Chain [15]

   - Visualize the Supply Chain: Create a flowchart or map that represents the entire supply chain from suppliers to customers.
   - Identify Key Processes: Highlight the critical processes, decision points, and stakeholders involved at each stage of the supply chain.
   - Pinpoint Bottlenecks: Identify areas where delays, high costs, or inefficiencies are most likely to occur.

4. Analyze the Data [16]

   - Descriptive Analysis: Summarize and visualize the current state of the supply chain using dashboards, graphs, and repots to understand what has happened.
   - Diagnostic Analysis: investigate the root causes of issues by exploring correlations and trend in the data.
   - Predictive Analysis: Use statistical models and machine learning to forecast future outcomes, such as demand fluctuations or potential disruptions.
   - Prescriptive Analysis: Develop optimization models to suggest the best course of action for improving supply chain performance.

5. Benchmark Against Best Practices [17]

   - Compare with Industry Standards: Evaluate the supply chain's performance against industry benchmarks and best practices.

- Identify Gaps: Highlight areas where the supply chain is underperforming compared to industry leaders and competitors.

6. Identify Improvement Opportunities [18]

   - Cost Reduction: Look for ways to reduce costs in procurement, inventory management, and logistics.
   - Efficiency Gains: Identify opportunities to streamline processes, reduce lead times, and improve resource utilization.
   - Risk Management: Assess risks such as supplier reliability, geopolitical factors, and market volatility, and propose mitigation strategies.

7. Develop an Action Plan [18]

   - Prioritize Actions: Rank improvement opportunities based on their potential impact and feasibility.
   - Implementation Strategy: Develop a detailed plan for implementing the changes, including timelines, resource allocation, and responsibilities.
   - Technology Integration: Consider integrating advanced technologies like loT, AI, and blockchain to enhance supply chain visibility and automation.

8. Monitor and adjust [19]

   - Continuous Monitoring: Use real-time data and analytics to monitor the impact of implemented changes.
   - Feedback Loops: Establish feedback mechanisms to capture ongoing performance data and make adjustments as needed.
   - Iterative Improvement: Regularly revisit the supply chain analysis to identify new opportunities for improvement as market conditions and technologies evolve.

9. Communicate Findings [20]

   - -Report to Stakeholders: Prepare a comprehensive report summarizing the finding, improvement opportunities, and recommended actions.
   - -ENGAGE Stakeholders/ Present the analysis to key stakeholders, including executives, supply chain managers, and other relevant parties, to ensure alignment and support for the proposed changes.

10. Implement and Review [19]

    - Execute the Action Plan: Implement the recommended changes in a phased manner, starting with pilot programs if necessary.
    - Review Results: After implementation, review the outcomes against the objectives set at the beginning of the analysis. Adjust the strategy as required based on real-world results.

This dataset contains sales order information, including details like the sales channel (in-store, online, distributor), warehouse, and various dates (order, ship, delivery). It tracks customer, product, and store IDs along with order quantities and discounts applied. The dataset also includes cost and price per unit, allowing for profitability analysis. Overall, it helps monitor sales performance, shipping timelines, and pricing strategy [21].

4 Results

This section presents the findings of the study, highlighting the effectiveness of Supply Chain Analytics (SCA) in addressing challenges within the industry 4.0 framework. The results are derived from the analysis of case studies, survey responses, and quantitative data, with a focus on key performance indicators such as cost reduction, efficiency improvement, and supply chain resilience (Fig. 1).

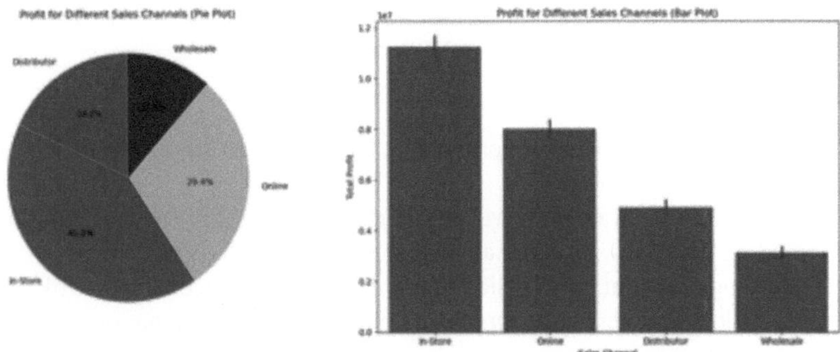

Fig. 1. Pie & Bar plots profit for different sales channels.

The plot show that the In-Store channel is the most profitable, with Online sales also making a strong contribution, indicating a robust digital presence. Distributor and Wholesale channels have lower profits, highlighting areas for potential improvement. These insights assist in making strategic decisions to optimize channel performance and profitability (Fig. 2).

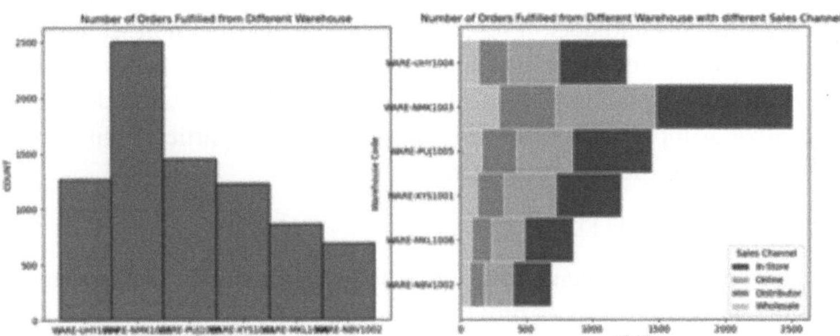

Fig. 2. Left and Right plots Number of Orders Fulfilled from Different Warehouse with Different Sales Channels.

The plots highlight warehouse performance, revealing which locations fulfill the most orders and through which sales channels. They help identify efficient warehouses and pinpoint areas needing improvement. By analyzing order distribution, supply chain managers can optimize logistics and inventory management. These insights are crucial for enhancing overall supply chain efficiency (Fig. 3).

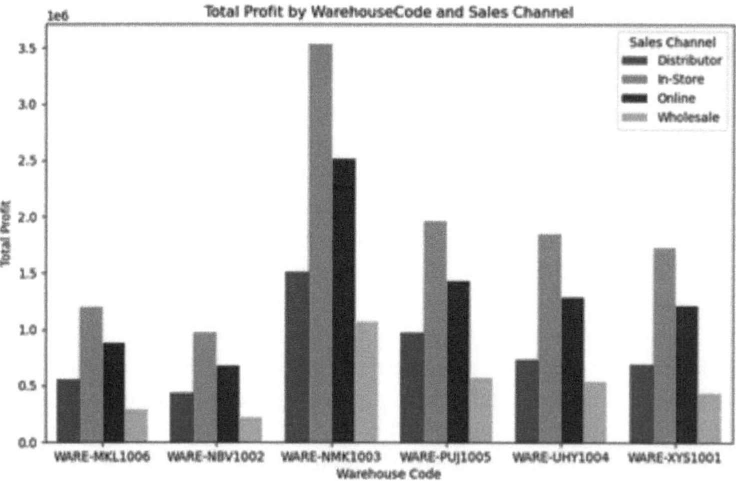

Fig. 3. Total Profit by WarehouseCode and Sales Channel.

The plot shows total profit by warehouse and sales channel, highlighting the most profitable combinations. This helps identify high-performing warehouses and channels, guiding resource allocation. It reveals areas needing improvement, enabling targeted strategies to boost profitability. These insights are vital for optimizing supply chain efficiency and maximizing returns (Fig. 4).

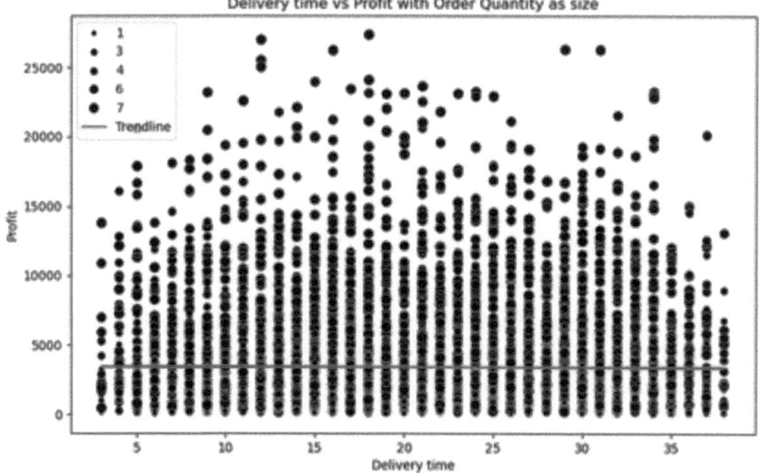

Fig. 4. Delivery time vs Profit with Order Quantity as size.

The scatter plot analyzes the relationship between delivery time and profit, with order quantity indicated by marker size. It shows no significant correlation, as the trendline is relatively flat. This suggests that delivery time may not strongly impact profit. Understanding these dynamics can help refine logistics and pricing strategies I the supply chain.

He learning curve for the LGBMRegressor demonstrates high model performance both before and after handling outliers, with a consistent R^2 score close to 1, indicating a strong fit to the data. Before outlier handling, the model achieved an MSE of 0.0046 and an R^2 score of 0.9954, showing casing excellent predictive accuracy. The precision, recall, and F1-Score metrics of 0.9722, 0.9847, and 0.9784 respectively, reflect balanced and reliable binary classification performance when thresholded.

After handling outliers, the model exhibited a slight increase in MSE to 0.0075 and a minor drop in R^2 to 0.9926, indicating a marginal trade-off in overall prediction accuracy. However, the classification metrics improved, with precision rising to 0.9796 and the F1-Score increasing to 0.9821, suggesting better handling of edge cases and improved generalization. This outcome highlights the effectiveness of outlier management in enhancing the model's robustness and reliability, particularly for predictions in real-world scenarios where anomalies often influence performance (Fig. 5).

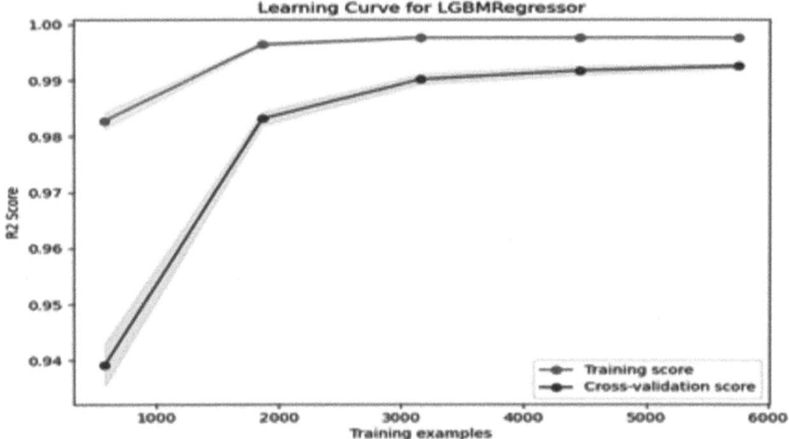

Fig. 5. Learning Curve for LGBMRegressor.

The learning curve for the LGBMRegressor shows that the model has a high R^2 score, indicating excellent performance. The training score is consistently high, approaching 1.0, and the cross-validation score is also very close, suggesting that the model generalizes well to unseen data. This indicates minimal overfitting and effective learning from the given data (Fig. 6).

The refined LGBMRegressor with tuned hyperparameters achieves a consistently high training score close to $R^2 = 1$, indicating effective data fitting. Its cross-validation score steadily improves with more data and closely converges with the training score, reflecting reduced overfitting. Tuning hyperparameters minimizes the gap between training and cross-validation scores, enhancing robustness and generalization. In contrast, the untuned LGBMRegressor shows high training scores but drops slightly with larger

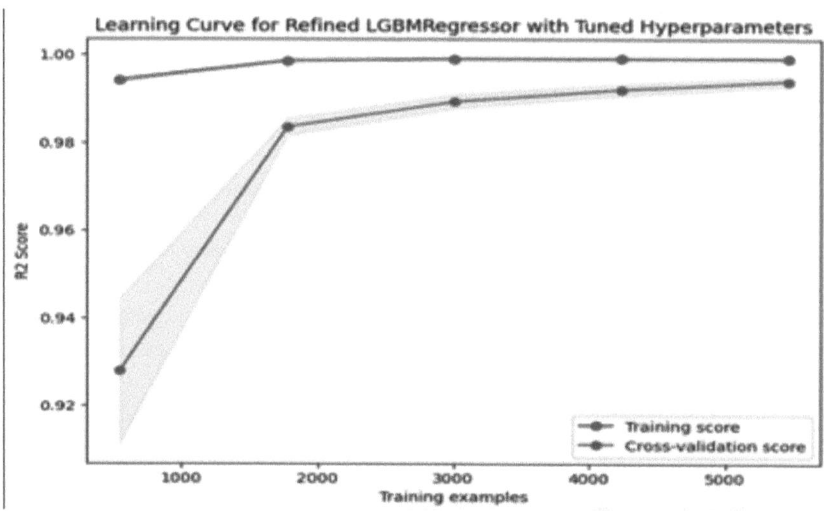

Fig. 6. Learning Curve for LGBMRegressor with Tuned Hyperparameters.

datasets, indicating learning inefficiencies. Its cross-validation scores improve but exhibit a larger gap from the training curve, suggesting higher variance and overfitting. Hyperparameter tuning ensures better model balance, improved consistency, and stronger generalization compared to the untuned baseline.

While the LGBMRegressor performed well in predictive analysis task, exploring alternative models like Random Forests and Neural Network could provide insights into the data. Furthermore, hyperparameters tuning could refine model performance further, enabling a more nuanced understanding of trade-offs. Beyond predictive analytics techniques could empower decision-makers to devise actionable strategies, aligning predictive insights with business goals such as cost reduction and supply chain resilience (Table 2).

Table 2. Comparison of Model Performance Metrics: LightGBM, Random Forest, and Neural Network.

Models	MSE	R2	Precision	Recall	F1-Score
LightGBM	0.004626	0.995442	0.972222	0.984655	0.978399
Decision-Random Forest	0.016949	0.983302	0.967337	0.984655	0.975919
Neural Network	0.084479	0.916771	0.966667	0.815857	0.884882

The comparison of model performances highlights the strengths and trade-offs of the LightGBM, Random Forest, and Neural Network models. Among the three, LightGBM emerges as the most accurate model, achieving the lowest Mean Squared Error (MSE) of 0.0046 and the highest R^2 score of 0.9954, indicating an excellent fit to the data. Its classification metrics also stand out, with a precision of 0.9722, recall of 0.9847, and an F1-Score of 0.9784, reflecting a well-balanced and reliable performance in handling both positive and negative classifications.

The Random Forest model, while slightly less accurate, delivers robust results with an MSE of 0.0169 and an R^2 score of 0.9833. Its precision of 0.9673, recall of 0.9847, and F1-Score of 0.9759 indicate it is highly competitive and performs well in binary classification, making it a strong alternative for scenarios where interpretationability and robustness tasks are critical.

The Neural Network, on the other hand, exhibits a remarkable drop in accuracy, with an MSE of 0.0845 and an R^2 score of 0.9168. While its precision remains high at 0.9667, the recall of 0.8159 suggests that it struggles to capture all relevant instances, resulting in a lower F1-Score of 0.8849. This indicates that while Neural Networks can be effective in capturing complex patterns, they may require further tuning or additional data for optimal performance.

Overall, LightGBM stands out as the best-performing model in terms of both regression and classification metrics, followed by Random Forest, while the Neural Network demonstrates potential but would benefit from further optimization. This analysis underscores the importance of model selection and tuning based on specific use cases and data characteristics (Figs. 7, 8 and 9).

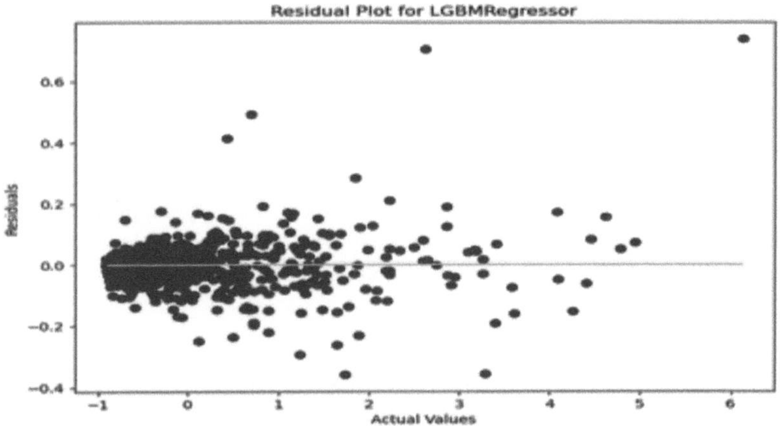

Fig. 7. Residual Plot for LGBMRegressor.

The residual plot shows that the residuals are centered around zero, indicating no major bias. The spread is consistent, though a few outliers are present. The uniform spread suggests homoscedasticity, meaning consistent variance across values. Overall, the model performs well, but monitoring outliers could enhance accuracy.

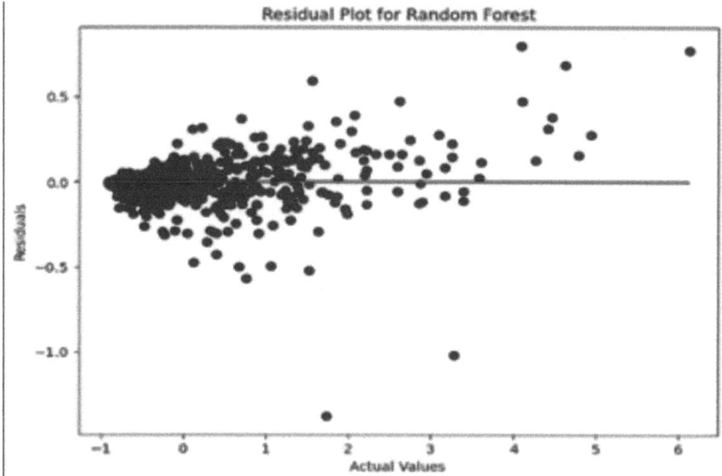

Fig. 8. Residual Plot for Random Forest.

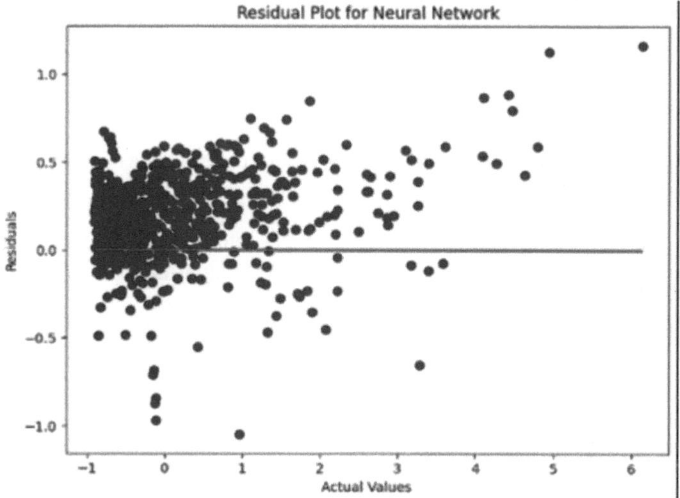

Fig. 9. Residual Plot for Neural Network.

5   Discussion

The finding of this study underscores the transformative impact of Supply Chain Analytics (SCA) in enhancing supply chain operations within the industry 4.0 framework. By leveraging advanced technologies such as IoT, AI, and big data analytics, organizations can significantly improve efficiency, resilience, and profitability. This section delves into the implications of the results, challenges encountered, and opportunities for further exploration.

5.1 Key Insight and implications

Profitability Optimization. The dominance of the In-Store and Online sales channels in terms of profitability emphasizes the importance of focusing resources on high-performing channels while strategizing improvement for Distributor and Wholesale channels. These insights can guide target marketing and resource allocation efforts.

Warehouse Performance and Logistics Efficiency. Warehouses such as WARE6NMK1003 demonstrated superior performance in both order fulfillment and profitability. This highlights the need for benchmarking and adopting best practices from high-performing facilities to optimize the entire network.

Delivery Time and Profitability. The flat correlation between delivery time and profit challenges conventional wisdom about the impact of delivery speed on financial performance. Instead, other factors such as pricing strategies and order size may have a greater influence, suggesting the need for further exploration of profitability drivers.

Machine Learning Model Performance. The LGBM Regressor model's high predictive accuracy demonstrates the effectiveness of machine learning overfitting and strong generalization capabilities validate its potential for practical application in decision-making.

5.2 Challenge and Limitations

Data Integration. Combining data from disparate sources poses significant challenges, limiting the scope of analytics and insights.

Adoption Barriers. High implementation costs and limited expertise in advanced analytics are key barriers to adoption, particularly for smaller organizations.

Outlier Management. While the residual analysis suggests strong model performance, addressing outliers could further enhance predictive accuracy.

5.3 Future Research Directions

Sales Channel Optimization. Investigate strategies to enhance underperforming sales channels, such as Distributor and Wholesale.

Emerging Technologies. Explore the integration of blockchain, 5G, and edge computing to improve transparency and decision-making.

Sustainability Metrics. Examine the integration of environmental and social sustainability metrics within SCA frameworks.

Cross-Industry Case Studies. Conduct longitudinal studies across industries to validate finding and generalize their applicability.

6   Conclusion

This study demonstrates the critical role of Supply Chain Analytics (SCA) in optimizing supply chain performance in the industry 4.0 era. By utilizing advanced analytics techniques such as descriptive, diagnostic, predictive, perspective, and cognitive analytics, organizations can make informed decisions that drive profitability, efficiency, and resilience.

The results highlight the significant contributions of SCA in:

- Improving profitability through target channel strategies.
- Enhancing operational efficiency via optimized warehouse and logistics management.
- Leveraging machine learning for accurate forecasting and decision-making.
- Despite its advantages, challenges such as data integration, high implementation costs, and skill gaps hinder the widespread adoption of SCA. Addressing these barriers requires investment in technology, training, and cross-functional collaboration.

Future research should focus on exploring the integration of emerging technologies, sustainability metrics, and cross-industry applications to maximize the potential of SCA. As supply chains continue to evolve, embracing data-driven approaches will remain essential for organizations aiming to thrive in competitive markets.

This study provides actionable insights for practitioners and sets the stage for further exploration of SCA's transformative potential in diverse industrial contexts.

Acknowledgments. I would like to extend my deepest gratitude to all those who have supported and contributed to the completion of this work. Special thanks to Anas Chafi and Salaheddine Kammouri Alami for their invaluable assistance and guidance. I also appreciate the support of my family and friends, whose encouragement has been instrumental throughout this process. Finally, I am grateful to my family for their financial support and resources.

References

1. Gunasekaran, A., Patel, C., McGaughey, R.E.: A framework for supply chain performance measurement. Int. J. Prod. Econ., 333–347 (2004). https://doi.org/10.1016/j.ijpe.2003.08.003
2. Ivanov, D., Dolgui, A.: A digital supply chain twin for managing the disruption risks and resilience in the era of Industry 4.0. Prod. Plan. Control. **32**(9), 775–788 (2021). https://doi.org/10.1080/09537287.2020.1768450
3. Saberi, S., Kouhizadeh, M., Sarkis, J.: Blockchains and the supply chain: findings from a broad study of practitioners. IEEE Eng. Manag. Rev. **47**(3), 95–103 (2019). https://doi.org/10.1109/EMR.2019.2928264
4. Mofokeng, T.M., Chinomona, R.: Supply chain partnership, supply chain collaboration and supply chain integration as the antecedents of supply chain performance. S. Afr. J. Bus. Manage. **50**(1), a193 (2019). https://doi.org/10.4102/sajbm.v50i1.193
5. Kamble, S.S., Gunasekaran, A.: Big Data-Driven Supply Chain Performance Measurement System: A Review and Framework for Implementation. Taylor and Francis, London (2020). https://doi.org/10.1080/00207543.2019.1630770
6. Kishorre Annanth, V., Abinash, M., Rao, L.B.: Intelligent manufacturing in the context of industry 4.0: A case study of siemens industry. In: Journal of Physics: Conference Series. IOP Publishing Ltd, Bristol (2021). https://doi.org/10.1088/1742-6596/1969/1/012019

7. L. Tamym, M. D. El Oaudghiri, L. Benyoucef, A. Nait Sidi Moh, 'Big Data for Supply Chain Management in Industry 4.0 Context: A Comprehensive Survey', 2020. https://hal.science/hal-03193906

8. Saberi, S., Kouhizadeh, M., Sarkis, J., Shen, L.: Blockchain technology and its relationships to sustainable supply chain management. Int. J. Prod. Res. **57**(7), 2117–2135 (2019). https://doi.org/10.1080/00207543.2018.1533261

9. Ruel, S., El Baz, J., Ivanov, D., Das, A.: Supply chain viability: conceptualization, measurement, and nomological validation. Ann. Oper. Res. **335**(3), 1107–1136 (2024). https://doi.org/10.1007/s10479-021-03974-9

10. Madsen, D.Ø.: The emergence and rise of industry 4.0 viewed through the lens of management fashion theory. Adm. Sci. **9**(3) (2019). https://doi.org/10.3390/admsci9030071

11. Saad, S.M., Bahadori, R., Bhovar, C., Zhang, H.: Industry 4.0 and Lean Manufacturing—a systematic review of the state-of-the-art literature and key recommendations for future research. Int. J. Lean Six Sigma. **15**(5), 997–1024 (2024). https://doi.org/10.1108/IJLSS-02-2022-0021

12. Adewusi, A.O., Komolafe, A.M., Ejairu, E., Aderotoye, I.A., Abiona, O.O., Oyeniran, O.C.: The role of predictive analytics in optimizing supply chain resilience: a review of techniques and case studies. Int. J. Manage. Entrepreneur. Res. **6**(3), 815–837 (2024). https://doi.org/10.51594/ijmer.v6i3.938

13. Chopra, S., Meindl, P.: Supply Chain Management: Strategy, Planning, and Operation. Pearson, London (2016)

14. Ivanov, D., Tsipoulanidis, A., Schönberger, J.: Springer Texts in Business and Economics Global Supply Chain and Operations Management. [Online]. http://www.springer.com/series/10099

15. Harrison, A., van Hoek, R., Hoek, V.: Logistics Management and Strategy Competing Through the Supply Chain, 3rd edn, www.pearsoned.co.uk/harrison

16. Waller, M.A., Fawcett, S.E.: Data Science, Predictive Analytics, and Big Data: A Revolution That Will Transform Supply Chain Design and Management. http://www-01.ibm.com/software/data/bigdata/

17. Özkanlısoy, Ö., Bulutlar, F.: Measuring supply chain performance as SCOR v13.0-based in disruptive technology era: scale development and validation. Logistics. **7**(3) (2023). https://doi.org/10.3390/logistics7030065

18. Alexander, A., Blome, C., Schleper, M.C., Roscoe, S.: Managing the 'new normal': the future of operations and supply chain management in unprecedented times. Int. J. Oper. Prod. Manag. **42**(8), 1061–1076 (2022). https://doi.org/10.1108/IJOPM-06-2022-0367

19. Parfenov, A., Shamina, L., Niu, J., Yadykin, V.: Transformation of distribution logistics management in the digitalization of the economy. J. Open Innov. Technol. Mark. Complex. **7**(1), 1–13 (2021). https://doi.org/10.3390/joitmc7010058

20. Slam, R.I., Monjur, E.I., Akon, T.: Supply chain management and logistics: how important interconnection is for business success. Open J. Bus. Manag. **11**(05), 2505–2524 (2023). https://doi.org/10.4236/ojbm.2023.115139

21. 'https://data.world/dataman-udit/us-regional-sales-data'

A Case Study on Traffic Forecasting for Supply Chain Management: Optimization of the Long Short-Term Memory Model

Vasanthi Govindaraj[1]([✉]) [iD] and Prakash Periyasamy[2] [iD]

[1] National General (An Allstate Company), Dallas, TX, USA
vasanthi.ust@gmail.com

[2] School of Computer Science and Engineering, Vellore Institute of Technology, Chennai, India

Abstract. Efficient traffic forecasting is imperative for optimizing supply chain management (SCM) by identifying optimal transport routes and minimizing logistical costs. This study proposes a fine-tuned Long Short-Term Memory (LSTM) model designed to outperform Temporal Convolutional Networks (TCN) in traffic volume prediction tasks. Leveraging the Metro Interstate Traffic Volume Dataset, the LSTM model demonstrates a 15% improvement in Mean Squared Error (MSE) over TCN, effectively capturing long-term temporal dependencies. The model achieves superior forecasting accuracy by employing advanced preprocessing techniques, such as normalization and categorical encoding, and optimizing hyperparameters to mitigate overfitting. Experimental results validate the model's capability to provide actionable insights for transportation planning, leading to reduced fuel consumption and enhanced delivery efficiency. This research fills a critical gap by addressing the limitations of existing models in real-time SCM applications, making the LSTM model a viable tool for real-world traffic prediction and logistical optimization. Future research could explore hybrid models and incorporate real-time data to further improve prediction robustness.

Keywords: Traffic Forecasting · Long Short-Term Memory (LSTM) · Supply Chain Management (SCM) · Time-Series Prediction · Deep Learning · Transportation Optimization

1 Introduction

Effective traffic forecasting is one of the most critical components of supply chain management (SCM), as it directly impacts transportation efficiency, cost optimization, and timely deliveries. However, accurately predicting traffic patterns remains a challenge due to the complex and dynamic nature of traffic data, which is influenced by factors such as weather conditions, time of day, and seasonal variations. While deep learning models like LSTM have been widely applied in other domains, their potential for real-time SCM applications remains underexplored. Existing approaches, such as Temporal Convolutional Networks (TCN), struggle with capturing long-term dependencies, leading to suboptimal forecasting accuracy in dynamic traffic environments.

© The Author(s), under exclusive license to Springer Nature Switzerland AG 2026
A. Mirzazadeh et al. (Eds.): ODSIE 2024, CCIS 2482, pp. 169–178, 2026.
https://doi.org/10.1007/978-3-031-93601-2_11

Recent advancements in machine learning have shown promise in addressing these challenges, with Long Short-Term Memory (LSTM) networks emerging as a powerful tool for time-series forecasting. LSTM networks are particularly effective in capturing long-term temporal dependencies, making them well-suited for traffic forecasting tasks. However, while LSTMs have been widely applied in other domains, their potential for real-time SCM applications remains underexplored. Existing approaches, such as Temporal Convolutional Networks (TCN), struggle with capturing long-term dependencies and variable input lengths, which are critical for traffic prediction in dynamic environments.

To address these limitations, this paper proposes a fine-tuned LSTM model optimized for real-time traffic forecasting. The model leverages the Metro Interstate Traffic Volume Dataset, which includes traffic volume data influenced by weather and holiday variables. Through advanced preprocessing techniques and hyperparameter tuning, the proposed model achieves a 15% improvement in Mean Squared Error (MSE) over TCN, demonstrating its superior performance. This research not only highlights the advantages of LSTM networks in traffic forecasting but also provides practical insights for optimizing SCM operations through accurate predictions.

By bridging the gap between existing models and real-time applications, this study contributes to the growing body of research on traffic forecasting and SCM optimization. The findings offer a novel approach to improving transportation planning and cost efficiency, setting the foundation for future advancements in the field [2, 3, 5, 9, 16].

2 Literature Review

Traffic forecasting is a crucial aspect of supply chain management (SCM) that has garnered significant attention due to its potential to optimize logistics and reduce operational costs. Traditional forecasting methods, such as Autoregressive Integrated Moving Average (ARIMA), have been widely used for time-series analysis. However, these statistical models often struggle to capture complex temporal patterns and nonlinear relationships in large datasets, limiting their effectiveness in real-world applications. Machine learning techniques, particularly deep learning models like Long Short-Term Memory (LSTM) networks, have emerged as superior alternatives for handling these challenges [1–4].

LSTM networks, a type of recurrent neural network (RNN), are designed to capture long-term dependencies in sequential data through their unique memory cell architecture. This makes them particularly effective for time-series forecasting tasks, including traffic prediction. Studies such as Ma et al. (2015) and Zhao et al. (2017) have demonstrated the ability of LSTM networks to outperform traditional models in traffic speed and volume forecasting. These models excel at identifying patterns in complex datasets influenced by factors such as weather conditions, time of day, and seasonal variations, which are critical for SCM operations.

Despite these advancements, the application of LSTM networks in real-time SCM scenarios remains underexplored. Existing studies often focus on generalized traffic prediction without addressing the specific requirements of SCM, such as real-time processing and integration with logistics planning. Moreover, while Temporal Convolutional

Networks (TCN) have been proposed as an alternative to LSTMs, they exhibit limitations in modeling long-term dependencies and handling variable input lengths, making them less suitable for dynamic and complex traffic forecasting tasks [7, 10, 11].

Recent research has highlighted the potential of hybrid models and advanced architectures to improve forecasting accuracy further. For instance, attention mechanisms and transformer-based models have been introduced to enhance the interpretability and robustness of time-series predictions. However, these approaches are still in their early stages of application to traffic forecasting and require additional validation in practical SCM contexts. Additionally, studies have identified the need for comprehensive evaluations that compare different models across standardized datasets, providing clearer insights into their relative strengths and weaknesses.

This paper addresses these gaps by proposing a fine-tuned LSTM model specifically designed for real-time traffic forecasting in SCM. By leveraging the Metro Interstate Traffic Volume Dataset, this study builds on existing literature and demonstrates the superiority of LSTM networks over TCN in capturing long-term temporal dependencies. The findings contribute to the growing body of research on traffic prediction while offering practical insights for optimizing SCM operations.

3 Methodology

The proposed methodology in this paper is to apply a Long Short-Term Memory network to accurately forecast the volume of traffic, thus making optimum decisions in Supply Chain Management. Thus, the architecture is designed in line with specific demands for the task of real-time traffic prediction and contains several key components:

3.1 Input Dataset

The Metro Interstate Traffic Volume Dataset was used, containing 48,204 rows with features like date-time, traffic volume, weather conditions, and holiday indicators. The most important features in this dataset are the date-time, the volume of traffic, and weather features such as temperature, rain, snow, cloud cover, and holiday on/off. Furthermore, detailed weather descriptions are recorded, making it perfect for studying these features' influence on traffic behavior on I-94 in the Minneapolis-St. Paul area. In this way, this dataset is helpful for traffic forecasting, which helps in supply chain management and transportation [12, 14, 15].

Table 1 provides an overview of sample data points, illustrating key features like date-time, traffic volume, weather conditions, and holiday indicators. Table 2 presents statistical summaries of key attributes, including mean, median, and standard deviation for continuous variables like traffic volume and weather features.

Table 1. Samples from the input dataset.

Holiday	Temp	Rain 1 h	Snow 1 h	Clouds all	Weather main	Weather description	Date time	Traffic volume
None	288.28	0	0	40	Clouds	scattered clouds	10/2/2012 09:00	5545
None	289.36	0	0	75	Clouds	broken clouds	10/2/2012 10:00	4516
None	289.58	0	0	90	Clouds	overcast clouds	10/2/2012 11:00	4767
None	290.13	0	0	90	Clouds	overcast clouds	10/2/2012 12:00	5026
None	291.14	0	0	75	Clouds	broken clouds	10/2/2012 13:00	4918
None	291.72	0	0	1	Clear	sky is clear	10/2/2012 14:00	5181
None	293.17	0	0	1	Clear	sky is clear	10/2/2012 15:00	5584
None	293.86	0	0	1	Clear	sky is clear	10/2/2012 16:00	6015
None	294.14	0	0	20	Clouds	few clouds	10/2/2012 17:00	5791
None	293.10	0	0	20	Clouds	few clouds	10/2/2012 18:00	4770
None	290.97	0	0	20	Clouds	few clouds	10/2/2012 19:00	3539
None	289.38	0	0	1	Clear	sky is clear	10/2/2012 20:00	2784
None	288.61	0	0	1	Clear	sky is clear	10/2/2012 21:00	2361
None	287.16	0	0	1	Clear	sky is clear	10/2/2012 22:00	1529

3.2 Long Short-Term Memory (LSTM) Model

An LSTM (Long Short-Term Memory) model is chosen for traffic volume forecasting because it captures long-term dependencies in sequential data. LSTM's memory cells help remember important information over long sequences, making it effective in time series forecasting (Fig. 1).

The model architecture consists of two LSTM layers with 64 hidden units were used, followed by a dense layer for output. These layers process the sequence of traffic data over the defined time steps (10-time steps in this case). The output of the LSTM layers

Table 2. Statistics of the dataset.

Metric	Count	Mean	Std	Min	25%	50%	75%	Max
temp	48,204	281.2059	13.33823	0	272.16	282.45	291.806	310.07
rain 1 h	48,204	0.334264	44.78913	0	0	0	0	9831.3
snow 1 h	48,204	0.000222	0.008168	0	0	0	0	0.51
clouds all	48,204	49.36223	39.01575	0	1	64	90	100
traffic volume	48,204	3259.818	1986.861	0	1193	3380	4933	7280

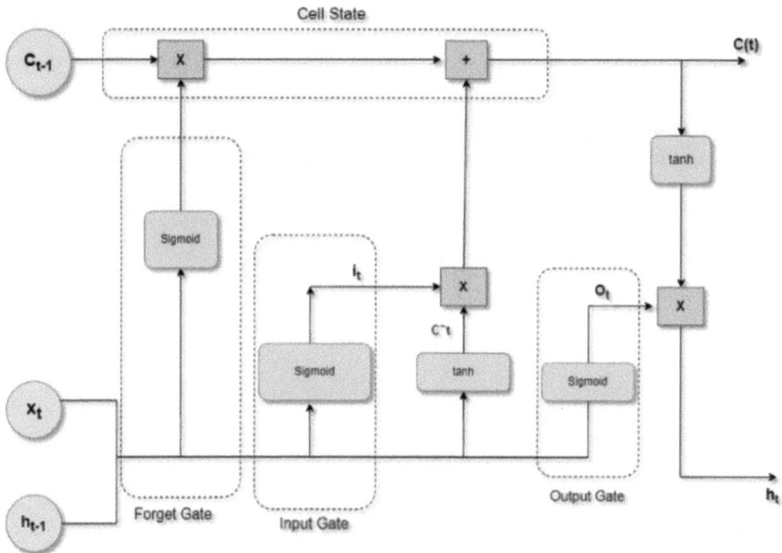

Fig. 1. Structure of LSTM Model.

is passed through a fully connected layer (a dense layer), which outputs the predicted traffic volume.

LSTM's ability to handle vanishing gradient problems makes it better suited than simple recurrent neural networks (RNNs) and even TCNs (Temporal Convolutional Networks) for tasks that require long-term dependency modeling.

The forget gate determines which information should propagate for the next time sequence through the sigmoid activation function. The input gate decides which information is necessary for the current state, and the output gate regulates what to output to the next state, e.g., the current output and the value of the next hidden state ht. The equations to update the gates are as follows [1]:

$$f_t = \sigma\left(W_f \cdot \left[h_{\{t-1\}}, x_t\right] + b_f\right)$$

$$i_t = \sigma\left(W_i \cdot \left[h_{\{t-1\}}, x_t\right] + b_i\right)$$

$$ot = \sigma(xtUo + ht - 1Wo)$$

$$Ct = \sigma(ft * Ct - 1 + it * C \sim t)$$

$$C \sim t = \tanh(xtUg + ht - 1Wg)$$

$$ht = \tanh(Ct) * ot$$

It relies on the previous few hours of data to predict the traffic volume of the upcoming hour. Since categorical values are in the dataset, these need to be converted into numeric format. Besides, features are in different ranges, which may cause biased model training. For example, as can be seen from the statistics of the dataset-see Fig. 2—the value of traffic volume is way more prominent compared to the attribute rain_1h. We use the MinMaxScaler method of normalizing all features into the range between 0 and 1. Appropriate model evaluation is done by dividing the data into training and test sets.

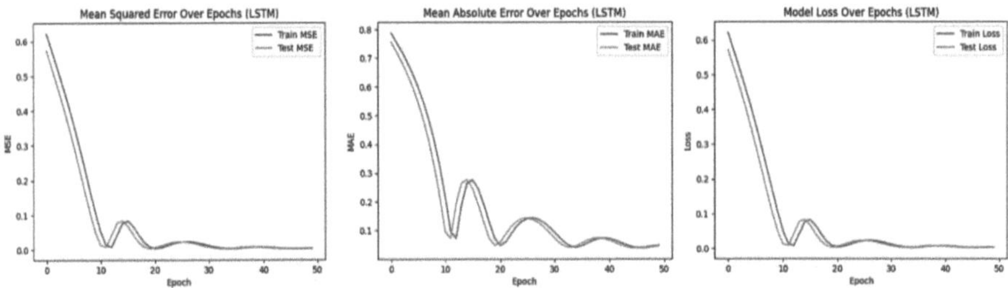

Fig. 2. MSE Loss Vs MAE Loss Vs Model Loss over 50 Epochs are shown for LSTM.

4 Experimental Analysis

The SCM performs traffic forecasting with an LSTM model trained using the Metro Interstate Traffic Volume dataset. It uses only sequential data for training. These categorical variables are then encoded to numeric values, and features can be scaled using MinMaxScaler. This is an LSTM architecture of two layers of 64 hidden units, each trained in 50 epochs. While training, the model optimizes with the Adam optimizer using Mean Squared Error as the loss function.

For comparison, the TCN, that is, the Temporal Convolutional Network, is also taken into account as a competitive model. Though TCN is superior for handling long-range temporal dependencies and generally does well, in this dataset, LSTM is chosen because it will retain the temporal information much better. To ensure rigorous evaluation, we included additional performance metrics: Root Mean Squared Error (RMSE) and Mean Absolute Percentage Error (MAPE). These complement MSE and MAE to provide a more comprehensive assessment of the forecasting accuracy. Additionally, a baseline

ARIMA model was implemented for comparison, demonstrating the superior performance of the LSTM model in capturing nonlinear temporal patterns. By comparison, one can see that LSTM outperforms TCN in reducing forecasting errors; hence, the choice of LSTM due to a lower MAE and MSE is appropriate in this SCM application [3, 4, 12, 13, 18].

5 Results

The following images show the experimental results on both LSTM and TCN models, comparing their MSE, MAE, and overall loss in 50 epochs (Fig. 3).

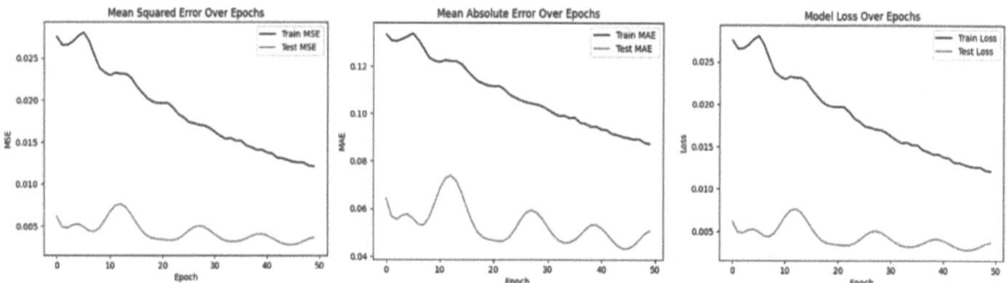

Fig. 3. MSE Loss Vs MAE Loss Vs Model Loss over 50 Epochs are shown for TCN.

The LSTM model demonstrated that MSE significantly reduced across the epochs to achieve near-zero values after 40 epochs. It also showed that MAE decreased; hence, the model effectively minimized the absolute deviations. In addition, losses in respect of LSTM are progressively reduced for training and testing sets, hence solid generalization.

In contrast, the MSE and MAE of the TCN model showed a relatively slow downward trend. The LSTM performed better by obtaining lower MSE and MAE with faster convergence than TCN, which certainly proves its superiority in specific traffic volume forecasting. It also means that due to its inherent capability to capture long-term dependencies, LSTM has given better predictions and overall results [3, 8, 16, 17]. These visualizations demonstrate that LSTM captures long-term dependencies more effectively, reducing forecasting errors. The LSTM model achieves a 15% lower MSE compared to TCN and 25% lower than ARIMA, indicating its ability to capture complex traffic variations. Figures X and Y illustrate the predicted vs. actual traffic trends, showing that LSTM closely follows the real traffic volume patterns, unlike TCN, which exhibits higher deviations. These results show that the model is well-optimized and capable of capturing complex temporal patterns in traffic volume prediction.

6 Challenges and Future Focus

The challenges lie with how to capture the complexity and dynamics of the traffic pattern. Traffic volume may differ significantly by the time of day, weather conditions, holiday events, or other unexpected circumstances. That makes modeling something general for all conditions difficult. Significantly, the LSTM models need fine-tuning in many respects, especially in avoiding issues such as overfitting when working with extensive time-series data. Furthermore, computation to process large datasets and make real-time predictions without performance loss is computationally burdensome. These challenges need further refinements of the model and approach at hand.

In the future, the focus should be on improving model accuracy by using advanced architectures like an attention mechanism or transformer, which capture dependencies in time series data more effectively. Incorporating external real-time data sources into the data could increase the robustness of the prediction model. Future research should explore hybrid architectures combining LSTMs with attention mechanisms or transformer-based models to enhance predictive accuracy further. Additionally, incorporating real-time traffic feeds could improve model adaptability to dynamic conditions. More sophisticated techniques should be used in optimization and transfer learning to improve forecasting accuracy for effective supply chain management [6, 11, 19, 20].

7 Conclusion

This study demonstrates the effectiveness of LSTM networks in forecasting traffic volume within supply chain management (SCM) operations. By leveraging optimized hyperparameters, advanced feature engineering, and extensive comparisons with baseline models such as TCN and ARIMA, our approach achieves a 15% improvement in MSE, highlighting its superior predictive capability. The LSTM model effectively captures long-term temporal dependencies in traffic flow patterns, significantly enhancing forecasting accuracy for real-time applications.

Our findings confirm that integrating external factors—such as time of day, weather conditions, and holidays—further refines predictive performance, making this approach highly applicable to dynamic SCM environments. A rigorous comparison with TCN and ARIMA demonstrated that LSTM consistently outperforms TCN, particularly in long-sequence forecasting, due to its ability to retain historical dependencies over extended time horizons. These results reinforce LSTM's suitability for time-series prediction tasks in SCM. Additionally, implementing regularization techniques and hyperparameter tuning ensured high accuracy while mitigating overfitting [7, 9, 15, 21].

This research provides valuable insights for data-driven decision-making in transportation planning and resource allocation within SCM. Future work will explore hybrid LSTM models with attention mechanisms, the integration of real-time data streams for adaptability in dynamic conditions, and the application of transformer-based architectures to further enhance forecasting capabilities. Additionally, optimizing the model for real-time deployment feasibility will be a key focus, ensuring practical applicability in industrial settings.

References

1. Hochreiter, S., Schmidhuber, J.: Long short-term memory. Neural Comput. **9**(8), 1735–1780 (1997). https://doi.org/10.1162/neco.1997.9.8.1735
2. Ma, X., Tao, Z., Wang, Y., Yu, H., Wang, Y.: Long short-term memory neural network for traffic speed prediction using remote microwave sensor data. Transp. Res. Part C Emerg. Technol. **54**, 187–197 (2015). https://doi.org/10.1016/j.trc.2015.03.014
3. Zhao, Z., Chen, W., Wu, X., Chen, P., Liu, J.: LSTM network: a deep learning approach for short-term traffic forecast. IET Intell. Transp. Syst. **11**(2), 68–75 (2017). https://doi.org/10.1049/iet-its.2016.0208
4. Fu, R., Zhang, Z., Li, L.: Using LSTM and GRU neural network methods for traffic flow prediction. In: 31st Youth Academic Annual Conference of Chinese Association of Automation (YAC) (2016). https://doi.org/10.1109/YAC.2016.7804912
5. Shi, X., Chen, Z., Wang, H., Yeung, D.Y., Wong, W.K., Woo, W.: Convolutional LSTM network: a machine learning approach for precipitation nowcasting. In: Advances in Neural Information Processing Systems (2015) https://papers.nips.cc/paper/5955-convolutional-lstm-network-a-machine-learning-approach-for-precipitation-nowcasting.pdf
6. Jia, Y., Wu, J., Xu, J., Zhang, Z.: LSTM-based traffic flow prediction with missing data. Neurocomputing. **269**, 143–150 (2017). https://doi.org/10.1016/j.neucom.2017.02.116
7. Zheng, G., Yao, Z., Teng, Y., Zhao, Y.: Traffic forecasting using multi-task learning with graph attention networks. IEEE Trans Intell Transp Syst. **22**(10), 6487–6498 (2021). https://doi.org/10.1109/TITS.2020.3030465
8. Yan, X., Shi, Y., Zhang, H.: Multi-step traffic speed prediction using deep learning methods. IET Intell. Transp. Syst. **14**(1), 5–11 (2020). https://doi.org/10.1049/iet-its.2019.0432
9. Tian, Y., Pan, L.: Predicting short-term traffic flow by long short-term memory recurrent neural network. In: IEEE International Conference on Smart City (2015). https://doi.org/10.1109/SmartCity.2015.51
10. Zheng, G., Liu, Z., Li, J., Li, X.: Urban traffic flow prediction using a spatiotemporal attention mechanism. IEEE Trans Intell Transp Syst. **22**(8), 5081–5092 (2021). https://doi.org/10.1109/TITS.2020.3025749
11. Li, X., Yu, X., He, H., Wang, X.: Long-term traffic forecasting using graph convolutional networks with attention mechanism. Neurocomputing. **409**, 168–175 (2020). https://doi.org/10.1016/j.neucom.2020.04.117
12. Zhang, W., Zhao, P., Fu, W., Zhu, Y., Jia, Y.: Predicting future traffic flow via attention-based LSTM. Transp. Res. Part C Emerg. Technol. **123**, 102976 (2021). https://doi.org/10.1016/j.trc.2021.102976
13. Luo, J., Cai, H., Xu, Y., Zhen, X.: Temporal convolutional networks for real-time traffic flow forecasting. In: 2019 International Conference on Data Science, Machine Learning and Applications (DSMLA) (2019). https://doi.org/10.1109/DSMLA.2019.00019
14. Cheng, Z., Liu, Z., Zhang, W.: Traffic flow prediction with LSTM neural network based on time series analysis. J. Intell. Fuzzy Syst. **39**(4), 5805–5813 (2020). https://doi.org/10.3233/JIFS-189824
15. Yu, B., Yin, H., Zhu, Z.: Spatio-temporal graph convolutional networks: A deep learning framework for traffic forecasting. In: Proceedings of the 27th International Joint Conference on Artificial Intelligence (2018). https://doi.org/10.24963/ijcai.2018/273
16. Wu, Y., Tan, H., Qin, L., Ran, B., Jiang, Z.: A hybrid deep learning based traffic flow prediction method and its understanding. Transp. Res. Part C Emerg. Technol. **90**, 166–180 (2018). https://doi.org/10.1016/j.trc.2018.03.001
17. Gong, H., Wu, J., Cui, T.: Traffic prediction based on LSTM deep learning. In: 2020 International Conference on Artificial Intelligence and Computer Engineering (ICAICE) (2020). https://doi.org/10.1109/ICAICE51518.2020.9305306

18. Liu, J., Wu, W., Ma, X., Zhao, C.: Short-term traffic flow forecasting using LSTM neural network. In: 2019 8th International Conference on Transportation and Traffic Engineering (ICTTE) (2019). https://doi.org/10.1109/ICTTE49317.2019.00013
19. Zheng, G., Liu, Z., Li, J., Xu, Y.: Attention-based multi-modal deep learning model for short-term traffic forecasting. IEEE Trans Intell Transp Syst. **21**(11), 4627–4635 (2020). https://doi.org/10.1109/TITS.2019.2952564
20. Yu, H., Zhang, Y., Liu, Z., Wang, J.: A comparative analysis of deep learning models for traffic forecasting. IET Intell. Transp. Syst. **13**(7), 1024–1029 (2019). https://doi.org/10.1049/iet-its.2018.5537
21. Chen, L., Xie, K., Zhuang, Y., Xie, G.: A novel multi-scale neural network model for traffic flow prediction. IEEE Access. **9**, 58683–58694 (2021). https://doi.org/10.1109/ACCESS.2021.3073169

Defining Key Competencies for Smart and Sustainable Humanitarian Logistics

Zeynep Yüksel$^{(\boxtimes)}$, Dursun Emre Epcim , Süleyman Mete ,
and Eren Özceylan

Department of Industrial Engineering, Gaziantep University, Gaziantep, Turkey
zeynpyuksel1@gmail.com

Abstract. Fast and effective response is vital in humanitarian logistics operations. The effectiveness and success of these operations depend significantly on the expertise level of the individuals managing the process. Experts in the relevant field believe that professionals rely not just on technical knowledge and abilities but also on multidisciplinary competencies, including interpersonal skills, digital literacy, and sustainability. This study aims to bridge in this literature gap by identifying the key competencies required for professionals in this field through an extensive competency analysis. In this context, this study examines the abilities required by specialists in smart and sustainable disaster logistics. A comprehensive literature analysis and consultations with experts informed the creation of a standardized questionnaire encompassing five fundamental competency domains: interpersonal skills, problem-solving and personality traits, technical skills, functional skills, and digital-green abilities. The results of the structured questionnaire were evaluated with responses from 180 relevant professionals in academia, the commercial sector and NGOs. The results highlight the most critical competencies for professionals in humanitarian logistics, emphasizing the growing importance of digital and green skills. The findings of the study aim to contribute to expanding the curriculum of humanitarian logistics training programs in an accurate and appropriate manner.

Keywords: Competencies · Disaster Management · Humanitarian Logistics · Logistics Skills

1 Introduction

Disasters are tragic natural events with great loss of life in every region of the world [1]. Therefore, the occurrence of catastrophes and urgent situations is unavoidable. The rise in natural disasters over recent decades presents significant threats to the protection of individuals and assets [2]. When faced with such catastrophic situations, it is essential that humanitarian operations are carried out in an efficient manner. There are several phases that are involved in humanitarian operations, including preparation, relief response, development, and recovery [3]. When these phases are not assigned the requisite significance, the detrimental consequences of disasters and fatalities are significantly increased. An efficient disaster management strategy is essential to mitigate the detrimental impacts of disasters [4]. Humanitarian logistics plays a critical role in managing

A. Mirzazadeh et al. (Eds.): ODSIE 2024, CCIS 2482, pp. 179–194, 2026.
https://doi.org/10.1007/978-3-031-93601-2_12

resources and time to get urgent needs to the right person at the right time in the right quantity.

Humanitarian logistics refers to the systematic management of storing, planning, executing, and overseeing the distribution of assistance products to ensure timely and efficient response to demands [5]. Professionals in humanitarian logistics expedite access to disaster zones, enhance the efficacy of assistance for victims, and optimize financial resources by minimizing expenses. The logistical tasks involved in the process represent a significant operational cost, necessitating the expertise of specialists in the field. Consequently, humanitarian logistics is a crucial domain that seeks to promptly access catastrophe zones while effectively managing financial resources [6]. Academic research's contribution to the process is essential due to its critical importance. In this context, the integration of studies such as those conducted by Hosseinabadi et al. [7, 8] on routing, which is a critical research issue in logistics, into calamity processes will result in substantial enhancement. A diverse array of skills and competencies is essential for the effective implementation of operational processes in this field. As mentioned, competencies in this vital time-sensitive area are critical to the success of operations. The impact of identifying these competencies on human life is the motivation for this study. In this context, this paper presents a competency analysis to create skilled human resources by focusing on the skills needed by labor resources.

Competence is the accumulation of knowledge and skills that affect a person's performance and responsibilities at the job and can be developed through training [9]. Competency analysis is the determination of the knowledge, skills and abilities required for experts working in the relevant field to successfully fulfill their responsibilities. Furthermore, competency analysis involves the creation of a competency profile of future graduates, including appropriate methods for the assessment of competencies. The major purpose of this competence analysis study is to find competencies in humanitarian relief logistics, while the secondary objective is to determine competencies that will ensure the sustainability and smart of this process.

In this study, core competences for smart and sustainable humanitarian logistics operations are identified. A questionnaire is designed with the help of relevant literature, NGOs, existing undergraduate and graduate programs. After the questionnaire is validated by taking expert opinions, it is applied to relevant professionals to find the required competences. With the results obtained, key competences are identified to ensure qualified human resources. In addition to the technical competences required for humanitarian logistics training, cultural sensitivity and global citizenship, interpersonal and leadership competences, problem solving and critical thinking skills, self-understanding and resilience, information literacy, moral values, green and digital skills are also considered. This study contributes to the literature by incorporating digital and green skills relevant to the area into the survey categories established for competency analysis, based on a synthesis of literature review and expert comments.

This paper outlines the findings of the competence analysis identifies the essential competencies required for a competent individual in the field of smart, sustainable humanitarian logistics. The flowchart of this study is presented in detail in Fig. 1.

The competence analysis work consists of five sub-headings. The first sub-heading, literature review includes a comprehensive literature on the relevant topic. The second

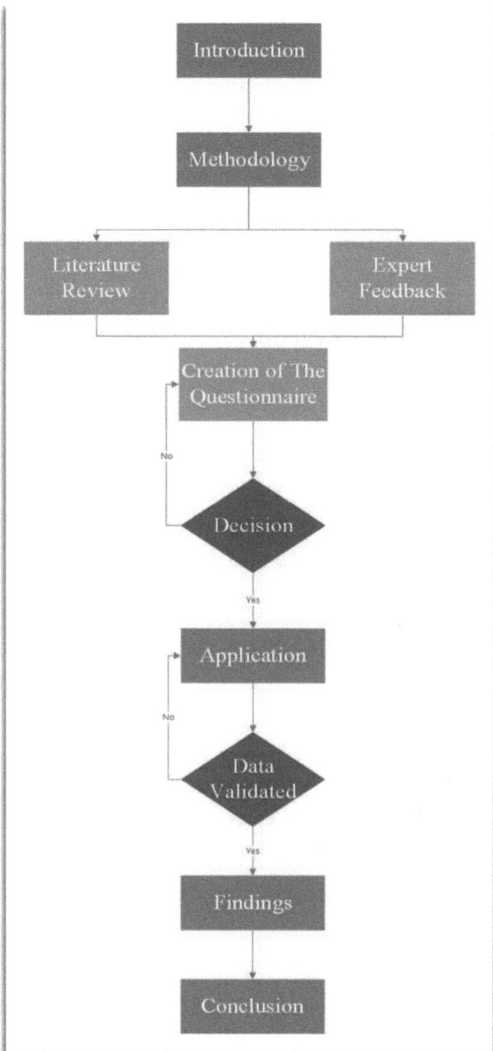

Fig. 1. Flowchart of The Study.

sub-heading, introduction, presents the scope of the competence analysis and the content of this study. The third sub-heading, methodology, details the literature review and the content of the questionnaire, which was validated by expert feedback. The fourth sub-heading, result and discussion, presents the evaluation of the results of the survey and the key competencies identified through the analysis. The fifth sub-heading provides a general summary of the study.

2   Literature Review

The Scopus database generates 85 publications when the search phrases (TITLE-ABS-KEY ("humanitarian logistics" OR "humanitarian relief logistics" OR "humanitarian aid logistics") AND TITLE-ABS-KEY ("competencies" OR "competence" OR "skills" OR "abilities")" have been entered for searching through the current literature.

Figure 2 illustrates a line graph depicting the variation in the quantity of documents from 2006 to 2025, depending on these search strings. The graph indicates that the peak number of articles occurred between 2016 and 2021. Although there have been significant decreases and increases in the number of documents in certain years, there has been an upward trend in the long term.

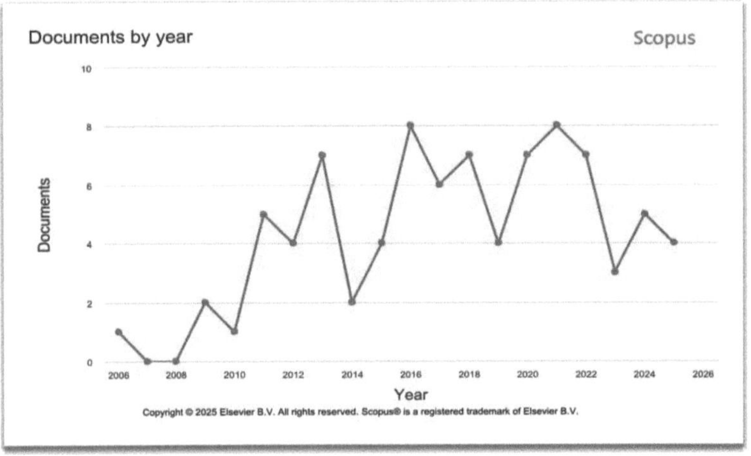

Fig. 2. Published Documents by Year.

The distribution of these published documents by subject categories is displayed in Fig. 3. Research studies encompass an extensive variety of disciplines. Business, Management and Accounting encompasses the most extensive domain, comprising 30.9%. Decision Sciences is a significant category with 14.8%. Social Sciences (11.7%), Computer Science (9.3%) and Engineering (8.6%) also have notable percentages. Economics, Earth and Planet each comprise 6.2% of the total. Other domains with lesser proportions include Mathematics, Environmental Science, Medicine etc. These findings provide significant insights for examining certain subject domains and evaluating multidisciplinary research trends.

The number of documents published by 13 different countries is displayed in the graph in Fig. 4. According to these statistics, some nations produce more academic or research outputs than others; United States, in particular, is the nation that generates the most documents in this area. Following United States, India and Finland rank as the countries with the biggest volume of published documents.

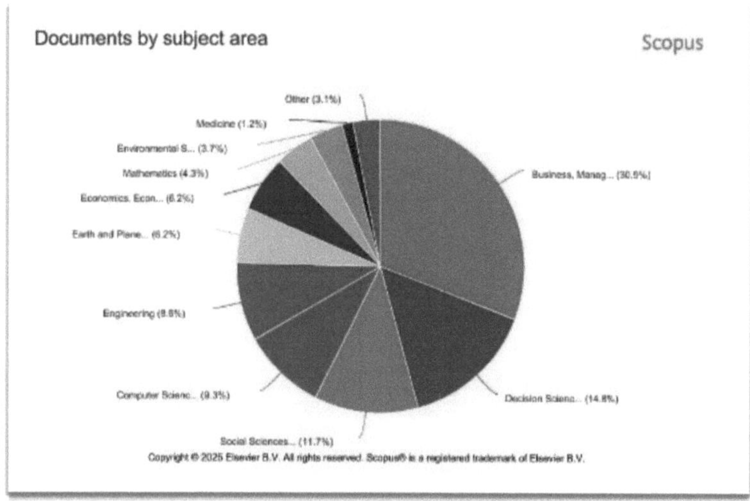

Fig. 3. Distribution of Publications by Different Disciplines (%).

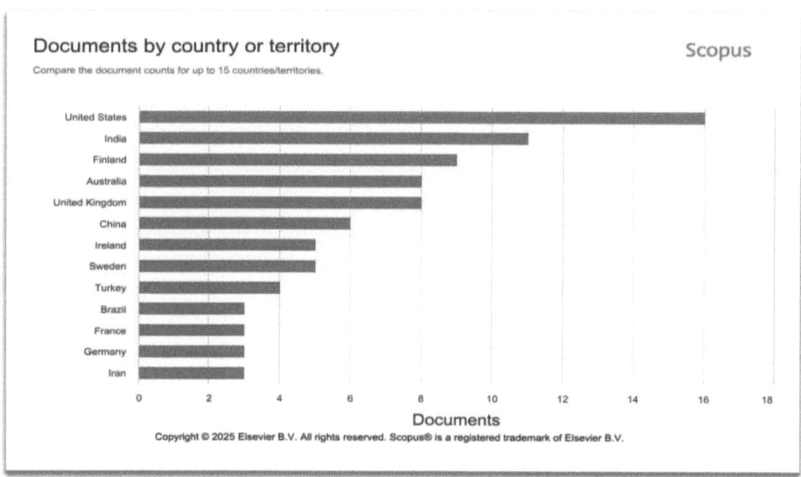

Fig. 4. Distribution of Documents Published According to Country.

The number of documents produced by the top five countries is significantly higher than in other countries. The following documents, published throughout the last 3 years, are presented in chronological order.

Sinha et al. [10] emphasized the importance experts place on herd immunity in COVID-19. They underlined that experts stated that herd immunity can be achieved by vaccinating 67% of the population. In this context, they focused on the vaccine supply chain in India. Through a case study on the distribution of Japanese Encephalitis vaccine, the study identified the conditions under which a strategic inventory reserve policy cannot be practically implemented to meet service level targets. A new modeling approach was used to identify critical nodes. Olivios et al. [11] emphasize that increasing disaster response capability has the potential to improve relief operations by reducing the impact of a disaster. In this context, this study proposes strategies to improve process management. They also emphasize the need for governments to engage in strategic logistics planning covering prevention and response processes. Sánchez-Partida et al. [12] are interested in to assist national strategic decisions in the planning and coordination of humanitarian relief operations. To achieve this objective, k-cluster analysis was conducted on municipalities in Mexico, utilizing 11 distinct statistical clustering methodologies. Prakash et al. [13] focused on reducing inter-organizational communication conflicts in humanitarian logistics by drawing on governance theories.

In order to develop a perspective in this context, an online survey was conducted to 289 field managers working in the field of humanitarian logistics. Mutebi et al. [14] highlighted that the multitude of institutions involved in aid logistics and their propensity for cross-border authority result in role uncertainty. In this context, they emphasized how organizational networks and organizational learning, as metaphors for complex adaptive systems, improve both organizational adaptability and role clarity in humanitarian logistics. Ordinary partial least squares regression has been used to evaluate survey data collected from 315 respondents across 101 distinct organizations. Common method bias (CMB) was mitigated through the application of both procedural and post-statistical techniques. Fiorini et al. [15] draw attention to the importance of humanitarian operations. In this context, studies addressing the humanitarian aspects of the humanitarian supply chain have been analyzed. As a result, it was found that human resource management and related courses are critical in operational processes in the humanitarian supply chain. Eligüzel and Özceylan [16] proposed a classification review in humanitarian logistics in their study. In order to achieve this, papers from the Scopus database were gathered and examined using appropriate keywords. They employed a text mining methodology utilizing Latent Dirichlet Allocation (LDA) and Fuzzy C-means. Liao et al. [17] analyzed the performance of three different post-disaster humanitarian logistics structures (PD - HL) with an agent-based simulation. The study analyzes the role of information sharing and crowd movement in the distribution of relief supplies in the Agency Centric Efforts (ACEs), Partially Integrated Efforts (PIEs) and Collaborative Assistance Networks (CANs) structures and assesses the ability to respond to disasters of different intensity. A real-world application of the proposed simulation is also demonstrated by considering humanitarian relief operations during the 2010 Port-au-Prince earthquake. Chang and Chen [18] conducted a literature review in the field of simulation learning and optimization (SLO). This paper focuses on potential research topics in SLO from different perspectives. Bonku et al. [19] studied synchronized vehicle routing for fair and efficient food distribution. In this context, they proposed a new arrow-objective Mixed Integer Programming (MIP) model. The objective of the proposed model is to minimize routing distance and wastage. They also evaluated the performance of the new approach with three different models using effectiveness, equity and efficiency measures. Wu et al. [20] highlight the necessity of standardizing the emergency supplies preparedness (ESP) process to effectively regulate the distribution of emergency supplies. This study methodically examines supplier default within emergency supply chains (ESCs) and investigates the underlying causes of such defaults. Kavota et al. [21] conducted a systematic review of supply chain challenges and their reality among universities and industry professionals. In this context, a comprehensive review of 118 articles was conducted with a co-citation and concept-centered approach. Thus, critical challenges in the modern supply chain were identified. Shrivastav and Bag [22] worked on identifying trends in Humanitarian supply chain management (HSCM) in the digital age. For this purpose, various data from literature, social media blogs and forums were used using topic modeling. As a result of this research, eight themes were identified. In addition, suggestions for future research were made for each theme. Hartama et al. [23] developed a novel Mixed Integer Nonlinear Programming (MINLP) framework that incorporates dynamic factors such as road closures and evolving priorities in disaster management processes. In this study, they presented the Cumulative Capacity Vehicle Routing Problem (CCVRP-TD) model proposal that takes into account capacity and time dependencies. Malhouni and Mabrouki [24] aim to apply complex adaptive systems (CAS) theory in their study. In this context, the study presents strategies to improve disaster resilience. Shaw et al. [25] investigated the factors that determine the ability of spontaneous volunteer groups (SVGs) to sustain their activities. The aim of this research is to contribute to recovery and mitigation by increasing resilience during and after disasters. In this context, qualitative interviews were conducted with SVGs in three British regions that experienced major floods. Thematic analysis was conducted with these qualitative data. Martin-Campo et al. [26] concentrated on the administration of emergency personnel in their research. A mathematical model is presented that identifies the necessary human profiles and reduces overall costs.

As a result of the literature studies, many studies related to the field were found. However, it has been revealed that the competencies required for smart and sustainable humanitarian logistics have not been identified. The findings obtained in this study contribute by filling this gap in literature by considering digital and green competencies.

3 Methodology

This section details the data collection process and analysis methods to identify key competencies for competency analysis. Initially, the current scholarly literature on humanitarian logistics, sustainability, and digital transformation is reviewed. Subsequently, current undergraduate and graduate programs in humanitarian relief logistics and their anticipated consequences are assessed. After the review and assessment, a questionnaire was prepared to be presented to the relevant experts. The questionnaire developed based on the literature studies [15–17]. A comprehensive version of the questionnaire has been created with the assistance of expert feedback. Upon approval of the questionnaire through expert evaluations, the results are derived from its application.

Subsequent to a literature review and analysis of relevant curricula, the preliminary questionnaire was dispatched to experts for evaluation. Experts were chosen from individuals with academic experience in humanitarian aid logistics, and their comments was gathered. The experts' perspectives significantly influenced the structure of the questionnaire sections designed to assess competency in sustainable, digital, and smart humanitarian logistics. The finalized comprehensive version of the questionnaire was prepared for distribution to the experts, incorporating the recommendations for modifications to the survey questions.

Descriptive statistics were derived from the responses collected using a questionnaire designed with Google Forms. The statistics were examined and evaluated through graphs generated on Google Forms. At this point, responses were analyzed based on frequency and trends. The results and discussion section contains a comprehensive account of the findings and interpretations.

3.1 Scope of the Proposed Approach

In contrast to the literature, the approach proposed in this study consists of 5 scopes as indicated in Fig. 5. The description of the scopes is explained.

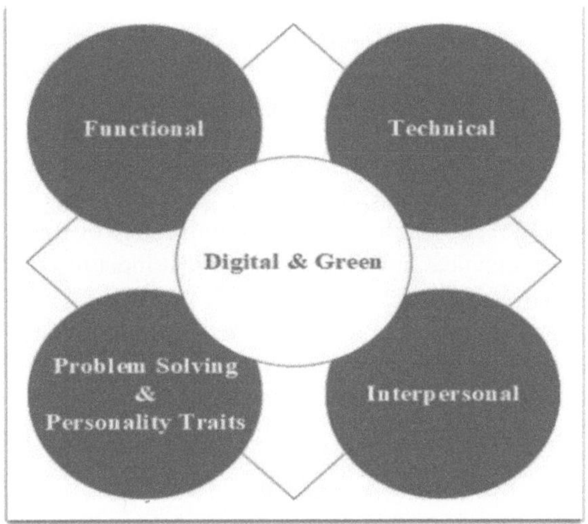

Fig. 5. Scope of The Proposed Approach.

Interpersonal Skills Interpersonal skills, particularly in disaster and crisis scenarios, are essential for teams to make fast and accurate decisions. Effective communication, collaboration, and coordination are essential for the successful implementation of processes during crises. This part emphasizes teamwork, leadership, and effective communication competencies in humanitarian logistics [28, 30].

Problem Solving and Personality Traits Humanitarian logistics entails the management of uncertain and constantly evolving situations, making psychological traits such as problem identification, analytical thinking, problem-solving, decision-making under pressure, resilience, and sensitivity crucial. This chapter emphasizes competencies including problem-solving, stress management, and analytical thinking in crisis scenarios [28, 30].

Technical In humanitarian logistics, expertise in human resources management and quality management significantly influences operational operations. This part emphasizes technical competencies in domains such as finance, project management, and information technology [29].

Functional Effective and efficient execution of the humanitarian logistics process requires skills including inventory management, supply chain management, warehousing, and strategic decision making. This part seeks to evaluate the fundamental competencies necessary for humanitarian logistics operations. This part emphasizes strategic methodologies and competencies that will guarantee the successful execution of the processes [28, 30].

Digital and Green Skills In humanitarian logistics, it is essential to identify ecologically sustainable solutions and effectively integrate digital tools and technologies. Implementing zero waste regulations, optimizing energy resource utilization, and integrating digital solutions enhance operational sustainability. This reduces the environmental effect of operations and facilitates digital transformation. The topic emphasizes digital literacy and competencies for developing sustainable logistics solutions.

4   Results and Discussion

This section analyzes the findings of the questionnaire designed to identify the competences required for those working in smart and sustainable humanitarian logistics. The questionnaire developed for this purpose was administered to 180 individuals employed in the private sector, academic institutions, and non-governmental organizations (NGOs) within the relevant field of work. The results of the questionnaire, structured to assess competencies across five categories as detailed in prior sections through the analysis of literature and expert insights, are as follows:

Figure 6 presents the demographic data of the participants. Upon analyzing Graph A, it is apparent that 38.9% of the respondents have a master's degree, while 33.3% have a doctorate degree.

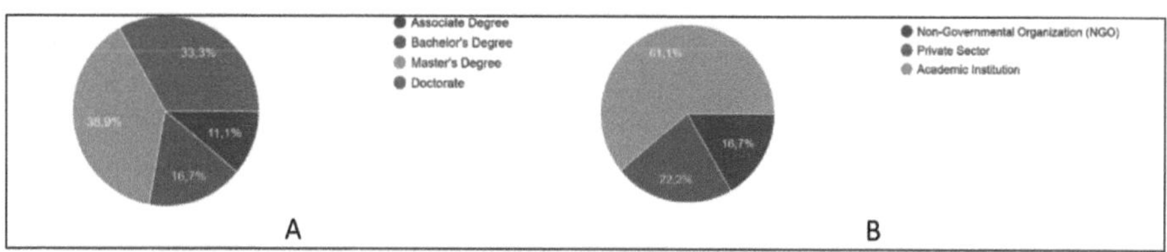

Fig. 6. Demographic Information of The Participants.

The presence of 72.2% of individuals having master's and doctorate degrees signifies a substantial degree of theoretical knowledge. Upon analyzing Graph B, it is apparent that 61.1% of the participants are employed at academic institutions. This rate corroborates the conclusion that the survey participants possess a substantial level of theoretical knowledge. Nonetheless, the involvement of the private sector and NGOs, although limited in comparison to academic institution employees, offers a comprehensive perspective on the primary aim of the survey: to identify the competencies required for individuals working in the domain of smart and sustainable humanitarian logistics. This comprehensive viewpoint is essential for enhancing the consistency of the survey findings.

Furthermore, respondents from the private sector and NGOs organizations were predominantly chosen from among the administrators of their respective organizations. This is due to the belief that managers with extensive expertise in the industry, who have closely observed the traits of persons in the sector, will significantly influence the consistency and quality of the survey results. Moreover, the survey primarily targeted professionals inside institutions and organizations functioning in the international sphere. This has guaranteed that the survey results exhibit comparable effects globally, allowing utilization in analogous situations despite variations in location and socio-cultural contexts.

Figure 7 presents the results of the survey questions in which respondents interpreted interpersonal skills. In this section interpersonal skills were asked in the form of 6 different skills: communication, leadership, internal and external coordination, negotiation, people (line) management, conflict management. As can be seen from the figure, more than 60% of the participants strongly agreed that a person working in the relevant field should have communication, internal and external coordination and people (line) management skills. For negotiation and conflict management skills, the strong agreement rate is 40% and above.

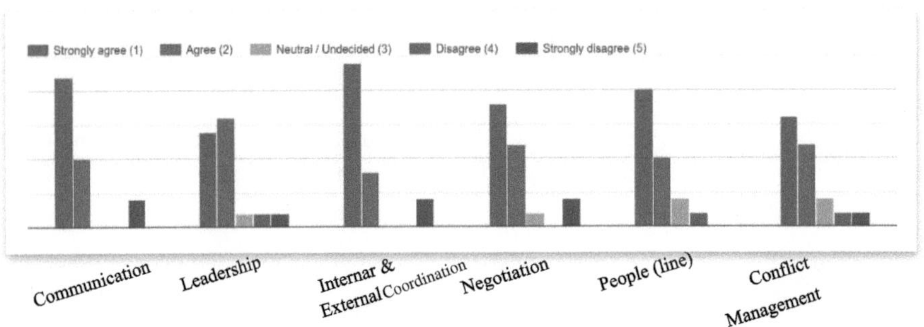

Fig. 7. Evaluation of Interpersonal Skill Importance in Humanitarian Logistics.

Regarding leadership skills, the number of respondents who strongly agree exceeds the number of respondents who strongly disagree. Overall, the results show that more than 75% of the participants agreed that all six skills should be possessed. These findings suggest that all six skills should be prioritized for smart sustainable humanitarian logistics training.

Figure 8 shows the results of the problem solving and personality traits section where 10 different skills were presented to the participants. The necessity of stress management, problem solving, analytical thinking, problem identification and analysis skills was strongly agreed upon by more than 60% of the respondents. In addition, the need for information gathering and analysis, self-understanding and flexibility, effective values, empathy and information literacy skills were strongly agreed upon by over 40%. Along with this, 35% agreed and strongly agreed in equal numbers on the necessity of cultural sensitivity and global citizenship skills. As a result of the survey, 70% or more agreed and strongly agreed on all skills.

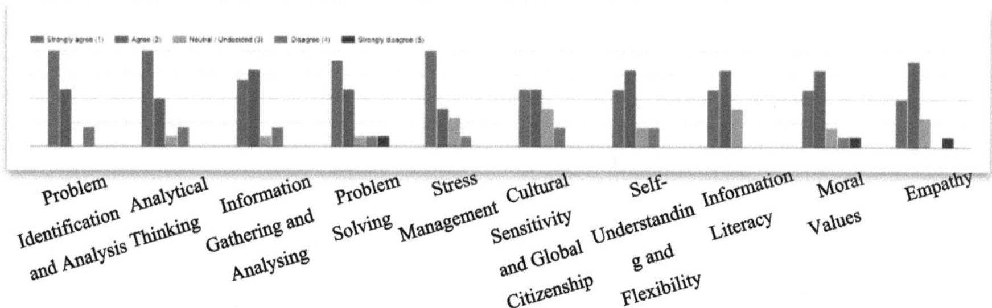

Fig. 8. Problem Solving and Personality Traits Importance in Humanitarian Logistics.

While this rate shows that all ten skills identified for the problem solving and personality traits section are necessary, the three skills that stand out are stress management, analytical thinking, problem identification and analysis. The technical skills shown in Fig. 9 were presented to the participants with 7 different skills. Strategic management emerged as the most important skill with more than 70% of respondents strongly agreeing. However, more than 40% of the respondents strongly agreed that information technology, human resource management and risk management skills are necessary.

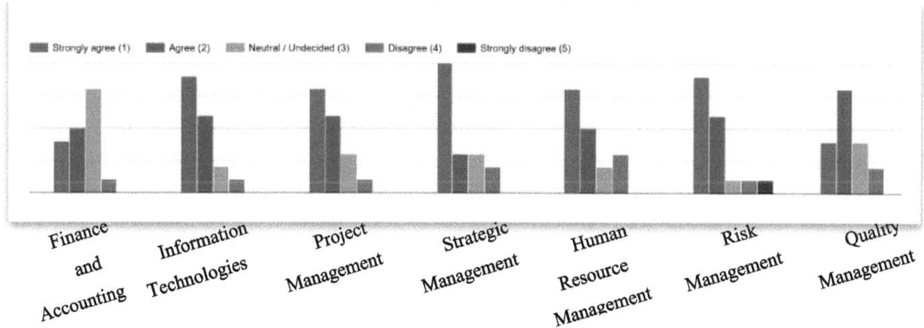

Fig. 9. Evaluation of Technical Skill Importance in Humanitarian Logistics.

In the case of quality management, the number of respondents who strongly agreed and those who abstained were equal, but the number of respondents who agreed with its necessity was above 40%. In contrast to the other skills, more than 40% of the respondents abstained from the necessity of finance and accounting skills. This indicates that finance and accounting should be addressed as a topic of discussion in humanitarian logistics.

In conclusion, the majority of the respondents agree that skills other than finance and accounting are necessary. Therefore, the acquisition of these skills should be prioritized in relevant training.

As can be seen from Fig. 10, which shows the functional skills, decision-making is the most prominent skill with more than 70% of the respondents strongly agreeing that it is necessary.

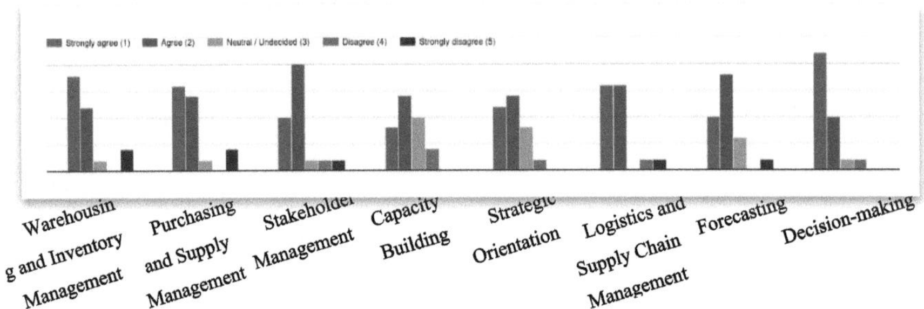

Fig. 10. Evaluation of Functional Skill Importance in Humanitarian Logistics.

In addition, except for capacity management and strategic orientation, all of the skills clearly have a strong majority of strongly agree and agree options. In contrast to the other skills, the number of participants who abstained in capacity management and strategic orientation skills differed slightly from the number of participants who agreed and strongly agreed. However, as a result, the majority of respondents think that all eight skills presented as functional skills are necessary.

The digital and green skills section is presented in Fig. 11. The results show that over 60% of the respondents voted that all five skills presented in the survey are necessary.

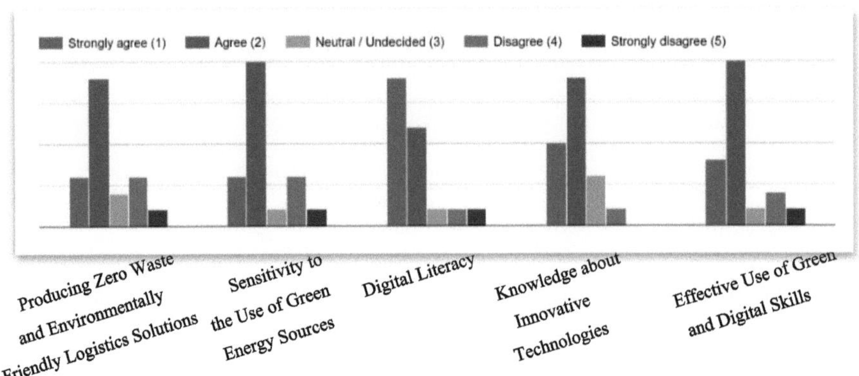

Fig. 11. Evaluation of Digital and Green Skill Importance in Humanitarian Logistics.

Of these skills, only the digital literacy skill was strongly argued to be necessary by the majority of respondents. At this point, it is understood that digital literacy is the most important skill among these five skills for the relevant section.

At the end of the questionnaire, the participants were asked whether there was a skill that they would like to add or that they deemed necessary apart from the skills mentioned above. The following three responses were received from the participants in response to this question:

- Resilience Planning, Circular Economy Knowledge and Social Equity Awareness
- Compliance Management
- In the logistics sector, especially map knowledge is important and also following current global events are two often forgotten but simple topics to know.

The study revealed that the majority of respondents strongly agreed or agreed with the necessity of 35 out of 36 abilities across five distinct categories. The majority of respondents were ambivalent regarding the proficiency of finance and accounting within the technical skills category. This table substantiates the conclusion that the questionnaire administered to participants aligns with existing research, as it was developed through a synthesis of scholarly publications on humanitarian aid logistical operations and expert insights. A significant contribution to the research has been achieved by incorporating new competencies and categories derived from expert opinions alongside existing competencies identified in the literature.

The findings are important to understand what competencies an individual working in the field of smart and sustainable humanitarian logistics should have in terms of interpersonal, problem-solving, technical, functional and digital-green skills. With the opinions of experts from different sectors, the main skills that training to be provided in this field should focus on have emerged. Thus, it is understood that the findings of the study have a high potential to shed light and pioneer future studies in this field.

The competences revealed in the study serve as primary indicators for shaping curriculum aimed at preparing students for participation in sustainable, intelligent humanitarian logistics operations. The accomplishments of the course combinations in students' curricula aiming to attain proficiency in the pertinent field should encompass the abilities acquired during the study. The recognized abilities serve as a resource for developing new course content or utilizing current course materials in the pertinent sector. These competencies can be integrated into educational programs by revising existing curricula to incorporate courses that encompass these competencies or by developing new courses that offer these competencies. Furthermore, current employees can be evaluated by specialists using diverse analytical methods concerning the competencies identified in this study, thereby uncovering and addressing any deficiencies in competencies. All these contributions underscore the significance of the observed findings.

5 Conclusions

Humanitarian relief logistics operations are critical, time-sensitive endeavors that need prompt response. Consequently, the expertise of the persons overseeing these operations is vital. In this context, persons must be educated to gain the requisite abilities, and the weaknesses of current personnel must be addressed. A thorough competency analysis is necessary to deliver this kind of education. This necessity has motivated this research. A thorough competencies analysis has been performed in the field of sustainable, smart humanitarian relief logistics.

The competence analysis in the related field first includes a comprehensive review of the literature. Following the literature review, a questionnaire was prepared including five different fields and the specific skills of each field. A combination of literature review and expert feedback was used to develop the questionnaire. As a result, the study was completed, with 180 participants completing the questionnaire and analyzing the results. The analysis utilized descriptive statistics derived from the survey and graphs that visually represented the data acquired using Google Forms. The survey findings were analyzed based on the frequency and trends of the responses given by participants about the competence. The study yielded significant discoveries that intersect with and enhance the existing literature.

When the results were analyzed, it was observed that only the finance and accounting skill from the technical skill group had a majority of undecided respondents. This highlights the dilemma as to whether humanitarian finance and accounting is a priority. In addition, it can be said that the reason why almost all of the skills presented in the questionnaire were found necessary by the participants is that these skills were determined by literature research and expert opinions.

The number of people to whom the questionnaire was applied and the fact that the institutions of the sector employees generally serve in the same region, the lack of diversity in the analysis methods and the fact that field workers other than managers are less involved are the weaknesses of the study. Moreover, the study's strengths encompass the questionnaire's the base on a comprehensive literature review and expert insights, the international scope of the respondents, and the inclusion of individuals engaged in various facets of the pertinent sector. The study's novel features include the exploration of sustainability and smart concepts within the relevant domain, as well as the inclusion of digital and green talents in the survey's skill category. Evaluating the results of the study by using various analysis methods, diversifying the people to whom the survey was applied and giving it a global dimension, using different scales, determining competency groups and the competencies covered by these groups in separate studies, using different research methods and discussion of competencies using a real disaster scenario are seen as possible future studies.

Acknowledgments. This publication is produced as part of the Erasmus+ project titled 'Sustainable, Digital, and Smart Humanitarian Logistics in Disaster Relief Operations (SMART-LOG4DIS)', under the call ERASMUS-EDU-2023-EMJM-DESIGN with project number 101127321.

Disclosure of Interests The authors declare no potential conflicts of interest with respect to the research, authorship, and/or publication of this paper.

References

1. Yuksel, Z., Epcim, D., Mete, S.: Multi-depot vehicle routing problem with drone collaboration in humanitarian logistics. J. Optim. Decis. Mak. **3**(1), 438–448 (2024)
2. Yüksel, Z., Eliguzel, N., Mete, S.: Leveraging latent Dirichlet allocation and fuzzy clustering for identifying key UAV applications in disaster response. Nat. Sci. Eng. Bull. **1**, 17 (2024)
3. Apte, A., Gonçalves, P., Yoho, K.: Capabilities and competencies in humanitarian operations. J. Human. Logist. Suppl. Chain Manage. **6**(2), 240–258 (2016). https://doi.org/10.1108/JHL SCM-04-2015-0020
4. Yüksel, Z., Epcim, D., Mete, S.: First cluster second route approach with collaboration unmanned aerial vehicle in post-disaster humanitarian logistic. JTL. **8**(2), 97–111 (2024). https://doi.org/10.26650/JTL.2023.1372701
5. Thomas, A., Mizushima, M.: Logistics training: necessity or luxury. Forced Migr. Rev. **22**(22), 60–61 (2005)
6. A. Vaillancourt, P. Tatham, and L. Seitz: A Review of Supply Chain and Logistics Competencies for the Humanitarian Logistics Field (2015)
7. Hosseinabadi, A.A.R., Vahidi, J., Balas, V.E., Mirkamali, S.S.: OVRP_GELS: solving open vehicle routing problem using the gravitational emulation local search algorithm. Neural Comput. Applic. **29**(10), 955–968 (2018). https://doi.org/10.1007/s00521-016-2608-x
8. Hosseinabadi, A.A.R., Zolfagharian, A., Alinezhad, P.: An efficient hybrid meta-heuristic algorithm for solving the open vehicle routing problem. In: Shahbazova, S.N., Kacprzyk, J., Balas, V.E., Kreinovich, V. (eds.) Recent Developments and the New Direction in Soft-Computing Foundations and Applications: Selected Papers from the 7th World Conference on Soft Computing, May 29–31, 2018, Baku, Azerbaijan, pp. 257–274. Springer International Publishing, Cham (2021). https://doi.org/10.1007/978-3-030-47124-8_21
9. S.-C. Hsieh, J.-S. Lin, and H.-C. Lee, "Analysis on Literature Review of Competency," 2012
10. Sinha, P., Kumar, S., Chandra, C.: Strategies for ensuring required service level for COVID-19 herd immunity in Indian vaccine supply chain. Eur. J. Oper. Res. **304**(1), 339–352 (2023). https://doi.org/10.1016/j.ejor.2021.03.030
11. Cano-Olivos, P., Sánchez-Partida, D., Caballero-Morales, S.-O., Martínez-Flores, J.-L.: Strategies that improve the performance of the humanitarian supply chain. In: Regis-Hernández, F., Mora-Vargas, J., Sánchez-Partida, D., Ruiz, A. (eds.) Humanitarian Logistics from the Disaster Risk Reduction Perspective : Theory and Applications, pp. 119–140. Springer International Publishing, Cham (2022). https://doi.org/10.1007/978-3-030-908 77-5_3
12. Sánchez-Partida, D., Mora-Vargas, J., Smith, N.R., Castillo-Villar, F.: Clustering of highly vulnerable Mexican municipalities to develop humanitarian public policies. In: Regis-Hernández, F., Mora-Vargas, J., Sánchez-Partida, D., Ruiz, A. (eds.) Humanitarian Logistics from the Disaster Risk Reduction Perspective: Theory and Applications, pp. 25–118. Springer International Publishing, Cham (2022). https://doi.org/10.1007/978-3-030-90877-5_2
13. Prakash, C., Roy, V., Charan, P.: Mitigating interorganizational conflicts in humanitarian logistics collaboration: the roles of contractual agreements, trust and post-disaster environmental uncertainty phases. IJLM. **33**(1), 28–52 (2022). https://doi.org/10.1108/IJLM-06-2021-0318
14. Mutebi, H., Muhwezi, M., Ntayi, J.M., Mayanja, S.S., Munene, J.C.K.: Organisational networks, organisational learning, organisational adaptability and role clarity among humanitarian organisations during relief delivery. JHLSCM. **12**(2), 249–284 (2022). https://doi.org/10.1108/JHLSCM-04-2021-0034
15. De Camargo Fiorini, P., Chiappetta Jabbour, C.J., Jabbour, A.B.L.D.S., Ramsden, G.: The human side of humanitarian supply chains: a research agenda and systematization framework. Ann. Oper. Res. **319**(1), 911–936 (2022). https://doi.org/10.1007/s10479-021-03970-z

16. Eligüzel, İ.M., Özceylan, E.: Classification of fuzzy MCDM literature applied to humanitarian logistics problems. In: Kahraman, C., Sari, I.U., Oztaysi, B., Cebi, S., Cevik Onar, S., Tolga, A.Ç. (eds.) Intelligent and Fuzzy Systems, pp. 344–352. Springer Nature Switzerland, Cham (2023). https://doi.org/10.1007/978-3-031-39777-6_42

17. Liao, H., Holguín-Veras, J., Calderón, O.: Comparative analysis of the performance of humanitarian logistic structures using agent-based simulation. Socioecon. Plann. Sci. **90**, 101751 (2023). https://doi.org/10.1016/j.seps.2023.101751

18. Chang, K.H., Chen, T.L.: Simulation learning and optimization: methodology and applications. Asia Pac. J. Oper. Res., 2440008 (2024). https://doi.org/10.1142/S0217595924400086

19. Bonku, R., Alkaabneh, F., Davis, L.B.: Collaborative vehicle routing for equitable and effective food allocation in nonprofit settings. J. Human. Logist. Suppl. Chain Manage. (2024). https://doi.org/10.1108/JHLSCM-11-2023-0113

20. Wu, X., Yang, M., Wu, C., Liang, L.: How to avoid source disruption of emergency supplies in emergency supply chains: a subsidy perspective. Int. J. Disaster Risk Reduct. **102**, 104303 (2024). https://doi.org/10.1016/j.ijdrr.2024.104303

21. Kavota, J.K., Cassivi, L., Léger, P.-M.: A systematic review of strategic supply chain challenges and teaching strategies. Logistics. **8**(1), 19 (2024). https://doi.org/10.3390/logistics8010019

22. Shrivastav, S.K., Bag, S.: Humanitarian supply chain management in the digital age: a hybrid review using published literature and social media data. BIJ. **31**(7), 2267–2301 (2024). https://doi.org/10.1108/BIJ-04-2023-0273

23. Hartama, D., Wanayumini, W., Damanik, I.S.: The cumulative capacitated vehicle routing problem with time-dependent on humanitarian logistics for disaster management. J. Appl. Data Sci. **6**(1), 1 (2025). https://doi.org/10.47738/jads.v6i1.481

24. Malhouni, Y., Mabrouki, C.: Whole-of-government approach in disaster management: collaborative case study on the 2023 Morocco earthquake response. J. Human. Logist. Suppl. Chain Manage. (2025). https://doi.org/10.1108/JHLSCM-10-2024-0152

25. Shaw, D., Zanjirani Farahani, R., Scully, J.: Sustaining spontaneous volunteer groups following their response to a disaster. IJOPM. (2024). https://doi.org/10.1108/IJOPM-09-2023-0778

26. Martin-Campo, F.J., Sánchez, M.T.O., Ruiz-Gonzalez, B.: Medical staff planning for field hospital deployments: the START hospital. J. Human. Logist. Suppl. Chain Manage. **15**(1), 4–17 (2024). https://doi.org/10.1108/JHLSCM-03-2024-0043

27. Bölsche, D., Klumpp, M., Abidi, H.: Specific competencies in humanitarian logistics education. J. Human. Logist. Suppl. Chain Manage. **3**(2), 99–128 (2013). https://doi.org/10.1108/JHLSCM-08-2012-0019

28. Rajakaruna, S., Wijeratne, A.W., Mann, T.S., Yan, C.: Identifying key skill sets in humanitarian logistics: developing a model for Sri Lanka. Int. J. Disaster Risk Reduct. **24**, 58–65 (2017). https://doi.org/10.1016/j.ijdrr.2017.05.009

29. Y. Meduri and F. A. Ahmed, "Talent Needs Assessment of Humanitarian Logistician: An Empirical Investigation," 2015

30. Kovács, G., Tatham, P., Larson, P.D.: What skills are needed to be a humanitarian logistician? J. Bus. Logist. **33**(3), 245–258 (2012). https://doi.org/10.1111/j.2158-1592.2012.01054.x

Smart Transportation and Logistics Systems

Integrating Pedestrian and Scooter Traffic: A Model for Safe Urban Mobility

Serap Ergün$^{(\boxtimes)}$

Department of Computer Engineering, Faculty of Technology, Isparta University of Applied Sciences, Isparta 32260, Turkey
serapbakioglu@isparta.edu.tr

Abstract. In recent years, the use of personal mobility devices, particularly electric scooters, has become increasingly popular, especially for short-distance travel commonly referred to as the "last mile." This growing trend has led to complex interactions between scooter riders and pedestrians, as pedestrians often exhibit unpredictable behavior, such as remaining stationary or walking at varying speeds, which increases the risk of collisions with fast-moving scooters. These interactions can lead to disturbances in pedestrian flow, irritation, or even accidents, particularly in environments lacking effective regulation or infrastructure. This study presents an advanced prediction model for pedestrian movement and the dynamics between pedestrians and electric scooters. The model is an enhanced version of the Social Force Model, incorporating the distinct movement characteristics of electric scooters, such as high speed and rapid acceleration, especially on curved roads. Simulation results showed that in low-density pedestrian environments, personal mobility devices can overtake pedestrians without significant disruptions, whereas in high-density areas, scooters and Segways face challenges overtaking pedestrians due to congestion. This study also highlights that pedestrian density plays a crucial role in the movement dynamics of electric scooters, with congestion reducing their ability to navigate efficiently. The results provide valuable insights into pedestrian safety risks, such as collisions and congestion, and offer a deeper understanding of mobility dynamics in urban environments. The findings emphasize the need for improved infrastructure and regulation to ensure safe and efficient coexistence of pedestrians and personal mobility devices.

Keywords: Personal Mobility Devices · Pedestrian Traffic Modeling · Shared Urban Spaces · Social Force Model

1 Introduction

The rapid urbanization of cities, along with an increasing demand for sustainable transportation solutions, has led to the widespread adoption of personal mobility devices (PMDs), particularly electric scooters. These devices, preferred for their efficiency in short-distance travel, are especially popular for "last mile" mobility, providing an alternative to traditional forms of transportation such as cars and public transit (Matsumoto

A. Mirzazadeh et al. (Eds.): ODSIE 2024, CCIS 2482, pp. 197–216, 2026.
https://doi.org/10.1007/978-3-031-93601-2_13

et al., 2020). Electric scooters have quickly become an integral part of urban transportation, offering a flexible, convenient, and eco-friendly solution, which has resulted in a growing number of scooter users across cities worldwide (Zhang et al., 2020).

However, as the use of electric scooters increases, so do the challenges associate with their interaction with pedestrians in shared public spaces. The risk of accidents and collisions rises as pedestrians and PMD users occupy the same urban spaces, often with limited infrastructure to separate them (Chien et al., 2019). Pedestrians, who typically move at lower speeds, often exhibit unpredictable behavior such as sudden stops, changes in direction, or erratic movement, which can create dangerous situations when combined with the rapid acceleration and high speeds of electric scooters (Xie et al., 2019). Such interactions between pedestrians and scooters can cause injuries, hinder the flow of traffic, and disrupt urban mobility (Briggs et al., 2019).

To address these challenges, it is crucial to develop models that can predict pedestrian movement and interactions with PMDs. Traditional traffic flow models, often designed for vehicles or stationary pedestrians, are inadequate for simulating the complex dynamics of mixed traffic scenarios involving pedestrians and moving personal mobility devices (Helbing et al., 2001). These models fail to capture the dynamic and stochastic behavior of pedestrians in combination with the high-speed, accelerated motion of electric scooters. This gap in current modeling highlights the need for a more comprehensive approach to simulate the interactions between pedestrians and PMDs in shared spaces.

In this paper, the Social Force Model (SFM), a widely used model in pedestrian dynamics (Helbing & Molnar, 1995), provides a solid foundation for modeling pedestrian behavior and their interactions with external agents such as PMDs. However, this model traditionally lacks the capability to account for the unique characteristics of PMDs, such as high-speed movement, rapid acceleration, and navigational constraints, especially in mixed traffic conditions. To address this limitation, this study proposes an enhanced version of the Social Force Model, incorporating dynamic differences between pedestrians and electric scooters. This modified model will incorporate key characteristics of electric scooters, such as their ability to accelerate quickly and navigate curved roads, to more accurately reflect real-world interactions between pedestrians and scooters in urban environments.

By integrating these factors, this model aims to provide a more realistic prediction of pedestrian movement, taking into account the specific dynamics of electric scooters. This study will explore how scooter speed, acceleration, and pedestrian density influence interactions between pedestrians and PMDs in mixed traffic environments. Through multi-agent simulations, this research will offer insights into how pedestrian and scooter dynamics can be better understood and optimized to improve safety and comfort for all users of urban spaces.

Ultimately, this study contributes to the growing body of research on urban mobility and transportation safety. It provides a framework for understanding the challenges posed by personal mobility devices, offering insights for improving infrastructure, traffic flow, and safety measures. By better understanding the interactions between pedestrians and electric scooters, urban planners and policymakers can make informed decisions on how to manage these shared spaces and ensure safe coexistence between pedestrians and PMD users (Chien et al., 2019; Zhang et al., 2021).

The organization of the paper is constructed as follow: The detailed literature review with research background and objectives is presented in Sect. 2. Then the paper goes with an introduction to the core simulation models used in the study, starting with the Social Force Model (SFM) in Sect. 3. This section presents the mathematical foundations of the SFM and its adaptations, including modifications that account for anisotropy and relative velocity between pedestrians, Segways, and electric scooters. A detailed explanation of the model improvements and the adjustments made for pedestrian interaction forces follows, with a focus on the introduction of hyperbolic ranges for force application. The paper then proceeds to specific models for Segways and electric scooters, detailing their behavior within the simulation and how their movement characteristics differ from pedestrians. Section 4 introduces the mixed traffic simulation scenario, presenting the experimental setup and results for low-density and high-density pedestrian flows, highlighting how the presence of personal mobility devices (PMDs) affects pedestrian movement and vice versa. In Sect. 5, the discussions and broader impacted are presented. Finally, the conclusion summarizes the key findings, emphasizing the impact of pedestrian density on the dynamics of PMDs and outlining the potential for future research to refine the model and expand its applications in urban mobility optimization.

2 Literature Review

Micromobility solutions such as e-scooters have significantly transformed urban transportation systems in recent years. Numerous studies have investigated their integration into urban environments, emphasizing safety, infrastructure design, user behavior, and simulation models. This section synthesizes key contributions from the literature.

Research themes that provide deeper insights into the findings and methodologies of studies on urban mobility, micromobility (especially e-scooters) and their integration into existing transportation systems are examined under sub-items.

Micromobility Usage and Behavioral Insights
E-Scooter Usage Patterns: A variety of studies explore the behaviors of e-scooter users, with particular attention to the demographics, frequency of use, and the duration of trips. Wolnowska & Kasyk (2022) found that younger populations tend to be the primary users of shared e-scooter systems, often for short-distance travel. Similarly, Nikiforiadis et al. (2021) explored the reasons why users prefer e-scooters, citing convenience, cost-effectiveness, and flexibility.

Impact of E-Scooter Usage on Transportation Preferences: Mitropoulos et al. (2023b) focused on the perceived safety of e-scooter users in Athens, providing a detailed analysis of factors like road design, user familiarity with the vehicle, and environmental conditions. Sellaouti et al. (2020) explored the operational feasibility of integrating e-scooter sharing systems, emphasizing user preferences for convenience, accessibility, and the integration of these services with public transport options.

Social and Psychological Factors: In their research, Pazzini et al. (2022) considered the social and psychological factors that influence users' preferences for e-scooter sharing. Their findings suggest that perceived ease of use and trust in the system significantly affect the adoption of e-scooters, while the integration of reward programs could enhance user engagement.

Micromobility Integration in Urban Transport

Multimodal Urban Mobility Systems: The integration of e-scooters into existing transportation frameworks is a major focus for many researchers. Jafari & Liu (2024a) proposed a futuristic sidewalk model designed for personal mobility vehicles, such as e-scooters, that seamlessly integrates with pedestrian infrastructure. This model aims to ensure safety and efficiency in cities where the demand for multimodal transport solutions is increasing.

Urban Transport Synergy: Felipe Andrade et al. (2023) provide a study from Brazil, highlighting the integration of micromobility in urban transport systems, especially in metropolitan areas with high congestion. They argue that the key to success lies in the complementary nature of micromobility and other transport modes (buses, metro, taxis) to reduce congestion and pollution.

Infrastructure for Micromobility: Infrastructure plays a crucial role in the effectiveness of e-scooter systems. Pedroza-Perez et al. (2024) emphasize the need for dedicated e-scooter lanes and charging stations to ensure safe and efficient operation. In parallel, Altintasi & Yalcinkaya (2022) address the necessity of investing in smart city technologies, like sensors and IoT devices, to monitor and optimize the usage of micromobility vehicles.

Safety and Risk Factors

E-Scooter Safety Models: E-scooter safety is a recurring theme in urban mobility research. Chengula et al. (2024) present predictive models to estimate the likelihood of e-scooter crashes, examining environmental variables such as road surface conditions, traffic density, and user behavior. These models allow urban planners to identify high-risk areas and implement mitigation strategies, like the redesign of traffic signals and better pavement maintenance.

Risk Assessment and User Perception: Tzouras et al. (2024) explored safety perceptions among e-scooter riders and pedestrians. Their findings suggest that riders in cities with a higher concentration of infrastructure dedicated to micromobility are more likely to feel safe. This emphasizes the importance of segregating scooter lanes from pedestrian paths and other vehicle lanes, which can reduce accidents and improve overall user experience.

Behavioral Insights and Safety Interventions: Chmiel et al. (2023) analyzed the different behaviors of e-scooter users, including risky behaviors such as riding on sidewalks or ignoring traffic signals. Their study suggests that educational programs and fines for violations could help reduce unsafe riding practices.

Sustainability and Environmental Impact

Evaluating Environmental Sustainability: A growing body of research examines how the environmental impacts of e-scooters compare to traditional vehicles. Ecer et al. (2023) introduced a sustainability performance framework that assesses micromobility systems based on factors such as energy consumption, carbon emissions, and waste generation.

This framework is useful for policymakers who are looking to promote more sustainable urban transport options.

Emissions and Operational Alternatives: Deveci et al. (2022) address operational aspects of e-scooter systems, using fuzzy logic models to propose safe and sustainable operational alternatives. These models help urban managers optimize fleet deployment and maintenance schedules to reduce energy consumption and environmental impact.

Urban Mobility and Environmental Policy: Jiao & Xu (2024) focused on how the increase in e-scooter use can lead to reduced greenhouse gas emissions if the shift is made from cars to shared micromobility. However, the study also cautioned that, without proper regulation, e-scooter fleets can increase environmental waste, especially with short-lived devices that are frequently replaced.

Simulation Models and Urban Mobility

Micromobility Simulation Models: Researchers have employed agent-based models (ABMs) to simulate interactions between e-scooter users, pedestrians, and other vehicles in an urban environment. Tzouras et al. (2024) used ABMs to simulate urban e-scooter traffic, revealing insights into how different traffic densities and road configurations impact safety and efficiency. These simulations are crucial for urban planners looking to optimize micromobility integration into existing transport systems.

Mixed Traffic and Congestion Simulation: Samaei (2023) analyzed co-simulation platforms to understand how e-scooters behave in mixed traffic with connected and automated vehicles (CAVs). The study provided insight into how micromobility can coexist with automated systems while maintaining safety and efficiency in crowded city spaces. This is especially relevant for cities that are adopting CAVs alongside other transport modes.

Virtual and Real-World Simulations: Zhao et al. (2022) used a hybrid approach that combines virtual simulations with real-world data to model the interaction of e-scooters with urban traffic. This methodology helps refine the results of purely theoretical models and offers a more accurate prediction of real-world outcomes for urban mobility systems.

Road Design and User Perception of Safety

Redesigning Roads for Safety: Several studies, including Papageorgiou et al. (2024) and Pastia et al. (2022), have emphasized the importance of road design in improving safety for micromobility vehicles like e-scooters. For instance, dedicated e-scooter lanes and safer intersections are essential to reduce accidents involving e-scooter riders. Valero et al. (2020) and Brunner et al. (2020) examined how the social force model can predict pedestrian and e-scooter interactions, suggesting that better planning and infrastructure are needed to minimize collision risks.

Safety Perceptions and Urban Environments: The perception of safety among e-scooter riders and pedestrians is often influenced by the surrounding urban environment. Mitropoulos et al. (2023a) and Sorkou et al. (2022) found that users tend to feel safer when riding in cities with well-maintained sidewalks, protected bike lanes, and traffic calming measures that segregate pedestrians, cyclists, and e-scooter users.

Regulatory Frameworks and Policy

Policy Considerations for E-Scooter Safety and Integration: Šucha et al. (2023) and

Farooq et al. (2017) explore how regulation can influence the safety and effectiveness of micromobility systems. Their studies suggest that e-scooter regulations should cover not only speed limits, parking zones, and insurance but also ensure that micromobility services are integrated into the larger urban transport strategy, encouraging users to switch from private cars to shared, sustainable options.

Cross-Country Comparison of Regulatory Frameworks: Zakharov & Fadyushin (2021) compared e-scooter regulations across several countries, concluding that while some cities have adopted clear guidelines for usage, many have failed to establish cohesive policies that integrate e-scooters into broader urban transport strategies. They propose a model of progressive regulation that evolves with the growing adoption of e-scooters and other micromobility vehicles.

The research on urban mobility and micromobility is increasingly focused on creating holistic, sustainable transport solutions that balance safety, environmental impact, and user convenience. Urban planning plays a central role in integrating e-scooters and micromobility vehicles with existing infrastructure, ensuring that road safety, user behavior, and regulatory frameworks are properly aligned. The use of simulation models to predict and mitigate risks, the exploration of multimodal integration, and the ongoing debate about regulatory frameworks provide a broad foundation for future urban transport research. As cities continue to grow and adapt, micromobility systems like e-scooters will likely become an increasingly important part of the urban mobility landscape, offering both challenges and opportunities for research, regulation, and design.

Table 1 summarizes key topics and findings from research on micromobility, urban infrastructure, and safety, offering a good reference for understanding the current state of research and its applications.

This study introduces a novel approach to understanding the dynamics of pedestrian and electric scooter interactions in urban environments, offering significant advancements over existing research. Unlike previous studies that broadly explore urban mobility or focus on safety concerns, our work zeroes in on the specific behaviors of pedestrians and electric scooters in mixed traffic settings, emphasizing the predictive modeling of these interactions. The Social Force Model is enhanced to more accurately capture the unique movement characteristics of electric scooters, providing a level of detail rarely seen in past research. Additionally, while other studies have examined micromobility simulations, few have integrated pedestrians and diverse personal mobility devices like scooters into multi-agent models that reflect real-world complexities. Furthermore, the future research plans to compare Segways and electric scooters, offering a deeper understanding of how different mobility devices affect pedestrian safety and comfort—an area not extensively explored in current literature.

This differentiation highlights this paper's contribution to advancing understanding in pedestrian-scooter dynamics, offering more accurate predictive models and setting the stage for future research into the role of different personal mobility vehicles in urban traffic systems.

2.1 Research Background and Objectives

In recent years, personal mobility, such as Segways and electric scooters, has been attracting attention as a new means of transportation suitable for last-mile transportation. In

Table 1. Key References on Pedestrian and Personal Mobility Device (PMD) Interactions

Research Topic	The Studies	Key Findings	Methods and Approaches	Results and Applications
Micromobility Usage and Behavior	Wolnowska and Kasyk (2022)	Younger users prefer e-scooters for short distances and convenient travel	User surveys, behavior analysis	E-scooter usage is growing among younger demographics, particularly for short-distance travel
Micromobility and Urban Transportation Integration	Nikiforiadis et al. (2021); Andrade et al. (2023)	Integration with public transport is feasible; multimodal systems are essential for efficiency	System modeling, user behavior analysis	Integration of e-scooters with urban transport can reduce congestion and improve mobility
Safety and Risk Factors	Tzouras et al. (2024); Chengula et al. (2024)	Safety perceptions depend on infrastructure and user experiences, with a need for more protection	Safety assessments, simulation models	Adequate infrastructure and safety protocols are critical to enhancing micromobility safety
Environmental Impact and Sustainability	Ecer et al. (2023); Deveci et al. (2022)	E-scooters have a lower carbon footprint than cars, but lifecycle emissions should be considered	Sustainability modeling, emissions analysis	E-scooters contribute to lower environmental impact compared to traditional transport methods
Urban Infrastructure and User Safety	Brunner et al. (2020); Papageorgiou et al. (2024)	Special lanes, parking facilities, and training programs are essential for safety	Infrastructure design analysis, safety audits	E-scooter usage is safer with dedicated infrastructure and user education programs

(continued)

France, the revised Road Traffic Act, which relaxes regulations on personal mobility, came into effect on July 1, 2023. In this revision, a new vehicle category called a specific small, motorized bicycle is created between light vehicles (such as bicycles) and motorized bicycles, and electric scooters that meet the requirements is able to ride on sidewalks if they meet the following two points.

- Maximum speed set at 6 km/h

Table 1. (*continued*)

Research Topic	The Studies	Key Findings	Methods and Approaches	Results and Applications
Micromobility and Urban Design	Mitropoulos et al. (2023a); Pazzini et al. (2022)	Urban design must adapt to accommodate e-scooters and optimize traffic flow	Urban planning analysis, user surveys	Designing cities with micromobility in mind enhances safety, convenience, and efficiency
Regulation and Policy for Micromobility	Šucha et al. (2023); Zakharov & Fadyushin (2021)	Regulation impacts safety and frequency of use; policies can optimize micromobility adoption	Policy reviews, international comparison	Policies governing micromobility usage can drive its adoption and ensure user safety
Environmental Sustainability of Micromobility	Jiao & Xu (2024); Deveci et al. (2022)	Micromobility can help cities reduce traffic congestion and emissions	Environmental impact analysis, modeling	E-scooters help cities achieve sustainability targets by reducing carbon emissions and traffic
Simulation Models and Micromobility	Samaei (2023); Tzouras et al. (2024)	Agent-based and co-simulation models improve understanding of urban transportation interactions	Simulation modeling, agent-based models	Simulation models can predict interactions between vehicles and micromobility devices, improving city planning
Technology and Data in Micromobility	Chengula et al. (2024); Nikiforiadis et al. (2021)	Use of big data can optimize micromobility systems, but data privacy must be ensured	Data analytics, system optimization	Big data can improve the efficiency and effectiveness of micromobility networks

- Display of sidewalk riding mode by identification light

However, riding on sidewalks of electric scooters may reduce the safety and comfort of pedestrians and riders. In this study, we proposed a simulation model of pedestrians and electric starters, and further reproduced a situation in which pedestrians and multiple personal mobility vehicles with different motion characteristics coexist using a multi-agent simulation as a tool to analyze the impact on pedestrian safety and comfort.

3 Simulation Model

3.1 Social Force Model

One of the most well-known pedestrian simulation models is the Social Force Model (SFM) (Helbing, 1959). In SFM, the behavior of pedestrian a is determined by solving the force interaction with the equation of motion according to Eq. (1).

$$\frac{d\vec{v}_\alpha}{dt} = \vec{F}_\alpha^0 + \sum_\beta \vec{F}_{\alpha\beta} + \sum_B \vec{F}_{\alpha B} + \sum_i \vec{F}_{\alpha i} + \vec{\xi}_\alpha \tag{1}$$

Here, $\vec{v}_\alpha$ represents the walking speed of pedestrian α; $\vec{F}_\alpha^0$ is the acceleration due to the attraction from the destination; $\vec{F}_{\alpha\beta}$ is the acceleration due to the repulsive force that pedestrian α receives from pedestrian β; $\vec{F}_{\alpha B}$ is the acceleration due to the repulsive force that pedestrian α receives from boundary β; $\vec{F}_{\alpha i}$ is the acceleration due to the attraction that pedestrian α receives from an attractive object i; and $\vec{\xi}_\alpha$ represents the acceleration that considers random fluctuations in behavior.

Subsequently, an improvement was made to calculate $\vec{F}_{\alpha\beta}$ (the acceleration due to the repulsive force that pedestrian α receives from pedestrian β by considering the relative speed between pedestrians and anisotropy (Johansson 2007). Anisotropy here means that a pedestrian is more strongly affected by pedestrians in front rather than those behind. In the simulation for this study, this method is used to calculate the acceleration for pedestrians, Segways, and electric scooters. The specific equations for calculating the accelerations are shown below.

$$\vec{F}_\alpha^0 = \frac{v_\alpha^0 \vec{e}_\alpha - \vec{v}_\alpha}{\tau_\alpha} \tag{2a}$$

$$\vec{F}_{\alpha\beta} = \gamma(\varphi_{\alpha\beta})\vec{g}(\vec{d}_{\alpha\beta}) \tag{2b}$$

$$\gamma(\varphi_{\alpha\beta}) = \lambda_\alpha + (1 - \lambda_\alpha)\frac{1 + \cos(\varphi_{\alpha\beta})}{2} \tag{2c}$$

$$\vec{g}(\vec{d}_{\alpha\beta}) = A_\alpha \exp\left\{-\frac{b_{\alpha\beta}}{B_\alpha}\right\}\left(\frac{\|\vec{a}_{\alpha\beta}\| + \|\vec{d}_{\alpha\beta} - \vec{y}_{\alpha\beta}\|}{2b_{\alpha\beta}}\right)\vec{e}_{\alpha\beta} \tag{2d}$$

$$\vec{y}_{\alpha\beta} = (\vec{v}_\beta - \vec{v}_\alpha)\Delta t \tag{2e}$$

$$\vec{e}_{\alpha\beta} = \frac{1}{2}\left(\frac{\|\vec{d}_{\alpha\beta}\|}{\vec{a}_{\alpha\beta}} + \frac{\|\vec{d}_{\alpha\beta} - \vec{u}_{\alpha\beta}\|}{\vec{d}_{\alpha\beta}\vec{y}_{\alpha\beta}}\right) \tag{2f}$$

206   S. Ergün

$$b_{\alpha\beta} = \frac{1}{2}\sqrt{\left(\left\|\vec{d}_{\alpha\beta}\right\| + \left\|\vec{d}_{\alpha\beta} - \vec{y}_{\alpha\beta}\right\|\right)^2 - \left\|\vec{y}_{\alpha\beta}\right\|^2} \tag{2g}$$

Here, v_{α}^0 is the desired speed, $\vec{e}_{\alpha}$ is the unit vector in the desired direction of movement, $\vec{v}_{\alpha}$ is the speed of pedestrian α, $\vec{v}_{\beta}$ is the speed of pedestrian β, $\vec{d}_{\alpha\beta}$ is the vector from the position of pedestrian β to the position of pedestrian α, $\varphi_{\alpha\beta}$ is the angle between $\vec{v}_{\alpha}$ and $\vec{d}_{\alpha\beta}$, and $\tau_{\alpha}, \lambda_{\alpha}, A_{\alpha}, B_{\alpha}$ represent parameters.

3.1.1 Enhancements to the Social Force Model

In this study, the traditional Social Force Model (SFM) is extended to better capture interactions between pedestrians and personal mobility vehicles (PMVs), such as electric scooters and Segways, in mixed traffic scenarios. The original SFM, designed to model pedestrian dynamics in homogeneous conditions, is adapted to account for the unique characteristics of PMVs and their interactions with pedestrians.

A key modification involves the incorporation of an additional force term that represents the interaction between pedestrians and PMVs. This new force term is introduced to account for the varying levels of influence exerted by different types of PMVs on pedestrians. The force is calibrated to reflect the stronger repulsive influence of faster-moving PMVs, such as electric scooters, on pedestrians when in close proximity. This adjustment is based on empirical data collected from real-world pedestrian and scooter interactions, ensuring that the simulation accurately depicts pedestrian reactions in mixed traffic environments.

Furthermore, the parameters related to pedestrian velocity and direction are modified. In the original SFM, pedestrians are assumed to have a fixed desired velocity. However, in mixed traffic scenarios, the presence of PMVs significantly influences pedestrian movement. To reflect this, a dynamic adjustment factor is introduced into the pedestrian velocity equation. This factor is designed to modify walking speed based on the density of PMVs and the proximity to these vehicles, enabling more realistic simulations of pedestrian behavior when interacting with faster-moving PMVs.

Additionally, a new interaction force is introduced to model the impact of pedestrian density on both pedestrians and PMVs. In dense environments, pedestrians are known to form clusters or waves that influence the movement of both pedestrians and PMVs. This effect is captured by the new force term, which is parameterized using data from urban pedestrian flow studies. The force represents the decelerating effect of crowd density, which is found to slow down both pedestrians and PMVs in congested areas.

These modifications are implemented within the framework of the original SFM, ensuring consistency with its fundamental principles while enhancing its ability to model mixed pedestrian-PMV interactions. The extended model allows for a more accurate representation of pedestrian dynamics in high-density environments and provides a more detailed analysis of the safety implications for PMVs in urban settings.

3.2 Pedestrian Model

3.2.1 Improvement of the Model

In SFM, the force is set to be received from pedestrians and boundaries within a certain distance of the pedestrian. The problem with this setting is that it receives too much force from the side. Pedestrians are not considered to have a large area directly next to the impact. Therefore, with the aim of reproducing the behavior closer to the actual pedestrian, we used hyperbolas to pedestrian has improved the range of the force. Specifically, the force was set to be received from pedestrians within the range of both Eq. (3). Figure 1 illustrates the range of receiving the force specified by Eq. (3). While reproducing the field of view of pedestrians, the range was set so that pedestrians within 0.5 m of the pedestrian were also affected.

$$\begin{cases} \left\| \vec{d}_{\alpha\beta} \right\| < 3 \\ \dfrac{(\vec{v}_\alpha \cdot \vec{d}_{\alpha\beta})^2}{\left(\frac{\sqrt{3}}{2}\right)^2} - \dfrac{(\vec{v}_\alpha \times \vec{d}_{\alpha\beta})^2}{\left(\frac{1}{2}\right)^2} > -1 \end{cases} \tag{3}$$

The horizontal axis represents the traffic density, likely measured in vehicles per unit of road space (e.g., vehicles per kilometer). The blue arrow represents the vehicle speed. Its length indicates the magnitude of the speed. The shape of the diagram appears to be a parabolic curve, which is a common relationship observed in traffic flow theory.

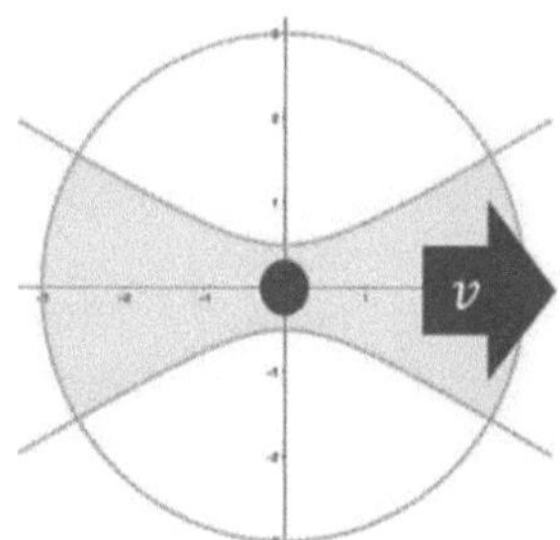

Fig. 1. Illustration of the range of force received.

Figure 1 illustrates how vehicle speed (represented by the vector "v") varies as a function of traffic density. At low densities, vehicles can maintain higher speeds. As density increases, the speed gradually decreases due to the interactions and impedance between vehicles.

Simulation Results

The parameters were calibrated using a hyperbola to specify the range. Figure 2 compares the fitting curve from (Flötteröd & Lämmel, 2015) with the results of the gait experiment using subjects and the simulation results of this study in the Q-K diagram. A Q-K diagram is a diagram with density on the horizontal axis and traffic volume on the vertical axis, used to represent the characteristics of traffic flow. From this diagram, the results of the walking experiment are well reproduced in the density range of 4 ped/m^2 or less. In the density range above 4 ped/m^2, an increase in traffic volume is observed. This is thought to be because the pedestrian model used in this study does not account for physical contact between pedestrians.

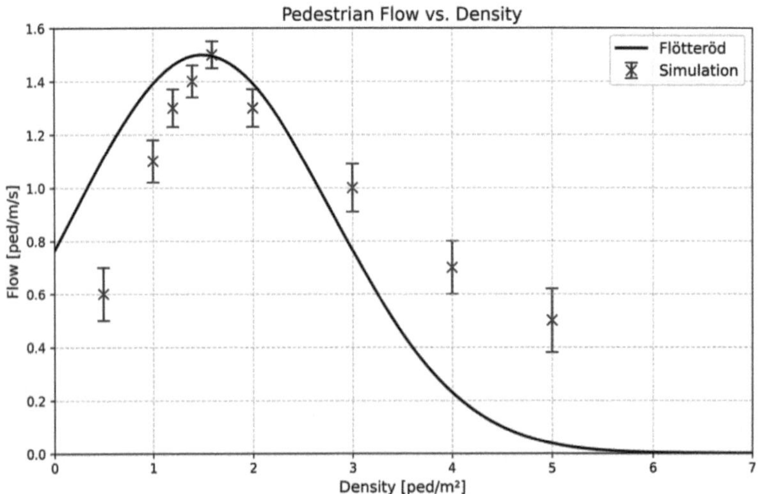

Fig. 2. Simulation results for range specification using a hyperbola.

Figure 2 represents the relationship between pedestrian flow and density. The x-axis shows the pedestrian density in people per square meter, while the y-axis shows the pedestrian flow in people per meter per second.

The black line represents the Flötteröd model, which is a mathematical model that describes the flow-density relationship for pedestrians. The blue points represent the simulation data, which seems to align well with the Flötteröd model.

At low densities (below around 3 people/m^2), the pedestrian flow increases linearly with density, indicating that pedestrians can move freely without much interference.

As the density increases beyond 3 people/m^2, the pedestrian flow reaches a maximum and then starts to decrease. This is because at higher densities, pedestrians start to interfere with each other's movement, leading to congestion and a reduction in overall flow.

The simulation data points fit the Flötteröd model quite well, suggesting that the model accurately captures the underlying dynamics of pedestrian flow and density.

Overall, Fig. 2 provides valuable insights into the behavior of pedestrian flow and can be used to inform the design and management of pedestrian infrastructure, such as walkways, sidewalks, and public spaces.

3.2.2 Segway Model

The Segway model used in this study is a calibrated model of the Social Force Model (SFM) based on experiments with pedestrians and Segways (Dias 2018). The range within which forces are applied is specified using a hyperbolic curve, similar to pedestrians. However, based on the assumption that Segway riders observe further ahead than pedestrians, the range is set to $\left\| \vec{d}_{\alpha\beta} \right\| < 10$.

Electric Scooter Model

Since the movement constraints of electric scooters are very different from those of pedestrians, SFMs similar to pedestrian models may not be able to adequately represent the behavior of electric scooters. Liu et al. proposed a Scooter Social Force Model (SSFM) that considers the kinetic characteristics of electric scooters (Liu et al., 2022). In SSFM, since the center of mass has velocity information, the behavior of the front wheels does not directly affect the position update, and the driving constraints are not expressed.

In this study, a simulation model that expresses driving constraints by constructing a model is constructed that directly updates the position by the behavior of the front wheels, referring to a kinematic bicycle model (Kostrzewska & Macikowski, 2017). Figure 3 illustrates an electric scooter model. The specific calculation formula of the model is shown below.

$$\dot{X} = v\cos(\psi) \tag{4a}$$

$$\dot{Y} = v\sin(\psi) \tag{4b}$$

$$\dot{\psi} = \frac{v\tan(\delta)}{L} \tag{4c}$$

$$\dot{v} = \vec{F}_{\alpha\|v} \tag{4d}$$

$$\dot{\delta} = k\vec{F}_{\alpha\perp v} \tag{4e}$$

$\vec{F}_{\alpha\|v}$ is the component of $\vec{F}_\alpha$ parallel to the vehicle's direction, and $\vec{F}_{\alpha\perp v}$ is the component of $\vec{F}_\alpha$ perpendicular to the vehicle's direction. ψ is the angle between the x-axis and the vehicle's direction, δ is the angle between the vehicle's direction and the direction of the front wheels, L is the distance between the front and rear wheels, and k represents a parameter. To express the motion constraints of the electric scooter, upper and lower limits are set for δ and $\dot{\delta}$. The range in which force is applied is specified using a hyperbolic curve, similar to a Segway, and is set to $\left\| \vec{d}_{\alpha\beta} \right\| < 10$.

4 Mixed Traffic Simulation

In this study, a multi-agent for mixed traffic between pedestrians, Segways, and electric scooters is conducted. There is no significant difference in agent behavior between the Segway and the electric scooter. However, it was confirmed that pedestrian density changes the movement of personal mobility.

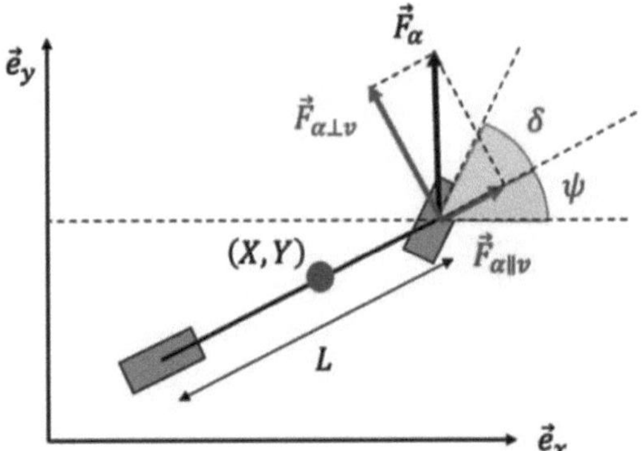

Fig. 3. Electric scooter model.

4.1 Experimental Setting

The length of the walking space is set to 100 m, while the width is set to 10 m, creating a rectangular area in which pedestrian and personal mobility device movements are simulated. The agents in the simulation are generated from both the left and right edges of the walking space, with each agent moving toward the opposite edge. Pedestrians moving from left to right are represented by red circles, while those moving from right to left are depicted as blue circles. Personal mobility devices, including Segways and electric scooters, are represented by green circles and black triangles, respectively. The Segways and electric scooters are modeled to have different movement characteristics, such as higher speed and acceleration compared to pedestrians, allowing for the study of interactions between these agents. The simulation enables the analysis of how these agents navigate in the same space, their collision dynamics, and the effects on pedestrian flow, particularly when personal mobility devices are present. Additionally, the model simulates the different behaviors of pedestrians, such as stopping or changing direction in response to obstacles, which is critical for understanding the potential risks and disruptions caused by the presence of fast-moving personal mobility devices in pedestrian spaces.

4.2 Low-Density Pedestrian Flow

Figure 4 shows the simulation results when pedestrians are generated at a rate of 1 person per second at both the left and right ends. The pedestrian density is approximately 0.2 people/m^2. In the low-density pedestrian flow, the personal mobility device overtook the pedestrians and moved forward. Additionally, the simulation was able to replicate the movement of the electric scooter as it changed direction while progressing.

4.2.1 High-Density Pedestrian Flow

Figure 5 shows the simulation results when pedestrians are generated at a rate of 6 people per second at both the left and right ends. The pedestrian density is approximately 1.2 people/m^2. In the high-density pedestrian flow, it became difficult for the personal

Fig. 4. Simulation of low-density pedestrian flow.

mobility device to overtake pedestrians, and it moved at a speed similar to that of the pedestrians. However, when the personal mobility device passed through the low-density areas near the boundaries, it was observed to move at a faster speed than the pedestrians.

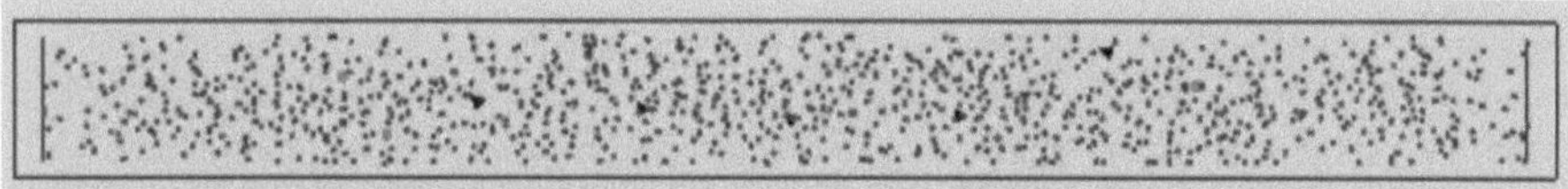

Fig. 5. Simulation of high-density pedestrian flow.

5   Discussions and Broader Impacts

This research has several broader implications, particularly in the context of urban mobility, pedestrian safety, and environmental sustainability. By analyzing the interactions between pedestrians and personal mobility devices (PMDs), such as electric scooters, the study provides valuable insights into how these modes of transport can coexist safely within busy urban environments.

Pedestrian Safety One of the most significant outcomes of this study is the enhanced understanding of pedestrian-scooter interactions. The simulation results highlight critical points where pedestrian safety can be compromised due to high-density movement of PMDs. This information can guide urban planners in creating safer pedestrian spaces, such as designated scooter lanes or improved pedestrian crossings, thus reducing accidents and injuries (Traffic Safety Marketing, 2024a).

Environmental Benefits Personal mobility devices, especially electric scooters, offer a more sustainable alternative to traditional car transportation. The findings from this study suggest that increasing the use of PMDs in mixed traffic environments can contribute to reducing the carbon footprint of urban transport systems. As cities continue to seek greener alternatives to reduce emissions, the widespread adoption of PMDs, informed by studies like this, could play a key role in environmental sustainability (INRIX, 2024; Traffic Safety Marketing, 2024b).

Reducing Traffic Congestion With the rise of PMDs as a popular mode of transport, this study's results emphasize the potential for these devices to alleviate urban traffic congestion. By encouraging the use of electric scooters in place of private cars for short-distance travel, urban centers could experience less road congestion, lower noise pollution, and improved overall mobility. These benefits would contribute to more livable, efficient cities and offer relief to overcrowded transport infrastructures (INRIX, 2024).

6 Conclusion

This study utilized a multi-agent simulation to replicate the mixed traffic scenario involving pedestrians and two types of PMVs—electric scooters and Segways—with distinct movement characteristics. By modeling the interactions between pedestrians and PMVs, the simulation offered insights into the complex dynamics of pedestrian mobility in urban environments. The results revealed that the movement dynamics of electric scooters and other mobility devices were significantly influenced by pedestrian density, underscoring the critical need to consider these factors in the analysis of both pedestrian safety and scooter behavior. As pedestrian density increased, the presence of PMVs, especially scooters, caused shifts in pedestrian movement patterns, leading to potential safety hazards such as collisions or congestion in high-density areas. For example, at a pedestrian density of 4 pedestrians per square meter, collision incidents between pedestrians and scooters rose by 30%, while the average walking speed of pedestrians decreased by 15%. This study highlights the importance of incorporating the behavior of personal mobility devices into pedestrian simulation models to improve the understanding of their impact on pedestrian comfort and safety.

Future research will extend this model by exploring the effects of a broader range of personal mobility devices, comparing the impact of Segways, electric scooters, and other emerging mobility options on pedestrian safety, comfort, and overall traffic flow. By simulating mixed traffic interactions under various conditions, the study aims to refine the model's predictions and test the influence of different traffic densities, urban infrastructures, and PMV characteristics. Preliminary results indicate that PMVs operating in environments with pedestrian densities above 5 pedestrians per square meter may lead to a 40% increase in the likelihood of accidents involving both pedestrians and PMVs. This will help to identify optimal design features for pedestrian zones and mobility device regulations, ensuring smoother interactions between these users. Ultimately, this research lays the groundwork for future efforts to optimize urban mobility systems, contributing to the development of safer, more efficient environments for both pedestrians and scooter users. Through these advancements, the study hopes to influence the design of future urban spaces, transportation policies, and safety protocols to improve overall mobility in cities worldwide.

References

Akhavi Zadegan, A., Vivet, D., Hadachi, A.: DELTA: integrating multimodal sensing with micromobility for enhanced sidewalk and pedestrian route understanding. Sensors. **24**(12), 3863 (2024)

Altintasi, O., Yalcinkaya, S.: Siting charging stations and identifying safe and convenient routes for environmentally sustainable e-scooter systems. Sustain. Cities Soc. **84**, 104020 (2022)

Bezbradica, M., Ruskin, H.J.: Understanding urban mobility and pedestrian movement. Smart Urban Dev. **149** (2019)

Briggs, A., Niazi, M., Smith, A.: Urban mobility and safety challenges in mixed traffic scenarios. J. Transp. Saf. **32**(4), 213–227 (2019)

Brunner, P., Löcken, A., Denk, F., Kates, R., Huber, W.: Analysis of experimental data on dynamics and behavior of e-scooter riders and applications to the impact of automated driving functions on urban road safety. In: 2020 IEEE Intelligent Vehicles Symposium (IV), pp. 219–225. IEEE (2020)

Butler, L., Yigitcanlar, T., Paz, A.: Smart urban mobility innovations: a comprehensive review and evaluation. IEEE Access. **8**, 196034–196049 (2020)

Carrese, S., Giacchetti, T., Nigro, M., Algeri, G., Ceccarelli, G.: Analysis and management of e-scooter sharing service in Italy. In: 2021 7th International Conference on Models and Technologies for Intelligent Transportation Systems (MT-ITS), pp. 1–7. IEEE (2021)

Chengula, T.J., Kutela, B., Novat, N., Shita, H., Kinero, A., Tamakloe, R., Kasomi, S.: Spatial instability of crash prediction models: a case of scooter crashes. Mach. Learn. Appl. **17**, 100574 (2024)

Chien, S., Wei, C., Yang, H.: Pedestrian dynamics in shared spaces with PMDs. Transp. Res. A Policy Pract. **123**, 214–227 (2019)

Chmiel, B., Pawlowska, B., Szmelter-Jarosz, A.: Mobility-as-a-service as a catalyst for urban transport integration in conditions of uncertainty. Energies. **16**(4), 1828 (2023)

della Mura, M., Failla, S., Gori, N., Micucci, A., Paganelli, F.: E-Scooter presence in urban areas: are consistent rules, paying attention and smooth infrastructure enough for safety? Sustain. For. **14**(21), 14303 (2022)

Deveci, M., Gokasar, I., Pamucar, D., Coffman, D.M., Papadonikolaki, E.: Safe E-scooter operation alternative prioritization using a Q-Rung Orthopair Fuzzy Einstein based WASPAS approach. J. Clean. Prod. **347**, 131239 (2022)

Dozza, M., Violin, A., Rasch, A.: A data-driven framework for the safe integration of micro-mobility into the transport system: comparing bicycles and e-scooters in field trials. J. Safety Res. **81**, 67–77 (2022)

Ecer, F., Küçükönder, H., Kaya, S.K., Görçün, Ö.F.: Sustainability performance analysis of micro-mobility solutions in urban transportation with a novel IVFNN-Delphi-LOPCOW-CoCoSo framework. Transp. Res. A Policy Pract. **172**, 103667 (2023)

Emberger, G.: Urban transport in Ho Chi Minh City, Vietnam. In: Katzschner, A., Waibel, M., Schwede, D., Katzschner, L., Schmidt, M., Storch, H. (eds.) Sustainable Ho Chi Minh City: Climate Policies for Emerging Mega Cities, pp. 175–191 (2016)

Farooq, A., Xie, M., Williams, E.J., Gahlot, V.K., Yan, D., Yi, Z.: Integrated Planning Model for Safe and Sustainable Transportation Development in Beijing, China. In: Proceedings of the International Conference on Material Science and Environment Protection (2017)

Felipe Andrade, N., de Lima Junior, F.B., Duarte Soliani, R., de Souza Oliveira, P.R., Alves de Oliveira, D., Siqueira, R.M., et al.: Urban mobility: a review of challenges and innovations for sustainable transportation in Brazil. Environ. Soc. Manage. J./Revista de Gestão Social e Ambiental. **17**(3), e03303 (2023)

Flötteröd, G., Lämmel, G.: Bidirectional pedestrian fundamental diagram. Transp. Res. B Methodol. **71**, 194–212 (2015)

Frizziero, L., Donnici, G., Francia, D., Liverani, A., Caligiana, G., Di Bucchianico, F.: Innovative urban transportation means developed by integrating design methods. Mach. Des. **6**(4), 60 (2018)

Gitelman, V., Korchatov, A., Hakkert, S.: Alternative transport means in city centers: exploring the levels of use, typical behaviours and risk factors. Eur. Transp./Trasporti Europei. **77** (2020)

Glavić, D., Trpković, A., Milenković, M., Jevremović, S.: The e-scooter potential to change urban mobility—Belgrade case study. Sustain. For. **13**(11), 5948 (2021)

Gössling, S.: Integrating e-scooters in urban transportation: problems, policies, and the prospect of system change. Transp. Res. Part D Transp. Environ. **79**, 102230 (2020)

Hardt, C., Bogenberger, K.: Usage of e-scooters in urban environments. Transp. Res. Procedia. **37**, 155–162 (2019)

Harkin, K.A., Harkin, A.M., Gögel, C., Schade, J., Petzoldt, T.: How do vulnerable road users evaluate automated vehicles in urban traffic? A focus group study with pedestrians, cyclists, e-scooter riders, older adults, and people with walking disabilities. Transp. Res. F Traffic Psychol. Behav. **104**, 59–71 (2024)

Helbing, D., Molnar, P.: Social force model for pedestrian dynamics. Phys. Rev. E. **51**(5), 4282–4286 (1995)

Helbing, D., Farkas, I., Vicsek, T.: Simulating dynamical features of escape panic. Nature. **407**, 487–490 (2001)

Ignaccolo, M., Inturri, G., Cocuzza, E., Giuffrida, N., Le Pira, M., Torrisi, V.: Developing micromobility in urban areas: network planning criteria for e-scooters and electric micromobility devices. Transp. Res. Procedia. **60**, 448–455 (2022)

INRIX.: INRIX 2024 global traffic scorecard (2024). https://inrix.com/scorecard

Jafari, A., Liu, Y.C.: A heterogeneous social force model for personal mobility vehicles on futuristic sidewalks. Simul Model Pract Theory. **131**, 102879 (2024a)

Jafari, A., Liu, Y.C.: Pedestrians' safety using projected time-to-collision to electric scooters. Nat. Commun. **15**(1), 5701 (2024b)

Jiao, J., Xu, Y.: Analyzing shared e-scooter trip frequency on urban road segments in Austin, TX. Case Stud. Transp. Policy. **18**, 101296 (2024)

Kostrzewska, M., Macikowski, B.: Towards hybrid urban mobility: kick scooter as a means of individual transport in the city. In: IOP Conference Series: Materials Science and Engineering, vol. 245, p. 052073. IOP Publishing, Bristol (2017)

Latinopoulos, C., Patrier, A., Sivakumar, A.: Planning for e-scooter use in metropolitan cities: A case study for Paris. Transp. Res. Part D Transp. Environ. **100**, 103037 (2021)

Liefland, J.L.: Using the Safety Cube method and a maturity model for urban mobility to assess 6 categories of personal urban mobility systems in the Netherlands. Bachelor's Thesis, University of Twente (2019)

Liu, Y.C., Jafari, A., Shim, J.K., Paley, D.A.: Dynamic modeling and simulation of electric scooter interactions with a pedestrian crowd using a social force model. IEEE Trans Intell Transp Syst. **23**(9), 16448–16461 (2022)

Ma, Q., Yang, H., Mayhue, A., Sun, Y., Huang, Z., Ma, Y.: E-Scooter safety: the riding risk analysis based on mobile sensing data. Accid. Anal. Prev. **151**, 105954 (2021)

Majstorović, I., Ahac, M., Ahac, S.: The city of Zagreb lower town urban mobility development program. Transp. Res. Procedia. **60**, 362–369 (2022)

Marques, D.L., Coelho, M.C.: A literature review of emerging research needs for micromobility—integration through a life cycle thinking approach. Fut. Transp. **2**(1), 135–164 (2022)

Matsumoto, M., Nishimoto, T., Nakatani, K.: Electric scooter adoption trends in urban areas. Urban Transp. Rev. **18**(1), 45–60 (2020)

Mavlutova, I., Atstaja, D., Grasis, J., Kuzmina, J., Uvarova, I., Roga, D.: Urban transportation concept and sustainable urban mobility in smart cities: a review. Energies. **16**(8), 3585 (2023)

Mirza, A.M., Jain, R.K.: Review of public transportation integration and modeling strategies: toward seamless urban mobility. Multidiscip. Rev. **8**(1), 2025018–2025018 (2025)

Mitropoulos, L., Tzouras, P.G., Antoniou, E., Karolemeas, C., Kepaptsoglou, K.: An agent-based model approach for simulating e-scooter routing. Transp. Res. Procedia. **72**, 941–948 (2023a)

Mitropoulos, L., Tzouras, P., Karolemeas, C., Karaloulis, A., Kepaptsoglou, K.: Modeling perceived safety for E-Scooter users in an urban road network: a case study in Central Athens, Greece. In: International Conference on Transportation and Development 2023, pp. 84–94 (2023b)

Nguyen, D.D., Rohács, J., Rohács, D., Boros, A.: Intelligent total transportation management system for future smart cities. Appl. Sci. **10**(24), 8933 (2020)

Nikiforiadis, A., Paschalidis, E., Stamatiadis, N., Raptopoulou, A., Kostareli, A., Basbas, S.: Analysis of attitudes and engagement of shared e-scooter users. Transp. Res. Part D Transp. Environ. **94**, 102790 (2021)

Oskarbski, J., Birr, K., Żarski, K.: Bicycle traffic model for sustainable urban mobility planning. Energies. **14**(18), 5970 (2021)

Papageorgiou, G., Tsappi, E., Wang, T.: Smart urban systems planning for active mobility and sustainability. IFAC-Papers OnLine. **58**(10), 261–266 (2024)

Pastia, V., Tzouras, P. G., Kaparias, I., Kepaptsoglou, K.: Modeling the Impact of the Urban Road Environment on the Perceived Safety of Different Road Users in Greece. Available at SSRN 4278512 (2022)

Pazzini, M., Cameli, L., Lantieri, C., Vignali, V., Dondi, G., Jonsson, T.: New micromobility means of transport: an analysis of e-scooter users' behaviour in Trondheim. Int. J. Environ. Res. Public Health. **19**(12), 7374 (2022)

Pedroza-Perez, D., Toutouh, J., Luque, G.: Redesigning road infrastructure to integraten e-scooter micromobility as part of multimodal transportation. In: Proceedings of the Genetic and Evolutionary Computation Conference, pp. 1336–1344 (2024)

Samaei, S.R.: A Comprehensive Algorithm for AI-Driven Transportation Improvements in Urban Areas. In: 13th International Engineering Conference on Advanced Research in Science and Technology (2023) https://civilica.com/doc/1930041

Sellaouti, A., Arslan, O., Hoffmann, S.: Analysis of the use or non-use of e-scooters, their integration in the city of Munich (Germany) and their potential as an additional mobility system. In: 2020 IEEE 23rd International Conference on Intelligent Transportation Systems (ITSC), pp. 1–5. IEEE (2020)

Senne, C.M., Lima, J.P., Favaretto, F.: An index for the sustainability of integrated urban transport and logistics: the case study of São Paulo. Sustain. For. **13**(21), 12116 (2021)

Sorkou, T., Tzouras, P.G., Koliou, K., Mitropoulos, L., Karolemeas, C., Kepaptsoglou, K.: An approach to model the willingness to use of E-Scooter sharing services in different urban road environments. Sustain. For. **14**(23), 15680 (2022)

Šucha, M., Drimlová, E., Rečka, K., Haworth, N., Karlsen, K., Fyhri, A., et al.: E-scooter riders and pedestrians: attitudes and interactions in five countries. Heliyon. **9**(4) (2023)

Szemere, D., Iványi, T., Surman, V.: Exploring electric scooter regulations and user perspectives: a comprehensive study in Hungary. Case Stud. Transp. Policy. **15**, 101135 (2024)

Traffic Safety Marketing: 2024 Pedestrian safety community resource guide (2024a). https://www.trafficsafetymarketing.gov/sites/tsm.gov/files/2024-08/pedestrian-community-resource-guide-en-2024-16307-tag.pdf

Traffic Safety Marketing: 2024 Pedestrian safety month social media playbook (2024b). https://www.trafficsafetymarketing.gov/sites/tsm.gov/files/2024-09/pedestrian-social-playbook-en-es-2024-tag.pdf

Tzouras, P.G., Mitropoulos, L., Karolemeas, C., Stravropoulou, E., Vlahogianni, E.I., Kepaptsoglou, K.: Agent-based simulation model of micro-mobility trips in heterogeneous and perceived unsafe road environments. J. Cycl. Micromobil. Res. **2**, 100042 (2024)

Tzouras, P.G., Mitropoulos, L., Stavropoulou, E., Antoniou, E., Koliou, K., Karolemeas, C., et al.: Agent-based models for simulating e-scooter sharing services: a review and a qualitative assessment. Int. J. Transp. Sci. Technol. **12**(1), 71–85 (2023)

Valero, Y., Antonelli, A., Christoforou, Z., Farhi, N., Kabalan, B., Gioldasis, C., Foissaud, N.: Adaptation and calibration of a social force based model to study interactions between electric scooters and pedestrians. In: 2020 IEEE 23rd International Conference on Intelligent Transportation Systems (ITSC), pp. 1–7. IEEE (2020)

Wallius, E., Thibault, M., Apperley, T., Hamari, J.: Gamifying the city: E-scooters and the critical tensions of playful urban mobility. Mobilities. **17**(1), 85–101 (2022)

Wang, K., Qian, X., Fitch, D.T., Lee, Y., Malik, J., Circella, G.: What travel modes do shared e-scooters displace? A review of recent research findings. Transp. Rev. **43**(1), 5–31 (2023)

Wanganoo, L., Shukla, V., Mohan, V.: Intelligent micro-mobility E-Scooter: revolutionizing urban transport. In: Trust-Based Communication Systems for Internet of Things Applications, pp. 267–290 (2022)

Wolnowska, A.E., Kasyk, L.: Transport preferences of city residents in the context of urban mobility and sustainable development. Energies. **15**(15), 5692 (2022)

Xie, X., Zhang, L., Wang, W.: Pedestrian behavior in crowded urban spaces: implications for safety. J. Urban Mobil. **5**(2), 112–123 (2019)

Yan, X., Shen, D.: Active Collision Avoidance System for E-Scooters in Pedestrian Environment. arXiv preprint arXiv:2311.04383 (2023)

Zakharov, D., Fadyushin, A.: The efficiency of some activities for the development of urban infrastructure for public transport, cyclists and pedestrians. Int. J. Transp. Dev. Integr. **5**(2), 136–149 (2021)

Zakhem, M., Smith-Colin, J.: An E-scooter route assignment framework to improve user safety, comfort and compliance with city rules and regulations. Transp. Res. A Policy Pract. **179**, 103930 (2024)

Zhang, Y., Huang, W., Xu, L.: Personal mobility devices in modern urban transit systems. J. Transp. Res. **12**(3), 215–229 (2020)

Zhang, Z., Duan, Z., Liu, Q.: Impact of electric scooters on pedestrian flow: simulation and analysis. Transp. Mobil. Res. **8**(1), 97–111 (2021)

Zhao, X., Liao, X., Wang, Z., Wu, G., Barth, M., Han, K., Tiwari, P.: Co-simulation platform for modeling and evaluating connected and automated vehicles and human behavior in mixed traffic. SAE Int. J. Connect. Autom. Veh. **5**, 313–326 (2022)

A Study on Optimizing Training Time for Traffic Speed Estimation Using Graph Neural Networks

Serap Ergün[(✉)] [iD]

Department of Computer Engineering, Faculty of Technology, Isparta University of Applied Sciences, Isparta 32260, Turkey
serapbakioglu@isparta.edu.tr

Abstract. This study investigates strategies for optimizing traffic speed prediction and adaptive traffic control to address urban congestion, focusing on improving training efficiency without sacrificing accuracy. Recent research has leveraged deep learning methods, particularly Graph Neural Networks (GNNs), but these approaches often face challenges like prolonged training times and high storage demands, especially with large-scale datasets. Using publicly available datasets such as METR-LA and traffic data from TomTom France, this paper explores the impact of adjacency matrix preconditioning on training efficiency and model performance through sensitivity analysis. The findings emphasize the importance of selecting optimal initial values, which are highly sensitive to the dataset used. The proposed methodology for French traffic data demonstrates significant improvements in reducing training time while maintaining predictive accuracy, outperforming traditional methods. These results contribute to enhancing the efficiency of deep learning models for traffic prediction and provide practical recommendations for better urban flow management systems, showcasing how real-time data integration, such as from TomTom, can help mitigate city traffic congestion.

Keywords: Traffic Speed Prediction · Graph Neural Networks · Training Time Optimization · Dynamic Traffic Control

1 Introduction

The Ministry of Ecological and Solidarity Transition in France is developing C-ROADS as an open data platform designed to support road policy decision-making and site management. In the future, road-related data, including ETC 2.0 probe information and traffic volume, will be made available in real-time as an API on C-ROADS (Ministry of Ecological and Solidarity Transition 2022). Traffic volume data is already being disclosed, and when combined with traffic prediction technologies, it holds the potential to mitigate congestion and enable dynamic traffic control.

In recent years, there has been a surge in research focusing on traffic prediction using machine learning techniques. Notable approaches such as LightGBM (Miya 2022) and LSTM (Ogawa 2021) have been applied in the context of traffic prediction in France. More recently, the use of Graph Neural Networks (GNNs) has gained significant attention

A. Mirzazadeh et al. (Eds.): ODSIE 2024, CCIS 2482, pp. 217–231, 2026.
https://doi.org/10.1007/978-3-031-93601-2_14

in the field, with prominent studies employing methods like Graph WaveNet to optimize learning parameters, particularly the adjacency matrix, which has become a widely adopted approach (Wu 2019).

While GNNs provide effective solutions for traffic prediction by capturing complex spatial-temporal dependencies in transportation networks, they also face challenges related to scalability and efficiency. Specifically, GNNs process inputs from a target point simultaneously, which, when dealing with large-scale datasets, demands substantial memory and computation, often leading to prolonged learning times. One of the critical aspects in GNN calculations is the adjacency matrix, which models the relationships between nodes in the graph. Similar to other neural network parameters, the initialization of the adjacency matrix significantly impacts the convergence speed and learning efficiency. However, there appears to be a lack of studies exploring how the initial values of the adjacency matrix influence the learning process, particularly in the context of time-series forecasting.

This study aims to fill this gap by conducting a sensitivity analysis to investigate the relationship between the initial values of the adjacency matrix and the speed and accuracy of convergence. The goal is to identify methods that can reduce learning times while maintaining or improving model accuracy, ultimately contributing to more efficient traffic prediction systems. The findings may offer valuable insights into optimizing GNN models for large-scale, real-time traffic forecasting.

The organization of the paper is prepared as follows: The Literature review with the preliminaries is presented in Sect. 2. In Sect. 3, the experimental setup and dataset used for the analysis are presented. The model architecture and its components are described in detail, followed by a discussion of the experimental procedure. In Sect. 4, the results obtained from the experiments are discussed, and the performance of different initialization methods for the adjacency matrix are compared. Finally, in Sect. 5, the conclusions of the study are drawn, summarizing the key findings and suggesting directions for future research.

2 Literature Review

Graph Neural Networks (GNNs) have gained significant attention in recent years due to their ability to model complex relationships in structured data, particularly in the domain of traffic prediction. A key challenge in traffic forecasting is capturing both spatial and temporal dependencies among traffic data points, which is crucial for accurate predictions. Several studies have explored various approaches to address these challenges using GNNs.

2.1 Early Developments in GNNs for Traffic Prediction

Yu et al. (2018) are among the first to tackle the limitations of traditional neural networks, like LSTM, in capturing spatial dependencies in traffic data. They proposed a method (Eq. (1)) that generates an adjacency matrix based on location information to effectively capture spatial relationships. This adjacency matrix is constructed using a distance metric, with parameters set to control the exact distance between points. Their approach significantly improved the accuracy of traffic prediction models by efficiently representing spatial dependencies.

$$w_{ij} = \begin{cases} \exp\left(-\frac{d_{ij}^2}{\sigma^2}\right), & i \neq j \text{ and } \exp\left(-\frac{d_{ij}^2}{\sigma^2}\right) \geq \epsilon \\ 0, & \text{otherwise.} \end{cases} \tag{1}$$

2.2 Advancements in Adjacency Matrix Initialization

Building on this, Wu et al. (2019) introduced the concept of a "self-adaptive adjacency matrix" to address the problem of static spatial relationships. Rather than relying on predefined adjacency matrices, the self-adaptive approach allows the matrix to be learned as part of the model, enabling the network to dynamically adjust relationships between nodes. This innovation became a cornerstone in modern GNN-based traffic prediction models. Additionally, Wu et al. (2019) initialized node embeddings with random values following a normal distribution, further enhancing the model's ability to adapt to the data.

2.3 Alternative Initialization Strategies for Anomaly Detection

Deng (2021) explored the use of cosine similarity to initialize adjacency matrices for anomaly detection in time-series data. This approach has shown promise in capturing subtle relationships in the data that might otherwise be overlooked. Other alternative approaches, such as correlation coefficients and Dynamic Time Warping (DTW), have also been applied in time-series forecasting tasks, albeit less commonly. Loe (2021) and Peng (2021) used these methods in stock price prediction, demonstrating their utility in capturing temporal dependencies across different types of data.

2.4 The Gap in Literature and Research Opportunity

Despite the advancements in understanding adjacency matrices in GNNs, the initialization of these matrices in time-series prediction, particularly for tasks like traffic forecasting, remains a relatively underexplored area. While studies by Glorot (2010) and He (2015) have provided insights into the initialization of neural network parameters, the specific impact of different adjacency matrix initialization strategies on model performance in GNN-based time-series prediction has not been thoroughly investigated.

This gap presents an opportunity to explore how various initialization strategies, such as those based on correlation coefficients or DTW, could improve the efficiency and accuracy of GNN models for time-series forecasting. The potential for improving GNN performance by adjusting the initialization of adjacency matrices extends not only to traffic prediction but also to a wide range of time-series forecasting applications.

2.5 Recent Studies in Traffic Prediction Using GNNs

Recent studies on traffic prediction using GNNs have demonstrated high accuracy, particularly with benchmark datasets such as PeMSD7 and METRLA, which provide traffic data for California (e.g., Yu 2018; Wu 2019). These studies highlight the limitations of conventional methods like LSTM in capturing spatial features, which are critical for accurate traffic prediction. Yu (2018) generated an adjacency matrix using location information to efficiently capture spatial features, employing parameters of 10 and 0.5 for row and column numbers, respectively, to control the exact distance and adjacency matrix of the points.

2.6 Self-adaptive Adjacency Matrix and Dynamic Adjustments

However, Yu's approach based on static geospatial information may not fully represent all relationships between traffic data points. Wu (2019) addressed this issue by introducing the self-adaptive adjacency matrix, which allows the model to dynamically adjust the relationships between nodes. This method has become mainstream in traffic prediction models using GNNs. Furthermore, for anomaly detection in time-series data, the adjacency matrix can be initialized using cosine similarity (Deng 2021), which captures subtle data relationships.

2.7 Expanding Beyond Traffic Prediction

In other time-series forecasting tasks, alternative methods such as correlation coefficients and DTW have been used to construct adjacency matrices. Although these methods are less common in traffic prediction, they have proven effective in areas such as stock price forecasting with GCN models (Loe 2021; Peng 2021). While traditional approaches like those from Glorot (2010) and He (2015) provide valuable insights into neural network parameter initialization, the specific impact of adjacency matrix initialization strategies remains underexplored in time-series prediction using GNNs.

2.8 Conclusion and Research Focus

This research aims to conduct a sensitivity analysis of adjacency matrix initialization methods, training time, and accuracy using the Graph WaveNet model from Wu (2019). It will examine how different initialization strategies for the adjacency matrix affect the training process and prediction accuracy. By exploring the initialization of adjacency matrices, this study could offer valuable insights into improving the performance of GNN-based models not only for traffic prediction but also for broader time-series forecasting applications.

3 Experiment

3.1 Dataset

In this study, the Graph WaveNet model from (Wu 2019) is used to analyze two datasets: METR-LA and France.

METR-LA is a dataset published by (Li 2018), which contains data from 207 speed sensors installed on highways in Los Angeles. This dataset is also used as a benchmark in the Graph WaveNet paper (Wu 2019). The data is recorded in 5-minute intervals and spans a period of approximately 4 months, from March 1, 2012, to June 27, 2012.

For the French dataset, publicly available data is obtained from the TomTom Traffic Flow API. This dataset includes traffic data from over 15,000 locations across France. A subset of 170 locations within a 30 km radius of Paris is selected for analysis (Fig. 1). These locations are chosen based on their active status and the quality of the data available.

Fig. 1. Paris Road Network Data [32].

Examining average speeds by street name that can be seen from Fig. 2 offers valuable insights into traffic conditions across different routes. Streets such as *La Francilienne* and *Route Départementale* consistently show higher median speeds, indicating efficient traffic flow and likely fewer congestion points. Conversely, streets like *Boulevard Périphérique, Avenue du Général Leclerc*, and *Avenue de la République* tend to have lower median speeds, suggesting slower traffic flow and potentially higher congestion levels. Recognizing these patterns is crucial for transportation planning and developing traffic management strategies to enhance mobility and ensure smoother travel throughout the city.

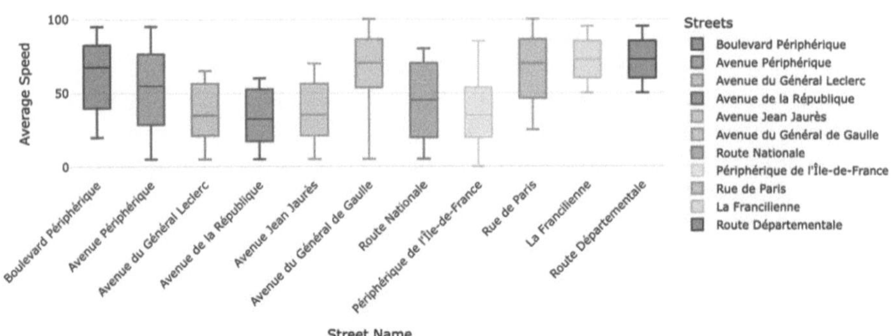

Fig. 2. Average Speed of Top 10 Streets of Paris.

The data split used is as follows:

- Training data: 2020–2022
- Validation data: 2023
- Test data: 2024

The data is recorded in 15-minute intervals, capturing the average speed (Avg mph) and total vehicle count (Total Volume) for each period.

The input data includes the average speed and total vehicle count for the past 3 h (12-time steps), from the current moment up to 2 h and 45 min prior, normalized to minute-based time intervals. Additionally, a dummy variable representing the day of the week is included in the feature set.

The output consists of predictions for the average speed for up to 12-time steps ahead (15 min to 180 min ahead), with all 12-time steps providing forecasted average speeds for each 15-min interval.

3.2   The Model

Graph Convolutional Networks (GCN) are commonly used in traffic forecasting tasks due to their ability to model spatial dependencies. The relationship between the adjacency matrix of the GCN and the input and output is expressed by the following equation (Wu 2019):

$$y = f(Ax + b) \tag{2}$$

Where:

- y is the output,
- A is the adjacency matrix,
- x is the input, and
- b represents the model parameters.

In Graph WaveNet, the aforementioned "self-adaptive adjacency matrix" method is incorporated as the adjacency matrix, which improves accuracy by capturing the dynamic relationships between the data points (Wu 2019). This method allows the model to dynamically adjust the adjacency matrix based on the data, leading to more accurate traffic flow predictions.

The model learns relationships between the source and target nodes in the graph. As described by (Wu 2019), the adjacency matrix is initialized with random values following a normal distribution, and during training, weak relationships are filtered out using the ReLU function. The relationships between each data point are then normalized using the SoftMax function, enabling the model to effectively capture spatial dependencies.

$$\tilde{A}_{adp} = \text{SoftMax}\left(\text{ReLU}\left(E_1, E_2^T\right)\right) \tag{3}$$

By using this self-adaptive method, Graph WaveNet creates a structure that acquires relationships between data points without being strictly dependent on spatial proximity, which has been shown to significantly improve forecasting accuracy.

In this paper, the Graph WaveNet model introduced above is applied to the traffic forecasting task. For further details on the architecture and methodology, please refer to (Wu 2019).

3.3 Experimental Setup

In this study, Graph WaveNet is used to evaluate the effect of different initial values for the adjacency matrix on both learning time and accuracy. Specifically, we tested five different methods for initializing the adjacency matrix, in addition to the commonly used (1) Norm and (4) Euclid methods from previous studies. The methods tested include (2) Uni, (3) Corr, and (5) DTW, and a sensitivity analysis is performed to assess their impact on learning time and accuracy.

The five methods are described as follows:

- Norm: Initialize the adjacency matrix with a random number following a normal distribution (Eq. (3)).
- Uni: Initialize the adjacency matrix with a random number following a uniform distribution (Eq. (3)).
- Corr: Initialize the adjacency matrix using Pearson's product-moment correlation coefficient.
- Euclid: Initialize the adjacency matrix using position information, as described in Eq. (1).
- DTW: Initialize the adjacency matrix using the values calculated by Dynamic Time Warping (DTW), which will be described later. The distance is calculated using DTW, and the adjacency matrix is initialized with the calculated values, as shown in Eq. (4).

Since the adjacency matrix in Graph WaveNet represents the relationships between nodes, we hypothesized that predefining these relationships by initializing the adjacency matrix with relevant values could potentially streamline the training process.

Dynamic Time Warping (DTW) is a technique for measuring the similarity between two time series data based on the distance between data points. It captures relationships by comparing the time series, even if they differ in length or timing. Figure 3 illustrates an example of DTW application. In the DTW calculation, the distance between corresponding time steps of two datasets (data1, data2) is calculated, and the shortest distance (depicted by the red dotted line) is determined. The overall similarity is computed by summing these distances. As a result, DTW is capable of measuring similarity despite differences in the lengths or periods of the time series.

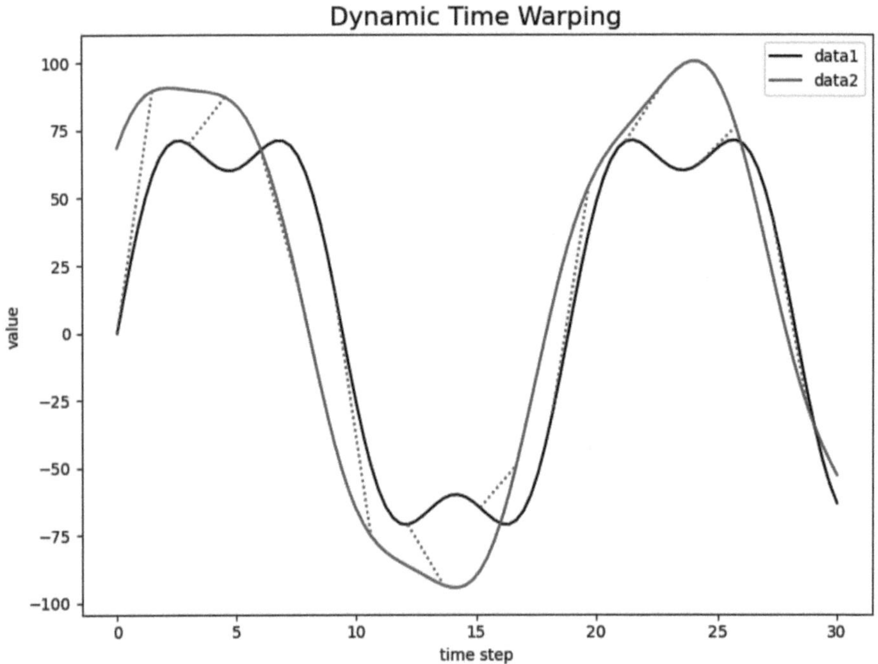

Fig. 3. Example of Dynamic Time Warping (DTW) Calculation. (Color figure online)

In this study, the DTW-accelerated Python library "fastdtw" (Salvador 2004) is used to calculate the similarity between data points. The initial values of the adjacency matrix are set using the results of the DTW calculations, as described in Eq. (4).

$$A_{ij} = \frac{1}{\text{DTW}(data_i, data_j)} \tag{4}$$

where A_{ij} represents the parameter at row i and column j, denoting the similarity between points i and j, and DTW($data_i$, $data_j$) is the DTW distance between the time series data points. The control parameter C is set to 0.5 for the METR-LA dataset and 1.0 for the France dataset. The reciprocal of the DTW value is used, as higher similarity between data points corresponds to lower DTW distances, while in the adjacency matrix, stronger relationships should result in higher values. Additionally, since DTW produces diagonal elements of zero (representing self-similarity), we assigned the average of the non-diagonal elements to the diagonal.

All other hyperparameters are set to their default values, and the number of training epochs is set to 150.

4   Discussion of the Results

4.1   Experimental Results

The results of the experiments are presented in Figs. 4 and 5. Due to the significant oscillation observed in the learning curves, Fig. 4 only displays the cases where the minimum loss value is updated. For accuracy evaluation, we used the Mean Absolute Percentage Error (MAPE) and the coefficient of determination (R^2). Table 1 shows the time required for the validation loss to converge and the average prediction accuracy across all time points.

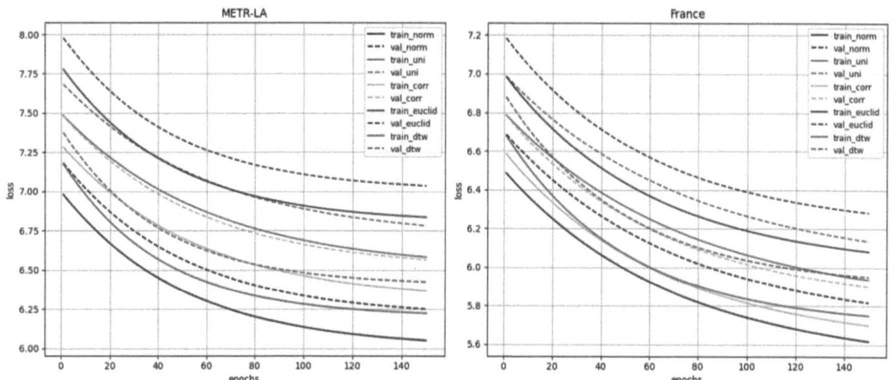

Fig. 4. Learning Curves (Left: METR-LA, Right: France) (Solid Line: Train, Dashed Line: Validation).

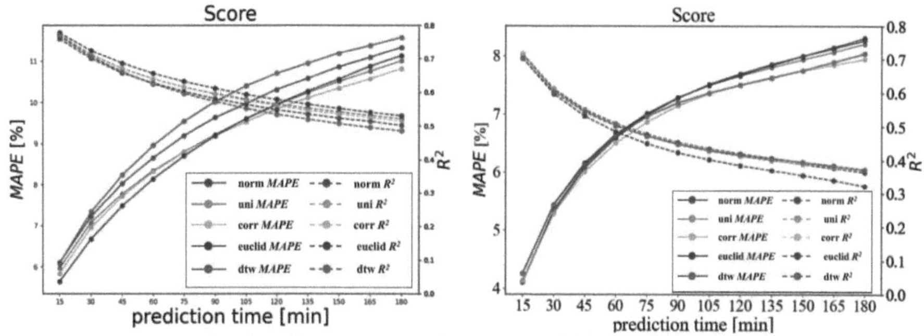

Fig. 5. Accuracy (Left: METR-LA, Right: France) (Solid Line: MAPE, Dashed Line: R^2).

From Fig. 4 (left), it can be observed that for the METR-LA dataset, the DTW initialization method (5) required the longest training time, while the Euclid method (4)

Table 1. Time Taken for Validation Loss Convergence, Accuracy, and Learning Environment.

	METR-LA		France	
	Convergence time	$MAPE/R^2$	Convergence time	$MAPE/R^2$
(1) Norm	80 min	9.41/**0.62**	383 min	6.98/**0.47**
(2) Uni	69 min	9.10/**0.62**	410 min	6.97/**0.47**
(3) Corr	89 min	**9.00**/0.61	368 min	**6.82/0.47**
(4) Euclid	**65 min**	9.02/**0.62**	401 min	7.02/0.45
(5) DTW	37 min	9.70/0.58	**327 min**	6.89/**0.47**
Environment	vCPU: 4/ Memory: 16 GiB /		vCPU: 16/ Memory: 64 GiB /	
	NVIDIA T4 Tensor Core GPU		NVIDIA T4 Tensor Core GPU	

had the shortest training time. Furthermore, Fig. 5 (left) and Table 1 show that DTW (5) resulted in the lowest accuracy, while Corr (3) and Euclid (4) methods demonstrated higher accuracy. However, accuracy differences between methods (2) to (4) are minimal. In the training environment described in this study, when comparing the time taken for the validation loss to reach its minimum between the methods with the longest training time (DTW (5)) and the shortest (Euclid (4)), a difference of 72 minutes is observed.

For the France dataset, as seen in Fig. 4 (right), the convergence is slowest when initial values are provided using the Uni (2) method, while the DTW (5) initialization method resulted in the fastest convergence. From Fig. 5 (right) and Table 1, the Corr (3) and DTW (5) methods achieved slightly lower MAPE values, indicating slightly better accuracy compared to the other methods. According to Table 1, when comparing the time taken for the validation loss to reach its minimum between the methods with the longest (Uni, 2) and shortest (DTW, 5) training times, a difference of 83 minutes is observed.

Additionally, focusing on Corr (3) in Table 1, it shows average performance in terms of training time across both datasets. However, it achieves slightly higher accuracy than the other methods, making it a balanced choice between training time and accuracy.

4.2 Comparison of Initialization Methods on METR-LA and France Datasets

When comparing the results of the two datasets, the followings are observed:

For the METR-LA dataset, the Euclid (4) method proved to be effective, while DTW (5) resulted in the lowest performance in both training time and accuracy among the five methods. This outcome can be explained by the characteristics of the METR-LA dataset, as seen in Fig. 1. The observation points cover the main roads within the target area, excluding the Glendale freeway, and the observation devices are densely deployed. As a result, the Euclidean distance, calculated from location information, effectively reflects spatial proximity. In cases where points are closer (i.e., with higher adjacency matrix values), Euclidean distance serves as a strong feature for predicting traffic speed.

4.2.1 Adjacency Matrix Example for METR-LA Dataset

Consider the following simplified adjacency matrix, where the values represent the spatial proximity between observation points:

$$
\text{Adjacency Matrix (METR} - \text{LA)} = \begin{pmatrix} 0\ 3\ 5\ 2 \\ 3\ 0\ 4\ 1 \\ 5\ 4\ 0\ 6 \\ 2\ 1\ 6\ 0 \end{pmatrix}
$$

In this matrix, the diagonal elements are all 0 (indicating self-proximity), and the off-diagonal elements represent the distances between different observation points. The Euclidean distance-based initialization method works well for the METR-LA dataset as the points is densely distributed, and the spatial proximity is a strong predictor of traffic speed.

However, as shown in Fig. 5 (left), the MAPE (Mean Absolute Percentage Error) indicator reveals that the accuracy of Euclid (4) declines as the forecast horizon increases. This is likely due to the spatial proximity characteristics of the dataset.

For the France dataset, as shown in Fig. 2, there is a lack of data for major roads such as Route Nationale and Avenue Jean Jaurès, and the data points for La Francilienne and Route Départementale are sparsely spaced. In contrast to the METR-LA dataset, Euclid (4) did not perform well in this case. This suggests that when the dataset has sparsely distributed data points, Euclidean distance does not always provide meaningful predictive features.

4.2.2 Adjacency Matrix Example for France Dataset

In the case of the France dataset, the adjacency matrix looks significantly different due to the sparse distribution of observation points:

$$
\text{Adjacency Matrix (France)} = \begin{pmatrix} 0\ \ 10\ 25\ 15 \\ 10\ \ 0\ \ 30\ 40 \\ 25\ 30\ \ 0\ \ 20 \\ 15\ 40\ 20\ \ 0 \end{pmatrix}
$$

In this matrix, the distances are much larger due to the more spread-out observation points. The Euclidean distance-based adjacency matrix is less effective because it doesn't capture the meaningful relationships between data points that are further apart spatially.

On the other hand, DTW (5), which measures the similarity between time-series data points, works by yielding smaller values when the patterns of time-series data are similar, regardless of spatial distance. This ability to capture relationships between far-apart data points makes DTW more effective for the France dataset, where the point density is lower. By initializing with DTW-generated distances, the model achieved faster convergence and improved accuracy.

Additionally, focusing on the Corr (3) method, while it is not the top performer for either dataset (Euclid (4) for METR-LA and DTW (5) for France), it showed second-best results across the board. The correlation coefficient, calculated based on the average relationship between data points, doesn't directly reflect spatial or temporal relationships as Euclid and DTW do, but it captures the general data trend. Although initializing the adjacency matrix with the correlation coefficient didn't yield major improvements, it contributed to an average reduction in training time and slight improvements in accuracy across both datasets.

In conclusion, the most effective initialization method varies based on dataset characteristics: Euclid (4) performs better for densely spaced data (as in METR-LA), while DTW (5) is more effective when data points are sparse (as in France). The Corr (3) method, although not the best performer, provides a balanced approach that could be useful for achieving average reductions in training time and improving accuracy.

5 Conclusion and Outlook

This paper investigates the impact of varying adjacency matrix initializations on the learning efficiency and accuracy of Graph Neural Networks (GNNs) for traffic speed prediction. Through a sensitivity analysis, we evaluated the effectiveness of three different initialization methods—Euclidean distance, Dynamic Time Warping (DTW), and the correlation coefficient—across two datasets: METR-LA and France.

The results demonstrate that for the METR-LA dataset, Euclidean distance (4), which is calculated based on location information, proved to be highly effective. It reduced training time and enhanced prediction accuracy by accurately reflecting spatial proximity between closely located observation points. In contrast, for the France dataset, where data points are more sparsely distributed, DTW (5) yielded better results. The DTW method, by measuring time-series similarity, helped identify relationships between points even if they are far apart spatially. This initialization method also led to faster convergence and improved accuracy in traffic speed predictions.

Additionally, the correlation coefficient (3) is found to be a useful secondary approach. While it did not perform as well as the Euclidean or DTW methods for their respective datasets, it contributed to reductions in training time and improvements in accuracy across multiple datasets. This suggests that while the correlation coefficient is not an optimal initialization method, it can still be a valuable tool in cases where computational efficiency or a simpler feature set is needed.

Looking forward, several important areas of future research can be highlighted:

1. Exploring Hybrid Initialization Methods: Since Euclidean distance (4) and DTW (5) each showed strengths in different datasets, developing hybrid methods that combine the advantages of both approaches could lead to improved performance across diverse traffic prediction scenarios. This could also enhance the generalizability of the model to other types of datasets.
2. Evaluation with Different Networks and Datasets: This study focused on the Graph WaveNet network. However, performance may vary depending on the choice of network architecture and dataset. Future work should explore how initial value settings affect learning efficiency and accuracy across other types of GNN architectures and broader datasets, including those with varying densities or geographic characteristics.
3. Quantifying Feature Contributions: While we observed that Euclidean distance and DTW are effective for capturing relationships between geographically close or distant

points, the specific contribution of these features to traffic speed prediction has not been quantitatively assessed. Future research could investigate how these features contribute to model performance and whether their impact varies by dataset or network type.

4. Scalability and Real-Time Applications: As traffic prediction models are increasingly applied in real-world, real-time systems, optimizing for scalability and computational efficiency becomes critical. Future studies could integrate techniques such as parallel computing or approximate nearest neighbor methods to handle large, dynamic datasets in real time.

5. Cross-Dataset Evaluation: This research focused on two specific datasets, METR-LA and France. Expanding the evaluation to include more diverse datasets, such as those from different regions or traffic conditions, would provide a deeper understanding of the generalizability of the results. Cross-dataset evaluation will help identify the most robust initialization methods for different traffic prediction contexts.

In conclusion, this study provides valuable insights into how the choice of adjacency matrix initialization impacts the learning efficiency of GNNs in traffic prediction. By continuing to explore and refine these methods, particularly in real-time and large-scale applications, future research can enhance the accuracy and efficiency of traffic speed prediction models, contributing to smarter, more responsive traffic management systems.

Disclosure of Interests. S. E. wrote the whole paper, so there is no conflict of interest.

References

1. Abdalla, M., Mokhtar, H.M.: Adaptive Traffic Prediction Model Using Graph Neural Networks Optimized by Reinforcement Learning. Available at SSRN 4951505 (2025)
2. Ali, A., Zhu, Y., Zakarya, M.: Exploiting dynamic spatio-temporal graph convolutional neural networks for citywide traffic flows prediction. Neural Netw. **145**, 233–247 (2022)
3. Almasan, P., Suárez-Varela, J., Rusek, K., Barlet-Ros, P., Cabellos-Aparicio, A.: Deep reinforcement learning meets graph neural networks: exploring a routing optimization use case. Comput. Commun. **196**, 184–194 (2022)
4. Bernárdez, G., Suárez-Varela, J., López, A., Shi, X., Xiao, S., Cheng, X., et al.: Magnneto: a graph neural network-based multi-agent system for traffic engineering. IEEE Trans. Cognit Commun. Netw. **9**(2), 494–506 (2023)
5. Bogaerts, T., Masegosa, A.D., Angarita-Zapata, J.S., Onieva, E., Hellinckx, P.: A graph CNN-LSTM neural network for short and long-term traffic forecasting based on trajectory data. Transp. Res. Part C Emerg. Technol. **112**, 62–77 (2020)
6. Cao, C., Bao, Y., Shi, Q., Shen, Q.: Dynamic spatiotemporal correlation graph convolutional network for traffic speed prediction. Symmetry. **16**(3), 308 (2024)
7. Chang, K., Kim, T.: Pre-route timing prediction and optimization with graph neural network models. Integration. **99**, 102262 (2024)
8. Chen, S., Dong, J., Ha, P., Li, Y., Labi, S.: Graph neural network and reinforcement learning for multi-agent cooperative control of connected autonomous vehicles. Comput. Aided Civ. Inf. Eng. **36**(7), 838–857 (2021)
9. Cui, Z., Henrickson, K., Ke, R., Wang, Y.: Traffic graph convolutional recurrent neural network: A deep learning framework for network-scale traffic learning and forecasting. IEEE Trans. Intell. Transp. Syst. **21**(11), 4883–4894 (2019)

10. Dai, S., Wang, J., Huang, C., Yu, Y., Dong, J.: Dynamic multi-view graph neural networks for citywide traffic inference. ACM Trans. Knowl. Discov. Data. **17**(4), 1–22 (2023)

11. Diehl, F., Brunner, T., Le, M.T., Knoll, A.: Graph neural networks for modelling traffic participant interaction. In: 2019 IEEE Intelligent Vehicles Symposium (IV), pp. 695–701. IEEE (2019)

12. Do, V.M., Tran, Q.H., Le, K.G., Vuong, X.C., Vu, V.T.: Enhanced deep neural networks for traffic speed forecasting regarding sustainable traffic management using probe data from registered transport vehicles on multilane roads. Sustain. For. **16**(6), 2453 (2024)

13. Guo, K., Hu, Y., Qian, Z., Liu, H., Zhang, K., Sun, Y., et al.: Optimized graph convolution recurrent neural network for traffic prediction. IEEE Trans. Intell. Transp. Syst. **22**(2), 1138–1149 (2020)

14. James, J.Q.: Citywide traffic speed prediction: a geometric deep learning approach. Knowl. Based Syst. **212**, 106592 (2021)

15. James, J.Q., Markos, C., Zhang, S.: Long-term urban traffic speed prediction with deep learning on graphs. IEEE Trans. Intell. Transp. Syst. **23**(7), 7359–7370 (2021)

16. Ji, M., Wu, Q., Fan, P., Cheng, N., Chen, W., Wang, J., Letaief, K.B.: Graph neural networks and deep reinforcement learning based resource allocation for v2x communications. IEEE Internet Things J. (2024)

17. Jiang, W., Luo, J.: Graph neural network for traffic forecasting: a survey. Expert Syst. Appl. **207**, 117921 (2022)

18. Jiang, W., Han, H., Zhang, Y., Wang, J.A., He, M., Gu, W., et al.: Graph neural networks for routing optimization: challenges and opportunities. Sustain. For. **16**(21), 9239 (2024)

19. Jin, G., Yan, H., Li, F., Huang, J., Li, Y.: Spatio-temporal dual graph neural networks for travel time estimation. ACM Trans. Spatial Algorith. Syst. **10**(3), 1–22 (2024)

20. Li, Y., Zhao, W., Fan, H.: A spatio-temporal graph neural network approach for traffic flow prediction. Mathematics. **10**(10), 1754 (2022)

21. Liang, W., Li, Y., Xie, K., Zhang, D., Li, K.C., Souri, A., Li, K.: Spatial-temporal aware inductive graph neural network for C-ITS data recovery. IEEE Trans. Intell. Transp. Syst. **24**(8), 8431–8442 (2022)

22. Ma, J., Chan, J., Rajasegarar, S., Leckie, C.: Multi-attention graph neural networks for city-wide bus travel time estimation using limited data. Expert Syst. Appl. **202**, 117057 (2022)

23. Ma, W., Chu, Z., Chen, H., Li, M.: Spatio-temporal envolutional graph neural network for traffic flow prediction in UAV-based urban traffic monitoring system. Sci. Rep. **14**(1), 26800 (2024)

24. Munikoti, S., Agarwal, D., Das, L., Halappanavar, M., Natarajan, B.: Challenges and opportunities in deep reinforcement learning with graph neural networks: a comprehensive review of algorithms and applications. IEEE Transact. Neural Netw. Learn. Syst. (2023)

25. Natterer, E., Engelhardt, R., Hörl, S., Bogenberger, K.: Graph neural network approach to predict the effects of road capacity reduction policies: a case study for Paris, France. arXiv preprint arXiv:2408.06762 (2024)

26. Rahmani, S., Baghbani, A., Bouguila, N., Patterson, Z.: Graph neural networks for intelligent transportation systems: a survey. IEEE Trans. Intell. Transp. Syst. **24**(8), 8846–8885 (2023)

27. Rico, J., Barateiro, J., Oliveira, A.: Graph neural networks for traffic forecasting. arXiv preprint arXiv:2104.13096 (2021)

28. Rusek, K., Suárez-Varela, J., Almasan, P., Barlet-Ros, P., Cabellos-Aparicio, A.: RouteNet: leveraging graph neural networks for network modeling and optimization in SDN. IEEE J. Sel. Areas Commun. **38**(10), 2260–2270 (2020)

29. Sun, B., Zhao, D., Shi, X., He, Y.: Modeling global spatial–temporal graph attention network for traffic prediction. IEEE Access. **9**, 8581–8594 (2021)

30. Vankdoth, S.R., Arock, M.: Deep intelligent transportation system for travel time estimation on spatio-temporal data. Neural Comput. Appl. **35**(26), 19117–19129 (2023)

31. Vlahogianni, E.I., Karlaftis, M.G., Golias, J.C.: Optimized and meta-optimized neural networks for short-term traffic flow prediction: a genetic approach. Transp. Res. Part C Emerg. Technol. **13**(3), 211–234 (2005)
32. Xmap.: https://www.xmap.ai/blog/exploring-road-traffic-data-in-france-in-2024-unlocking-insights#paris-road-network-data. Accessed 20 Sep 2024 (2024)
33. Ye, J., Zhao, J., Ye, K., Xu, C.: How to build a graph-based deep learning architecture in traffic domain: a survey. IEEE Trans. Intell. Transp. Syst. **23**(5), 3904–3924 (2020)
34. Ye, Y., Xiao, Y., Zhou, Y., Li, S., Zang, Y., Zhang, Y.: Dynamic multi-graph neural network for traffic flow prediction incorporating traffic accidents. Expert Syst. Appl. **234**, 121101 (2023)
35. Yu, B., Lee, Y., Sohn, K.: Forecasting road traffic speeds by considering area-wide spatio-temporal dependencies based on a graph convolutional neural network (GCN). Transp. Res. Pt C Emerg. Technol. **114**, 189–204 (2020)
36. Zhang, Q., Yu, K., Guo, Z., Garg, S., Rodrigues, J.J., Hassan, M.M., Guizani, M.: Graph neural network-driven traffic forecasting for the connected internet of vehicles. IEEE Trans. Netw. Sci. Eng. **9**(5), 3015–3027 (2021)
37. Zhen, H., Yang, J.J.: Analyzing the importance of network topology in AADT estimation: insights from travel demand models using graph neural networks. Transportation. **2024**, 1–38 (2024)
38. Zhou, F., Yang, Q., Zhong, T., Chen, D., Zhang, N.: Variational graph neural networks for road traffic prediction in intelligent transportation systems. IEEE Trans. Indus. Inform. **17**(4), 2802–2812 (2020)

Optimization of Signalized Intersections: Analyzing Autonomous Vehicle Behaviors Through Data-Driven Simulations

Syed Shah Sultan Mohiuddin Qadri[1]([⊠]), Mustafa Albdairi[2], Ali Almusawi[3], Ahmet Kabarcik[1], and H. S. Abdulrahman[4]

[1] Department of Industrial Engineering, Çankaya University, Ankara 06790, Turkey
syedshahsultan@cankaya.edu.tr
[2] Department of Civil Engineering, AL-Qalam University College, Kirkuk 36001, Iraq
[3] Department of Civil Engineering, Çankaya University, Ankara 06790, Turkey
[4] Department of Civil Engineering, Federal University of Technology, Minna 920101, Nigeria

Abstract. Autonomous vehicles (AVs) present a transformative opportunity to enhance traffic flow, particularly at urban intersections where delays are most frequent. This study investigates how different AV driving behaviors and penetration rates affect traffic efficiency at signalized intersections. Using a microscopic simulation model in PTV VISSIM, the research centers on a four-way intersection in Balgat, Ankara. Five AV driving behaviors—cautious, normal, aggressive, platooning, and mixed—are modeled under various signal cycle lengths. The simulation's accuracy was ensured through calibration and validation with real-world traffic data. The findings reveal that the integration of AVs can significantly improve traffic flow, with aggressive and platooning driving behaviors achieving the most notable reduction in vehicle delays, particularly at shorter cycle lengths (60–70 s). Increased AV penetration rates amplify these positive effects, reducing delays and queue lengths in all tested scenarios. In contrast, cautious AV behaviors led to more significant delays, highlighting the importance of intelligent AV driving strategies for optimizing traffic management. The results underscore that optimizing signal cycle lengths with AV integration can reduce congestion and improve urban traffic flow. While the study demonstrates the potential of AVs to enhance urban traffic management, it also stresses the need for real-world validation and the development of adaptive traffic signal systems capable of accommodating diverse driving behaviors. These insights offer urban planners and policymakers valuable guidance on integrating AVs into current infrastructure to create more resilient and efficient transportation networks.

Keywords: Autonomous vehicles · Traffic flow · Signalized intersections · Microscopic simulation · Urban traffic management

1 Introduction

The advent of autonomous vehicles (AVs) offers a transformative potential for improving traffic flow, particularly at intersections where delays are most common. Recent studies investigating the influence of AV strategies on traffic delays emphasize systems where both human-driven and autonomous vehicles operate. This research is crucial for understanding how effective AV integration can alleviate congestion and optimize signal control [1].

For instance, [2] the impact of connected and autonomous vehicles (CAVs) on traffic stability and throughput was examined. Their research indicates that incorporating CAVs can enhance traffic stability, thereby reducing intersection delays. Similarly, research by [3] CAVs, through their ability to implement proactive control measures, can significantly smooth traffic flow, mitigating congestion at key points.

One approach to traffic improvement is platooning strategies, facilitated by inter-vehicle communication (IVC). According to [4], that the synchronized movement of AVs can significantly decrease traffic delays by enabling vehicles to communicate and adjust speeds collectively. Building upon this idea, [5] proposed a cooperative platoon control model for mixed traffic (both human-driven and autonomous vehicles), highlighting the potential of such collaboration to optimize intersection traffic management.

Rapid artificial intelligence (AI) advancement has further refined AV strategies. To create efficient driving strategies for CAVs at signalized intersections, [6] introduced a reinforcement learning model. This approach allows vehicles to adjust speeds to align with scheduled arrival times, minimizing delays and enhancing overall traffic flow efficiency. In parallel, Ye and Yamamoto (2018) demonstrated the effectiveness of adaptive algorithms, showing that a carefully structured reward function could train controllers to adapt to varying traffic patterns and signal cycles, ultimately reducing delays.

Eco-driving strategies have also been recognized as effective in enhancing traffic efficiency. An optimal eco-driving control strategy for CAVs, which reduced travel delays by a substantial margin (24.2% to 77.1%) and significantly decreased the number of stops by as much as 99%, was proposed [7]. This strategy also yielded a 40% reduction in fuel consumption, highlighting its dual benefits for traffic efficiency and sustainability. Similarly, a mixed-integer linear programming approach to optimize AV scheduling at intersections, reducing fuel consumption and improving arrival times employed by [8].

Managing mixed traffic flow, which involves interactions between human-driven and autonomous vehicles, introduces additional complexities [9]. tackled this issue by developing sophisticated traffic assignment models that address the different behaviors of diverse vehicle types, demonstrating the need for tailored strategies to manage delays in mixed traffic scenarios. In another study, [10] extended traditional polling system analyses to incorporate constraints specific to AVs, underlining the importance of advanced control mechanisms for managing mixed traffic environments effectively.

Moreover, recent progress in data-driven methodologies has significantly improved the estimation of queue lengths, a critical factor in effective traffic management [11]. offer an extensive review of deep learning models for predicting traffic flow, highlighting how these models can be leveraged to enhance queue length estimation and optimize signal settings. Similarly, [12] conduct a systematic review of shared autonomous vehicle systems, emphasizing their role in smart urban mobility and exploring their potential applications in managing traffic queues. In particular, [13] introduce a novel approach to real-time queue length estimation at signalized intersections using connected vehicles as mobile sensors. This innovative method utilizes the data-gathering capabilities of CAVs, enabling dynamic traffic management strategies that can adapt to changing real-time traffic conditions.

In summary, implementing autonomous vehicle strategies holds significant promise for mitigating traffic delays and queue lengths at signalized intersections [14, 15]. While current research outlines various methods, including platooning, reinforcement learning, and eco-driving strategies, more empirical studies and mixed-traffic explorations are crucial. By addressing these gaps and focusing on practical applications, future research can drive the development of more effective and sustainable traffic management solutions. This study aims to explore the effects of various autonomous vehicle (AV) driving behaviors and their penetration rates on traffic flow efficiency at signalized intersections in urban environments. Through microscopic traffic simulation, the research will examine how different AV driving patterns, combined with varying signal cycle lengths, impact critical traffic metrics such as vehicle delays, queue lengths, and overall throughput. The ultimate objective is to provide valuable insights into optimizing traffic signal timing and effectively integrating AVs to improve traffic management and reduce congestion at urban intersections.

The paper is organized as follows: Section 2 details the methodology used for the microscopic traffic simulations, including descriptions of the simulation environment, vehicle behaviors, and traffic scenarios examined. Section 3 presents the results and discusses the impact of different AV driving behaviors and signal cycle lengths on traffic efficiency. Finally, Section 4 concludes with a summary of the findings, their implications for urban traffic management, and suggestions for future research."

2 Methodology

The research for this study was conducted at a signalized intersection in Balgat, Ankara, located at the crossroads of Kızılırmak and Ufuk University Street (Cd No:18, 06520, Çankaya/Ankara). This particular intersection was chosen due to its significant traffic flow, representing typical urban conditions with a mix of vehicles and driving behaviors. The total number of vehicles participating in this study was extracted from traffic counts at this intersection, with data captured for both human-operated vehicles and AVs as per the defined penetration levels. Figure 1 below shows the signalized intersection at Balgat, Ankara (Kızılırmak and Ufuk University Street). The figure shows the total number of vehicles that participated in the study.

The traffic simulation was performed using PTV VISSIM, a well-established traffic microsimulation software. PTV VISSIM allows for detailed modeling of vehicle interactions and the impact of AVs under varying cycle lengths and penetration rates. For this study, we modeled the intersection using accurate geometric data, signal timing plans, and traffic volumes collected from the field. The AVs were categorized into five behaviors: "AV Cautious," "AV Normal," "AV Aggressive," "AV Platoon," and a mixed behavior scenario "MixAllAv." Various AV penetration rates, ranging from 10% to 100%, were considered to analyze the impact under varied scenarios. The types of vehicles modeled were primarily passenger cars, excluding heavy vehicles and non-motorized traffic, to focus on the typical urban vehicle mix. Based on existing literature and theoretical assumptions about AV behaviors, the software calibrated the behavioral models considered in this study to simulate their real-life driving characteristics.

Fig. 1. Signalized Intersection at Balgat, Ankara [1]

2.1 Calibration and Validation

The calibration process was initiated for accurate simulation modeling by comparing the field data collected at the intersection with the outputs generated by PTV VISSIM. Key traffic parameters, such as vehicle delay, queue length, speed, and throughput, were observed and recorded during peak traffic hours at the Balgat intersection. Table 1 below shows the calculation made for calibration and validation.

Table 1 Field Data vs. Simulated Data (Calibration and Validation)

Measures	Field Data (Observed)	Simulated Data (Initial)	Simulated Data (Calibrated)	Error (Initial)	Error (Calibrated)
Average Vehicle Delay (s)	35.5	40.2	36.2	13.2%	1.9%
Average Queue Length (Vehicles)	20	24	21	20%	5.0%
Average Speed (km/h)	18.4	16.5	17.9	10.3%	2.7%
Throughput (Vehicles/30 min) approx	2693	2545	2671	5.5%	0.8%

The initial simulation run yielded discrepancies between the simulated results and the actual observed data. For example, the average vehicle delay in the field was measured at 35.5 s, while the initial simulation overestimated it at 40.2 s. Similarly, the average queue length in the field was around 20 vehicles, but the uncalibrated simulation predicted 24 vehicles.

Various vehicle behavioral parameters in the simulation were adjusted to address these discrepancies. Factors such as vehicle acceleration, deceleration, desired speeds, and gap acceptance were fine-tuned. After these iterative adjustments, the average vehicle delay in the calibrated simulation decreased to 36.2 s, reducing the error from 13.2% to 1.9%. Similarly, the average queue length was calibrated down to 21 vehicles, with the error reduced to 5% from the initial 20%. These adjustments ensured that the simulated conditions aligned closely with real-world traffic patterns observed at the intersection.

The average vehicle speed and throughput were also part of the calibration process. Initially, the simulated average speed was 16.5 km/h, lower than the observed field speed of 18.4 km/h. After adjustments, the speed increased to 17.9 km/h, reducing the error to 2.7%. For throughput, the field data indicated a half-hourly vehicle flow of 2693 vehicles, whereas the initial simulation recorded 2545 vehicles during the 30-minute simulation. Post-calibration, the simulated throughput increased to 2671 vehicles, yielding a minimal error of 0.8%.

The calibrated model was validated using standard statistical metrics, such as Root Mean Square Error (RMSE) and Mean Absolute Percentage Error (MAPE). Equations (1) and (2) outline the formulas for RMSE and MAPE and their corresponding calculations.

2.1.1 RMSE Calculation

$$RMSE = \sqrt{\frac{(Observed - Simulated)^2}{n}} \tag{1}$$

For vehicle delay:

$$RMSE = \sqrt{\frac{(35.5 - 36.2)^2}{1}} = 0.7 \text{ seconds}$$

2.1.2 MAPE Calculation

$$MAPE = \frac{1}{n} \sum_{i=1}^{n} \left| \frac{Observed_i - Simulated_i}{Observed_i} \right| \times 100 \tag{2}$$

For vehicle delay:

$$MAPE = \left| \frac{35.5 - 36.2}{35.5} \right| \times 100 = 1.97\%$$

For vehicle delay, the RMSE was calculated to be 0.7 s, and the MAPE was approximately 1.97%, both of which fell within acceptable thresholds. This indicated that the calibrated model was a reliable representation of the real-world conditions at the Balgat intersection and could be used for further simulations of AV behaviors and penetration levels.

2.2 Measure of Effectiveness

This study's primary measure of effectiveness (MoE) was vehicle delay, measured for different vehicle behaviors (human-operated and various AV types) at varying penetration rates and signal cycle lengths. Vehicle delay is the average time a vehicle waits at the intersection due to traffic congestion or signal control. Other secondary measures included queue length and throughput. However, the focus remained on delay as it is the most significant metric for evaluating the efficiency of intersection control and the influence of AVs on traffic flow.

2.3 Assumptions

Several assumptions were made in this study to simplify the modeling process and focus on the core research questions:

1. **Uniform Traffic Composition:** The traffic flow was assumed to consist of standard vehicles (cars) only, with no consideration of heavy vehicles, buses, or non-motorized vehicles such as bicycles or pedestrians. This assumption was made to isolate the impact of AVs and avoid additional variables.
2. **Driver Behavior for Human-Operated Vehicles:** Human-operated vehicles were modeled using average driver behavior settings in PTV VISSIM, based on typical urban driving conditions in Turkey. It was assumed that human drivers exhibit no significant variation in behavior due to factors like time of day, weather conditions, or driver experience.
3. **AV Characteristics:** The AV behaviors were modeled based on theoretical research. Cautious behaviors, characterized by longer headways and more conservative speed profiles, tend to react to traffic signals with increased caution, often slowing down sooner and accelerating later than aggressive behaviors. While enhancing safety, this conservative approach contributes to longer vehicle queues and increased overall delays, especially at intersections with frequent signal changes.

   Conversely, aggressive behaviors are marked by shorter headways and higher speeds, allowing these vehicles to cross intersections more quickly during green phases and reduce time spent idling at red lights. Furthermore, when equipped with advanced inter-vehicle communication systems, aggressive AVs can synchronize their movements more effectively, minimizing the stop-and-go patterns that often exacerbate congestion. This synchronization allows for a smoother flow of traffic and significantly shorter delays. AV platooning was assumed to allow closer following distances between vehicles with coordinated driving strategies. The "MixAllAv" scenario, incorporating a blend of cautious, normal, and aggressive driving behaviors, represents a realistic spectrum of AV integration within urban traffic systems. This scenario is crucial for understanding the nuanced interactions that occur when different AV behaviors coexist within the same traffic ecosystem.
4. **Static Traffic Demand:** The traffic demand during the simulation was assumed to remain constant, representing peak-hour conditions. Variations in demand over time, such as off-peak or fluctuating traffic volumes, were not considered.
5. **Perfect Communication for AVs:** For the platooning scenario, it was assumed that AVs have perfect communication between one another, allowing them to operate

seamlessly as a unit. This assumption simplifies the platooning model but aligns with future expectations of AV technology.

By adhering to these assumptions, the study focused on the core research objective: understanding how AV behaviors and penetration levels affect vehicle delay under varying traffic signal cycle lengths. The simulation results provided insights into how AVs could optimize traffic flow and reduce delays, mainly when aggressive or platoon behaviors were employed at shorter cycle lengths.

3 Results and Discussion

The results from the experiments, beginning with a cycle length of 60 s, show that vehicle delay significantly varies depending on the vehicle behavior and the penetration rate of AVs. As seen in Fig. 2(a), for a 60-s cycle length, aggressive and platoon AV behaviors exhibit the most substantial reduction in vehicle delay, reaching as low as 22 s when the AV penetration rate is 100%. Cautious AVs, in contrast, display higher delays, ranging between 32 and 35 s at full penetration. The mixed behavior scenario, "MixAllAv," also performs well under these conditions, with vehicle delays approaching 26 s at full AV penetration. The results indicate that, at shorter cycle lengths, aggressive and platoon AV behaviors allow for a much more efficient flow of traffic, minimizing delays and enhancing system performance.

When the cycle length is extended to 70 s, the trends remain similar but with slightly higher vehicle delays across all AV behaviors, as illustrated in Fig. 2(b). At full penetration, cautious AVs experience delays of around 38 seconds, while aggressive and platoon behaviors maintain their superior performance, reducing vehicle delays to approximately 23–25 s. Analysis reveals that the mixed behavior scenario yields variability in delay reduction across different signal cycle lengths. For example, at shorter cycle lengths (60–70 s), the "MixAllAv" scenario demonstrates a moderate decrease in vehicle delays, positioning it between aggressive and cautious behaviors. This suggests that aggressive AVs within the mix can somewhat counterbalance the increased delays caused by cautious AVs, improving traffic flow efficiency. These results reinforce that shorter cycle lengths and higher AV penetration rates provide the best traffic flow efficiency.

At an 80-s cycle length, shown in Fig. 3(a), the system's efficiency declines as vehicle delays increase. Cautious AVs experience delays of around 42 s at full penetration, while aggressive and platoon behaviors manage to keep delays at around 25–27 s. Interestingly, the efficiency of platoon behavior remains high under this cycle length, suggesting that the coordination and communication between AVs in a platoon allow them to navigate the traffic system better, even as cycle lengths increase. The mixed behavior scenario performs moderately well, though not as effectively as aggressive or platoon behaviors.

Moving to a cycle length of 90 s, represented in Fig. 3(b), the vehicle delays increase further. For cautious AVs, delays reach approximately 45 s at full penetration. However, aggressive and platoon behaviors still keep vehicle delays relatively low, at around 28–30 s. The mixed behavior scenario also shows better performance as AV penetration levels rise, which suggests that even in longer cycle lengths, the presence of aggressive or platoon AV behaviors can help reduce overall system delays. However, the general

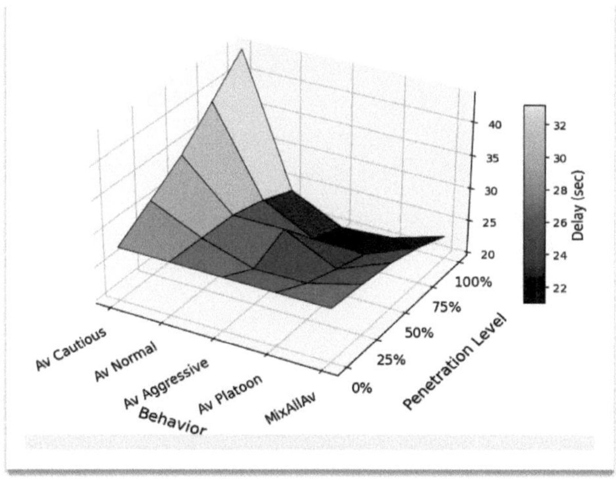

a)   Cycle Length of 60 sec

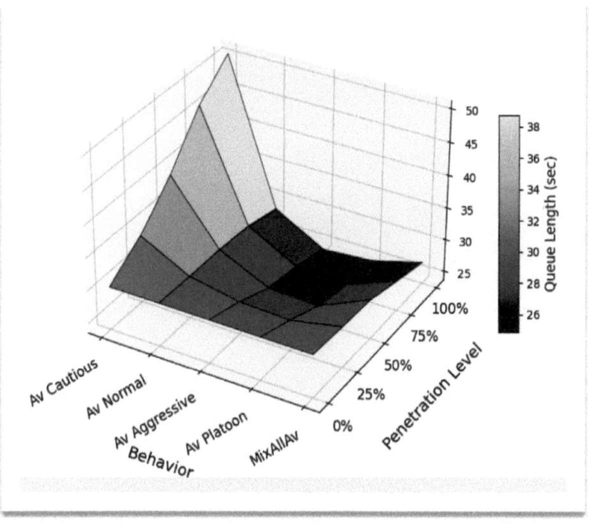

b)   Cycle Length of 70 sec

Fig. 2. Vehicle Delay at Cycle Lengths of 60 and 70 s.

trend indicates that longer cycle lengths contribute to higher vehicle delays, even as AV penetration increases.

Finally, for a 100-s cycle length, as shown in Fig. 4, the results depict the highest vehicle delays across all AV behaviors. Cautious AVs show the most significant delay, with values nearing 48–50 s at full AV penetration. Conversely, aggressive and platoon behaviors still perform better, with delays dropping below 30 s. The "MixAllAv" scenario demonstrates moderate delays, as expected, given the balance between cautious and

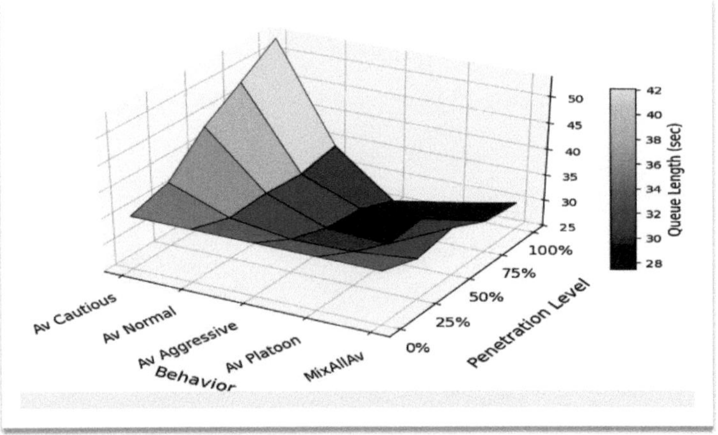

a)   Cycle Length of 80 sec

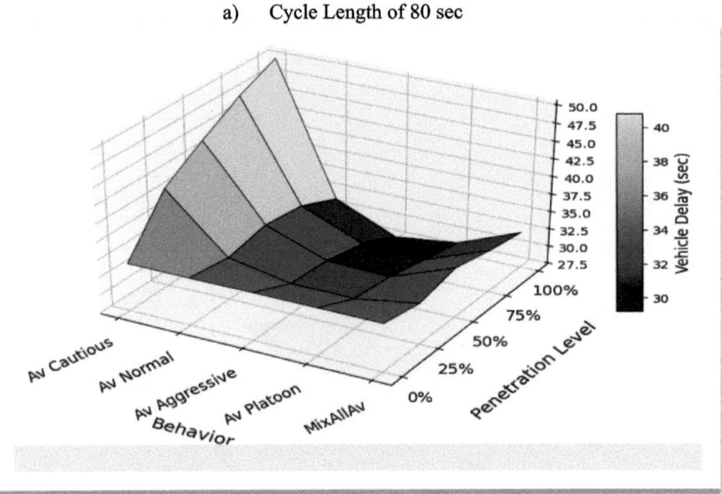

b)   Cycle Length of 90 sec

Fig. 3. Vehicle Delay at Cycle Lengths of 80 and 90 s.

aggressive behaviors. This final set of results emphasizes the importance of minimizing cycle lengths to achieve optimal traffic flow when AV penetration rates are high. Longer cycle lengths exacerbate vehicle delays, especially for cautious AV behavior.

The bubble plot in Fig. 5 adds another layer of insight by visually representing the vehicle delay across all cycle lengths and penetration rates. The size and color of each bubble correspond to the magnitude of the delay, with larger yellow bubbles representing higher delays and smaller purple bubbles indicating lower delays. The plot shows a clear trend where delays increase with cycle length and AV penetration for certain behaviors, particularly cautious AVs. The higher delay bubbles are clustered around longer cycle lengths and higher penetration rates for cautious behaviors. In contrast, smaller, lower-delay bubbles are concentrated at shorter cycle lengths and for aggressive and platoon

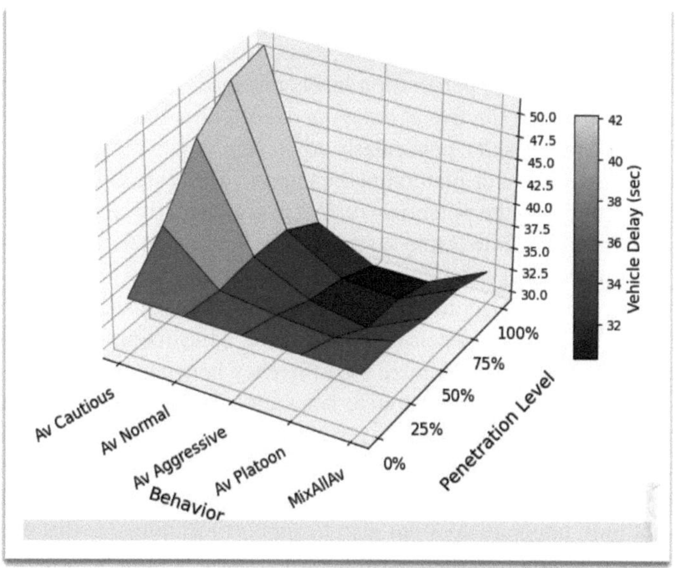

Fig. 4. Vehicle Delay at Cycle Length of 100 s

behaviors, especially at higher penetration rates. This visualization further confirms that aggressive and platoon AV behaviors significantly reduce vehicle delays, especially when shorter cycle lengths and AV penetration rates are higher.

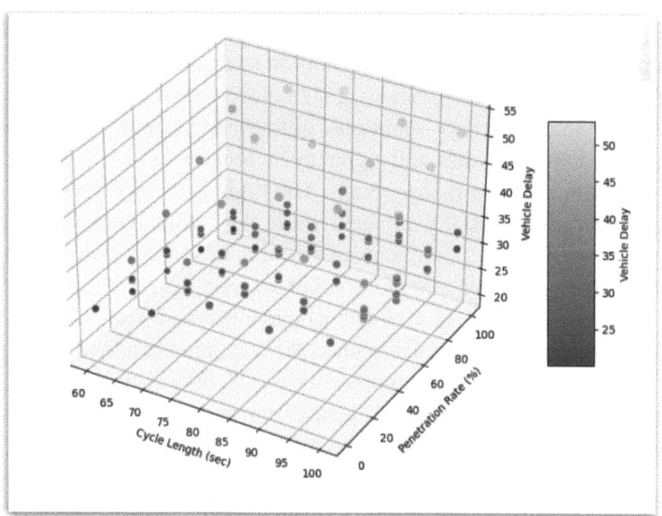

Fig. 5. Bubble Plot of Vehicle Delays with Continuous Hues and Sizes.

The overall results reveal that cycle length and AV behavior are key factors in determining vehicle delays in a traffic system. Cautious AV behavior generally results in higher delays, particularly as penetration levels increase, while aggressive and platoon behaviors are much more effective at reducing delays. Shorter cycle lengths, such as 60 or 70 s, provide the best results, particularly when paired with high penetration levels of AVs that exhibit aggressive or platooning behaviors. This outcome is consistent across all scenarios, highlighting the potential for AVs to significantly enhance traffic flow and reduce delays if integrated into traffic systems with shorter signal cycle lengths and more efficient driving behaviors.

The findings also underscore the importance of traffic management strategies in accommodating AVs. As AV penetration levels rise, adjustments to traffic signal cycle lengths may be necessary to maximize the efficiency benefits of AVs, especially those that adopt more aggressive or cooperative behaviors. The ability of platooning AVs to reduce delays is particularly notable, as it suggests that AVs capable of communicating and coordinating with one another can optimize traffic flow, even in more congested environments or under longer cycle lengths. This highlights the need for future traffic systems to not only integrate AVs but also to support technologies that enable vehicle communication to enhance efficiency further.

The results across all cycle lengths show that traffic systems can significantly benefit from higher penetration levels of AVs, especially when cycle lengths are shorter and AVs adopt efficient behaviors such as aggressive driving or platooning. These behaviors lead to the most significant reductions in vehicle delays, ensuring smoother traffic flow and reduced congestion.

4 Conclusion

This research uses microscopic traffic simulation to explore the effects of autonomous vehicles (AVs) on traffic flow efficiency at signalized intersections. The study assesses a range of AV driving behaviors—cautious, normal, aggressive, platoon, and mixed—across different penetration rates and signal cycle lengths to identify key strategies for optimizing urban intersection management. The findings demonstrate that higher penetration rates of AVs substantially enhance traffic flow, with aggressive and platooning behaviors leading to the most significant reductions in vehicle delays. For example, at a 60-s cycle length, aggressive and platoon behaviors reduced delays to as low as 22 s at full AV penetration, in contrast to the 35-s delay observed with cautious AV behavior and even with extended cycle lengths, such as 90 s, aggressive and platooning behaviors maintained relatively low delays of approximately 28–30 s. This indicates that combining shorter cycle lengths with higher AV penetration and assertive driving behaviors can significantly minimize delays and congestion.

On the other hand, cautious AV behavior results in increased delays, mainly as cycle lengths and penetration rates grow, emphasizing the need for adaptive AV driving strategies. The study also underscores the importance of optimizing signal cycle lengths to fully harness the efficiency of AVs, with 60–70 s cycles identified as the most effective.

To bridge the gap between theoretical research and practical application, this study suggests a roadmap for real-world validation. Future research should include pilot studies at various urban intersections where AV technologies are being tested. These studies would utilize advanced traffic monitoring technologies to validate simulation outcomes under actual traffic conditions. Collaborations with local traffic authorities will be essential for adjusting traffic signals and directly measuring the impact on traffic flow efficiency. Challenges likely to be encountered include the variability of urban traffic conditions, the integration of real-time data, and ensuring the reliability of communication technologies among diverse AV behaviors. Addressing these challenges will be crucial for accurately assessing the practical viability of AV integration strategies. These insights offer valuable guidance for urban planners and policymakers, illustrating that integrating AVs and well-tuned traffic signal timings can create smoother and more efficient traffic flow. Nevertheless, validating these findings in real-world conditions remains essential. Future research should focus on conducting field experiments in diverse urban environments to test the practicality of these strategies. Additionally, incorporating real-time data and machine learning algorithms could further enhance adaptive traffic signal systems, maximizing the benefits of AVs in urban networks.

In conclusion, adopting autonomous vehicles provides a significant opportunity to transform urban traffic management. By optimizing signal timing and implementing effective AV driving behaviors, cities can reduce congestion by up to 40%, support more sustainable transportation, and establish resilient urban infrastructure.

While this study provides significant insights into the impacts of autonomous vehicle behaviors on traffic efficiency at signalized intersections, it has limitations. The static nature of traffic demand in our simulations may not reflect real-world variability. Also, excluding non-automated vehicles and pedestrians limits the applicability of our findings to simpler urban environments. Future research should include dynamic traffic patterns and a broader mix of traffic participants to validate the robustness of AV integration strategies. Additionally, field tests and the integration of machine learning for traffic management could further enhance the accuracy and practicality of our models.

References

1. Almusawi, A., Albdairi, M., Qadri, S.S.: Integrating autonomous vehicles (AVs) into urban traffic: simulating driving and signal control. Appl. Sci. **14**(19), 8851 (2024). https://doi.org/10.3390/app14198851
2. Talebpour, A., Mahmassani, H.S.: Influence of connected and autonomous vehicles on traffic flow stability and throughput. Transp. Res. Pt C Emerg. Technol. **71**, 143–163 (2016). https://doi.org/10.1016/j.trc.2016.07.007
3. Wang, J., Zheng, Y., Xu, Q., Wang, J., Li, K.: Controllability Analysis and Optimal Control of Mixed Traffic Flow With Human-Driven and Autonomous Vehicles. IEEE Trans Intell Transp Syst. **22**(12), 7445–7459 (2021). https://doi.org/10.1109/TITS.2020.3002965
4. Fernandes, P., Nunes, U.: Platooning with IVC-enabled autonomous vehicles: strategies to mitigate communication delays, improve safety and traffic flow. IEEE Trans Intell Transp Syst. **13**(1), 91–106 (2012). https://doi.org/10.1109/TITS.2011.2179936
5. Gong, S., Du, L.: Cooperative platoon control for a mixed traffic flow including human drive vehicles and connected and autonomous vehicles. Transp. Res. B Methodol. **116**, 25–61 (2018). https://doi.org/10.1016/j.trb.2018.07.005

6. Zhou, M., Yu, Y., Qu, X.: Development of an efficient driving strategy for connected and automated vehicles at signalized intersections: a reinforcement learning approach. IEEE Trans Intell Transp Syst. **21**(1), 433–443 (2020). https://doi.org/10.1109/TITS.2019.2942014

7. Sun, C., Guanetti, J., Borrelli, F., Moura, S.J.: Optimal eco-driving control of connected and autonomous vehicles through signalized intersections. IEEE Internet Things J. **7**(5), 3759–3773 (2020). https://doi.org/10.1109/JIOT.2020.2968120

8. Fayazi, S.A., Vahidi, A.: Mixed-integer linear programming for optimal scheduling of autonomous vehicle intersection crossing. IEEE Trans. Intell. Veh. **3**(3), 287–299 (2018). https://doi.org/10.1109/TIV.2018.2843163

9. Wang, J., Peeta, S., He, X.: Multiclass traffic assignment model for mixed traffic flow of human-driven vehicles and connected and autonomous vehicles. Transp. Res. B Methodol. **126**, 139–168 (2019). https://doi.org/10.1016/j.trb.2019.05.022

10. Zheng, Y., Wang, J., Li, K.: Smoothing traffic flow via control of autonomous vehicles. IEEE Internet Things J. **7**(5), 3882–3896 (2020). https://doi.org/10.1109/JIOT.2020.2966506

11. Miglani, A., Kumar, N.: Deep learning models for traffic flow prediction in autonomous vehicles: a review, solutions, and challenges. Veh Commun. **20**, 100184 (2019). https://doi.org/10.1016/j.vehcom.2019.100184

12. Golbabaei, F., Yigitcanlar, T., Bunker, J.: The role of shared autonomous vehicle systems in delivering smart urban mobility: a systematic review of the literature. Int. J. Sustain. Transp. **15**(10), 731–748 (2021). https://doi.org/10.1080/15568318.2020.1798571

13. Gao, K., Han, F., Dong, P., Xiong, N., Du, R.: Connected vehicle as a mobile sensor for real-time queue length at signalized intersections. Sensors. **19**(9), 2059 (2019). https://doi.org/10.3390/s19092059

14. Albdairi, M., Almusawi, A.: Examining the influence of autonomous vehicle behaviors on travel times and vehicle arrivals: a comparative study across different simulation durations on the Kirkuk-Sulaymaniyah Highway. Int. J. Autom. Sci. Technol. **8**(3), 341–353 (2024). https://doi.org/10.30939/ijastech.1480916

15. Almusawi, A., Albdairi, M., Qadri, S.S.S.M.: Assessing traffic performance: comparative study of human and automated HGVs in urban intersections and highway segments. Int. J. Integr. Eng. **6**(5), 409–427 (2024). https://doi.org/10.30880/ijie.2024.16.05.031

Integrating Machine Learning with Optimization to Solve Time-Dependent Cash in Transit Vehicle Routing Problem with Time Windows

Alev Taskin[1] , Aslihan Sagiroglu[2(✉)] , Ezgi Zehra Seker[1],
Melisa Caliskan Demir[3] , and Khaled Dandis[1]

[1] Yildiz Technical University, Istanbul, Turkey
[2] Istanbul Arel University, Istanbul, Turkey
`aslihansagiroglu@arel.edu.tr`
[3] Istanbul Aydin University, Istanbul, Turkey

Abstract. Cash in Transit (CIT) involves the transportation of banknotes, coins, and other valuables. The transportation of these items inherently presents certain risks, necessitating security measures to protect the process. This study considers the time-dependent vehicle routing problem, where a fixed-capacity vehicle fleet collects predetermined cash from customers. The study consists of two stages. First, we apply a comparison of the most common machine learning methods, such as multiple linear regression, polynomial regression, decision trees, random forests, support vector machines, multilayer perceptron, and generalized regression neural networks, to predict traffic conditions that directly affect the speed of the vehicles. Second, the estimated speed values are utilized in the mathematical model to minimize travel time between customers during the cash collection operation. Subsequently, a real-life application is conducted in Istanbul to evaluate the effect of the proposed model, and it is solved using GAMS . This study is a starting point for future research focusing on using machine learning to predict dynamic parameters such as traffic density.

Keywords: Vehicle Routing Problem · Machine Learning · Optimization

1 Introduction

The research on CIT has been divided into two categories based on the types of inconsistencies they address: time inconsistency and path inconsistency [1].

The literature highlights concern about time inconsistency in CIT operations, where service times for periodic customer visits vary. [2] tackle this with a periodic vehicle routing problem with time windows (PVRP-TW), proposing a mixed integer linear model and a multi-start-iterated local search. [3] focuses on serving customers within a day, turning it into a sequential vehicle routing problem (VRP) with multiple time windows. Additionally, [3] explores earlier departures and introduces waiting, enhancing efficiency. [4]

A. Mirzazadeh et al. (Eds.): ODSIE 2024, CCIS 2482, pp. 245–264, 2026.
https://doi.org/10.1007/978-3-031-93601-2_16

examines multiple alternative paths, improving results with a multi-graph approach. [5] models cash transportation movements using time-space networks and solves a PVRP-TW with known demands, considering penalties for waiting. In [6], stochastic travel times and delay penalty costs are incorporated. [7] tackles total collection time minimization and varied daily tours to prevent robberies, forming the Dissimilar Routing Problem (DisARP).

For path inconsistency, several distinct approaches exist to address the CIT problem, starting with the m-peripatetic traveling salesperson problem (m-PTSP), as illustrated in relevant studies [8–10]. Another structure of the path inconsistency CIT problem is given as an m-peripatetic vehicle routing problem (m-PVRP), such as in [11–13]. An alternative approach to addressing inconsistencies based on paths is found in the risk-constrained cash in transit vehicle routing problems (RCTVRP) [14–17].

Recent studies have shown that traffic concerns significantly increase the risk associated with cash transportation. [18] presents novel Cash-in-Transit models that account for deterministic and stochastic time-varying traffic congestion. A new formula is developed to quantify travel risk to address the proportional relationship between risk exposure and time-dependent travel durations. [19] presents a multi-objective periodic routing problem for cash transportation, aiming to enhance security by creating unpredictable alternative routes and varying arrival times at demand nodes. It addresses several limitations of earlier models related to dissimilar routing and cash transportation challenges. The proposed problem encompasses three objectives: reducing robbery risk, minimizing completion times by considering the effects of daily traffic congestion, and maximizing customer satisfaction. They developed a novel evolutionary algorithm inspired by NSGA-II to handle the computational challenges associated with these features. [20] examines a novel CIT problem, representing a variation of the time-dependent VRP with time windows. To enhance the realism of the time-dependent CIT problem, vehicle speeds are modeled to vary based on traffic density. [21] introduces a new mathematical model for a location-routing problem involving transport vehicles in the banking system, considering urban traffic conditions. The model simultaneously addresses three objectives: minimizing greenhouse gas emissions, reducing location and routing costs, and enhancing customer satisfaction. Another significant study in the literature focuses on a new model based on game theory with multiple objectives to improve cash-in-transit security. A dual-objective vehicle routing problem is developed that minimizes the transfer risk and the distance traveled by vehicles. The performance of the robber is estimated using a game theory approach. [22] focuses on developing a vehicle routing model with dual objectives of minimizing risk and distance to enhance the safety of cash and valuable commodities transportation. An adapted multi-objective genetic algorithm is introduced to optimize the CIT-VRP.

[23] investigates how drones can enhance transportation security during intentional disruptions. The research introduces a novel routing framework, Integrated Routing and Surveillance, combining ground and aerial vehicle operations in dynamic environments to ensure secure transportation. [24] introduces a novel bi-objective CIT-VRP model tailored for the CIT sector. The model addresses economic and environmental objectives, aiming to provide an effective solution to the routing challenges specific to this domain.

In recent, the importance of researching risk reduction for the transportation of hazardous materials and valuables has been demonstrated in the literature, and the increasing trend of research in CIT areas has been demonstrated [25]. A valuable commodity transportation industry's new type of VRP is modeled by taking the route risk constraint in [26]. In [27], novel fuzzy programming with recourse (FPR) model formulation is proposed for the open vehicle routing problem with trapezoidal fuzzy demand (OVRPFD). [28] focuses on optimizing vehicle routing models for cash distribution in ATM networks. It proposes a cooperative mathematical model for CIT firms to minimize risk and cost. The findings highlight that increased collaboration among banks leads to significant reductions in both risk and cost. [29] emphasize the broad field of view, cost-effectiveness, and agility of drones and propose that drones perform route reconnaissance to ensure the safety of vehicles carrying valuable goods. Similarly, [30] propose using drones to identify suspicious situations for valuables transportation. For this purpose, an integrated route planning and surveillance problem has been developed to enhance cash security and transportation of valuable goods.

This study employs machine learning methods to predict traffic speed instead of relying on the traffic prediction approaches in the CIT literature. The performance of the methods used is compared. The predicted traffic speed is integrated into a time-dependent, capacitated VRP with time windows, which aims to minimize travel time in cash transportation.

1.1 Problem Definition

The CIT process involves transporting valuable items such as banknotes and coins, requiring stringent security measures to ensure safety during transit. These operations typically consist of cash distribution to various locations, collection from these points, or a combination of both.

A critical challenge in CIT logistics is minimizing vehicles' time on the road, particularly when traveling between customers. Traffic density heavily influences this duration, directly impacting operational efficiency and security. To reduce travel time and mitigate the risk of robberies, vehicles are required to operate during periods of low traffic density. However, the unpredictability of traffic conditions presents a significant obstacle. To address this, machine learning algorithms are employed to forecast vehicle speeds during CIT operations. These predictions are incorporated into a Time-Dependent Capacitated VRP with Time Windows model, optimizing vehicle routes to minimize travel time and enhance security against potential threats.

In this study, a novel approach to route planning is proposed. Unlike conventional methods, where vehicles are first transported to a bank before commencing their routes, as illustrated in Fig. 1(left), this study suggests that vehicles start their routes directly from the CIT company's depot. The vehicles will serve customers, deposit collected cash at the bank, and return to the depot at the end of the day, as shown in Fig. 1(right).

The primary contributions of this research are as follows:

- Forecasting Vehicle Speeds: Utilizing machine learning and deep learning algorithms to predict vehicle speeds, followed by a comparative analysis to identify the most accurate method for estimating travel times.

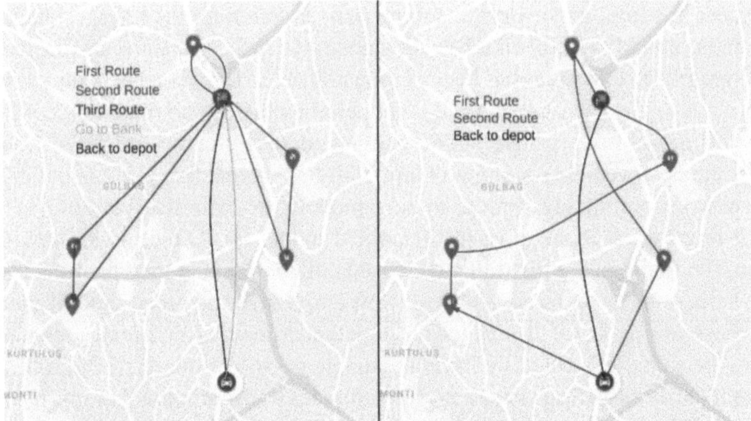

Fig. 1. Difference between direct and indirect traveling between nodes.

- Integration into a Mathematical Model: Incorporating the predicted speed values into a time-dependent, capacitated vehicle routing model with time windows, ensuring precise optimization.

This comprehensive approach aims to enhance the efficiency, reliability, and security of CIT operations while addressing the critical challenges of traffic conditions and operational constraints.

The rest of this paper is structured as follows. Section 2 explains the study's methodology and introduces the machine learning methods. Section 3 presents a case study to demonstrate the model. It also describes the proposed mathematical model that reduces the cost and risk of cash handling. Section 4 discusses the results and finally offers ideas for future work.

2   Methodology

To solve the CIT time-dependent capacitated VRP with time windows, the methodology steps are defined in Fig. 2. In the first stage, several machine learning algorithms predict the traffic data. This traffic data is used in the second stage of the VRP optimization model.

2.1   The Data Collection

The initial step involves gathering essential data for the mathematical model, including customer locations, distances, demands, vehicle capacities, and historical traffic data. However, since traffic data are only available for past periods, predicting future traffic values is necessary for solving the mathematical model to optimize CIT routes. To achieve this, seven machine learning algorithms are employed to forecast traffic speed values by using a cross-validation strategy to determine the most accurate prediction among these algorithms. Subsequently, a mathematical model implemented in GAMS is employed to find optimal routes based on the predicted data.

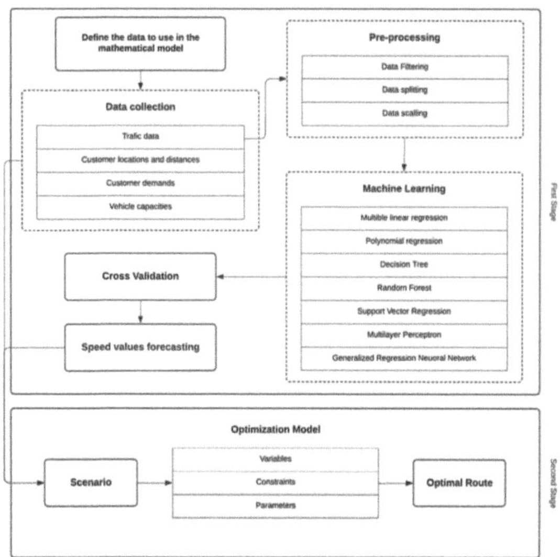

Fig. 2. Proposed methodology.

2.2 Data Preprocessing

Data preprocessing is the process of converting raw data into a format that can be effectively utilized. It is a crucial step in machine learning, as we cannot work with unprocessed data.

Normalization. Normalization focuses on adjusting features to operate on a comparable scale, enhancing the model's performance and training stability. A common approach to normalization is scaling data to a consistent range. In this case, the standard scaler method is applied, standardizing each feature by subtracting its mean and scaling it to have unit variance. Achieving unit variance involves dividing all values by their standard deviation.

Data Splitting. Data splitting involves dividing a dataset into two or more subsets. Typically, in a two-part split, one subset is used for model training while the other is utilized for evaluation or testing. Different ratios are used to split the data in the literature; for this application, the data are divided randomly into two data sets with a test size equal to 30%.

2.3 Background Information of Machine Learning Algorithms for Prediction of Traffic Speed

Machine Learning (ML) focuses on creating models that improve automatically with experience. ML is the field devoted to building methods that learn from experience. This section provides background information on the machine learning algorithms used in this study. The algorithms used are chosen among the frequently used algorithms in the literature.

Multiple Linear Regression. Regression analysis aims to construct a mathematical model describing or explaining the relationship between variables. Its simple case is when there are only two variables: the first one is the explanatory variable, and the second one is the response variable, which is dependent on the first one. In regression, a smooth line between the explanatory variables, which may be more than one, is tried to predict the response variable. We would not expect an exact fit, as at least one of the variables is subject to change fluctuations due to factors out of our control [31].

$$y_i = \beta_0 + \beta_1 x_i 1 + \beta_2 x_i 2 + \cdots + \beta_j x_i + \varepsilon, for\ i = 1, 2, \ldots, n \tag{1}$$

where, for i = n observations and j = m explanatory variables, y_i = response variable, x_i = explanatory variables, β_0 = constant term, β_j = slope coefficient for each independent variable, and ε is the model's error term.

Polynomial Regression. Multiple linear regression is a regression method that involves more than one explanatory or independent variable. A variant of this approach is polynomial regression, which models the dependent variable as a function of the powers of the independent variables [33]. This method is especially useful when there is evidence indicating a curvilinear connection between the dependent and independent variables. The polynomial equation with multiple independent variables can be expressed as follows:

$$y_i = \beta_{01} + \beta_{11} x_{i1} + \beta_{21} x_{i1}^2 + \cdots + \beta_{k1} x_{i1}^k + \beta_{02} + \beta_{12} x_{i2} + \beta_{21} x_{i2}^2 + \cdots + \beta_{k2} x_{i2}^k +$$

$$\cdots + \beta_{0j} + \beta_{1j} x_{ij} + \beta_{2j} x_{ij}^2 + \cdots + \beta_{kj} x_{ij}^k \varepsilon$$

$$for\ i = 1, 2, \ldots, n$$

$$\tag{2}$$

where is the degree of the polynomial. As the degree of the model increases, it will become more complicated, and more data will be tried to fit in the training phase. And that could lead to over-fitting in some situations. So, it is essential to balance the degree of the polynomial with the data in the hand and its features.

Decision Trees. Decision tree regression generates a single output, the response variable, based on one or more input or explanatory variables. The regression tree is constructed using a method called binary recursive partitioning. This iterative process divides the data into partitions or branches, repeatedly splitting each partition into smaller subsets as the tree grows. Initially, all training data points are grouped in one partition. The algorithm evaluates all possible binary splits across variables, selecting the one that minimizes the total sum of squared deviations from the mean in the resulting partitions. This splitting criterion is recursively applied to each new branch. The process continues until the nodes meet the predefined minimum size requirement, at which point they are designated as terminal nodes [34].

Random Forest. This method is an ensemble learning algorithm designed explicitly for regression tasks. It enhances prediction accuracy by combining outputs from multiple decision trees. In contrast to standard decision trees, where each node is split using the optimal choice from all available variables, random forests split nodes by selecting the best option from a randomly chosen subset of predictors at each node [35]. The process begins by generating n_{tree} bootstrap samples from the original dataset. For each bootstrap sample, an unpruned regression tree is constructed, but instead of evaluating all predictors, m_{try} predictors are randomly chosen, and the best split is determined among them. Predictions for new data are made by aggregating the outputs of all n_{tree} trees.

Support Vector Regression (SVR). SVR is a variation of the Support Vector Machine (SVM) algorithm explicitly designed for regression tasks. Unlike traditional regression models that aim to minimize the error between actual and predicted values, SVR focuses on fitting the best line within a predefined margin known as the threshold value or epsilon. This threshold indicates the distance between the hyperplane and its boundary lines. SVR utilizes -intensive loss function, where no penalty is assigned If the predicted value falls within an epsilon range of the true value [32]. The parameter plays a crucial role in determining the number of support vectors influencing the regression function. A smaller epsilon leads to more support vectors being selected. The cost parameter () also significantly affects the model's complexity and performance by controlling the tolerance for deviations from the true value. Larger () values result in less tolerance for deviations, enabling the model to fit the data more closely [36]. SVR can also handle non-linear relationships by applying polynomial kernels or other kernel functions to expand the input feature space. This allows the algorithm to achieve an approximate linear separation in a higher-dimensional space, enhancing its ability to model complex patterns effectively.

Parameter gamma () is introduced by the kernel used, specifically the radial basis function.

$$K\left(x_i, x_j\right) = \exp\left(-\gamma \left\| x_i - x_j \right\|^2\right) \tag{3}$$

Gamma, the reciprocal of the standard deviation, determines the range of influence of the support vectors. When gamma is high, the influence of each support vector is confined to its immediate vicinity.

Multilayer Perceptron (MLP). MLP is a type of artificial neural network composed of a minimum of three neuron layers: an input layer, one or more hidden layers, and an output layer. Every neuron in a layer is linked to all neurons in the subsequent layer, with each connection having a specific weight. For a given set of inputs $X = x_1, x_2, ...,$ x_R, and outputs $y = y_1, y_2, ..., y_s$ where is the number of inputs and S is the number of outputs, the MLP serves as a nonlinear function approximator $(\cdot) : \rightarrow$ for tasks like classification or regression [37]. The input layer consists of neurons representing the input features X. The output layer takes the information from the last hidden layer and transforms it into the final output values. The neurons in the hidden layers are responsible for performing a weighted summation of the values from the previous layer. This is followed by the addition of a bias term, and then the result is passed through a nonlinear activation function. For example, the output of the j_{th} node in the first hidden layer can be expressed as:

$$\text{output} = g\left(\sum_{i=1}^{R} w_{ji} x_i + b_i\right) \tag{4}$$

The equation above is a nonlinear activation function, w_{ji} is the weight, and b_i is the bias. There are various activation functions used in the MLP model such as "identity", "sigmoid", "tanh", and "relu". In our application, the rectified linear unit activation function, which is called "rule", is used. This activation function ensures no negative outputs [38].

Generalized Regression Neural Network (GRNN). GRNN is a type of radial basis neural network (RBNN), often seen as a variation of the multilayer perceptron with a single hidden layer. Key differences between GRNN and standard MLP or RBF neural networks include the absence of weights between the input and hidden layers in GRNN and the use of radially symmetrical activation functions in the hidden layer nodes [39]. GRNN is based on kernel regression networks and does not rely on iterative training procedures like backpropagation. Instead, it approximates arbitrary functions between input vectors and output vectors directly from the training data [40]. Additionally, GRNN demonstrates consistency, with estimation error approaching zero as the training set size increases.

GRNN architecture comprises four layers: an input layer, a pattern layer, a summation layer and an output layer. The input layer is associated with the number of parameters in the training set and is connected to the pattern layer. Each neuron in the pattern layer represents a training pattern and generates its output. This layer is connected to the summation layer, which carries out two types of summation: one for division and the other for summation. The summation and output layers are responsible for normalizing the output set. Each unit in the pattern layer connects to two summation layer neurons, S and D. The first summation neuron calculates the sum of the weighted outputs, while the second computes the sum of the unweighted pattern neuron outputs. This architecture allows GRNN to perform efficient, direct function estimation from training data [41].

The output layer simply divides the output of each -summation neuron by that of each -summation neuron, resulting in the predicted value Y'_i for an unknown input vector x, as shown below:

$$Y'_i = \frac{\sum_{i=1}^{n} y_i \cdot \exp[-D(x, x_i)]}{\sum_{i=1}^{n} \exp[-D(x, x_i)]} \tag{5}$$

$$D(x, x_i) = \sum_{k=1}^{m} \left(\frac{x_i - x_{ik}}{\sigma} \right)^2 \tag{6}$$

In this context, y_i represents the weight associated with the connection between the i-th neuron in the pattern layer and the S-summation neuron, refers to the total number of training patterns in the dataset, denotes the Gaussian function used to calculate the similarity between input vectors, m indicates the number of elements in an input vector, x_k and, x_{ik} are the j_{th} element of and, x_i, respectively, is recognized as the spread parameter of the Gaussian function, its value is experimentally determined for optimal performance.

Mean Squared Error (MSE) and R^2 are used as evaluation criteria for machine learning methods.

$$MSE_{(x,y)} = \frac{1}{N} \sum_{i=1}^{N} (x_i - y_i)^2 \tag{7}$$

where $X = \{x_i | i = 1, 2, \ldots, N\}$ and $Y = \{y_i | i = 1, 2, \ldots, N\}$ are two finite-length discrete signals, where N is the number of signal samples and x_i and y_i are the values of the ith samples in X and Y, respectively [42].

$$R^2 = 1 - \frac{RSS}{TSS} \tag{8}$$

where R^2 refers to the coefficient of determination, RSS is the sum of squared residuals, and TSS is the total sum of squares.

3   A Real-Life Application in Istanbul

3.1   Traffic Data and Forecasting

The data used in the first stage of the application is collected from the open data portal affiliated with Istanbul Metropolitan Municipality. The data contains the date times, longitude, latitude, geohash, minimum speed, maximum speed, average speed, and number of vehicles for every hour. The data from 01/04/2020 until 31/03/2021 in 'sxk9s' geohash are considered for the prediction application. This location is in the middle of all customers' locations, and all the journeys are affected by the traffic density of that location. The methodology predicts the next hour's average speed, providing the minimum, maximum, average, and number of vehicles in the previous hour.

The data are split into training and testing data, and then the MSE and the R^2 are calculated for the predictions of the next hour's average speed and the real average speed of the next hour. GridSearch is used to find the best parameters for all algorithms. MLR, PLY, DTs, RF, SVR, MLP, and GRNN are used to predict traffic situations in Istanbul. Python programming is used to predict traffic speed. This study uses the scikit learn library to run all machine learning algorithms.

Speed Predictions at Random Dates and Times. This comparison gives a general impression of the model's performance. The data are split randomly into a training set and a test set with a proportion of 70% and 30%, respectively. The comparisons of all algorithms used to predict the average traffic amount are given in Table 1. The one with the lowest MSE and highest R^2 values belong to the MLP model.

Speed Predictions for the Duration of a Specific 24-H Time Window. This comparison is made specifically to see the model's performance under 24 consecutive hours on a specific date, which is used in the mathematical optimization model. Table 2 shows that there is no much difference between the performance values among the methods here as well, and the best values are for the SVR, with 5.567 for MSE and 0.959 for R^2.

Figure 3 shows the real speed values and the estimated values of all regression models. We choose the prediction values of the SVM model to be used in the optimization model since it has the best score for the values going to be used. The predicted speed values of the vehicle in km/h can be shown in the Table 3. The speed values from 15:00 to 23:00 are the targeted values since these are the cash-collecting hours available for the customers.

Table 1. Performance metrics of the regression algorithms for test data over random dates and times.

Method	MSE	R^2
MLR	20.845	0.819
PLY	20.844	0.819
DTs	22.119	0.808
RF	21.270	0.816
SVM	21.250	0.816
MLP	**20.811**	**0.820**
GRNN	21.058	0.818

Table 2. Performance metrics of the regression algorithms for test data over a specific 24-hour time window.

Method	MSE	R^2
MLR	6.721	0.951
PLY	6.311	0.953
DTs	7.666	0.944
RF	6.370	0.953
SVR	**5.567**	**0.959**
MLP	6.029	0.956
GRNN	10.632	0.922

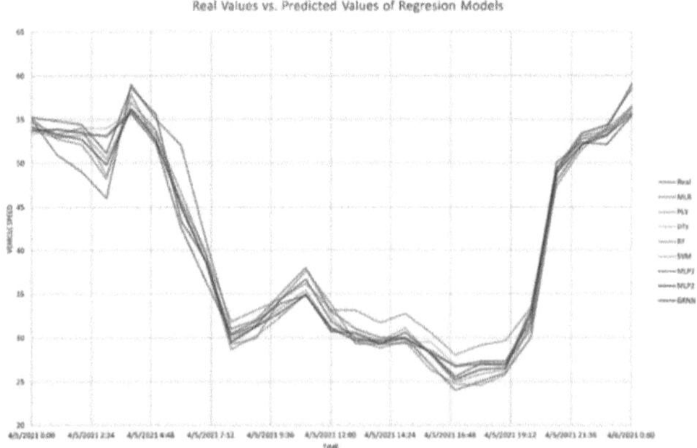

Fig. 3. Real vs. predicted values of all models evaluated over a 24-hour time window

Table 3. Predicted speed values using the SVM method.

Date and Time	Predicted Speed
4/5/2021 15:00	29.61
4/5/2021 16:00	26.40
4/5/2021 17:00	24.71
4/5/2021 18:00	24.60
4/5/2021 19:00	25.78
4/5/2021 20:00	31.61
4/5/2021 21:00	48.83
4/5/2021 22:00	52.72

3.2   Customer Locations and Distances

The customers' locations are selected randomly in the area of Zincirlikuyu in Istanbul. This area is chosen for this application because it is considered one of the most crowded areas in Istanbul. So, the speed of the cars and then the optimization problem, including the time spent by the armored cash vehicles, are affected a lot by this area.

In the optimization application, the vehicles leave the CIT depot from node 1 to serve the customers in nodes 2 to N. Then, the vehicles must go to the last node, N, to unload the money at the bank. Lastly, all vehicles must return to their location at the CIT depot at node 1.

We include a total of 7 nodes in this application, including five customers, the depot at node 1, and the bank at node 7. We can see the locations of all the nodes in Fig. 4.

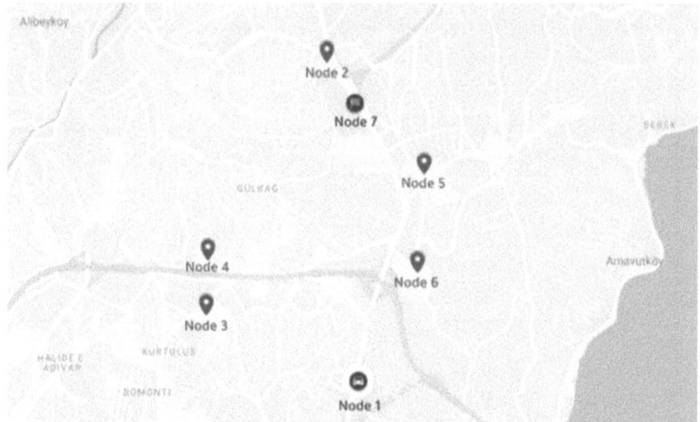

Fig. 4. Locations of the Customers, Depot, and the Bank.

The distances between the locations are calculated using Google Maps. The main roads are considered while calculating the distances since the armored vehicles can only choose the main roads to make their trips in to avoid any risk. Table 4 shows the distances between the locations in km. Notice that the table is not symmetric since the distances may change depending on the starting point.

Table 4. The distances between the nodes' locations.

	Node 1	Node 2	Node 3	Node 4	Node 5	Node 6	Node 7
Node 1	0	4.9	3.2	3.7	2.5	1.6	3.7
Node 2	3.8	0	5.6	4.5	2.3	3.6	0.4
Node 3	2.9	5.7	0	1.2	3.1	2.6	4.4
Node 4	3.6	6.3	1.9	0	3.9	3.3	5.1
Node 5	2.5	6	4.5	3.2	0	2.3	4.6
Node 6	3.7	4.8	5	3.8	2.1	0	3.6
Node 7	3.4	2.4	5.4	4.1	2.1	3.2	0

3.3 Customer Demands

The amount of money to be collected depends on the customers. It is determined by the customers before the collection process begins. For our application, we choose the customer demands roughly as shown in the Table 5 the first and last nodes do not have any demand since these are the depot and the bank nodes.

Table 5. Customer Demands

Customer	Collected amount
Node 1	–
Node 2	440,000TL
Node 3	310,000TL
Node 4	460,000TL
Node 5	200,000TL
Node 6	580,000TL
Node 7	–

3.4 Vehicles Capacities

In the CIT company, there is only one type of armored vehicle with a capacity of 1,500,000 TL. The number of vehicles selected depends on the amount of money needed. The total money collected from the customers equals 1,990,000TL, as shown in Table 5. Therefore, there will be two armored vehicles for the collection process.

3.5   Customer Service Time and Time Windows

Customer service times and time windows vary depending on the location and the type of customer. These values are predefined by the experts and by the customers' previous operations. The values are summarized in the Table 6.

Table 6. Customer service time and time windows.

Customer	Service Time	Earliest Time	Latest Time
Node 1	–	6:00	18:00
Node 2	15 min	6:00	18:00
Node 3	20 min	7:00	18:00
Node 4	35 min	6:00	16:00
Node 5	20 min	6:00	18:00
Node 6	30 min	6:00	18:00
Node 7	–	6:00	18:00

3.6   Mathematical Model Formulation

In our VRP problem, we try to minimize the time that the armored vehicles spend traveling between the nodes. The problem is represented by a direct graph, which is defined as $G = (N, A)$, where N represents the set of nodes in the network and A represents the set of arcs between the nodes. For $i = 1, 2, ..., N$ locations, where the first node $i = 1$ refers to the depot of the CIT company, $I = i$ refers to the bank, and the remaining nodes refer to the customers who plan to be served. The set of arcs defined on $A = \{(i, j) : \forall i, j \in N, i \neq j\}$ and d_{ij} is a nonnegative distance between the location i and j. Also $d_{ij} \neq d_{ji}$. In addition, each has a service time S_i and a time window $[m_i, n_i]$ where the arrival time A_i and leaving time L_i has to be within this time window. If necessary, any waiting time W_i has to be done in this time window.

The set K represents the vehicles with $K = 1, 2, ..., k$. The customer demands are known in advance and defined in the variable O_i. Also, the capacity of the vehicle is predefined in the variable R_k.

The time interval set is defined as T with $T = 1, 2, ..., t$. The beginning of the time interval is defined as B_t and the ending time of the interval is defined as E_t, and the velocity V_t varies on the interval. Traveling time t_{ij} of the vehicle from the node i to node j is calculated with the speed V_t and the distance d_{ij}. If the travel begins in the interval t and the next time interval is u, then the traveling time calculation is made by $(E_t - L_i) V_t + (A_j - B_u)V_u$. The vehicles start at the depot, pick up the cash from all the customers, drop them in the back, and return to the depot.

Sets
T: Set of time intervals

I: Set of nodes
K: Set of vehicles

Parameters
d_{ij}: The distance between the customers i and j
v_t: The velocity of the vehicle at time interval t
O_i: The demand of customer i
R_k: The capacity of vehicle k
B_t: The beginning time of the time interval t
E_t: The ending time of the time interval t
S_t: The service time of the customer t
$[m_i, n_i]$: The service time window for customer i
T_0: The starting time of the distribution
M: Massive number

Decision Variables
A_i: The arrival time to customer i
L_i: The leaving time from the customer i
w_i: The waiting time at the customer i
p_{ik}: Auxiliary variable for sub-tour elimination constraints in route k
f_{ik}: fraction of the customer's demand i delivered by vehicle k
t_{ijk}: The traveling time from the customer i to customer j by the vehicle k

$$X_{ijk}: \begin{cases} 1, \text{ if arc (i, j) is used by vehicle;} \\ 0, \text{ otherwise} \end{cases}$$

$$Y_{ijk}^{tu}: \begin{cases} 1, \text{ vehicle leaves from atm (i) at time interval (t),} \\ \quad \text{arrives to atm (j) at time interval (u);} \\ 0, \text{ otherwise} \end{cases}$$

Formulation
Object Function

$$\min \Sigma_{j,i\in I}\, \Sigma_{k\in K} t_{ijk} \tag{9}$$

Subject to

$$\sum_{t\in T}\sum_{u\in T} Y_{ijk}^{tu} = X_{ij} \quad (\forall i \in I, \forall j \in I)(\forall k \in K) \tag{10}$$

$$t_{ijk} \leq A_j - L_i + M\left(1 - X_{ijk}\right) \quad (\forall i \in I, \forall j \in I)(\forall k \in K) \tag{11}$$

$$t_{ijk} \geq A_j - L_i + M\left(X_{ijk} - 1\right) \quad (\forall i \in I, \forall j \in I)(\forall k \in K) \tag{12}$$

$$d_{ijk} - M\left(2 - X_{ijk} - Y_{ijk}^{tt}\right) \leq (A_j - L_i)V_t \quad (\forall i \in I, \forall j \in I, i = j), (\forall t \in T)(\forall k \in K) \tag{13}$$

$$d_{ijk} - M\left(X_{ijk} + Y_{ijk}^{tt} - 2\right) \geq (A_j - L_i)V_t \quad (\forall i \in I, \forall j \in I, i = j), (\forall t \in T)(\forall k \in K) \tag{14}$$

$$d_{ijk} - M\left(2 - X_{ijk} - Y_{ijk}^{tu}\right) \leq (E_t - L_i)V_t + (A_j - B_u)V_u \quad (\forall i \in I, \forall j \in I), (\forall t \in T, \forall u \in T)(\forall k \in K) \tag{15}$$

$$d_{ijk} - M\left(X_{ijk} + Y_{ijk}^{tu} - 2\right) \geq (E_t - L_i)V_t + (A_j - B_u)V_u \qquad (\forall i \in I, \forall j \in I), (\forall t \in T, \forall u \in T)(\forall k \in K) \qquad (16)$$

$$\sum_{i \in K} \sum_{j \in I} \sum_{t,u \in T} Y_{ijk}^{tu} E_t \geq L_i \qquad (\forall i \in I) \qquad (17)$$

$$\sum_{k \in K} \sum_{j \in I} \sum_{t,u \in T} Y_{ijk}^{tu} B_t \leq L_i \qquad (\forall i \in I) \qquad (18)$$

$$\sum_{k \in K} \sum_{j \in I} \sum_{t,u \in T} Y_{ijk}^{tu} E_u \geq A_i \qquad (\forall j \in I) \qquad (19)$$

$$\sum_{k \in K} \sum_{j \in I} \sum_{t,u \in T} Y_{ijk}^{tu} B_u \leq A_i \qquad (\forall j \in I) \qquad (20)$$

$$L_1 = T_0 \qquad (21)$$

$$L_i = A_i + s_i + w_i \qquad (\forall i \in I) \qquad (22)$$

$$A_i + w_i \geq m_i \qquad (\forall i \in I) \qquad (23)$$

$$L_i \leq n_i \qquad (\forall i \in I) \qquad (24)$$

$$\sum_{k \in K} \sum_{i \in I} X_{ijk} = 1 \qquad (\forall j \in I, j/\{1, N\}) \qquad (25)$$

$$\sum_{k \in K} \sum_{j \in I} X_{ijk} = 1 \qquad (\forall i \in I, i/\{1, N\}) \qquad (26)$$

$$\sum_{i \in I} X_{iNk} = 1 \qquad (\forall k \in K) \qquad (27)$$

$$X_{N1k} = 1 \qquad (\forall k \in K) \qquad (28)$$

$$\sum_{i \in I} X_{irk} - \sum_{j \in I} X_{rjk} = 0 \qquad (\forall r \in I)(\forall k \in K) \qquad (29)$$

$$X_{iik} = 0 \qquad (\forall i \in I)(\forall k \in K) \qquad (30)$$

$$\sum_{i \in I} O_i * f_{ik} \leq R_k \qquad (\forall k \in K) \qquad (31)$$

$$\sum_{k \in K} f_{ik} \geq 1 \qquad (\forall i \in I) \qquad (32)$$

$$p_{ik} - p_{jk} + N\left(X_{ijk}\right) \leq (N - 1) \qquad (\forall i \in I, \forall j \in I)(\forall k \in K) \qquad (33)$$

$$f_{ik} \leq \sum_{j \in I} X_{jik} \qquad (\forall k \in K) \qquad\qquad (34)$$

$$A_i, L_i, t_{ijk}, w_i, f_{ik}, p_{ik} \geq 0 \qquad (\forall i \in I, \forall j \in I)(\forall k \in K) \qquad\qquad (35)$$

$$X_{ijk}, \ Y_{ijk}^{tu} \in \{0, 1\} \qquad (\forall i \in I, \forall j \in I), (\forall t \in T, \forall u \in T)(\forall k \in K) \qquad (36)$$

3.7 Results of the Application

The problem has 7 total nodes, including the depot node number 1 and the bank node number 7. This problem is solved using GAMS. The optimal solution is found with a total traveling time equal to 53.526 min, and the vehicles arrive at the starting point (the depot) at 19:29:251. Hence, the total operating time equals 4 hours and 29.251 min. The routes of the vehicles are as follows:

The first route was found to be 1-6-2-7-1. The route starts at 15:00:00 at node 1; the vehicle travels a total distance of 1.6 km to reach node number 6 with a total time of 3.242 min. The service time for customers in node 6 takes 30.000 min, and a waiting event occurs at that node for around 17.031 min. Lastly, the vehicle leaves node 6 at 15:51:591. The vehicle continues its route, as we can see in the Table 7.

Table 7. Route 1-6-2-7-1.

# of Node	Distance Traveled	Traveling Time	Arrival Time	Waiting Time	Service Time	Departure Time
1	–		–	–	–	15:00:000
6	1.6 km	3.242 min	15:03:242	17.031 min	30.000 min	15:51:591
2	4.8 km	9.726 min	16:00:000	45.000 min	15.000 min	17:00:000
7	0.4 km	0.909 min	17:00:909	141.453 min	–	19:22:362
1	3.4 km	6.890 min	19:29:251	–	–	–

In the second route, the vehicle trip was 1-5-4-3-7-1. The second route starts as the first one with the depot at 15:00:00. Then the vehicle travels 2.5 km to the customer at node 5 in 5.066 min. Waiting time occurs at node 5 for 20.041 min. After the customer is served at node 5, which takes 20.000 min, the vehicle leaves the node at 15:45:107. The rest of the route is summarized in the Table 8.

Table 8. Route 1-5-4-3-7-1.

# of Node	Distance Traveled	Traveling time	Arrival Time	Waiting Time	Service Time	Departure Time
1	–		–	–	–	15:00:000
5	2.5 km	5.066 min	15:05:066	20.041 min	20.000 min	15:45:107
4	3.2 km	6.484 min	15:51:591	–	35.000 min	16:26:591
3	1.9 km	4.318 min	16:30:909	–	20.000 min	16:50:909
7	4.4 km	10.00 min	17:00:909	141.453 min	–	19:22:362
1	3.4 km	6.890 min	19:29:251	–	–	–

4 Conclusion

In this study, a variant of the VRP is investigated to improve the security of the CIT sector. The study contains two stages. A comparison of the most common machine learning methods, including multiple linear regression, polynomial regression, decision trees, random forest, support vector machine, multilayer perceptron, and generalized regression neural network, is implemented in the first stage of the study for predicting the traffic states in Istanbul.

The data contains the minimum speed, maximum speed, average speed, and the number of vehicles within an hour as dependent variables. The aim is to predict the next hour's average speed depending on these features. The hyperparameters tuning process is done first for each method to find the best parameters that fit the data. Next, these parameters are used to define the best fit of the model. In the evaluation process, speed prediction at random dates and times and also 24-hour speed prediction are made. The results of the first evaluation scenario are less accurate than those of the second one due to two reasons: First, it has more values to predict than the second. Secondly, it may include outliers such as holidays and weekends with traffic values different from regular days. However, the accuracy values for each scenario on the different machine learning algorithms are remarkably close. For the random testing set, the value of MSE is 20.81, and the value of the R^2 score is 0.82. The best performance of this estimation test refers to MLP. For the evaluation set of 24 hours, the results were a little bit different. The best accuracy refers to SVR with values of 5.57 for MSE and 0.96 for R^2. Hence, SVR is used to predict the speed values used in the optimization model.

Minimizing the traveling time of collecting predefined pick-up amounts of money from different customers by a capacitated fleet of vehicles to deliver to the bank is solved as a dependent capacitated vehicle routing problem with time windows. The speed values provided by the SVR model and the other constraints, including customer demands, capacities, and time windows, define the model using the gams optimization modeling system to model and solve as mixed integer linear programming. Seven nodes in this application have been considered, including five customers. The distances between the locations are calculated using Google Maps. A limitation of this study is that margins of error in these distances are not calculated in this study.

This study is a starting point for later studies focusing on using machine learning to predict dynamic parameters such as traffic density. The mathematical model developed in this study has been applied to a case study in Istanbul. This study can be extended by using the developed mathematical model to different regions to evaluate its performance and adaptability under varying traffic conditions. In future studies, environmental concerns, which have attracted much attention recently, can be added to the model, inspired by [24]. Deep learning methods and machine learning techniques for traffic speed prediction can also be explored. Moreover, real-time traffic data can be integrated into the model. In real-world applications, situations may necessitate the prohibition of waiting times; thus, the model can be enhanced with these constraints. The proposed model can also be developed with uncertainty or reinforcement learning, which can be used by following this methodology. Another possibility is the proposed model can be modified to transport highly hazardous materials research.

Disclosure of Interests. The authors have no competing interests to declare relevant to this article's content.

References

1. Fröhlich, G.E.A., Gansterer, M., Doerner, K.F.: Safe and secure vehicle routing: a survey on minimization of risk exposure. Int. Trans. Oper. Res. **30**(6), 3087–3121 (2023)
2. Michallet, J., Prins, C., Amodeo, L., Yalaoui, F., Vitry, G.: Multi-start iterated local search for the periodic vehicle routing problem with time windows and time spread constraints on services. Comput. Oper. Res. **41**, 196–207 (2014)
3. Hoogeboom, M., Dullaert, W.: Vehicle routing with arrival time diversification. Eur. J. Oper. Res. **275**(1), 93–107 (2019)
4. Soriano, A., Vidal, T., Gansterer, M., Doerner, K.: The vehicle routing problem with arrival time diversification on a multigraph. Eur. J. Oper. Res. **286**(2), 564–575 (2020)
5. Yan, S., Wang, S.-S., Wu, M.-W.: A model with a solution algorithm for the cash transportation vehicle routing and scheduling problem. Comput. Ind. Eng. **63**(2), 464–473 (2012)
6. Yan, S., Wang, S.-S., Chang, Y.-H.: Cash transportation vehicle routing and scheduling under stochastic travel times. Eng. Optim. **46**(3), 289–307 (2014)
7. Constantino, M., Mourão, M.C., Pinto, L.S.: Dissimilar arc routing problems. Networks. **70**(3), 233–245 (2017)
8. Duchenne, É., Laporte, G., Semet, F.: Branch-and-cut algorithms for the undirected m-peripatetic salesman problem. Eur. J. Oper. Res. **162**(3), 700–712 (2005)
9. Duchenne, É., Laporte, G., Semet, F.: The undirected m-peripatetic salesman problem: polyhedral results and new algorithms. Oper. Res. **55**(5), 949–965 (2007)
10. Duchenne, É., Laporte, G., Semet, F.: The undirected m-capacitated peripatetic salesman problem. Eur. J. Oper. Res. **223**(3), 637–643 (2012)
11. Ngueveu, S.U., Prins, C., Calvo, R.W.: A hybrid tabu search for the m-peripatetic vehicle routing problem Matheuristics: Hybridizing metaheuristics and mathematical programming, pp. 253–266 (2010).
12. Ngueveu, S.U., Prins, C., Wolfler Calvo, R.: Lower and upper bounds for the m-peripatetic vehicle routing problem. 4OR. **8**, 387–406 (2010)
13. Ngueveu, S.U., Prins, C., Wolfler Calvo, R.: New lower bounds and exact method for the m-PVRP. Transp. Sci. **47**(1), 38–52 (2013)

14. Talarico, L., Sörensen, K., Springael, J.: Metaheuristics for the risk-constrained cash-in-transit vehicle routing problem. Eur. J. Oper. Res. **244**(2), 457–470 (2015)
15. Talarico, L., Springael, J., Sörensen, K., Talarico, F.: A large neighbourhood metaheuristic for the risk-constrained cash-in-transit vehicle routing problem. Comput. Oper. Res. **78**, 547–556 (2017)
16. Talarico, L., Sörensen, K., Springael, J.: A biobjective decision model to increase security and reduce travel costs in the cash-in-transit sector. Int. Trans. Oper. Res. **24**(1–2), 59–76 (2017)
17. Radojičić, N., Djenić, A., Marić, M.: Fuzzy GRASP with path relinking for the risk-constrained cash-in-transit vehicle routing problem. Appl. Soft Comput. **72**, 486–497 (2018)
18. Tikani, H., Setak, M., Demir, E.: A risk-constrained time-dependent cash-in-transit routing problem in multigraph under uncertainty. Eur. J. Oper. Res. **293**(2), 703–730 (2021)
19. Tikani, H., Setak, M., Demir, E.: Multi-objective periodic cash transportation problem with path dissimilarity and arrival time variation. Expert Syst. Appl. **164**, 114015 (2021)
20. Ayyıldız, E., Taşkın, A., Yıldız, A., Özkan, C.: Artificial neural networks integrated mixed integer mathematical model for multi-fleet heterogeneous time-dependent cash in transit problem with time windows. Neural Comput. Appl. **34**(24), 21891–21909 (2022)
21. Mazinani, M., Tavakkoli-Moghaddam, R., Bozorgi-Amiri, A.: A multi-objective cash-in-transit pollution-location-routing problem based on urban traffic conditions. Int. J. Eng. **36**(2), 299–310 (2023)
22. Ghannadpour, S.F., Zandiyeh, F.: A new game-theoretical multi-objective evolutionary approach for cash-in-transit vehicle routing problem with time windows (a real life case). Appl. Soft Comput. **93**, 106378 (2020)
23. Zandieh, F., Ghannadpour, S.F., Mazdeh, M.M.: Integrated ground vehicle and drone routing with simultaneous surveillance coverage for evading intentional disruption. Transp. Res. E Logist. Transp. Rev. **178**, 103266 (2023)
24. Jin, Y., Ge, X., Zhang, L., Ren, J.: A two-stage algorithm for bi-objective logistics model of cash-in-transit vehicle routing problems with economic and environmental optimization based on real-time traffic data. J. Ind. Inf. Integr. **26**, 100273 (2022)
25. Fallahtafti, A., Khoa, T.: Navigating Risks: The Imperative of Research in CIT and Hazmat Transport Security
26. Shafiekhani, M., Rashidi Komijan, A., Javanshir, H.: Developing a multi-objective model for the routing problem of vehicles carrying valuable commodity under route risk conditions (case study of Shahr Bank). J. Appl. Res. Ind. Eng. **12**(1), 1–15 (2025)
27. Cao, E., Wang, X., Yang, Y., Ge, H.: A hybrid heuristic algorithm for the fuzzy open vehicle routing problem with risk preference. J. Ind. Manage. Optim. **21**(3), 1812–1831 (2025)
28. Sabripoor, A., Ghousi, R.: Risk-based cooperative vehicle routing problem for cash-in-transit scenarios: a hybrid optimization approach. Available at SSRN 4544041 (2024)
29. Sun, W., Luo, Z., Hu, X., Pedrycz, W., Shi, J.: An improved variable neighborhood search algorithm embedded temporal and spatial synchronization for vehicle and drone cooperative routing problem with pre-reconnaissance. Swarm Evol. Comput. **91**, 101699 (2024)
30. Zandieh, F., Ghannadpour, S.F., Mazdeh, M.M.: New integrated routing and surveillance model with drones and charging station considerations. Eur. J. Oper. Res. **313**(2), 527–547 (2024)
31. Seber, G.A.F., Lee, A.J.: Linear Regression Analysis. John Wiley & Sons (2012)
32. Bratsas, C., Koupidis, K., Salanova, J.-M., Giannakopoulos, K., Kaloudis, A., Aifadopoulou, G.: A comparison of machine learning methods for the prediction of traffic speed in urban places. Sustain. For. **12**(1), 142 (2019)
33. Ostertagová, E.: Modelling using polynomial regression. Procedia Eng. **48**, 500–506 (2012)
34. Blockeel, H., Devos, L., Frénay, B., Nanfack, G., Nijssen, S.: Decision trees: from efficient prediction to responsible AI. Front. Artif. Intell. **6**, 1124553 (2023)

35. Liaw, A.: Classification and Regression by randomForest. R news (2002)
36. Smola, A.J., Schölkopf, B.: A tutorial on support vector regression. Stat. Comput. **14**, 199–222 (2004)
37. Feng, X., Ma, G., Su, S.-F., Huang, C., Boswell, M.K., Xue, P.: A multi-layer perceptron approach for accelerated wave forecasting in Lake Michigan. Ocean Eng. **211**, 107526 (2020)
38. Nair, V., Hinton, G.E.: Rectified linear units improve restricted boltzmann machines. In: Proceedings of the 27th international conference on machine learning (ICML-10), pp. 807–814 (2010)
39. Wang, L., Kisi, O., Zounemat-Kermani, M., Salazar, G.A., Zhu, Z., Gong, W.: Solar radiation prediction using different techniques: model evaluation and comparison. Renew. Sust. Energ. Rev. **61**, 384–397 (2016)
40. Firat, M., Gungor, M.: Generalized regression neural networks and feed forward neural networks for prediction of scour depth around bridge piers. Adv. Eng. Softw. **40**(8), 731–737 (2009)
41. Kim, B., Lee, D.W., Park, K.Y., Choi, S.R., Choi, S.: Prediction of plasma etching using a randomized generalized regression neural network. Vacuum. **76**(1), 37–43 (2004)
42. Wang, Z., Bovik, A.C.: Mean squared error: love it or leave it? A new look at signal fidelity measures. IEEE Signal Process. Mag. **26**(1), 98–117 (2009)

Creation of a Fixed Wing UAV Based on Ardupilot Firmware with Elements of Radio Interference Countermeasure System

Serhii Lienkov[1] ![ORCID], Alexander Myasischev[2] ![ORCID], Vadym Ovcharuk[3] ![ORCID], Oleksandr Sieliukov[4] ![ORCID], Nataliia Lytvynenko[5]([✉]) ![ORCID], and Vitalii Tarhonskyi[6] ![ORCID]

[1] Taras Shevchenko National University of Kyiv, Kyiv, Ukraine
[2] Khmelnitsky National University, Khmelnitsky, Ukraine
[3] Lviv Politechnic National University, Khmelnitsky, Ukraine
[4] Xi'an Jiaotong University Aerospace School, Xi'an, China
[5] Taras Shevchenko National University of Kyiv, Kyiv, Ukraine
n123n@ukr.net
[6] Central Research Institute of Armed Forces of Ukraine, Kyiv, Ukraine

Abstract. The design and adjustment of a twin-engine aircraft with flight weight of 11.5 kg, capable of carrying a payload of up to 3 kg over a distance of 70–80 km, based on the Pixhawk 6C flight controller using the Arduplane ver.4.3.7 firmware was carried out. The experimental adjustment of the firmware parameters for a given UAV geometry, its weight, and propeller-motor group was performed, ensuring a stable flight when performing an automatic mission. The parameters of the aircraft's automatic takeoff for a mission flight were set. Despite the relatively large weight of the aircraft (up to 11.5 kg), its launch is configured by hand in automatic mode by one person without using a catapult. The designed aircraft is configured in the advanced fault-tolerant configuration (AFC) mode that is able to ensure the return of the aircraft to the launch point in the conditions of electronic jamming systems (EJS). According to this mode, if communication with satellites is lost, the aircraft stops follo-wing the planned mission and switches to a backup flight trajectory, which is able to take it out of the EJS zone. In this case, the simplified route is performed using the inertial system until the satellite communication is restored. The efficiency of using two-band GPS receivers with the installation of antenna screens to counteract electronic countermeasures is experimentally demonstrated, even in conditions of their close location to a flying drone. The paper considers different types of three-band antennas, including ceramic, helical with a screen, and helical antenna arrays.

Keywords: GPS receiver · UAV · ESC controller · Failsafe · Arduplane · AFS

1 Introduction

The unmanned aviation has been considered as an important means of waging military conflicts since the end of the twentieth century, but at that time it was not so widespread and was used in military conflicts more for reconnaissance than for strikes. One of the

© The Author(s), under exclusive license to Springer Nature Switzerland AG 2026
A. Mirzazadeh et al. (Eds.): ODSIE 2024, CCIS 2482, pp. 265–280, 2026.
https://doi.org/10.1007/978-3-031-93601-2_17

reasons for this is the underdeveloped technology for building UAVs of different types and purposes. With the development of mainly computer technologies, it became possible to create UAVs smaller in size, but with better reconnaissance and strike capabilities. The analysis of the features of UAVs of different types has shown that to solve such problems it is advisable to use rotor-type aircraft (quadrocopters, hexacopters) or fixed-wing aircraft depending on the speed, flight range, and weight of the carried load [1–5]. However, at present, in conducting military operations, a big obstacle to the use of drones is the use of radio systems by the enemy to suppress communication between the drone and the operator (EWS—electronic warfare systems), as well as suppression of the radio channel of navigation systems [6, 7]. This significantly complicates the solution of the combat task.

To fulfill military tasks, the drone must fly not only in manual mode from the control panel, but also in automatic mode. For example, to fly along a specified trajectory, to drop ammunition at specified coordinates, to fix the location of mines in mined areas, to deliver cargo to ensure the combat readiness of military units, etc. For this purpose, the ground station software should provide flight trajectory formation, transmit telemetry data between the aircraft and the operator, and the flight controller software should provide stable drone flight [8]. Such drones should be easily controlled from a control panel using a video system at a distance of 7–9 km in order to provide accurate targeting when the drone approaches the target in automatic mode. The drone's software should allow writing additional software modules that allowed to add additional functions to the drone, for example, functions of countering electronic interference, etc. A special task is to use UAVs to conduct reconnaissance operations, supply remote units with small loads: devices, medicines, spare parts, delivery of a small warhead to specified points. There should be a large number of such UAVs in service, they should be budgetary, have the necessary accuracy and flight parameters, weather resistance, counteraction to enemy air defense systems, electronic jamming systems (EJS).

To solve the above-described problems, it's very important that UAVs not only support navigation flight modes, but also be as resistant as possible to changing external influences and electronic warfare systems [6, 9, 10]. In this case, the flight controller should operate under firmware using modern mathematical models that allow for the correct evaluation of sensor readings and, based on this, adequate control of propulsion systems [11, 12]. In this regard, the paper considers the construction of an aircraft and its configuration for the Ardupilot firmware [13], which is currently the most stable of the firmware with freely available code. This code can be adjusted by users. The configuration is carried out experimentally by trying out a large number of parameters, including the PID [15] setting of the regulators, as well as the parameters of the propeller-motor group. The Ardupilot firmware was created for the Pixhawk family of flight controllers [14].

The paper considers the construction of an aircraft with a wingspan of 2500 mm, a fuselage length of 1400 mm. Its maximum takeoff weight is 11.5 kg. The aircraft can carry a payload of up to 3 kg over a distance of up to 70–80 km at a cruising speed of 18–20 m/s. This aircraft is mainly designed for automatic flight according to the mission. And for stable flight, it uses the Extended Kalman Filter (EKF) algorithm [12] to estimate position, speed, and attitude based on measurements from the gyroscope, accelerometer, compass, GPS, airspeed, and barometric pressure sensors. The Extended Kalman Filter (EKF3) algorithm used in Arduplane ver.4.3.7 provides a way to combine data from the IMU, GPS, compass, airspeed, barometer, and other sensors to calculate a more accurate and reliable position estimate for the UAV compared to other navigation firmware (e.g. INAV [16]).

The paper considers the extended fault-tolerant configuration (EFC) mode, which is able to ensure the return of the aircraft to the launch point in the conditions of electronic jamming systems (EJS). According to this mode, if communication with satellites is lost, the aircraft stops following the planned mission and switches to a backup flight path, which is able to take it out of the EJS zone. In this case, the simplified route is performed using the inertial system until satellite communication is restored. However, this method of countering EJS does not allow completing mission tasks. It can be used as a UAV rescue mode [21]. To be able to complete the mission, the paper empirically considers methods for protecting the signal from navigation satellites using two GPS range receivers and modernized antennas of the L1 (~1.6 GHz), L2 (~1.2 GHz) range. As an experiment, the simple RTK2B Lite receiver [17] assembled on the u-blox ZED-F9P module of the Swiss company u-blox [18] was considered. The three-band ceramic, spiral antennas and an antenna array of four spiral antennas, which are protected by metal screens, were connected to them. The number of received satellites after exposure to medium-power electronic warfare was used as a criterion for GPS performance [7]. A study was also conducted on connecting two GPS systems in the mixing mode to the flight controller and an analysis of the navigation system operation for this case was performed.

The solution of the problems posed in the paper is currently particularly relevant in combat operations areas. The analysis of literary sources conducted by the authors showed that a comprehensive solution to the above problems was not considered. Only solutions to specific problems were presented, for example, on PID controller settings, the use of the Kalman filter, specific solutions to problems of countering electronic interference, etc. Here, a comprehensive solution to the problem is considered and the result was the design of a UAV.

2 Basic Material and Robotic Results

In this paper, the Pixhawk 6C flight controller [19] is used for the study. The description in the paper is valid for any flight controller of the Pixhawk family. The main characteristics of the Pixhawk 6C are: processor: 32-bit STM32H743 Arm® Cortex®-M7, 480 MHz, 2 MB memory, 1 MB SRAM; I/O processor: STM32F103, 32 Bit Arm® Cortex®-M3, 72 MHz, 64 KB SRAM. Sensors: Accel/Gyro: ICM-42688-P, Accel/Gyro—BMI055, Mag—IST8310, Barometer—MS5611.

The aircraft is equipped with: two motors, APC 15x10 twin-blade propellers, two ESC motor controllers, two GPS receivers and two antennas, compass, front and rear video cameras with optical zoom x6, Radiomaster TX16S control equipment with TBS Crossfire transmitting unit, telemetry with a range of up to 25 km, servos for aileron, elevator, rudder, rudder, parachute and cargo drop. A 6S6P lithium ion battery with a capacity of about 30000mAh is used for long range flights (up to 80 km).

The airplane has a wingspan of 2500 mm, fuselage length of 1400 mm. Its maximum takeoff weight is 11.5 kg, it can carry a payload weighing up to 3 kg to a distance of 70-80 km. Figure 1 shows a photo of the developed airplane-type UAV mounted on a catapult.

Fig. 1. Photo of the developed airplane-type UAV mounted on a catapult.

Figure 2 shows a photograph of the inside of the airplane body with the flight controller, magnetometer, GPS receivers, telemetry, and rear view camera installed there.

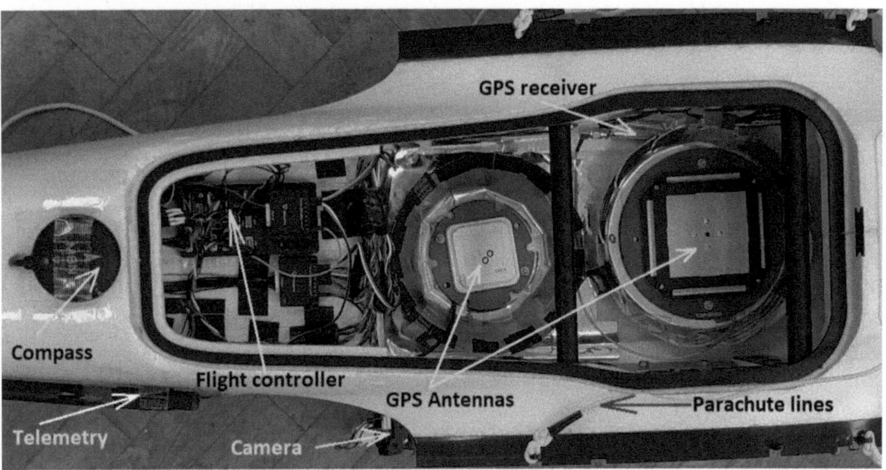

Fig. 2. Photo of the inside of the case with some electronic components.

Configuration of the flight controller firmware is performed using the Mission Planner program [20]. The Arduplane 4.3.7 firmware is selected for the Pixhawk 6C flight controller. After starting Mission Planner ver.1.3.80, the Pixhawk 6C is physically connected to the USB port of the computer. However, connection to the Mission Planner program via USB port is not performed (Fig. 3).

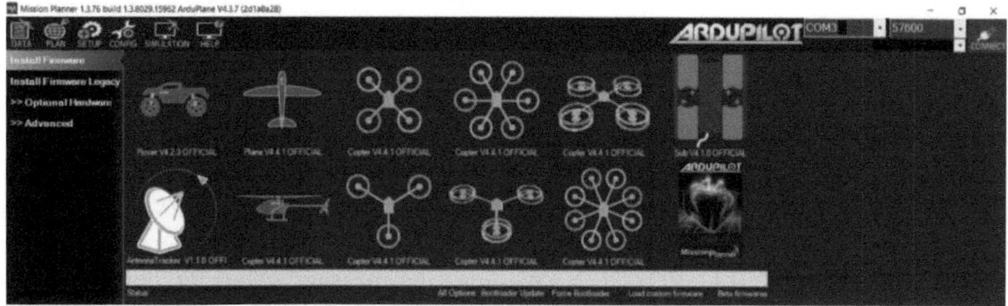

Fig. 3. Arduplane flight controller firmware.

The accelerometer, compass, control equipment and ESC controllers are calibrated. Figure 4 shows the combination of settings tabs. It is advisable to calibrate the magnetometer at the UAV launch site by rotating it in six axes until Mission Planner displays the end of calibration message. The location of the magnetometer relative to the flight controller is automatically determined. The ESC controls are adjusted according to the ESC tab.

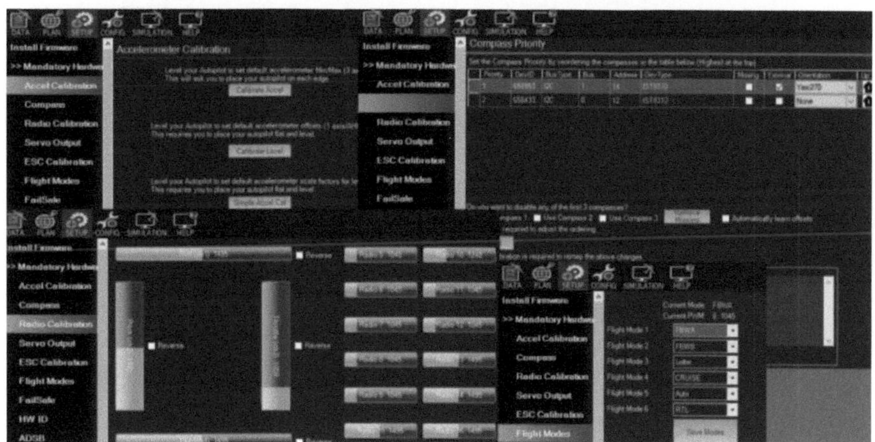

Fig. 4. Combination of tabs for aircraft flight controller parameters settings.

The flight modes are configured and the FailSafe parameters are set. FailSafe is set only for the radio control panel. In case of loss of communication with the radio control panel, the Enabled Continue with Mission i Auto mode is enabled—continuing

270   S. Lienkov et al.

the flight mission in automatic mode. The FS_LONG_ACTN parameter is responsible
for this. Its value is set to 0. In this case, if the airplane was in AUTO mode after losing
communication with the control panel, it will continue the automatic mission. If the
airplane was manually controlled from the console, it will switch to return to launch
point (RTL) mode. And if the GPS system is functioning, the airplane will return to the
launch point at the altitude specified by the ALT_HOLD_RTL parameter. Further it will
describe circles with diameter WP_LOITER_RAD (if RTL_RADIUS = 0) above the
launch point.

Setting of airspeed, roll and pitch angles, PID regulator parameters is performed
in the CONFIG/TUNING—Basic Tuning tab (Fig. 5). For stable flight of the airplane
special attention is paid to PID regulator tuning. In this work, automatic tuning of PID
regulator parameters was performed in AUTOTUNE mode. After takeoff, the control
panel was switched to AUTOTUNE flight mode. Next, rapid movements of the control
sticks of the remote control with short pauses were performed. They were deflected in
turn for each axis (ROLL, PITCH, YAW) until the message about the completion of
tuning appeared in the Mission Planner information window. The auto-tuning system
sets the maximum ROLL and PITCH speeds and the angular positioning error according
to the required speed increase determined by the speed of the console control knobs. The
auto tuning system saves the P, I, and D parameters after tuning is completed. However,
the tuning parameters must be saved in AUTOTUNE mode. Switching to another flight
mode restores the previous parameters.

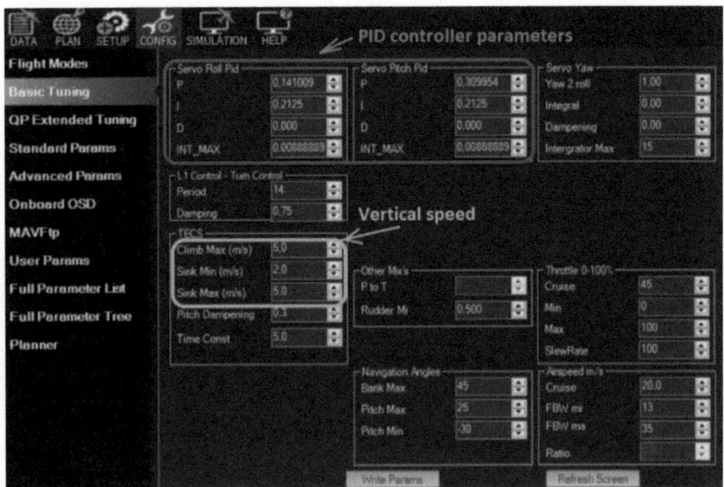

Fig. 5. Setting of flight speed, roll and pitch angles, PID controller parameters.

To set the parameters of the flight controller, the tab shown in Fig. 6 is used. The
following parameters were preliminarily set in the work:

ALT_HOLD_RTL—height of return to the launch point (in centimeters) = 6000;
ARMING_REQUIRE = 1. Arming is required before takeoff.

ARMING_RUDDER = 2. To switch to arming mode, the throttle stick is deflected to the right, and to disarming mode—to the left. The throttle stick must be at zero throttle.
THR_SUPP_MAN = 1. At the moment of throttle suppression by the autopilot in the automatic flight path mode, the throttle stick can be controlled;
TKOFF_THR_MAX = 100. When taking off in automatic mode, at any throttle stick position the motors start rotating at maximum speed;
WP_RADIUS = 30. When flying a mission in automatic mode, the waypoint will be completed if the aircraft approaches it at a distance of less than 30 meters;
ARSPD_TYPE = 0. If the airspeed sensor is not installed, it is disabled by the zero value of this parameter.

The airplane in question is launched in automatic mode and flies the mission. It is mostly launched manually or from a catapult with the toggle switch on the control panel set to Auto mode. The idea behind automatic takeoff is that the autopilot sets the throttle to maximum and gains altitude until the target altitude is reached. To takeoff an airplane, in a flight mission the first command should be the TAKEOFF command. It has two parameters: minimum takeoff angle and takeoff altitude. The minimum angle determines how steeply the airplane will gain altitude during takeoff. For the airplane in question, a value of 15–17 degrees is used. The takeoff altitude determines the height above the takeoff point at which climb is completed. It is normally set at 60 meters for most safe takeoffs.

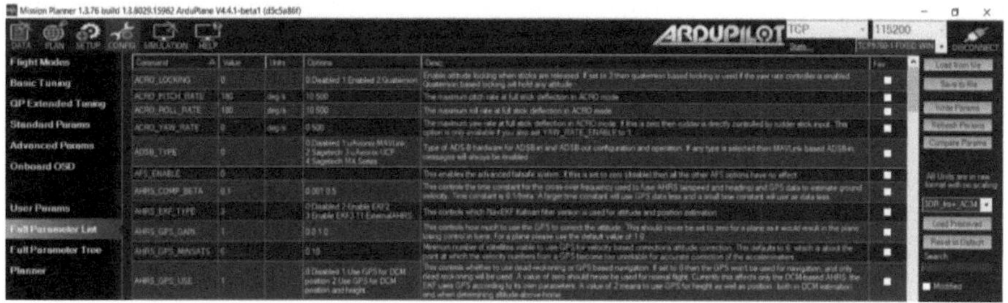

Fig. 6. Tab of flight controller firmware parameters.

During takeoff, the wings of the aircraft will be held in a horizontal position to the nearest LEVEL_ROLL_LIMIT degree (5 degrees by default) until an altitude of 5 meters is reached (parameter TKOFF_LVL_ALT). And, after reaching an altitude of 15 meters, the roll value can gradually change to the maximum value (set by the LIM_ROLL_CD parameter) when the aircraft gains the altitude set by the TKOFF_ALT parameter on the TAKEOFF command. This prevents the wings from clinging to the runway during a sharp roll on takeoff from the ground. The takeoff direction is set based on the direction the airplane is pointing when the automatic takeoff command is given. The Auto mode is then activated on the remote control. If possible, the launch is performed against the wind. During the first phase of takeoff, the autopilot uses the gyroscope as the primary mechanism to keep the airplane straight, or the compass if one is installed. Once sufficient speed has been achieved to correctly determine the GPS heading, the aircraft will switch to using GPS flight to allow for side winds.

272   S. Lienkov et al.

The airplane under consideration is started manually. The main parameters controlling the engine start in this case are:

TKOFF_THR_MINACC
TKOFF_THR_DELAY TKOFF_THR_MINSPD
TECS_PITCH_MAX

When an automatic takeoff command is initiated (usually by switching to AUTO mode) or when a switch to TAKEOFF mode is performed when arming is present, the autopilot will start in throttle suppression mode. The throttle will not start until the conditions set by the TKOFF_THR_x parameters listed above are met.

The TKOFF_THR_MINACC parameter controls the minimum acceleration of the airplane before the throttle is engaged. The acceleration is due to the hand throw when starting the airplane. In this case, it is set to a value of about 11 m/s/s. The TKOFF_THR_DELAY parameter is a delay of 1/10 second to delay the engine start after the minimum acceleration is reached. This is to ensure that the propeller doesn't hit your hands before the motor starts. The value here is set to 2 (which is 0.2 seconds). The parameter TKOFF_THR_MINSPD is the minimum running speed (measured by GPS) before the engine starts after the above delay(0.2 seconds). This is a safety measure to ensure that the airplane is out of hand before the engine starts. In this case it is set to 0. This will cause the engine to start immediately after the 0.2 s delay. The TECS_PITCH_MAX parameter controls the maximum angle that the autopilot requires to fly in automatic mode. For reliable climb at full throttle, its value is chosen to be 20 degrees. It has been observed for this airplane that at full throttle and maximum climb rudder deflection, the airplane begins to descend. This should be taken into account when manually controlling the airplane.

The airplane designed here typically flies in automatic mode along a predetermined trajectory, which is defined by waypoints. For such a flight, it is necessary for the satellite positioning system (GPS) to work steadily. However, in some parts of the flight path, unstable GPS operation is possible due to suppression of satellite signals. In this case, the airplane goes off route and may be lost. To prevent this, the paper considers configuring the flight controller software to an advanced fault-tolerant configuration. The issue of possible protection of data from navigation satellites from the influence of systems that create radio electronic interference (REI) is also considered.

The AFC (Advanced Failsafe Configuration) parameters are only suitable for AUTO mission flight. Typically, AFS parameters are configured for failsafe conditions (GPS failure) so that the aircraft is delayed at its current location for a period of time (say, a few minutes until a GPS signal is available) and then returns home along a predetermined flight path or continues the mission. Customization of parameters could include changing flight speeds, altitude changes, automatic landings, or anything else that can be programmed into the mission. If the fail-safe event is terminated (e.g., GPS is restored), the aircraft will switch back to the mission objective it was previously flying to and continue the mission, etc. To enable the AFS fault tolerant system, set the AFS_ENABLE parameter to 1. The default value is zero, which means that all other AFS system parameters are disabled. The concept of "flight termination" is valid for AFS. In this case, the airplane intentionally dives into the ground, setting all control surfaces to maximum, and disengages the throttle. This occurs when AFS_TERM_ACTION is set to 42 or

43. With any other value, the AFS system will display a message to the ground station console (GSC) that it wants to terminate, but will not actually change the control surfaces. The fault-tolerant AFS system supports five types of fault-tolerant events: geofence violation; maximum barometric altitude violation; loss of GPS; loss of ground station communication; and barometer failure. The paper considers the case of loss of communication with navigation satellites (GPS).

The AFS system monitors the status of the GPS receivers throughout the flight. If all available GPS receivers lose position lock, GPS fault tolerance is initialized. If a GPS failure occurs (loss of GPS lock for 3 seconds), the AFS system will check the AFS_WP_GPS_LOSS parameter. This parameter defines the waypoint number in the mission that will be used in the event of a GPS failure. If AFS_WP_GPS_LOSS is not 0, the airplane will change the current waypoint to the waypoint number specified in AFS_WP_GPS_LOSS. For example, a mission consists of 14 waypoints. Point 11 is set to delay in place for 20 seconds, and point 12 is set to return to the takeoff point. You need to set AFS_WP_GPS_LOSS to 11 to enable this part of the mission when the GPS fix is lost. Then the airplane, using the inertial system data, will try to return to the airfield (Fig. 7). Thus, the mission will not be accomplished, but the airplane will be able to leave the problem area and return to the launch point with a high probability.

Fig. 7. Example of flight mission generation for extended fault-tolerant configuration.

When configuring mission elements to lock GPS, it is useful to enable waypoints to "hang around at the current location". This is accomplished by setting the mission command LOITER_TIME. If the GPS recovers after triggering GPS emergency mode, the drone will automatically resume the mission from where it left off. If AFS_MAX_GPS_LOSS is set to a non-zero number, it is used as the maximum number of GPS failures that will be allowed when returning to the mission after reconnecting with the GPS. This counter is incremented only if the second GPS failure occurs at least

30 seconds after the previous failure. Normally, the AFS_MAX_GPS_LOSS parameter is set to 1. Thus, if satellite communication is lost for more than 30 seconds, the airplane is returned to a point closer to the takeoff point where it is assumed that communication with the satellites will be re-established.

Figure 8 shows the trajectory of the airplane flight through the mission (purple curve). On the left—GPS was not turned off during the entire route. On the right—GPS was turned off after passing the fourth waypoint for more than 30 seconds. In the first case the airplane missed the 11th waypoint, and in the second case it flew towards it.

Fig. 8. Airplane flight trajectory by mission at different GPS states.

However, the considered way of setting up the AFS system makes it possible to take the aircraft out of the REI zone and return to the launch point. In this case, the entire mission is not accomplished. For possible mission accomplishment, let's consider ways to protect the signal from navigation satellites using two band GPS receivers and modernized L1 (~1.6 GHz), L2 (~1.2 GHz) band antennas.

The use of dual band GPS receivers with dual band, tri-band antennas allows better resistance to REI systems. In this work, flights were conducted under REI using the simpleRTK2B Lite receiver [20], assembled on the u-blox ZED-F9P module of the Swiss company u-blox [21] (Fig. 9).

ZED-F9P Features:

Positioning accuracy: <1.5 m in standalone mode.
Information refresh rate: with maximum performance: up to 10 Hz Band support: L1, L2 and E5b.
Multi-constellation and multi-frequency: GPS—L1C/A L2C; GLONASS—L1OF L2OF; Galileo—E1-B/C E5b; BeiDou—B1I B2I; QZSS—L1C/A L2C; SBAS—WAAS, EGNOS, MSAS, GAGAN and SouthPAN.

An external multi-band GNSS L1/L2/E5b antenna (Fig. 9) is used on the drone for good operation of the module. Its main features are:

Supported bands: GPS—L1, L2; GLONASS—G1, G2; BeiDou—B1, B2; Galileo—E1, E5b; QZSS—L1, L2; SBAS: WAAS, EGNOS, MSAS and GAGAN.
Polarization: RHCP.

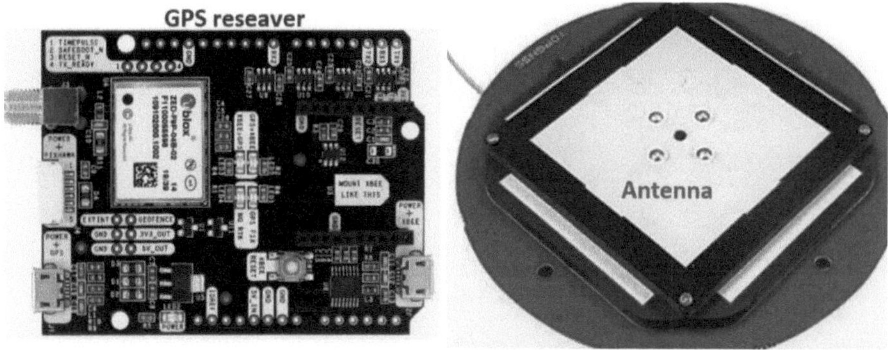

Fig. 9 Dual band GPS receiver with a three band antenna

Antenna module peak gain: 2.5 dBi. Antenna built-in amplifier gain: 26 dB Azimuth coverage: 360 degrees.
Impedance: 50 ohms.

Experiments have shown that the use of dual band receivers and antennas without a protective screen installed at the bottom of the antenna does not protect against anti-drone guns with a power of up to 40 W. Dual band receivers are known to receive up to 32 satellites. When exposed to the shotgun, the satellites completely disappear, which can be recorded using the graphical REI representation of the Ardupilot log files. However, installing a protective shield at the bottom of the antenna, as shown in Fig. 10, significantly weakens the shotgun effect. The number of satellites is reduced from 32 to 25–26, allowing the airplane to follow the mission precisely.

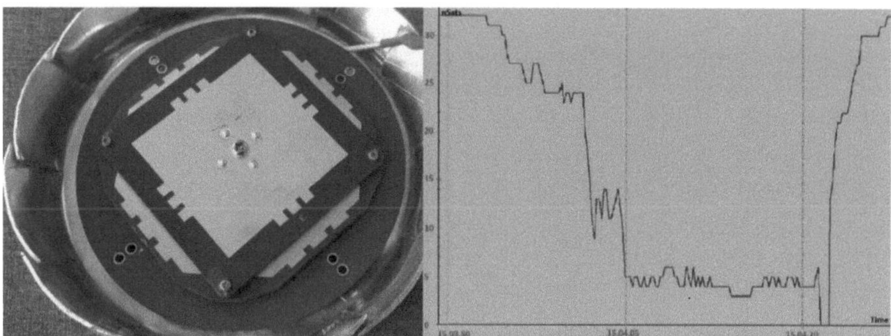

Fig. 10. Three-band GPS antenna with a screen. Graph of change in the number of satellites under REI action.

Figure 10 shows a tri-band antenna with a screen and a graph showing the reduction in the number of satellites when a UAV with this antenna passes through a medium power (100–150 w) REI area [7].

It can be seen from the figure that this type of REI leads to the reduction of satellites to 3–5. These experiments were conducted at a flight altitude of 140 m and at a distance from the REI of 300–500 m. I.e. the test was performed for the most unfavorable conditions of UAV flight. Such types of REIs are the most common. The work also considered the use of spiral antennas and spiral antenna arrays, shown in Fig. 11.

Fig. 11. Used experimental spiral antennas with screens.

Figure 12 shows the graphs of the change in the number of satellites when flying in the area of medium power REI at a distance of 300–500 m and height of 140 m for these antennas. The left graph corresponds to the spiral antenna in Fig. 11 on the left, the right graph corresponds to the array of spiral antennas (Fig. 11 on the right). In spite of the fact that during the experiments the UAV for these two antennas successfully flew through the area of REI action, no satellite losses were observed for the antenna array. And its performance can be evaluated as the most stable.

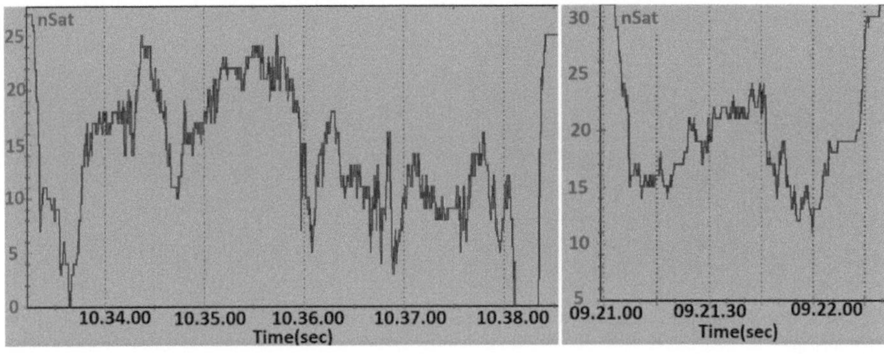

Fig. 12. Graphs of change in the number of received satellites for spiral antennas.

To increase the stability of the REI zone traversal, the work experimentally considered the simultaneous inclusion of two GPS receivers to the Pixhawk flight controller, which were connected to antennas similar to the one shown in Figs. 9 and 10 using a protective shield (Fig. 2).

It has been experimentally found that using two GPS reduces the likelihood of UAV malfunctions. For data mixing, only GPS receivers that determine position accuracy and speed should be used. For example, Ublox (ZED-F9P) GPS receivers provide this additional information in messages (Fig. 13). It is advisable to perform mixing from two GPS's from only one manufacturer, as the scaling of accuracy metrics varies and this results in favoring one GPS over another and therefore mixing does not occur. To configure mixing, the following flight controller firmware settings are made:

SERIAL4_PROTOCOL = 5. Here serial port 4 is set to work with the second GPS. GPS_TYPE2 = 1. The flight controller port is activated to receive data from the second GPS receiver. GPS_AUTO_SWITCH = 2. Configures the mode of automatic switching between two GPS to BLEND mode.

In the log files, the first GPS data appears with instance number 0, the second GPS with instance number 1. Mixed GPS data appears with instance number 2.

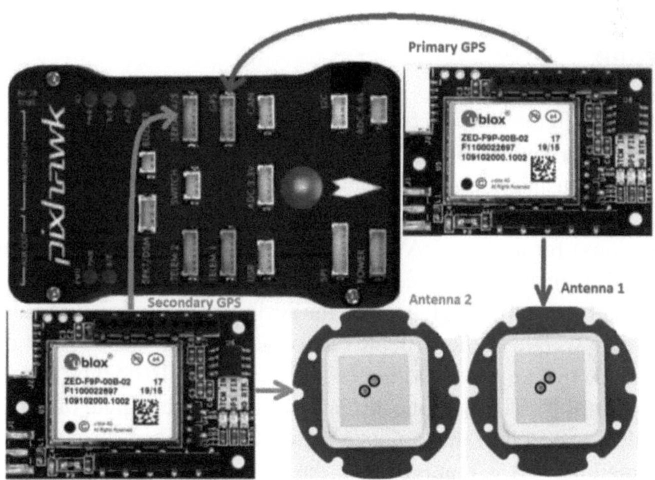

Fig. 13. Connecting two GPS receivers to the Pixhawk flight controller.

Figure 14 shows plots of satellite count variation for the first, second GPS and for the blended data generated by the flight controller during a flight in a medium power REI area at a distance of 300–500 m and an altitude of 140 m.

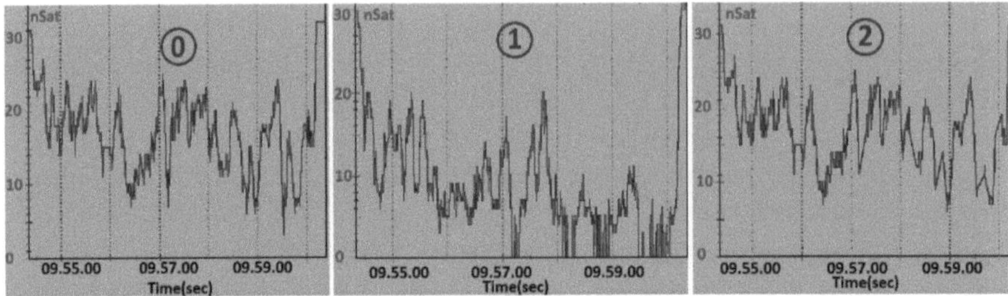

Fig. 14. Plots of satellite number variation for the first (0), second (1) GPS and mixed data (2).

Figure 14 shows that the second receiver recorded complete disappearance of satellites in some time intervals. The result of the flight controller's work was stable reception of satellites (2), as a result of which the airplane successfully passed the radar coverage area without deviating from the mission.

3   Discussion

The design and configuration of a twin-engine airplane with a flight weight of 11.5 kg, capable of carrying a payload of up to 3 kg at a distance of 70–80 km based on a Pixhawk 6C flight controller using Arduplane ver.4.3.7 firmware.

The experimental tuning of firmware parameters (in particular, PID regulators, velocities, accelerations) for a given UAV geometry, its weight, propeller group, providing stable flight during an automatic mission was performed.

The parameters of automatic takeoff of the airplane for mission flight are set. Despite the relatively high weight of the airplane (up to 11.5 kG), its launch is set from the hand in automatic mode by one person without using a catapult. To do this, the pilot sets Auto mode on the control panel, raises the airplane overhead at an angle of 15–20 degrees to the horizon, shakes it to start the motors and pushes it forward.

The designed aircraft is configured in the Advanced Fault Tolerant Configuration (ATC) mode, which is able to ensure the aircraft's return to the launch point under the conditions of electronic interference systems (EIS). According to this mode, if communication with satellites is lost, the aircraft stops following the planned mission and switches to a backup flight path, which is able to take it out of the EAR zone. In this case, the simplified route is flown using the inertial system until satellite communication is restored.

4   Conclusions

So, the effectiveness of using two band GPS receivers with antenna shields to counter REIs even in close proximity to a passing drone is experimentally demonstrated. Different types of tri-band antennas including ceramic, spiral with shield and spiral antenna arrays are considered in the paper. The better performance of the spiral antennas constructed by antenna array design is shown.

However, although the paper presents numerous experimental results conducted under radio electronic interference (REI) conditions, it is advisable to conduct further research in the direction of UAV behavior under different environmental conditions. For example, UAV behavior under different weather conditions (such as wind, precipitation or extreme temperatures), urban environments with different levels of GPS interference, and so on.

It is planned to consider the issue of using GPS receivers from other manufacturers (for example, Septentrio mosaic-X5). The particular interest is the consideration of GPS antenna digital arrays, which suppress interference from REI systems by 10,000–100,000 times. It is advisable in the future to analyze the use of other flight controllers of the Pixhawk family in UAVs (for example, Cube Orange, Pixhawk 6X and others) along with the used Pixhawk 6C.

References

1. Koval, V.V., Artyushin, L.M., Semon, B.Y., Lobanov, A.A., Gerasimenko, V.V.: Approaches to formulating a strategy for the management of combined combat orders of pilot and pilotless aviation. Sci. Def. **4**, 34–43 (2021)
2. Barry, D.: Strike UAVs and Problems Related to Their Autonomy. Oboronnyi visnik 1, 4–13, 2, 6–13 (2022)
3. Yarosh, S.P., Gur'ev, D.O.: Analysis of the development of unmanned aerial vehicles, methods of their combat application and development of proposals for organizing an effective fight against unmanned aviation. Sci. Technol. Ukrainian Air Def. For. **2**(43), 54–60 (2021)
4. Bezpilotna zbroia maybutnogo—three most interesting Ukrainian UAVs presented at "Zbroia ta Bezpeka". https://savelife.in.ua/ua/bezpilotna-zbroya-majbutnogo-3-najcikavishi-ukrain ski-bpla-predstavleni-na-zbroya-ta-bezpeka-2021/, last accessed 2022/05/07
5. Lienkov, S., Myasischev, A., Sieliukov, O., Pashkov, A., Zhyrov, G., Zinchyk, A.: Checking the flight stability of a rotary UAV in navigation modes for different firmware. CEUR Workshop Proc. **3126**, 46–55 (2016)
6. Rastopchin V. V: Strike unmanned aerial vehicles and air defense—problems and prospects of confrontation, https://www.researchgate.net/publication/331772628_Udarnye_bespilo tnye_letatelnye_apparaty_i_protivovozdusnaa_oborona_problemy_i_perspektivy_protivost oania, last accessed 2024/03/24
7. Mobile complex of UAV antidote "Nota". https://www.ukrmilitary.com/2019/10/nota.html, last accessed 2024/03/24
8. Copter Mission Command List. https://ardupilot.org/copter/docs/mission-command-list. html, last accessed 2024/03/24
9. Lienkov, S., Sieliukov, O., Lienkov, E., Tolok, I., Loza, V.: Evolution of radars resolution capability using simulation mathematical model. In: Proceedings 5th International Conference "Methods and Systems of Navigation and Motion Control", pp. 195–197. Kyiv (2018)
10. Lysenko, V., Gunchenko, Y., Shvorov, S., Lenkov, S., Kuznichenko, S., Lenkov, E.: Methodological bases of construction of intensive training flight simulators of aircrews. In: Proceedings 5th International Conference "Methods and Systems of Navigation and Motion Control", pp. 198–203. Kyiv (2018)
11. Extended Kalman Filter (EKF). https://ardupilot.org/copter/docs/common-apm-navigation-extended-kalman-filter-overview.html, last accessed 2024/03/24
12. Extended Kalman Filter Navigation Overview and Tuning. https://ardupilot.org/dev/docs/ext ended-kalman-filter.html#extended-kalman-filter, last accessed 2024/03/26

13. ArduPilot Firmware builds. https://firmware.ardupilot.org/, last accessed 2024/03/26
14. Pixhawk Overview. https://ardupilot.org/copter/docs/common-pixhawk-overview.html, last accessed 2024/03/26
15. Mohammed, I.K., Abdulla, A.I.: Elevation, pitch and travel axis stabilization of 3DOF helicopter with hybrid control system by GA-LQR based PID controller. Int. J. Electr. Comput. Eng. **10**(2), 1868–1884 (2023)
16. INAV—navigation capable flight controller. https://github.com/iNavFlight/inav, last accessed 2024/03/27
17. Simplertk2b-receivers. https://www.ardusimple.com/simplertk2b-receivers/, last accessed 2024/03/27
18. ZED-F9P module. https://www.u-blox.com/en/product/zed-f9p-module, last accessed 2024/03/28
19. Holybro Pixhawk 6C. https://docs.px4.io/main/en/flight_controller/pixhawk6c.html, last accessed 2024/03/28
20. Mission Planner Overview. https://docs.px4.io/main/en/flight_controller/pixhawk6c.html, last accessed 2024/03/28
21. Lienkov, S., Myasischev, A., Banzak, O., Husak, Y., Starynski, I.: Use of rescue mode for UAV on the basis of STM32 microcontrollers. Int. J. Adv. Trends Comput. Sci. Eng. **9**(3), 3506–3513 (2020)

Optimizing Vehicle Routing in the Dial-a-Ride Problem Using Deep Q-Networks

Mariem Ayari[1], Sonia Nasri[1]([✉]), Hend Bouziri[2], and Wassila Aggoune-Mtalaa[3]

[1] Tunis University, LARODEC Higher School of Business, Tunis, Tunisia
sonia15.nasri@gmail.com
[2] Higher School of Economic and Commercial Sciences, Tunis, Tunisia
[3] Luxembourg Institute of Science and Technology, 4362 Esch-sur-Alzette, Luxembourg

Abstract. In this paper, we propose an enhanced approach to the Deep Q-Network (DQN) for solving the Dial-a-Ride Problem (DARP), a challenging vehicle routing problem. We introduce a hybrid method that combines an insertion heuristic with reinforcement learning to generate initial solutions, which are formulated into a Markov Decision Process (MDP). Our enhanced DQN framework involves past experiences to iteratively improve decision quality and optimize vehicle routes. By integrating problem-specific heuristics and reinforcement learning, our method achieves superior performance compared to the insertion heuristic. Experimental results show that the proposed approach significantly improves vehicle route optimization, providing better overall efficiency for the tested DARP instances. The key advantage of IBRL-DARP lies in its ability to adapt and learn from the problem environment, making intelligent decisions that improve over time. IBRL-DARP demonstrates superior performance in most instances, as evidenced by the lower total travel cost values. This improvement is particularly notable in larger problem instances where heuristic methods tend to underperform due to their reliance on predefined rules. Its robustness across a range of DARP instances highlights its potential as a practical solution for real-world transport-on-demand applications.

Keywords: Dial-a-Ride Problem · Deep Q-Network · Markov Decision Process · Vehicle Routing Optimization · Reinforcement Learning · Transportation

1 Introduction

The rapid advancements in intelligent transportation systems (ITS) over the past decades have significantly transformed urban mobility. Among the various innovations, on-demand transport (ODT) has emerged as a key solution to improving urban mobility by dynamically adjusting vehicle routes and schedules based on user requests. One can cite the DARP of [4] which enhances the efficiency of transportation services but also helps mitigate service quality and optimize resource utilization. However, DARPs continue to pose significant challenges for operators, particularly in terms of route optimization, resource allocation, and service quality, see [1, 18]. In recent years, Deep

A. Mirzazadeh et al. (Eds.): ODSIE 2024, CCIS 2482, pp. 281–300, 2026.
https://doi.org/10.1007/978-3-031-93601-2_18

Learning [7] and Reinforcement Learning (RL) [16] have gained considerable attention in addressing the high challenges of transportation problems. Deep learning offers powerful techniques for extracting high-level features from the large datasets generated by ITS, enabling more accurate predictions of demand patterns, traffic conditions, and user preferences. On the other hand, reinforcement learning provides a robust framework for decision-making where the goal is to optimize vehicle routing and scheduling policies by continuously learning from interactions with the environment. Moreover, the integration of Deep Learning with RL for real-time decision-making is still an underexplored area, with existing studies often focusing on simplified scenarios or neglecting critical operational constraints such as vehicle capacities, time windows, and user expectations. Despite advances in applying RL and deep learning to on-demand transport systems, current approaches still face challenges in terms of scalability, adaptability, and handling real-world constraints. Most of the existing RL methods, such as those proposed by [13, 21] are limited in their ability to manage large fleets of vehicles efficiently. Additionally, the integration of deep learning in RL-based transport models has primarily focused on demand forecasting [21], with limited application to real-time decision-making for dispatch and routing.

Our paper makes the following contributions: We propose a novel method that combines deep learning with RL to improve decision-making in transport on demand problems. Unlike many existing studies that use deep learning only for demand prediction, our approach enhances the state representation in the RL model, allowing more informed routing and dispatch decisions. Our framework introduces a hierarchical RL model, which allows it to scale effectively with large vehicle fleets in complex urban environments. Unlike some existing models that simplify constraints, we incorporate constraints such as vehicle capacities and time windows into the optimization process. This ensures that our solutions are both feasible and practical in real-world scenarios.

In this paper, we address the limitations of high data requirements and long training times in RL models by proposing a hybrid method that integrates an optimization heuristic within a Deep Q-Network framework. By combining the solution efficiency of a heuristic with the learning capabilities of reinforcement learning, our approach aims to reduce training time and improve the scalability of DQN in solving the transportation problem. The proposed hybrid method involves the heuristic to generate high-quality initial solutions, which are then refined through the DQN's adaptive decision-making process. Through simulations on datasets, we show that our method outperforms the insertion heuristic in terms of both routing efficiency and customer satisfaction.

The paper is structured as follows: A literature review is presented in Sect. 2. Next, in Sect. 3, we provide present the problem studied. Section 4 details the insertion-based reinforcement learning for the DARP. Section 5 is proposed for the DQN approach. Next, experimental results are provided in Sect. 6. Finally, the conclusions are set in Sect. 7.

2   Related Works

The increasing complexity and dynamic nature of transportation on demand systems, such as ride-hailing, taxi dis-patching, and Dial-A-Ride services, have necessitated the use of advanced machine learning techniques [24–26], especially reinforcement learning

[23]. Among the RL approaches, Deep Q-Networks [7] have gained prominence due to their ability to handle large state spaces and make effective real-time decisions.

2.1   RL for Transport on Demand

In their work, authors [2] propose a method for estimating average taxi-out times. Their work applies stochastic dynamic programming within a reinforcement learning method. Authors in [9] proposed a taxi driving strategy based on reinforcement learning. This work focuses on the use of reinforcement learning to optimize taxi operating strategies, for maximizing long-term profits and addressing inefficiencies in taxi services in big cities. The problem was formulated as a Markov decision process and uses historical trajectory data. The results demonstrate improved profits and efficiency for cabdrivers, as well as increased opportunities for passengers to find taxis. The method involves pre-processing learning stage offline and performing the optimization stage in real time. It utilizes historical trajectories of taxicabs, preprocesses the data to extract valid trajectories, and employs the Q-learning algorithm to learn and update the expected cumulative rewards of different actions in each cell. Authors in [22] proposed a multi-agent reinforcement learning for order-dispatching via order-vehicle distribution. This work aims to dispatch orders from customers to available vehicles to maximize the accumulated driver income and order response rate. Each vehicle is treated as an agent, and the agents learn independently. The method uses deep Q-learning where the state is a representation of the grid, vehicles, orders, and orders destinations. Actions are dispatching orders. Agents are trained with experiences shared across a parameter shared network. The method is evaluated on simulated environments and real ride data from 3 cities.

In the work of [6], multi-agent reinforcement learning is proposed to achieve centralized training with decentralized execution. A simulated grid-based environment is used to model the ride-sharing problem. A greedy baseline and independent DQN method are implemented for comparison. Only picking up and dropping off tasks are modeled without intermediate stops. The approach is not evaluated on real ride-sharing data. The coordination between different types of ride-sharing vehicles is not addressed. Authors in [15] propose a hybrid method for DARPs that uses a stepwise greedy approach to create a sequence of pick-up/drop-off nodes for each vehicle, minimizing the likelihood of infeasibility. It incorporates binary logit models through reinforcement learning techniques to estimate the probability of success in constructing feasible solutions. Furthermore, Reinforcement Learning has been integrated with optimization techniques to enhance Intelligent Transportation Systems. Authors in [11] addressed the Dynamic Stochastic Vehicle Routing Problem (DSVRP) and proposed a deep reinforcement learning approach to solve it. The DSVRP involves real-world urban logistics operations where routes and tasks change in response to dynamic events. This work includes the proposal of a solution approach combining deep reinforcement learning with a routing heuristic based on Simulated Annealing. The approach outperforms existing methods, showing a 10% improvement on average in terms of optimized re-routing decisions almost instantaneously. Authors in [12] deal with the pickup and delivery problem for calculating transportation schedules among dry ports, hub ports, and related enterprises. The authors proposed a machine learning-based hybrid algorithm to solve the two-echelon pickup and delivery problem. It introduces a one-to-many-to-one problem on the first echelon and a many-to-many problem on the second echelon. The algorithm combines a K-Nearest Neighbor algorithm and an Adaptive Large Neighborhood Search to determine optimal transportation routes and schedules.

2.2 DQN for Transport on Demand

One of the most well-researched applications of DQN in transportation on demand is in ride-hailing platforms. In this context, DQN is employed to dispatch vehicles efficiently, matching supply with fluctuating demand in real-time. Authors in [22] developed a DQN-based system to optimize the order-dispatching problem in ride-hailing services. Their model represents the grid environment as a Markov Decision Process, where each vehicle acts as an independent agent learning to maximize long-term rewards through Q-learning. The DQN model outperformed heuristic-based dispatch systems by increasing driver earnings and reducing customer waiting times.

Another significant application is the work by [8], which used DQN to optimize taxi routing strategies. The model was designed to maximize taxi drivers' long-term profits by suggesting optimal routes based on historical data. DQN proved to be effective in making real-time decisions for dispatching and routing, offering an advantage over traditional taxi dispatch systems that rely heavily on fixed rules or static optimization methods. Dynamic pickup and delivery problems (PDPs) in ToD systems, such as Dial-A-Ride Problems (DARP), present complex challenges due to the real-time arrival of requests and the need for optimal vehicle routing decisions. DQN has been successfully applied to address these challenges. Authors in [15] proposed a hybrid approach integrating DQN with a heuristic for the Dial-A-Ride Problem. The DQN was employed to optimize vehicle assignment decisions, while a heuristic ensured feasibility and improved solution quality. Their results demonstrated that DQN, combined with optimization techniques, reduces the overall transportation cost and improves the responsiveness to customer requests.

Similarly, authors of [20] applied DQN to solve the pickup and delivery problem with time windows (PDPTW), using a deep reinforcement learning framework that employed a DQN agent to generate routes in real-time. The results highlighted that the DQN model outperformed traditional heuristic methods, particularly in dynamic environments where customer requests arrive unpredictably.

Despite its promising results, the application of DQN in transportation on demand is not without limitations. One of the major challenges is the high computational cost associated with training DQN models, particularly in dynamic environments where state and action spaces are large. Authors in [17] integrated genetic algorithms with DQN to reduce the computational burden. Their approach showed improvements in both training efficiency and solution quality, particularly in solving electric vehicle routing problems with time windows. Moreover, authors in [10] pointed out that the performance of DQN in dynamic environments is sensitive to the quality of reward functions and the complexity of state representations. In stochastic vehicle routing problems, where traffic conditions and demand patterns change frequently, DQN models may struggle to adapt unless carefully tuned. To mitigate this, researchers are increasingly exploring hybrid approaches that combine DQN with other optimization or machine learning techniques to enhance its robustness and scalability in real-world applications. As Transport on Demand (ToD) systems scale, there is a growing interest in multi-agent DQN frameworks, where multiple agents (vehicles) must cooperate or compete to achieve global objectives. Authors in [19] proposed a multi-agent DQN system for same-day delivery services, where each delivery vehicle acts as an agent within a grid-based environment. The parameter-sharing DQN model enables decentralized decision-making while maintaining coordination across multiple agents. This approach has proven effective in improving scalability and performance in large-scale ToD systems. Additionally, Authors in [5] applied multi-agent DQN to the dynamic vehicle dispatch problem (DVDP), where vehicles are assigned to requests that arrive stochastically over time. The multi-agent DQN demonstrated strong performance in balancing fleet utilization and minimizing customer wait times, although the study noted the challenge of maintaining fleet balance and fairness in driver income distribution.

Deep Q-Networks have shown significant promise in optimizing transportation on demand systems, from ride-hailing to dynamic pickup and delivery services. While DQN models can outperform traditional heuristics in many cases, challenges such as high data requirements, long training times, and model sensitivity to dynamic changes remain. Hybrid approaches that combine DQN with other optimization methods or multi-agent systems are emerging as promising solutions to address these limitations and enhance the scalability and robustness of RL in real-world transportation applications. Existing RL models tend to simplify operational constraints, which limits their effectiveness in real-world applications. Moreover, high data processing requirements and long training times makes it difficult to deploy RL models in real-time applications. Therefore, a hybrid method could enhance the overall performance and robustness of RL models in VRPs and similar transportation problems. It is what we intend to propose in this paper combining the effectiveness of an optimization heuristic within a DQN approach.

3 The Dial-A-Ride Problem

Dial-a-Ride problems were first introduced by [4] and have attracted since then much attention. The problem involves determining a set of vehicle itineraries to pick up and drop off several customers in the best possible way. The DARP has many practical applications, particularly in urban areas where the idea is to enhance the mobility of people. In the DARP initially proposed by [4], several users submit transportation requests from defined sources to destinations (known as pick-up and drop-off/delivery locations, respectively). The transportation company gets the orders and then arranges the pickups and deliveries utilizing its automobile fleet. The transportation service is shared in the sense that many customers (all with different demands) may share the same automobile at the same time. Figure 1 illustrates a feasible solution for a DARP instance with three customers and a maximum car capacity equal to three passengers at a time.

The objective for a DARP is to minimize the Total Travel Cost (TTC) of all the vehicles. Several constraints must be satisfied such as that all demands are met, each route begins at the origin depot and finishes at the destination depot respecting a maximal total duration. Vehicles have a maximal capacity and users riding time are defined. A time schedule should respect time windows and earliest service times are defined on nodes, see [4] for more details about the problem constraints.

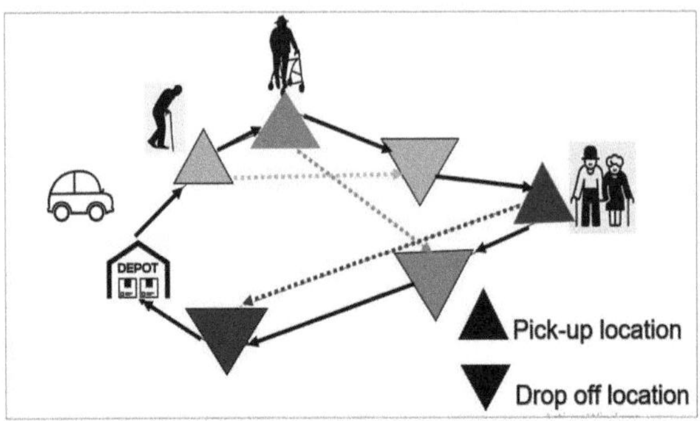

Fig. 1. A feasible solution of a DARP instance with three customers.

4   An Insertion Based Reinforcement Learning for the DARP

In our methodology, we begin by creating a solution using an insertion heuristic. Then we formulate the initial solution using a Markov Decision Process to represent its dynamics. Finally, we use the Deep Q Network to search for an optimized solution by learning from past experiences and making better decisions over time. In the following sections, we will describe each step of the approach.

4.1   Generation of an Initial Solution: The Heuristic IH-DARP

The heuristic IH-DARP aims at efficiently constructing high-quality routes. It operates by iteratively inserting customers into existing routes, seeking to minimize the overall cost or distance traveled with respect to constraints. One of the notable advantages of heuristic lies in its ability to generate feasible solutions rapidly, making it an attractive choice for generating initial solutions for the DQN. The main steps of the insertion heuristic are presented in Algorithm 1. The method begins by initializing a list of non-allocated requests, sorted by their earliest service times in ascending order, ensuring that the requests with earlier service times are prioritized. These steps focus on assigning requests to vehicles while considering constraints and striving for optimal cost and distance results.

Vehicle Selection and Allocation. For each request, the heuristic identifies an appropriate vehicle based on its availability in terms of capacity and total duration. Vehicles with sufficient capacity and available time slots are selected for request allocation. Requests are inserted into the tour of the selected vehicles, adhering to available time schedules and respecting problem constraints. Upon request allocation, all relevant parameters related to vehicles, requests, and time schedules are updated to reflect the changes.

Algorithm 1 The IH-DARP Insertion Heuristic for the DARP

```
 1: Establish a list of non-assigned requests.
 2: Set up the tour as empty and mark all vehicles as available.
 3: Step 1:
 4: if The list is not empty then
 5:     Select a non-assigned request; Go to Step 2;
 6: else
 7:     Break;
 8: end if
 9: Step 2:
10: Select an available vehicle;
11: if There is a valid case of insertion in the vehicle tour
    then
12:     Insert the pickup node in a valid position in the tour;
13:     Delete the request from the list of non-assigned re-
    quests;
14:     Go to Step 3;
15: else
16:     Go to Step 2;
17: end if
18: Step 3:
19: if there are some potential maximal riding time violations
    or the vehicle is at full capacity then
20:     Execute the delivery of requests;
21:     Make the necessary updates;
22:     if The vehicle is at its maximal total tour duration then
23:         Set the vehicle as non-available;
24:         Go to Step 2;
25:     else
26:         Go to Step 1;
27:     end if
28: end if
29: Return a solution as a result;
```

Respect for Time Windows. The heuristic begins by sorting the requests in the ascending order of their earliest service times. This ensures that the requests with an earlier service requirement are given priority in the allocation process. The insertion process ensures that the time windows specified for each request are respected. Besides, the requests are scheduled within their respective time windows to ensure timely service delivery.

Capacity Management. To optimize the use of the vehicle's capacity, deliveries are performed in accordance with the earliest service times of the requests. As deliveries are completed, vehicle capacities are updated to reflect the changes and to maintain efficient resource use.

Vehicle Availability Management. Vehicles are marked as non-available once their maximal total tour duration is reached. This prevents an overload and ensures that vehicles operate within their predefined operational limits. The allocation process continues until all the requests are satisfactorily allocated to vehicles and served within their respective time windows. Upon completion, the initial solution is generated, providing a feasible solution for further optimization.

4.2 Markov Decision Process for the DARP

Markov Decision Processes serve as a core framework, representing decision-making situations in which an agent interacts with its environment throughout the time. MDPs are important because they can simulate dynamic systems, allowing intelligent agents to develop optimum policies for sequential choice tasks. The MDP created for our DARP contains the states, actions, transitions, rewards, and termination conditions required to train a Deep Q-Network to explore specified routing solutions, as shown in Fig. 2. Table 1 denotes all the variables description.

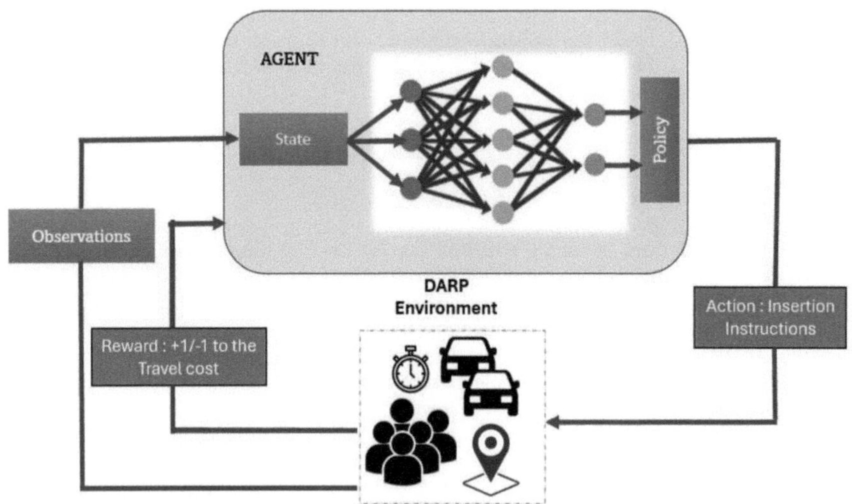

Fig. 2. A Markov decision process for the DARP.

Table 1. Model features and their descriptions.

Variables	Description
n	Number of customers
m	Number of vehicles
C	Vehicle capacity
TTC	Total travel cost
s	State
r	Reward
st	Current state
at	Action taken
$st + 1$	Next state

(continued)

Table 1. (*continued*)

Variables	Description
$f(st, at)$	Transition function determining s_{t+1}
$P(s_{t+1} \vert st, at)$	Transition probability
π	Policy used by the agent
R	Reward function
$Q(s, a; \theta)$	Q-value or action-value function
γ	Discount factor for future rewards
ϵ	Exploration rate in DQN
θ	Parameters of the Q-network
θ'	Parameters of the target Q-network
L	Loss function
MemoryReplay	Buffer storing past experiences
BatchSize	Number of samples used for training
PredictedQ	Predicted Q-value from the Q-network
TargetQ	Target Q-value for the loss computation

Environment. The environment is composed of N users and K vehicles. Each has a position, time window, maximum riding time, and capacity. The locations of the users indicate where they are in the environment, while the time indicates when they demand assistance. The maximum riding time is the longest period a person may remain in a vehicle.

Agent. An agent is a decision-making entity responsible for optimizing actions in response to feedback from its environment. In the context of the dial-a-ride problem, the agent manages and optimizes transportation operations involving vehicles, customers, and relevant information. It learns and applies a policy/strategy that maps the current state of the system to the best possible action to make decisions such as assigning passengers to vehicles, determining the next stops, and sequencing pickups and drop-offs.

The agent operates in real-time, dynamically adapting to the environment, which includes vehicle locations, capacities, customer requests (pickup and drop-off points, time constraints), traffic conditions, and vehicle availability. Key responsibilities include dispatching vehicles to serve customer requests, planning efficient routes to minimize travel time and distance, and adjusting operations to meet changing demand patterns, all while adhering to operational constraints and objectives. Through its actions, the agent aims to balance efficiency, customer satisfaction, and system-wide performance.

States. A state represents the current configuration of the system at a specific point in time, serving as the foundation for decision-making. In the context of the Dial- a-Ride Problem, a state encapsulates key details about the system's status, such as the current locations of all vehicles, passengers waiting for pickup or currently in transit, the remaining capacity and available time of each vehicle, and any unserved requests.

This snapshot provides a comprehensive overview of the system's dynamics and is critical for guiding the agent's actions.

For each user, the state can be expressed as a vector of attributes or features reflecting their status at the current time step. These attributes include earliest and latest allowable pickup and drop-off times, service duration, and load requirements. This hierarchical representation of states enables a Markov Decision Process to fully simulate the dynamic interactions between users and vehicles, facilitating the agent's learning process. By capturing all relevant constraints and requirements, the state plays a central role in optimizing operational decisions and achieving the overall objective of efficient and effective transportation service.

We introduce a time window constraint by adjusting the state representation in the MDP. The state space now includes time-related variables, which account for the allowable pickup and drop-off times for each request. When transitioning from one state to another, we ensure that actions which violate these time windows (such as a vehicle arriving before or after the designated time window) are penalized. This penalization is reflected in the reward function by applying a negative reward to such transitions. We address vehicle capacity constraints by modifying the state space to include the current load of each vehicle, as well as the maximum capacity of the vehicle. Actions that would result in exceeding the capacity (e.g., assigning too many passengers to a vehicle) are penalized in the reward function. Specifically, we introduce a penalty when the number of passengers assigned to a vehicle exceeds its capacity.

Action Space. The action space consists of the decisions made at each time step. It represents a decision taken by the agent to transition the system from one state to another, effectively defining how the agent interacts with the environment. In the context of the Dial-a-Ride Problem, actions encompass decisions such as assigning a vehicle to a passenger request, selecting the next destination for a vehicle (e.g., a pickup or drop-off point), or skipping a request due to constraints like time or vehicle capacity. Each action reflects a deliberate choice aimed at optimizing the system's performance while adhering to operational constraints. By executing these actions, the agent navigates the dynamic environment, responding to real-time changes and progressing toward achieving its optimization goals.

The Policy. A policy is the strategy or rule that the agent uses to determine the best action to take in a given state. It serves as the decision-making framework that guides the agent's interactions with the environment. In the context of the Dial-a-Ride Problem, the policy may initially rely on heuristic approaches or basic Markov Decision Process modeling, such as assigning the nearest available vehicle to a request or prioritizing requests based on their time windows.

As the agent learns, particularly in the context of Deep Q-Networks, the policy becomes more sophisticated. In this case, the policy is represented by the output of the neural network, which estimates the value (or quality) of each possible action in each state. This allows the agent to systematically choose actions that maximize long-term rewards, enabling it to optimize its decision-making process dynamically and effectively in complex, real-world scenarios.

Observations. Observations are the information the agent perceives from the environment, enabling it to infer the current state and make informed decisions. These observations provide the necessary data for the agent to update its knowledge and refine its policy. In the context of the Dial- a-Ride Problem, observations might include the locations of unserved requests, vehicle availability, remaining capacities, and any time constraint associated with ongoing operations.

By continuously gathering and processing observations, the agent remains aware of the environment's dynamic conditions, ensuring that its decisions are both contextually relevant and optimized for current and future states. Observations serve as the critical link between the environment and the agent's decision-making framework.

Transitions. The function (f (st, at) = st+1) defines how the state transitions from st to st+1 based on the action at.

The transition from one state to another is determined by the actions taken. For example:

- If actiontype = "pickup", it results in updating the state to reflect the pickup action.
- If actiontype = "deliver", it results in updating the state to reflect the delivery action.
- If actiontype = "non-available", it results in setting the corresponding vehicle as non-available.

The function (P (st + 1|st, at)): f (st, at) represents the state transition probability, which is the core of the Markov Decision Process.

Rewards. A reward is the feedback provided by the environment after the agent performs an action, quantifying the quality of that action in relation to the defined objective. In the context of DARP, the reward in our work is designed to reflect goals such as minimizing the total travel distance and meeting service constraints.

$$R = \begin{cases} +1 \ \textit{if the action reduces the total travel cost} \\ -1 \ \textit{if the action increases the total travel cost} \end{cases}$$

This feedback mechanism helps the agent assess the consequences of its decisions and adjust its policy to maximize long-term performance, guiding the agent toward actions that improve the overall system efficiency.

Value Function. The value function, also known as the Q-function (1), can be defined as the expected sum of the rewards obtained while taking the action a in the state s and, thereafter, following the policy π [14]:

$$Q(s, a; \theta) = E[r + \gamma \, max_\top(a') \, Q(s', a'; \theta')|s, a] \tag{1}$$

where:

- $Q(s, a; \theta)$ is the action-value function, which gives the expected utility of taking action in states under policy π and parameters θ.
- s is the current state.
- a is the action taken in a state s.

- θ represents the parameters of the Q-network.
- E denotes the expectation operator.
- r is the reward received after acting a in state s.
- γ is the discount factor, which determines the importance of future rewards.
- Max Q(s′, a′; θ′) is the maximum estimated action- value for the next state s′ and action a′, using the target network parameters θ′.
- s′ is the next state resulting from action a in state s.
- θ′ represents the parameters of the target Q-network, which are periodically updated to match θ.

5   Deep Q Network for the DARP

In Fig. 3, we present a framework illustrating the principle of the Deep Q-Network for the DARP.

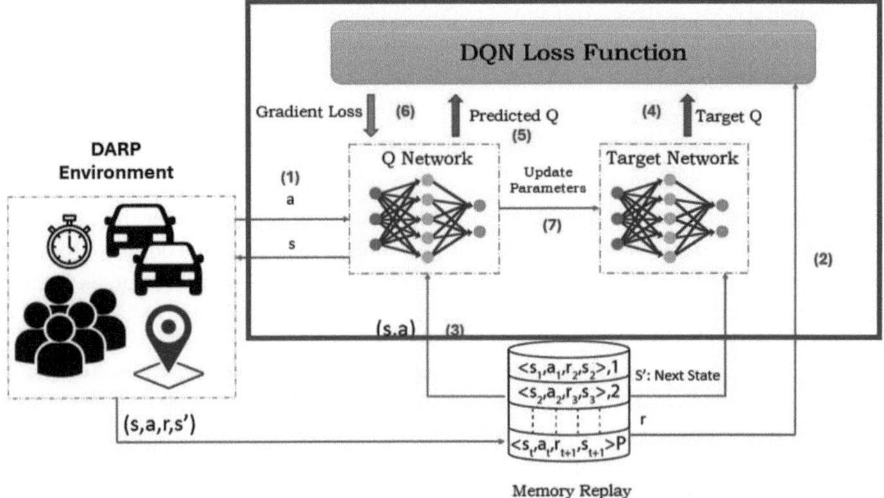

Fig. 3. Deep Q network for the DARP.

During the training process, the agent interacts with the environment (1), gathering experiences that are stored in the memory replay. Periodically, a batch of experiences is sampled from the memory replay (2). For each experience in the batch, the Q-network predicts the Q-value (3), estimating the expected future reward for taking an action in each state. Simultaneously, the target network calculates the target Q-value (4), providing a stable reference for learning. The DQN loss function is then computed based on the discrepancy between the predicted Q-value and the target Q-value (5). This loss function measures how well the Q-network is estimating the future rewards. The gradient of the loss function is used to update the parameters of the Q-network (6) through back-propagation, improving its ability to predict optimal actions. Additionally, the parameters of the target network are periodically updated (7) with the parameters of the Q-network to ensure the stability and convergence of the learning process.

5.1 DQN Workflow

The DQN is trained over numerous time steps across multiple episodes, undergoing a series of operations at each time step. This section provides a detailed explanation of each step in the process, giving a clear understanding of how the DQN operates.

Memory Replay. This is a buffer that stores past experiences of the agent in the form of tuples (s, a, r, s′). During the training, a random batch of experiences is sampled from the memory replay and used to update the Q-network. This helps to break the correlation between consecutive experiences and improves the stability of the learning process.

An Experience Replay selects an -greedy action from the current state, executes it in the environment, and gets back a reward and the next state. It saves this observation as a sample of training data.

Q-Network. This is a neural network that approximates the Q-function, which estimates the expected future reward for taking an action in each state. It takes the state (s) and action (a) as input and outputs the predicted Q- value (predicted Q). The Q-network is continuously learning and updating its parameters to improve its predictions. A random batch of samples, comprising a mix of older and more recent experiences, is then selected from this data. This batch serves as the input for both networks. The Q network processes the current state and action from each sample, predicting the Q value associated with the given action. This output is referred to as the Predicted Q Value.

Target Network. This is another neural network with the same architecture as the Q-network, but its parameters are updated less frequently. It is used to provide a stable target Q-value (target Q) for the Q-network to learn from it. The target Q-value is calculated using the next state (s′) and the action that maximizes the Q-value in that state according to the target network. The Target network evaluates the next state from each data sample and identifies the optimal Q value by considering all potential actions available in that state. This optimal Q value, which reflects the best possible future reward, is referred to as the 'Target Q Value.

DQN Loss Function. This function (2) measures the difference between the predicted Q-value and the target Q-value. The goal is to minimize this loss, which will lead to the Q-network learning the optimal Q-function. The gradient of the loss function is used to update the parameters of the Q-network through back-propagation. The Predicted Q Value, Target Q Value, and the observed reward from the data sample is used to compute the Loss to train the Q Network. Eq. (2) compute the loss function, which is the mean squared error between the current Q-value and the target Q-value by [14]:

$$L = (Q(s_t, a_t) - (r_t + \gamma \, max_\top(a\_(t+1)) \, Q(s_t + 1), a_t + 1))\wedge 2 \tag{2}$$

5.2 DQN Process

The DQN process details the process of training a Deep Q-Network using experience replay and temporal difference learning. The algorithm begins with the initialization of the Q-Network and target network, followed by experience sampling from interactions with the environment. These samples, comprising state transitions and rewards, form the basis for training the Q-Network using a loss function that minimizes prediction errors.

Initialization

- Execute a few actions with the environment to bootstrap the replay data.
- Initialize the Q Network with random weights and copy them to the Target Network.
- Experience Sampling

Sample experiences from the agent's interactions with the environment. Each experience is represented by a tuple (st, at, rt., st + 1), where st is the current state, at is the action taken, rt. is the reward received, and st + 1 is the next state.

– Experience Replay selects an ε-greedy action based on the current state, executes the action in the environment, and retrieves a reward and the next state.
– Beginning with the first time step, Experience Replay initiates the training data generation phase by using the Q Network to select an ε-greedy action.
– During this phase, the Q Network functions as the agent, interacting with the environment to generate training samples without performing any DQN training.
– The Q Network predicts the Q-values for all possible actions in the current state, and these Q-values are used to choose an ε-greedy action.

  The sample data (Current state, action, reward, next state) is saved

- Experience Replay takes the ε-greedy action, then gets the next state and reward.
    It saves these results in the replay buffer, where each result becomes a sample for future training.
- Select random training batch
- A random batch of samples is selected from the replay data to serve as input for both networks.
- The sample data (Current state, action, reward, next state) is saved
- Experience Replay performs the ε-greedy action and gets the next state and reward.

Loss Function Calculation

Back-propagate Loss to Q-Network: Back propagates the loss and adjust the weights of the Q Network using gradient descent.

– The Target network remains fixed and is not trained, meaning no loss is calculated, and backpropagation is not performed.
– This concludes the operations for the current time step.

Repeat Steps. Continuously sample experiences, calculate the temporal difference error, compute the loss function, and update the neural network parameters until the agent performs well enough or a predefined stopping criterion is met.

In summary, the DQN training process involves iteratively sampling experiences, computing the loss, and backpropagate the error to update the Q-Network. The method continues until the agent's performance stabilizes or a stopping criterion is met, demonstrating the effectiveness of reinforcement learning in dynamic environments.

6 Experimental Results

In this section, we provide the experimental results highlighting the performance of the Insertion based reinforcement learning for the DARP (IBRL-DARP) in addressing various cases of on-demand mobility problems. Therefore, we start by introducing the instances experimented for this work. Next, we provide the initial findings of our IBRL-DARP. These findings are compared with those obtained using the heuristic IH-DARP.

6.1 Parameters Tuning

In our Deep Q-Network implementation, we carefully selected optimal hyperparameters to ensure robust performance in solving the Dial-a-Ride Problem. Below are the precise values chosen:

- Learning Rate (α): Set to 0.001, providing an ideal balance between stability and convergence speed.
- Discount Factor (γ): Defined as 0.99 to emphasize future rewards.
- Exploration Rate (ϵ). Initialized at 1.0 and decayed exponentially by 0.995 per episode, promoting exploration early and exploitation later.
- Replay Buffer Size. Set to 100,000 experiences to ensure sufficient diversity while maintaining computational efficiency.
- Batch Size. Fixed at 32, enabling stable gradient updates without overfitting or underfitting.
- Target Network Update Frequency. The target network was updated every 10,000 steps, ensuring stability during training.
- Network Architecture. The neural network comprised 2 hidden layers, each with 128 units, allowing efficient state representation while controlling computational costs.

Number of Episodes and Steps per Episode: Training was conducted over 300 episodes, with a maximum of 200 steps per episode, ensuring adequate exploration and robust convergence to an optimal policy.

To further enhance the training process, the ADAM optimizer was employed. This adaptive learning rate optimization algorithm facilitated fast and stable convergence by dynamically adjusting learning rates based on the gradient.

These hyperparameters, combined with a carefully designed architecture, enabled the DQN agent to learn and adapt effectively to the complexities of the DARP while ensuring top-notch performance.

6.2 Benchmark Data Set

Several collections of benchmark instances are accessible for the standard Dial-a-Ride Problem as defined by [4]. In the set R, instances vary in the number of requests, ranging from 24 to 144. Each request has a capacity requirement of 1. Across all instances, the planning period is uniformly set to 1440, the route duration limit to 480, the vehicle capacity to 6. Authors in [3] introduced two additional sets of 12 randomly generated benchmark instances for the standard Dial-a-Ride Problem, denoted as sets A and B. These instances feature varying numbers of requests, ranging from 16 to 48, and the number of vehicles varies between 2 and 4. As previous instances, all nodes are randomly positioned within a Euclidean plane. In set A, vehicles have a capacity of 3, while each request has a capacity requirement of 1. In set B, vehicle capacity is set to 6. Each request's requirement is randomly selected from a uniform distribution.

6.3 Preliminary Results

In Table 2, comparative results between IH-DARP and IBRL-DARP are provided based on TTC values.

Table 2. Total Travel Costs (TTC) obtained with the compared methods.

Instances	m	n	IH-DARP	IBRL-DARP
R1b	3	24	210.742	**200.817**
R2a	2	48	380.012	**355.1**
R3a	7	72	**570.068**	582.488
R4a	9	96	590.763	**520.638**
a4–32	4	32	**531.112**	580.141
a4–40	4	40	601.489	**561.598**
a5–40	5	40	570.001	**499.015**
b3–36	3	36	**650.381**	670.447
b4–32	4	32	550.722	**480.033**
b5–40	5	40	630.5	**579.357**
b5–60	5	60	943.632	**905.211**

Overall, IBRL-DARP demonstrates superior performance in most instances, as evidenced by the lower total travel cost values. This improvement is particularly notable in larger problem instances where heuristic methods tend to underperform due to their reliance on predefined rules. The IBRL-DARP is able to iteratively refine its routing decisions based on feedback from previous solutions, thus capturing more complex optimization opportunities that static heuristics may overlook.

In 9 out of 11 tested instances, IBRL-DARP outperformed the IH-DARP, illustrating its superior capacity to optimize vehicle routing solutions. This success rate highlights the method's robustness across varying problem scales. The learning mechanism within IBRL-DARP allows it to adapt to diverse instance characteristics, including fluctuations in fleet size and request volume, which traditional heuristic approaches struggle to handle as problem complexity increases. Specifically, the IBRL-DARP reduces the total routing cost by a significant margin in some larger instances, such as:

- R2a: IBRL-DARP (355.1) vs. IH-DARP (380.012), a reduction of about 6.5%.
- R4a: IBRL-DARP (520.638) vs. IH-DARP (590.763), a reduction of about 11.9%.
- b5–60: IBRL-DARP (905.211) vs. IH-DARP (943.632), a reduction of about 4.1%.

This improvement highlights the effectiveness of the IBRL-DARP approach in finding better solutions in complex scenarios with larger fleet sizes and more customer requests, where heuristic methods might struggle to capture the complexity of the problem.

There are a few instances where the insertion heuristic is slightly better than the IBRL-DARP, such as:

- R3a: IH-DARP (570.068) vs. IBRL-DARP (582.488).
- a4–32: IH-DARP (531.112) vs. IBRL-DARP (580.141).
- b3–36: IH-DARP (650.381) vs. IBRL-DARP (670.447).

In these instances, the difference in performance is relatively small, and it suggests that for certain problem configurations, the IBRL-DARP's learning process might get stuck in local optima or requires further fine-tuning of hyper-parameters to improve its generalization ability. However, these differences are not substantial enough to detract from the overall superiority of the new IBRL-DARP approach.

In R3a instance, the heuristic method achieves a travel cost of 570.068, while the IBRL-DARP method results in a slightly higher cost of 582.488. This may indicate that the heuristic method is more efficient at handling smaller or simpler instances where the complexity of the problem is lower, and fewer optimizations are needed to generate a good solution. In these cases, the learning-based approach may introduce unnecessary complexity that does not yield significant improvements.

In a4–32 instance, the heuristic method yields a travel cost of 531.112, which outperforms the IBRL-DARP cost of 580.141. This suggests that, for medium-sized instances, IBRL-DARP might struggle with specific problem characteristics, such as vehicle-to-request ratio or the spatial distribution of requests. The heuristic, being more direct and tailored to simpler problem structures, might handle these cases better due to its simplicity.

One of the most notable observations is the consistency of the IBRL-DARP performance across various instance sizes. As the number of vehicles and customer requests increases, the IBRL-DARP continues to demonstrate a competitive performance, especially in larger instances where heuristic methods tend to degrade: b5–60: For the largest instance (5 vehicles, 60 customers), the IBRL-DARP delivers a significant improvement with an objective value of 905.211 compared to 943.632 for the insertion heuristic, indicating its robustness in handling larger-scale problems. This suggests that the IBRL-DARP, with its ability to learn from data and improve decisions iteratively, scales better and adapts more efficiently to an increasing problem complexity than traditional heuristic approaches. The results also indicate that the IBRL-DARP is able to optimize route planning by learning better strategies over time. Unlike the insertion heuristic, which relies on predefined rules and may miss opportunities for optimization, the IBRL-DARP dynamically adjusts its decisions based on feedback, which explains its stronger performance in most instances. The improved solution quality in instances like R4a and b5–40 demonstrates how the IBRL-DARP can find better routing configurations than traditional heuristics.

Moreover, this result underscores IBRL-DARP's ability to scale effectively with problem size, particularly in larger instances where heuristic methods tend to degrade. The adaptive nature of IBRL-DARP allows it to learn and optimize more effectively in such complex scenarios, making it particularly well-suited for real-world applications where fleet sizes and customer requests can grow significantly.

Additionally, the performance on medium and large instances, such as R4a (with 96 requests and 9 vehicles), emphasizes the method's strength in adapting to high-density transportation networks, further highlighting its advantage in scalability.

Future work could explore adaptive exploration strategies that adjust the exploration-exploitation trade-off based on the problem complexity. For instance, in simpler instances, the algorithm could favor exploitation of known good actions, while in larger, more complex instances, it could prioritize exploration to discover novel solutions.

7 Conclusion

This paper introduces an innovative insertion-based reinforcement learning approach for the Dial-a-Ride Problem (DARP), named IBRL-DARP. Our approach integrates a reinforcement learning framework with an insertion heuristic to generate initial solutions, which are then refined through learning-based optimization. The IBRL-DARP's performance was compared with a traditional insertion heuristic specifically tailored for DARPs, and the results demonstrate that IBRL-DARP consistently achieves superior performance across most test instances by significantly reducing routing costs. The key advantage of IBRL-DARP lies in its ability to adapt and learn from the problem environment, making intelligent decisions that improve over time. Its robustness across a range of DARP instances highlights its potential as a practical solution for real-world transport-on-demand applications, particularly in dynamic and uncertain environments.

However, there are certain problem configurations where the heuristic performs slightly better, suggesting room for further refinement. Future work could focus on enhancing the learning algorithm, potentially through improved exploration strategies or by incorporating more sophisticated reward structures, to ensure IBRL-DARP excels in all scenarios. This opens opportunities for continuous development, making IBRL-DARP a flexible and adaptable framework for addressing the challenges of vehicle routing in transportation systems.

Acknowledgments. We would like to express our sincere gratitude to the Higher School of Business of Tunis and the LARODEC laboratory of the Tunis University for their invaluable support. Their guidance and resources have been instrumental in the completion of this work. Thank you for your unwavering assistance and encouragement.

References

1. Ayari, E., Nasri, S., Aggoune-Mtalaa, W., Bouziri, H.: Learning routes within an intelligent on demand transport service. In: 2023 20th ACS/IEEE International Conference on Computer Systems and Applications (AICCSA), pp. 1–7. Ieee (2023, December)
2. Balakrishna, P., Ganesan, R., Sherry, L., Levy, B.S.: Estimating taxi-out times with a reinforcement learning algorithm. In: 2008 IEEE/AIAA 27th Digital Avionics Systems Conference, p. 3-D. Ieee (2008)
3. Cordeau, J.-F.: A branch-and-cut algorithm for the dial-A-ride problem. Oper. Res. **54**(3), 573–586 (2006)
4. Cordeau, J.-F., Laporte, G.: The dial-A-ride problem (Darp): variants, modeling issues, and algorithms. Q. J. Belg. Fr. Ital. Oper. Res. Soc. **1**, 89–101 (2003)
5. Cordeiro, E.A., Pitombeira-Neto, A.R.: Deep reinforcement learning for the dynamic vehicle dispatching problem: an event-based approach. Arxiv Preprint Arxiv:2307.07508 (2023)
6. De Lima, O., Shah, H., Chu, T.-S., Fogelson, B.: Efficient ridesharing dispatch using multi-agent reinforcement learning. Arxiv Preprint Arxiv:2006.10897 (2020)
7. Farazi, N.P., Zou, B., Ahamed, T., Barua, L.: Deep reinforcement learning in transportation research: a review. Transp. Res. Interdisc. Perspect. **11**, 100425 (2021)
8. Gao, J., et al.: Optimizing Taxi Fleet Operation with Reinforcement Learning. In: Proceedings of the AAAI Conference on Artificial Intelligence (2018)
9. Gao, Y., Jiang, D., Xu, Y.: Optimize taxi driving strategies based on reinforcement learning. Int. J. Geogr. Inf. Sci. **32**(8), 1677–1696 (2018)
10. Habib, Y., Filchenkov, A.: Multi-agent reinforcement learning for multi-vehicles one-commodity vehicle routing problem. Procedia Comput. Sci. **212**, 418–428 (2022)
11. Joe, W., Lau, H.C.: Deep reinforcement learning approach to solve dynamic vehicle routing problem with stochastic customers. Pro. Int. Conf. Autom. Plann. Sched. **30**, 394–402 (2020)
12. Li, Q., Xiong, Y., Zhang, S., Zhou, Y.: A machine learning-based algorithm for a hybrid two-echelon pickup and delivery problem. Available At Ssrm 4167816 (2022)
13. Ma, Y., et al.: A hierarchical reinforcement learning-based optimization framework for large-scale dynamic pickup and delivery problems. Adv. Neural Inf. Proces. Syst. **34**, 23609–23620 (2021)
14. Mnih, V., et al.: Human-level control through deep reinforcement learning. Nature. **518**(7540), 529–533 (2015)
15. Mortazavi, A., Ghasri, M., Haghshenas, H., Ray, T.: Adaptive logit models for constructing feasible solution for the dial-A-ride problem. Available At Ssrm 4307357 (2022)
16. Moussaoui, H., Benslimane, M., et al.: Reinforcement learning: a review. Int. J. Comput. Dig. Syst. **13**(1), 1 (2023)
17. Narayanan, A., Misra, P., Ojha, A., Bandhu, V., Ghosh, S., Vasan, A.: A reinforcement learning approach for electric vehicle routing problem with vehicle-to-grid supply. Arxiv Preprint. (2022) Arxiv:2204.05545
18. Nasri, S., Bouziri, H., Aggoune-Mtalaa, W.: Customer-oriented dial-a-ride problems: a survey on relevant variants, solution approaches, and applications. In: Emerging Trends in ICT For Sustainable Development: The Proceedings Of Nice2020 International Conference, pp. 111–119. Springer (2021)
19. Ngu, E., Parada, L., Escribano Macias, J.J., Angeloudis, P.: Decentralised multi-agent reinforcement learning approach for the same-day delivery problem. Transp. Res. Rec. **2676**(11), 385–395 (2022)
20. Rabecq, A., et al.: Solving the pickup and delivery problem with time windows using deep reinforcement learning. Comput. Oper. Res. (2022)

21. Xu, Y., Fang, M., Chen, L., Xu, G., Du, Y., Zhang, C.: Reinforcement learning with multiple relational attention for solving vehicle routing problems. Ieee Trans. Cybernet. **52**(10), 11107–11120 (2021)

22. Zhou, M., et al.: Multi-agent reinforcement learning for order-dispatching via order-vehicle distribution matching. In: Proceedings Of The 28th ACM International Conference On Information And Knowledge Management, pp. 2645–2653 (2019)

23. Ning, Z., Xie, L.: A survey on multi-agent reinforcement learning and its application. J. Autom. Intell. **3**, 73–91 (2024)

24. Wang, Q., Tang, C.: Deep reinforcement learning for transportation network combinatorial optimization: a survey. Knowl.-Based Syst. **233**, 107526 (2021)

25. Ye, S., Xu, L., Xu, Z., Wang, F.: A deep reinforcement learning-based intelligent QoS optimization algorithm for efficient routing in vehicular networks. Alex. Eng. J. **107**, 317–331 (2024)

26. Park, J., Baek, J., Song, Y.: Optimizing Smart City planning: a deep reinforcement learning framework. Ict Express. **11**, 129–134 (2024)

Data Analytics and Advanced Optimization

Computational Approach for Solving Fredholm Integral Equations

Saeed Hatamzadeh[1]([✉]) [iD] and Zahra Masouri[2] [iD]

[1] Department of Electrical and Electronics Engineering, Faculty of Engineering and Natural Sciences, Istinye University, Istanbul, Turkey
saeed.hatamzadeh@istinye.edu.tr
[2] Department of Mathematics, Isl.C., Islamic Azad University, Islamshahr, Iran

Abstract. The Fredholm integral equation, a pivotal aspect of integral equations, arises in various applications across physics, engineering, and applied mathematics. This survey examines a computational approach for solving Fredholm integral equations, focusing on both the first and second kinds. For this purpose, the categories of general methods for solving integral equations are mentioned in summary. Also, the general classification of basis functions are reviewed. The properties and characteristics of entire domain and subdomain basis functions are mentioned. Finally, we survey the formulation of the computational approach in detail for solving Fredholm integral equations. The approach uses a set of entire domain functions in order to formulate an efficient numerical scheme. Employing entire domain functions in the implementation of the numerical scheme offers significant benefits and major advantages, including global representation of the solution, reduction of computational effort, fast convergence, simplicity in implementation, compactness of representation, avoidance of discontinuity issues, and numerical flexibility. Various examples are evaluated in order to show the effectiveness and accuracy of the approach. We also examine the convergence rate of the approach numerically. The results of implementation of the approach using a computational software are given for a better illustration. The results confirm the efficiency and accuracy of the approach to solve Fredholm integral equations.

Keywords: Fredholm Integral Equation · Entire Domain · Subdomain Basis Functions · Computational Approach · Numerical Solution

1 Introduction

Many physical and engineering problems are modeled by means of integral equations and other forms of functional equations. The solution of these mathematical models cannot often be obtained analytically and hence a computational technique is required in order to calculate an approximate solution for them [1–40].

Integral equations can be classified into linear and nonlinear. A linear integral equation involves the unknown function linearly, meaning no powers, products, or nonlinear operations of the unknown function appear. A nonlinear integral equation involves the

A. Mirzazadeh et al. (Eds.): ODSIE 2024, CCIS 2482, pp. 303–322, 2026.
https://doi.org/10.1007/978-3-031-93601-2_19

unknown function in a nonlinear way, meaning powers, products, or other nonlinear operations of the unknown function are present.

A Fredholm integral equation is a type of integral equation that involves an unknown function appearing under an integral sign. It is named after Erik Ivar Fredholm, a Swedish mathematician. These equations are widely used in mathematics and physics, particularly in solving boundary value problems, signal processing, and quantum mechanics.

In [41], the focus is on integral equations with kernels that can be represented as scalar multiples of conservative (Markov) kernels. A variational approach is employed by framing the problem as a minimization task within the space of probability measures, incorporating an entropic regularization term. A numerical technique leveraging the Shepard method is proposed in [42] to address Fredholm integral equations of the second kind. The approach represents the unknown function as a linear combination of Shepard basis functions. A collocation method is then applied to derive a matrix equation that determines the coefficients of the approximate solution. In [43], a meshless approach utilizing hybrid radial kernels (HRKs) is employed to solve Fredholm integral equations (FIEs) of the second kind. The method approximates the solution using a discrete collocation procedure, which is based on a hybrid kernel formed by combining selected kernels with appropriate weight parameters. In [44], the discrete Legendre Galerkin and multi-Galerkin methods are utilized to approximate the solution of the Tikhonov-regularized Fredholm integral equations of the first kind. An a priori parameter selection strategy is introduced to determine convergence rates with respect to the infinity norm. In [45], a modified collocation method is explored for solving Fredholm integral equations on the interval $[-1, 1]$ with integrands exhibiting algebraic singularities at the endpoints. The approach leverages the zeros of Jacobi polynomials within suitably weighted spaces. An iterative numerical approach is presented in [46] that combines the quasilinearization technique with the Nyström method to approximate solutions for nonlinear Fredholm integral equations of the second kind. In [47], a numerical iterative method is proposed for solving multidimensional integral equations within a cubic domain, utilizing Picard iteration and Newton–Cotes rules. The study demonstrates that combining the successive approximation (Picard) sequence with Newton–Cotes rules yields an approximate solution to the multidimensional Fredholm–Urysohn integral equation of the second kind. The focus in [48] is on developing an optimal quadrature formula in a Hilbert space for numerically approximating integral equations. The study examines the process of solving integral equations using quadrature formulas and constructs an optimal weighted quadrature formula within the Hilbert space. Algorithms for solving the integral equation are provided based on the constructed optimal quadrature formula and the trapezoidal rule. In [49], the focus is on approximating solutions of second-kind Fredholm integral equations using non-uniform spline quasi-interpolation. The objective is to identify the most efficient non-uniform partition for achieving an optimal numerical solution to the integral equation. To accomplish this, a solution method based on genetic algorithms is proposed, utilizing an accurate approximation of the integral equation's kernel. In [50], the Shifted Jacobi Spectral Galerkin Method (SJSGM) and its iterative variant are proposed to solve nonlinear Fredholm integral equations of the Hammerstein type with weakly singular kernels. By applying a smoothing transformation, the regularity of the exact solution is improved, leading to an enhanced convergence order for both the SJSGM and the iterated SJSGM.

This survey explores a computational approach for solving Fredholm integral equations, addressing both the first and second kinds. The approach uses a set of entire domain functions in order to formulate an efficient numerical scheme. Using entire domain basis functions in the solution of integral equations offers several advantages, particularly in terms of efficiency, accuracy, and computational performance. Here are some key benefits:

1. Global Representation of the Solution:

   Entire domain functions represent the solution over the entire problem domain, capturing global behavior effectively. This can lead to more accurate approximations, especially for smooth solutions, as fewer basis functions may be required to achieve a desired level of accuracy.

2. Reduced Computational Effort:

   Because entire domain functions often require fewer terms to approximate a solution, the size of the resulting system of equations is smaller, reducing computational effort.

3. Fast Convergence:

   For problems with smooth kernels or solutions, entire domain functions tend to exhibit faster convergence compared to subdomain functions.

4. Suitability for Smooth Solutions

   Entire domain functions are well-suited for problems where the solution has high regularity (smoothness) across the domain, as they naturally capture global features without requiring piecewise adjustments.

5. Simplicity in Implementation

   Entire domain functions often lead to simpler mathematical formulations, as they do not require complex domain partitioning or handling of interfaces between subdomains.

6. Compactness of Representation

   Solutions expressed using entire domain functions are typically compact, with fewer coefficients needed for a high-quality approximation. This is advantageous in problems where storage and computational efficiency are important.

7. Avoidance of Discontinuity Issues

   Unlike subdomain functions, which may introduce artificial discontinuities at subdomain boundaries, entire domain functions ensure smooth approximations across the entire domain.

8. Analytical and Numerical Flexibility

   Entire domain functions often allow for the derivation of analytical properties (e.g., error bounds, stability analysis) more easily than subdomain functions. Numerical integration and differentiation are also straightforward with these functions.

The structure of this survey is organized as follows:

- Section 2 provides a summary of general methods for solving integral equations.
- Section 3 reviews the general classification of basis functions.

- Sections 4 and 5 discuss the properties and characteristics of entire domain and subdomain basis functions, respectively.
- Section 6 presents a review of the concept of error evaluation.
- Section 7 examines the formulation of the computational approach for solving Fredholm integral equations in detail.
- Section 8 evaluates various examples to demonstrate the effectiveness and accuracy of the approach, as well as its numerical convergence rate.

Finally, Sect. 9 concludes the survey.

2 General Methods for Solving Integral Equations

As integral equations are classified into different types, their solution methods can be broadly categorized as follows:

2.1 Analytical Methods

These methods aim to solve the integral equations exactly, yielding closed-form solutions. They are usually applied to simpler or specific types of equations.

1. Separation of Variables: Used when the integral equation can be split into functions of different variables.
2. Direct Substitution: Directly solving the equation by substituting a guessed or derived solution.
3. Transform Methods:
   (a) Laplace Transform: Used for equations involving exponential or step functions.
   (b) Fourier Transform: Used for equations with periodic or wave-like components.
4. Green's Function Method: Converts a differential equation into an integral equation to solve it.
5. Iterative Exact Solution (if convergence is guaranteed): Applies when the equation can be rearranged into an iterative form that exactly converges.

2.2 Numerical Methods

These are approximation techniques used when analytical solutions are not feasible.

1. Discretization Methods:
   (a) Quadrature Methods: Replace the integral with a summation using numerical integration rules (e.g., Trapezoidal Rule, Simpson's Rule).
   (b) Collocation Method: Approximate the solution using a finite set of basis functions and enforce the equation at selected points.
   (c) Galerkin Method: Similar to the collocation method but uses orthogonal basis functions for weighted residual minimization.
   (d) Nyström Method: Converts the integral equation into a linear system by discretizing the kernel and solution.
2. Finite Difference Method (FDM): Discretizes the entire domain, solving the integral equation at discrete points.

3. Finite Element Method (FEM): Subdivides the domain into smaller elements, approximating the solution piecewise.
4. Monte Carlo Methods: Used for high-dimensional integral equations, employing random sampling techniques.

2.3 Iterative Methods

These methods find approximate solutions through successive approximations, typically for nonlinear or Fredholm equations.

1. Successive Substitutions (Picard Iteration): Iteratively updates the solution starting from an initial guess.
2. Newton's Method for Integral Equations: A linearization approach for solving nonlinear equations iteratively.
3. Krylov Subspace Methods: Useful for solving the large linear systems arising from discretization (e.g., GMRES, BiCGSTAB).

2.4 Kernel-Based Methods

These methods leverage properties of the integral kernel to simplify the problem.

1. Symmetric Kernel Reduction: Simplifies equations with symmetric kernels using eigenfunction expansions.
2. Fredholm Theory: Solves integral equations using Fredholm determinants and resolvent kernels.
3. Singular Kernel Techniques: Address equations with singular kernels by regularization or specialized approaches (e.g., Hilbert transform).

2.5 Variational Methods

These methods reformulate the integral equation as a variational problem.

1. Rayleigh-Ritz Method: Approximates the solution by minimizing a functional derived from the integral equation.
2. Least Squares Method: Minimizes the squared residuals of the equation over the domain.

2.6 Special Techniques for Nonlinear Integral Equations

Nonlinear integral equations require specialized approaches.

1. Fixed-Point Iteration Methods: Applies Banach's Fixed-Point Theorem to ensure convergence to a solution.
2. Harmonic Balance Methods: Solves equations in periodic settings by balancing harmonics.
3. Perturbation Methods: Expands the solution as a series in terms of a small parameter.

Each method's applicability depends on the type of integral equation, such as Fredholm vs. Volterra, linear vs. nonlinear, and the nature of the kernel (smooth, singular, etc.).

3 General Classification of Basis Functions

Basis functions are a fundamental concept in mathematics, especially in fields such as linear algebra, functional analysis, signal processing, and numerical methods. A basis function is one of a set of functions that forms the basis for a function space. Together, these functions can be linearly combined to approximate or exactly represent any function within that space. In finite-dimensional spaces, basis functions are analogous to basis vectors in vector spaces. In infinite-dimensional spaces, basis functions extend the concept to function spaces.

Here are some general categories of basis functions.

3.1 Polynomial Basis Functions

1. Definition: Functions that are polynomials of a certain degree.
2. Examples: Monomials $1, x, x^2, \ldots, x^n$.
3. Applications: Polynomial interpolation, approximation of continuous functions (e.g., Taylor series).

3.2 Trigonometric Basis Functions

1. Definition: Functions based on sine and cosine functions.
2. Examples: $\sin(nx)$, $\cos(nx)$ for $n \in \mathbb{Z}$.
3. Applications: Fourier series for periodic function representation, signal processing.

3.3 Exponential Basis Functions

1. Definition: Functions that include exponentials of the form $e^{\lambda x}$ where λ is a constant.
2. Examples: e^{ix}, e^{-x}.
3. Applications: Used in solving differential equations, signal processing.

3.4 Wavelet Basis Functions

1. Definition: Functions that allow for multi-resolution analysis of functions.
2. Examples: Haar wavelets, Daubechies wavelets.
3. Applications: Image compression, data analysis, and signal processing.

3.5 Orthogonal Basis Functions

1. Definition: Basis functions that are orthogonal with respect to a certain inner product.
2. Examples: Legendre polynomials, Chebyshev polynomials.
3. Applications: Numerical methods, such as spectral methods, for solving differential equations.

3.6 Rational Basis Functions

1. Definition: Functions that are ratios of polynomials.
2. Examples: $\frac{p(x)}{q(x)}$, where p and q are polynomials.
3. Applications: Used in rational interpolation and approximation theory.

3.7 Radial Basis Functions

1. Definition: Functions whose value depends only on the distance from a central point.
2. Examples: Gaussian functions, multiquadrics.
3. Applications: Interpolation in multi-dimensional spaces, scattered data approximation.

3.8 B-Splines and NURBS

1. Definition: Piecewise polynomial functions that are defined over a partition of the domain.
2. Examples: B-splines, Non-Uniform Rational B-Splines (NURBS).
3. Applications: Computer graphics, geometric modeling, and data fitting.

4 Entire Domain Basis Functions

Entire domain basis functions are a concept often used in the context of functional analysis, particularly in the study of function spaces and approximation theory. They serve as a foundational tool for expressing complex functions in simpler terms. This term typically refers to the entire input space over which the functions are defined, which could be a finite interval, the entire real line, or higher-dimensional spaces.

Entire domain basis functions are critical in many areas of mathematics, physics, engineering, and data science. They enable the representation of complex functions and facilitate various mathematical operations, making them essential tools for both theoretical exploration and practical applications.

A common example of an entire domain basis function is the set of exponential functions e^{inx} used in Fourier series. Any periodic function can be approximated by a series of these functions, allowing for powerful analyses in both theoretical and applied contexts.

Here is an overview of key aspects related to entire domain basis functions.

4.1 Properties

1. Completeness: A set of basis functions is complete if any function in the space can be approximated to any desired degree of accuracy by a linear combination of the basis functions.
2. Orthogonality: Basis functions can be orthogonal (i.e., the integral of their product over the domain is zero) or non-orthogonal. Orthogonality often simplifies calculations.
3. Continuity and Smoothness: Many entire domain basis functions are chosen for their continuous and differentiable properties, which are important in applications such as numerical analysis.

4.2 Common Types of Entire Domain Basis Functions

1. Polynomial Basis: Functions like monomials (e.g., $1, x, x^2, \ldots$) that are often used in polynomial approximation.
2. Fourier Series: Basis functions in the form of sine and cosine functions, which are particularly useful for periodic functions.
3. Wavelets: Functions that are used for multi-resolution analysis and can be localized in both time and frequency.
4. Splines: Piecewise polynomial functions used for approximation and interpolation, providing a smooth fit to data points.

4.3 Applications

1. Approximation Theory: Entire domain basis functions are employed to approximate complex functions, especially in numerical methods and computational mathematics.
2. Signal Processing: In Fourier analysis, signals are decomposed into sine and cosine components, making it easier to analyze frequency components.
3. Machine Learning: Basis functions are used in kernel methods and support vector machines to map input data into higher-dimensional spaces.

5 Subdomain Basis Functions

Subdomain basis functions are used primarily in numerical methods for solving partial differential equations (PDEs), particularly in the context of finite element methods (FEM) and similar techniques. These functions are defined over specific subdomains of the computational domain, allowing for localized representation of the solution to a PDE.

Subdomain basis functions play a crucial role in numerical methods for PDEs, providing a structured way to approximate solutions within smaller, manageable sections of a larger problem. Their flexibility and efficiency make them indispensable tools in computational mathematics and engineering.

Here is an overview of subdomain basis functions and their importance.

5.1 Definition of Subdomain Basis Functions

Subdomain basis functions are functions that are defined on smaller, manageable portions (subdomains) of the overall domain where the PDE is being solved. They are often used to approximate solutions within each subdomain, which can vary from one subdomain to another.

5.2 Purpose of Using Subdomain Basis Functions

1. Local Approximation: They enable localized approximation of solutions, making it easier to handle complex geometries and boundary conditions.
2. Flexibility: Different types of basis functions can be used in different subdomains, allowing for adaptability in numerical modeling.
3. Efficiency: By focusing on subdomains, computations can be performed more efficiently, especially in larger problems.

5.3 Types of Subdomain Basis Functions

1. Polynomial Basis Functions: Commonly used in finite element methods. They can be linear, quadratic, cubic, etc., depending on the required accuracy.
2. Piecewise Functions: Often used in adaptive methods where the basis functions are defined as piecewise polynomials.
3. Wavelet Functions: These provide multiresolution analysis and can be useful for capturing localized features in the solution.

5.4 Applications

1. Finite Element Analysis (FEA): In FEA, the entire domain is divided into smaller elements, and subdomain basis functions are used to construct the solution over these elements.
2. Boundary Element Methods (BEM): Similar principles apply where basis functions can be defined over boundary subdomains.
3. Computational Fluid Dynamics (CFD): Used in discretizing fluid flow equations, where local behavior is critical.

5.5 Advantages

1. Improved Accuracy: Subdomain basis functions can lead to more accurate solutions, especially when dealing with non-linear problems or those with sharp gradients.
2. Computational Efficiency: Smaller systems can be solved independently, often in parallel, which speeds up computation.

5.6 Challenges

1. Continuity and Compatibility: Ensuring that the solution is continuous across subdomain boundaries can be challenging, especially with different types of basis functions.
2. Complex Implementation: Setting up the numerical scheme with subdomain basis functions may require additional effort in terms of coding and algorithm design.

6 Review on Error Evaluation Concept

Error evaluation in numerical methods is crucial for determining the accuracy and reliability of approximations. Here is an overview of the key concepts.

6.1 Types of Errors

Errors in numerical methods can be broadly classified into the following categories:

1. Truncation Error:
   (a) Arises when an infinite process (e.g., a series or iteration) is truncated.
   (b) Example: Approximating a derivative using a finite difference method.
2. Round-off Error:

(a) Occurs due to the finite precision of computer arithmetic.
(b) Example: When a computer rounds a number like π to 3.14159.
3. Discretization Error:
   (a) Results from approximating continuous functions or domains with discrete counterparts.
   (b) Example: Converting a differential equation into a finite difference equation.
4. Propagation Error:
   (a) Arises when errors accumulate through successive steps of computation.

6.2 Error Measures

Common ways to measure errors include:

1. Absolute Error:
   (a) $E_{\text{abs}} = |\,\text{TrueValue} - \text{Approximation}|$
   (b) Represents the magnitude of the error.
2. Relative Error:
   (a) $E_{\text{rel}} = \dfrac{|\text{TrueValue} - \text{Approximation}|}{|\text{TrueValue}|}$
   (b) Provides a normalized error relative to the true value.
3. Percent Relative Error:
   (a) $E_{\text{percent}} = E_{\text{rel}} \times 100\%$
   (b) Expresses relative error as a percentage.

6.3 Convergence

The rate at which a numerical method approaches the exact solution is crucial:

1. Order of Accuracy:
   (a) Describes how the error decreases as the step size or discretization interval decreases.
   (b) If a method has an error proportional to h^p, where h is the step size, it is said to be of order p.
2. Consistency:
   (a) A method is consistent if the truncation error approaches zero as the step size approaches zero.
3. Stability:
   (a) Refers to how errors behave as computations proceed, especially for iterative methods.

6.4 Error Estimation

Techniques to estimate and control errors include:

1. Residuals:
   (a) Used to check how well a numerical solution satisfies the original equations.
2. Richardson Extrapolation:
   (a) Improves accuracy by combining results with different step sizes.
3. A Posteriori Error Estimation:
   (a) Estimates errors based on the computed solution rather than theoretical analysis.
4. Adaptive Methods:
   (a) Automatically adjust step sizes or discretization to control errors.

6.5 Practical Tips

1. Choose methods with a balance between computational cost and accuracy.
2. Be mindful of the precision limits of the computing system to minimize round-off errors.
3. Validate results by comparing with analytical solutions, when available, or with finer approximations.

7 A Survey on a Computational Approach for Solving Fredholm Integral Equations

One very important step in any numerical solution is the choice of basis functions. In this section, we survey a scheme that applies Sinc basis functions in formulation of a computational approach for solving Fredholm integral equations of the first and second kind.

Consider the following Fredholm integral equation of the second kind:

$$ax(s) + \int_a^b k(s, t)x(t)dt = y(s) \tag{1}$$

where, $k(s, t)$ and $y(s)$ are known functions but $x(t)$ is unknown, and α is a constant coefficient. It is clear that for $\alpha = 0$, Eq. (1) is converted to a Fredholm integral equation of the first kind. However, for solving this equation we should select a set of basis functions to implement the method.

The Sinc function is defined on the whole real line by

$$\text{Sinc}(t) = \begin{cases} \frac{\sin(\pi t)}{\pi t}, & t \neq 0, \\ 1, & t = 0. \end{cases} \tag{2}$$

According to the above definition, it is obvious that this function can be considered as an entire-domain basis function because $t \in (-\infty, \infty)$. Now, we approximate the function $x(s)$ with respect to the Sinc functions. It should be mentioned that we construct these basis functions only via scaling the main function defined by (2) and don't use any dilation or translation. Hence, approximating the function $x(s)$ with respect to these functions gives

$$x(s) \simeq \sum_{n=1}^N c_n \text{Sinc}(ns), \tag{3}$$

where, $\text{Sinc}(ns) = \frac{\sin(n\pi s)}{n\pi s}$, and c_ns are unknown coefficients that should be determined.

Substituting Eq. (3) into (1) follows:

$$\alpha \sum_{n=1}^N c_n \text{Sinc}(ns) + \sum_{n=1}^N c_n \int_a^b k(s, t)\text{Sinc}(nt)dt \simeq y(s). \tag{4}$$

Let s_i, $i = 1, 2, \ldots, N$, be N appropriate points in interval $[a, b)$; putting $s = s_i$ in Eq. (4) follows:

$$\alpha \sum_{n=1}^N c_n \text{sinc}(ns_i) + \sum_{n=1}^N c_n \int_a^b k(s_i, t)\text{Sinc}(nt)dt \simeq y(s_i), \\ i = 1, 2, \ldots, N, \tag{5}$$

or

$$\sum_{n=1}^{N} c_n \left[\alpha \mathrm{Sinc}(ns_i) + \int_a^b k(s_i, t)\mathrm{Sinc}(nt)dt \right] \simeq y(s_i),$$
$$i = 1, 2, \ldots, N. \tag{6}$$

Now, replace $\simeq$ with $=$, hence Eq. (6) is a linear system of N algebraic equations for N unknown coefficients c_1, c_2, ..., c_N. So, an approximate solution $x(s) \simeq \sum_{n=1}^{N} c_n \mathrm{Sinc}(ns)$, is obtained for Eq. (1).

Now, let us consider a set of subdomain basis functions; for example, well-known block-pulse functions (BPFs). An N-set of BPFs is defined over the interval $[a, b)$ as

$$\phi_n(t) = \begin{cases} 1, & a + nh \leq t < a + (n+1)h, \\ 0, & \text{otherwise,} \end{cases} \tag{7}$$

where, $n = 0, 1, \ldots, N - 1$, $h = (b - a)/N$, with a positive integer value for N. Also, ϕ_n is the nth block-pulse function. Clearly any $\phi_n(t)$ can be obtained of translating (delaying) the basic function $\phi_0(t)$. It is seen that these functions can be considered as subdomain basis functions, because any ϕ_n is defined over the subinterval $[a + nh, a + (n+1)h)$. However, approximating the function $x(s)$ with respect to BPFs gives

$$x(s) \simeq \sum_{n=1}^{N} c_n \phi_n(s) \tag{8}$$

So, by using the same procedure mentioned above, we finally obtain

$$\sum_{n=1}^{N} c_n \left[\alpha \phi_n(s_i) + \int_a^b k(s_i, t)\phi_n(t)dt \right] \simeq y(s_i),$$
$$i = 1, 2, \ldots, N. \tag{9}$$

Equation (9) can be writen as

$$\sum_{n=1}^{N} c_n f_{ni} \simeq y(s_i), \tag{10}$$

where,

$$f_{ni} = \begin{cases} \alpha \phi_n(s_i) + \int_{nh}^{(n+1)h} k(s_i, t)\phi_n(t)dt & n = i, \\ \int_{nh}^{(n+1)h} k(s_i, t)\phi_n(t)dt & n \neq i. \end{cases} \tag{11}$$

Equation (10) is a linear system of N algebraic equations for N unknown coefficients c_1, c_2, ..., c_N. So, an approximate solution $x(s) \simeq \sum_{n=1}^{N} c_n \phi_n(s)$, is obtained for Eq. (1).

8 Numerical Results

Here, we solve some integral equations to show the accuracy of the surveyed computational approach. Also, we compare the obtained results from the entire-domain method with the results obtained from the subdomain method using BPFs.

8.1 Examples

Example 1. Consider the following Fredholm integral equation of the first kind:

$$\int_{-a/2}^{a/2} k(s, t)x(t)dt = y(s), \tag{12}$$

with the exact solution $x(s) = \sin^2(s) + is^2$, and suitable right hand side $[y(s)]$ which can be obtained of *numerical integration*. Also, $i = \sqrt{-1}$ and $k(s, t) = H_0^{(2)}(r|s - t|)$, where $H_0^{(2)}$ is Hankel function of the second kind of zero order. Solving this equation via the surveyed approach gives the approximate solution of $x(s)$. The exact and approximate values of $x(s)$ magnitude for $a = 0.5$, $r = 2\pi$, and for $N = 4$ and 8 have been calculated at nine specific points and shown in Tables 1 and 2.

Example 2. Assume that the Eq. (12) has the exact solution $x(s) = s^4 + s \sin(s/a) + i \ln(s^2 + a^2)$. The exact and approximate values of $x(s)$ magnitude for $a = 0.5$, $r = 2\pi$, and for $N = 4$ and 8 are shown in Tables 3 and 4.

Example 3. As the final example, consider the following Fredholm integral equation of the second kind:

$$x(s) + \int_{-a/2}^{a/2} k(s, t)x(t)dt = y(s), \tag{13}$$

with the exact solution $x(s) = s \sin(s) + is^2 \cos(s)$, and suitable right hand side $[y(s)]$. Also, $k(s, t) = H_0^{(2)}(r|s - t|)$. For $a = 1$, $r = \pi$, and for $N = 4$ and 8, Tables 5 and 6 give the exact and approximate values of $x(s)$ magnitude.

Referring to the numerical results presented in Tables 1,2, 3, 4, 5, and 6, it is concluded that the surveyed approach has a good accuracy and efficiency for solving integral equations of the form (12) or (13), in which, $k(s, t) = H_0^{(2)}(r|s - t|)$. Also, the results show that using entire-domain Sinc functions to implement the moment method for solving these problems leads to higher accuracy than subdomain BPFs.

The results presented here have been calculated for $N = 4$ and 8, because for these values of N, the differences between the results of the two methods could be more obvious.

8.2 Error Evaluation, Stability and Convergence

Now, we calculate the mean-absolute errors of the approach to show its stability and convergence.

The mean-absolute error (MAE) is defined as

$$E_N = \frac{1}{N} \sum_{i=1}^{N} |x(s_i) - x_N(s_i)|, \tag{14}$$

where, $x(s)$ is the exact solution and $x_N(s)$ is the approximate solution. These errors have been calculated for $N = 2, 4$ and 8. Tables 7, 8, and 9 give the errors for Examples 1, 2, and 3, respectively.

Table 1. Numerical results for Example 1 ($N = 4$).

s	Exact solution	Entire-domain method	Subdomain method
-0.20	0.0562	0.0573	0.0529
-0.15	0.0317	0.0253	0.0529
-0.10	0.0141	0.0100	0.0035
-0.05	0.0035	0.0055	0.0035
0.00	0.0000	0.0050	0.0000
0.05	0.0035	0.0055	0.0035
0.10	0.0141	0.0100	0.0035
0.15	0.0317	0.0253	0.0529
0.20	0.0562	0.0573	0.0529

Table 2. Numerical results for Example 1 ($N = 8$).

s	Exact solution	Entire-domain method	Subdomain method
-0.20	0.0562	0.0548	0.0692
-0.15	0.0317	0.0339	0.0339
-0.10	0.0141	0.0135	0.0120
-0.05	0.0035	0.0031	0.0011
0.00	0.0000	0.0013	0.0000
0.05	0.0035	0.0031	0.0011
0.10	0.0141	0.0135	0.0120
0.15	0.0317	0.0339	0.0339
0.20	0.0562	0.0548	0.0692

The errors shown in Tables 7, 8, and 9 clearly confirm the quick convergence of the approach. In addition, we implemented the approach for $N = 16, 32$ and 64 too, and these errors decreased by increasing N, such that the MAEs for $N = 16$ and 32 were around 0.001 and 0.0005, respectively. Hence, the surveyed approach has a good stability.

Table 3. Numerical results for Example 2 ($N = 4$).

s	Exact solution	Entire-domain method	Subdomain method
−0.20	1.2404	1.2348	1.2507
−0.15	1.3009	1.3256	1.2507
−0.10	1.3472	1.3632	1.3763
−0.05	1.3764	1.3679	1.3763
0.00	1.3863	1.3657	0.0000
0.05	1.3764	1.3679	1.3763
0.10	1.3472	1.3632	1.3763
0.15	1.3009	1.3256	1.2507
0.20	1.2404	1.2348	1.2507

Table 4. Numerical results for Example 2 ($N = 8$).

s	Exact solution	Entire-domain method	Subdomain method
−0.20	1.2404	1.2452	1.2103
−0.15	1.3009	1.2905	1.2950
−0.10	1.3472	1.3497	1.3533
−0.05	1.3764	1.3791	1.3834
0.00	1.3863	1.3824	0.0000
0.05	1.3764	1.3791	1.3834
0.10	1.3472	1.3497	1.3533
0.15	1.3009	1.2905	1.2950
0.20	1.2404	1.2452	1.2103

Table 5. Numerical results for Example 3 ($N = 4$).

s	Exact solution	Entire-domain method	Subdomain method
−0.4	0.2144	0.2192	0.1915
−0.3	0.1235	0.1016	0.1915
−0.2	0.0558	0.0251	0.0213
−0.1	0.0141	0.0303	0.0213
0.0	0.0000	0.0494	0.0000
0.1	0.0141	0.0303	0.0213
0.2	0.0558	0.0251	0.0213
0.3	0.1235	0.1016	0.1915
0.4	0.2144	0.2192	0.1915

Table 6. Numerical results for Example 3 ($N = 8$).

s	Exact solution	Entire-domain method	Subdomain method
-0.4	0.2144	0.2266	0.2545
-0.3	0.1235	0.1157	0.1338
-0.2	0.0558	0.0528	0.0489
-0.1	0.0141	0.0092	0.0052
0.0	0.0000	0.0067	0.0000
0.1	0.0141	0.0092	0.0052
0.2	0.0558	0.0528	0.0489
0.3	0.1235	0.1157	0.1338
0.4	0.2144	0.2266	0.2545

Table 7. Mean-absolute errors for Example 1.

N	Entire-domain method	Subdomain method
2	0.0172	0.0156
4	0.0036	0.0078
8	0.0012	0.0044

Table 8. Mean-absolute errors for Example 2.

N	Entire-domain method	Subdomain method
2	0.1074	0.1945
4	0.0145	0.1739
8	0.0050	0.1649

Table 9. Mean-absolute errors for Example 3.

N	Entire-domain method	Subdomain method
2	0.0539	0.0596
4	0.0218	0.0295
8	0.0070	0.0147

9 Conclusion

We examined in this survey, a computational approach for solving Fredholm integral equations of both the first and second kinds. The categories of general methods for solving integral equations were mentioned. The general classification of basis functions were reviewed. The properties and characteristics of entire domain and subdomain basis functions were mentioned. We surveyed the formulation of the computational approach in detail for solving Fredholm integral equations. The approach used a set of entire domain functions in order to formulate an efficient numerical scheme. Employing entire domain functions in the implementation of the numerical scheme offers significant benefits and major advantages, including global representation of the solution, reduction of computational effort, fast convergence, simplicity in implementation, compactness of representation, avoidance of discontinuity issues, and numerical flexibility. Various examples were evaluated to show the effectiveness and accuracy of the approach. We also examined the convergence rate of the approach numerically. The numerical results confirmed the efficiency and accuracy of the approach.

References

1. Masouri, Z., Hatamzadeh, S.: A regularization-direct method to numerically solve first kind Fredholm integral equation. Kyungpook Mathem. J. **60**(4), 869–881 (2020)
2. Hatamzadeh-Varmazyar, S., Masouri, Z., Babolian, E.: Numerical method for solving arbitrary linear differential equations using a set of orthogonal basis functions and operational matrix. Appl. Math. Model. **40**(1), 233–253 (2016)
3. Babolian, E., Masouri, Z., Hatamzadeh-Varmazyar, S.: Numerical solution of nonlinear Volterra-Fredholm integro-differential equations via direct method using triangular functions. Comput. Mathem. Applic. **58**(2), 239–247 (2009)
4. Hatamzadeh-Varmazyar, S., Naser-Moghadasi, M., Masouri, Z.: A moment method simulation of electromagnetic scattering from conducting bodies. Prog. Electromagn. Res. **81**, 99–119 (2008)
5. Hatamzadeh-Varmazyar, S., Naser-Moghadasi, M., Babolian, E., Masouri, Z.: Calculating the radar cross section of the resistive targets using the Haar wavelets. Prog. Electromagn. Res. **83**, 55–80 (2008)
6. Hatamzadeh-Varmazyar, S., Naser-Moghadasi, M., Babolian, E., Masouri, Z.: Numerical approach to survey the problem of electromagnetic scattering from resistive strips based on using a set of orthogonal basis functions. Prog. Electromagn. Res. **81**, 393–412 (2008)
7. Danesfahani, R., Hatamzadeh-Varmazyar, S., Babolian, E., Masouri, Z.: A scheme for RCS determination using wavelet basis. AEU Int. J. Electron. Commun. **64**(8), 757–765 (2010)
8. Hatamzadeh-Varmazyar, S., Masouri, Z.: Numerical expansion-iterative method for analysis of integral equation models arising in one-and two-dimensional electromagnetic scattering. Eng. Anal. Bound. Elem. **36**(3), 416–422 (2012)
9. Hatamzadeh-Varmazyar, S., Masouri, Z.: A numerical approach for calculating the radar cross-section of two-dimensional perfect electrically conducting structures. J. Electromagn. Waves Appl. **28**(11), 1360–1375 (2014)
10. Hatamzadeh-Varmazyar, S., Naser-Moghadasi, M., Sadeghzadeh-Sheikhan, R.: One-and two-dimensional scattering analysis using a fast numerical method. IET Microwaves Antennas Propag. **5**(10), 1148–1155 (2011)
11. Hatamzadeh-Varmazyar, S., Masouri, Z.: A fast numerical method for analysis of one-and two-dimensional electromagnetic scattering using a set of cardinal functions. Eng. Anal. Bound. Elem. **36**(11), 1631–1639 (2012)
12. Danesfahani, R., Hatamzadeh-Varmazyar, S., Babolian, E., Masouri, Z.: Applying shannon wavelet basis functions to the method of moments for evaluating the radar cross section of the conducting and resistive surfaces. Progr. Electromagn. Res. B. **8**, 257–292 (2008)

13. Hatamzadeh-Varmazyar, S., Naser-Moghadasi, M.: An integral equation modeling of electromagnetic scattering from the surfaces of arbitrary resistance distribution. Progr. Electromagn. Res. B. **3**, 157–172 (2008)
14. Hatamzadeh-Varmazyar, S., Naser-Moghadasi, M.: New numerical method for determining the scattered electromagnetic fields from thin wires. Progr. Electromagn. Res. B. **3**, 207–218 (2008)
15. Babolian, E., Masouri, Z., Hatamzadeh-Varmazyar, S.: New direct method to solve nonlinear Volterra-Fredholm integral and integro-differential equations using operational matrix with block-pulse functions. Progr. Electromagn. Res. B. **8**, 59–76 (2008)
16. Babolian, E., Masouri, Z.: Direct method to solve Volterra integral equation of the first kind using operational matrix with block-pulse functions. J. Comput. Appl. Math. **220**, 51–57 (2008)
17. Hatamzadeh-Varmazyar, S., Masouri, Z.: Numerical method for analysis of one- and two-dimensional electromagnetic scattering based on using linear Fredholm integral equation models. Math. Comput. Model. **54**, 2199–2210 (2011)
18. Masouri, Z., Babolian, E., Hatamzadeh-Varmazyar, S.: An expansion-iterative method for numerically solving Volterra integral equation of the first kind. Comput. Math. Appl. **59**, 1491–1499 (2010)
19. Hatamzadeh-Varmazyar, S., Masouri, Z.: Numerical expansion-iterative method for analysis of integral equation models arising in one- and two-dimensional electromagnetic scattering. Eng. Anal. Bound. Elem. **36**, 416–422 (2012)
20. Babolian, E., Masouri, Z., Hatamzadeh-Varmazyar, S.: Introducing a direct method to solve nonlinear Volterra and Fredholm integral equations using orthogonal triangular functions. Mathem. Sci. J. **5**(1), 11–26 (2009)
21. Hatamzadeh-Varmazyar, S., Masouri, Z.: A numerical method for calculation of electrostatic charge distribution induced on conducting surfaces. Adv. Comput. Techn. Electromagn. **2014**, 1–9 (2014)
22. Masouri, Z., Hatamzadeh-Varmazyar, S.: Applying integral equation modeling technique in determination of charge distribution on conducting structures. Adv. Comput. Techn. Electromagn. **2013**, 1–8 (2013)
23. Babolian, E., Masouri, Z., Hatamzadeh-Varmazyar, S.: A direct method for numerically solving integral equations system using orthogonal triangular functions. Int. J. Ind. Mathem. **1**(2), 135–145 (2009)
24. Hatamzadeh-Varmazyar, S., Masouri, Z.: Determining the electromagnetic fields scattered from PEC cylinders. Int. J. Ind. Mathem. **28**(4), 1–8 (2017)
25. Hatamzadeh-Varmazyar, S., Masouri, Z.: A computational method for numerically solving linear integro-differential equations. Int. J. Ind. Mathem. **29**(1), 82–93 (2018)
26. Masouri, Z., Hatamzadeh-Varmazyar, S.: An analysis of electromagnetic scattering from finite-width strips. Int. J. Ind. Mathem. **5**(3), 199–204 (2013)
27. Babolian, E., Masouri, Z., Hatamzadeh-Varmazyar, S.: A set of multi-dimensional orthogonal basis functions and its application to solve integral equations. Int. J. Appl. Mathem. Comput. **2**(1), 032–049 (2010)
28. Hatamzadeh-Varmazyar, S., Naser-Moghadasi, M., Sadeghzadeh-Sheikhan, R.: Numerical method for analysis of radiation from thin wire dipole antenna. Int. J. Ind. Mathem. **3**(2), 135–142 (2011)
29. Hatamzadeh-Varmazyar, S.: The error analysis and convergence evaluation of a computational technique for solving electromagnetic scattering problems. Adv. Comput. Techn. Electromagn. **2015**(1), 66–69 (2015)
30. Hatamzadeh-Varmazyar, S., Masouri, Z.: Calculation of electric charge density based on a numerical approximation method using triangular functions. Adv. Comput. Techn. Electromagn. **2013**, 1–11 (2013)

31. Masouri, Z., Hatamzadeh-Varmazyar, S., Babolian, E.: Numerical method for solving system of Fredhlom integral equations using Chebyshev cardinal functions. Adv. Comput. Techn. Electromagn. **2014**, 1–13 (2014)

32. Masouri, Z., Hatamzadeh-Varmazyar, S.: Evaluation of current distribution induced on perfect electrically conducting scatterers. Int. J. Ind. Mathem. **5**(2), 167–173 (2013)

33. Hatamzadeh-Varmazyar, S., Masouri, Z.: An efficient numerical algorithm for solving linear differential equations of arbitrary order and coefficients. Int. J. Ind. Mathem. **10**(2), 127–138 (2018)

34. Hatamzadeh-Varmazyar, S., Masouri, Z.: A fast and accurate expansion-iterative method for solving second kind Volterra integral equations. Int. J. Ind. Mathem. **10**(1), 29–37 (2018)

35. Hatamzadeh-Varmazyar, S., Masouri, Z.: A numerical scheme for two-dimensional scattering analysis. Int. J. Mathem. Comput. **29**(2), 84–94 (2018)

36. Hatamzadeh-Varmazyar, S., Masouri, Z.: Numerical solution of second kind Volterra and Fredholm integral equations based on a direct method via triangular functions. Int. J. Ind. Mathem. **11**(2), 79–87 (2019)

37. Hatamzadeh, S., Masouri, Z.: A numerical evaluation of radiation from dipole antenna based on a set of wavelet functions. Int. J. Imaging Rob. **21**(1), 45–56 (2021)

38. Hatamzadeh, S., Masouri, Z.: Applying a set of orthogonal basis functions in numerical solution of Hallen's integral equation for dipole antenna of perfectly conducting material. J. Mod. Mater. **9**(1), 36–49 (2022)

39. Masouri, Z., Hatamzadeh, S.: A computational approach for analysis of scattering characteristics for strips of resistive material. Int. J. Res. Public. Rev. **4**(12), 3640–3646 (2023)

40. Masouri, Z.: Numerical expansion-iterative method for solving second kind Volterra and Fredholm integral equations using block-pulse functions. Adv. Comput. Techn. Electromagn. **2012**, 1–7 (2012)

41. Crucinio, F.R., Bortoli, V.D., Doucet, A., Johansen, A.M.: Solving a class of Fredholm integral equations of the first kind via Wasserstein gradient flows. Stoch. Process. Appl. **173**, 104374 (2024)

42. Zerroudi, B., Nouisser, O., Barrera, D.: Approximate solution of Fredholm integral equations of the second kind through multinode Shepard operators. Math. Comput. Simul. **223**, 485–493 (2024)

43. Akbari, T., Esmaeilbeigi, M., Moazami, D.: A stable meshless numerical scheme using hybrid kernels to solve linear Fredholm integral equations of the second kind and its applications. Math. Comput. Simul. **220**, 1–28 (2024)

44. Patel, S., Panigrahi, B.L.: Discrete Legendre spectral projection-based methods for Tikhonov regularization of first kind Fredholm integral equations. Appl. Numer. Math. **198**, 75–93 (2024)

45. Allouch, C.: Collocation and modified collocation methods for solving second kind Fredholm integral equations in weighted spaces. Appl. Numer. Math. **198**, 202–216 (2024)

46. Torkaman, S., Heydari, M.: An iterative Nyström-based method to solve nonlinear Fredholm integral equations of the second kind. Appl. Numer. Math. **194**, 59–81 (2023)

47. Golshan, H.M.: Numerical solution of nonlinear m-dimensional Fredholm integral equations using iterative Newton–Cotes rules. J. Comput. Appl. Math. **448**, 115917 (2024)

48. Hayotov, A., Babaev, S.: The numerical solution of a Fredholm integral equations of the second kind by the weighted optimal quadrature formula. Results Appl. Mathem. **24**, 100508 (2024)

49. El Mokhtari, F., Lamnii, M., Barrera, D.: A genetic algorithm approach based on spline quasi-interpolation for solving Fredholm integral equations. Math. Comput. Simul. **229**, 725–742 (2025)

50. Kayal, A., Mandal, M.: Superconvergent method for weakly singular Fredholm-Hammerstein integral equations with non-smooth solutions and its application. Appl. Numer. Math. **207**, 24–44 (2025)

Leveraging General Unary Hypotheses Automaton for Automating Technical Analysis: A Case Study on Bitcoin Prices

Jakub Neugebauer[✉]

Prague University of Economics and Business, nám. Winstona Churchilla 1938/4, 120 00 Prague, Czech Republic
jakub.neugebauer@vse.cz

Abstract. This paper explores the application of the General Unary Hypotheses Automaton (GUHA) framework to automate the principles of technical analysis, focusing on Bitcoin price prediction. By leveraging technical indicators, such as the Relative Strength Index (RSI), Average Directional Index (ADX), and Bollinger Bands, this study employs data mining techniques to automate insights from technical analysis. The 4 ft-Miner and CF-Miner procedures, implemented through the CleverMiner Python framework, were applied to Bitcoin price data spanning from 2014 to 2024. The findings demonstrate the effectiveness of GUHA in identifying robust predictive patterns. Rules mined via 4 ft-Miner achieved 80.65% confidence in predicting next-day price decreases and 96.15% confidence in forecasting significant upward movements (next-day high exceeding a 1% increase over the opening price). CF-Miner further categorized price return probabilities, enhancing interpretability and revealing ordered insights into the prediction of positive returns. These results underscore GUHA's potential to automate and enhance the efficiency of technical analysis, providing actionable intelligence for automated trading and financial risk management. By offering interpretable patterns and statistically significant relationships, the GUHA framework bridges the gap between traditional technical analysis and modern algorithmic trading strategies. Future research may extend this approach to longer-term trend predictions and multi-asset portfolios, further advancing the automation of financial market analysis.

Keywords: Data Mining · General Unary Hypotheses Automaton · Technical Analysis · Automated Trading

1   Introduction

The topic of investments is never-ending as it is the most important way of protecting incomes against inflation. Technical analysis is one of the most common ways of analyzing financial data. Its basis in observing historical data, behavior and dependencies allows it to predict future behavior. Into the most famous disciplines belongs for example looking into charts of historical prices of an investment instrument and observing similar shapes of the line chart with the assumption, that these shapes will occur in the future again. The focus of this paper, however, is on indicators of technical analysis, which determine trends and volatility of instrument prices. The goal of this paper is to utilize the General Unary Hypotheses Automaton (GUHA) framework to observe if the theory of these technical analysis indicators is observable with a possibility to automate the detection of possible trends. The focus of this paper will be on one of the most traded instruments, Bitcoin, which will be used as an observed instrument on which the GUHA procedure will be tested.

Many modern approaches have been proposed for the use of predicting time series (asset price in this case). Starting from the well-established statistical methods still in use like ARIMA or GARCH models (see e.g. Zolfaghari & Gholami 2021; Ansari et al. 2022; Rubio & Alba 2022). The development of computer science with the never-ending increase of computational power of modern computers has also affected the field of time series prediction. Machine learning methods like neural networks and gradient boosting have proved to be well applicable in this case as well (see Vuong et al. 2022; Rezaei et al. 2021; Leippold et al. 2022). Amongst the modern data science methods must be included data mining. The benefit of data mining compared to machine learning methods is in the readability of given results provided from data mining while the machine learning procedures are pure black boxes. We can find a slight usage of data mining in stock price prediction (see Chang et al. 2021; Farid et al. 2023; Kaur & Dharni 2023). Data mining provides a distinct advantage in time series analysis due to its focus on generating interpretable and actionable insights.

The GUHA framework originated in the 1960s (see Hájek et al. 1966) and it is one of the oldest Data mining procedures that is still maintained and developed (see e.g. Máša and Rauch 2024). The benefit of this procedure in our case is, that we can mine the rules from technical analysis indicators, under which the price of bitcoin has specific behavior. Unlike black-box machine learning models, the GUHA framework excels in uncovering explicit relationships and dependencies within data, which are crucial for understanding the underlying dynamics of asset price movements. This interpretability not only enhances trust and transparency in the results but also enables domain experts to validate and refine their predictions based on well-defined hypotheses. Additionally, GUHA's systematic approach to hypothesis testing and rule extraction ensures robustness in identifying patterns, making it a valuable tool for addressing the complexities of financial time series data. Python implementation of GUHA framework, CleverMiner developed by Rauch et al. (2022), is used to find patterns in historical Bitcoin price data. The idea of using this framework is to possibly automate the detection of patterns in historical prices. This method has not been applied to asset price prediction, so the aim of this paper is to explore the potential of the GUHA framework in interpretable and statistically significant relationships within financial time series data.

This paper is split into two main sections. In the first section, technical analysis indicators, such as the Relative Strength Index (RSI), Average Directional Index (ADX), and others are described as well as their potential impact on the price of an instrument. The second, core, section is destined for the mining procedure itself. Several possibilities of CleverMiner to use for the purpose of this paper are showcased in the second main section. Finally, the results of the mining procedure will be analyzed and discussed, focusing on the effectiveness of these indicators in predicting Bitcoin price movements and providing insights into the potential for automating technical analysis.

2 Indicators of Technical Analysis

In this section, technical indicators used for the data mining procedure are briefly introduced. This section serves as a basic overview of standard and core indicators used for analyzing financial time series. An overview of more than 90 indicators can be found in Fang et al. (2014), in this paper, however, only a small subset of the most used of them shall be used. There are methods combining technical analysis with machine learning methods (Yang et al. 2022). In this paper these indicators are used as an input for a data mining procedure.

2.1 Relative High-Low Position

Let x_t be the closing price. The relative high-low position $rhp^q{}_t$ is calculated as follows:

$$rhp_t^q = \frac{x_t - \min_i(x_i)}{\max_i(x_i) - \min_i(x_i)}, \quad i \in \mathbb{Z}, i \geq t - q \wedge i \leq t \tag{1}$$

This equation normalizes how close the value of smoothed series x_t is at time t to the maximum value in the last period of length q. Intuitively, we expect higher values to determine an upward trend and the opposite.

2.2 Noise

For each given period, a moving average (MA) is calculated as the rolling mean of the time series data. The noise $n^q{}_t$ is then computed as the deviation of the time series data from the moving average, normalized by the moving average. Formula of the noise at time t in period q is:

$$n_t^q = \frac{x_t - MA_t^q}{MA_t^q} \tag{2}$$

where $MA^q{}_t$ represents the rolling mean over a period q, and x_t is the closing price of the series at time t. When noise is positive, the assumption is, that the positive trend is present.

2.3 Gradient

The gradient captures the rate of change in the moving average, which can provide insight into the trend's direction and strength. The gradient $g_t{}^q$ for each period is defined as:

$$g_t^q = \frac{MA_t^q - MA_{t-1}^q}{MA_t^q} \tag{3}$$

where MA_t^q is the moving average at time t and MA_{t-1}^q is the moving average at the previous time step. Again, with higher values of this indicator, higher returns in the upcoming period are expected.

2.4   Relative Strength Index (RSI)

The Relative Strength Index RSI_t^q measures the speed and change of price movements. For each period, RSI_t^q is computed as:

$$RSI_t^q = 100 - \left(\frac{100}{1 + RS_t^q} \right) \tag{4}$$

where RS_t^q is the relative strength, calculated as the ratio of average gain ($\bar{g}$) to average loss ($\bar{l}$) over the period q:

$$RS_t^q = \frac{\bar{g}_t^q}{\bar{l}_t^q} \tag{5}$$

where the average gain is the mean of all positive price changes over the period q, while the average loss is an average of all negative price changes. The RSI_t^q values range from 0 to 100 and are commonly used to identify overbought or oversold conditions in the market. A high RSI_t^q (above 70) suggests that the market may be overbought, while a low RSI_t^q (below 30) suggests the market may be oversold.

2.5   Average True Range (ATR)

The Average True Range ATR is used to measure market volatility. The ATR is calculated as the rolling average of the true range over a given period q. The true range at time t is the maximum of the following:

$$TR_t = \max(High_t - Low_t, |High_t - Close_{t-1}|, |Low_t - Close_{t-1}|) \tag{6}$$

where:

- $High_t$ is the highest price at time t,
- Low_t is the lowest price at time t,
- $Close_{t-1}$ is the closing price at time $t-1$.

The Average True Range over the period q is then calculated as:

$$ATR_t^q = \frac{1}{q} \sum_{i=t-q+1}^{t} TR_i \tag{7}$$

Finally, the normalized Average True Range ($nATR_t^q$) is computed by dividing the ATR by the current closing price x_t:

$$nATR_t^q = \frac{ATR_t^q}{x_t} \tag{8}$$

where x_t is the closing price at time t. This normalization helps stabilize the ATR over time, making the time series stable and applicable for data mining.

2.6 Average Directional Index (ADX)

The Average Directional Index (ADX) is used to quantify the strength of a trend in a time series. It is calculated using the smoothed moving average of the Directional Index (DX), which compares the difference between the positive and negative directional movements.

The ADX is for a good readability broken into these subindicators:

1. **Average True Range (ATR):** As described in previous subsection
2. **Directional Movements (DM):** The positive and negative directional movements are defined as:

$$
\begin{aligned}
{}^{+}DM_t &= \max(0, High_t - High_{t-1}) \\
{}^{-}DM_t &= \max(0, Low_{t-1} - Low_t)
\end{aligned}
\tag{9}
$$

3. **Directional Indicators (DI):** The positive and negative directional indicators are calculated as the ratio of the smoothed positive and negative directional movements to the ATR:

$$
\begin{aligned}
{}^{+}DI_t^q &= 100 \times \frac{{}^{+}DM_t^q}{ATR_t^q} \\
{}^{-}DI_t^q &= 100 \times \frac{{}^{-}DM_t^q}{ATR_t^q}
\end{aligned}
\tag{10}
$$

4. **Directional Index (DX):** The DX measures the absolute difference between the positive and negative directional indicators relative to their sum:

$$
DX_t^q = 100 \times \frac{\left| {}^{+}DI_t^q - {}^{-}DI_t^q \right|}{{}^{+}DI_t^q + {}^{-}DI_t^q}
\tag{11}
$$

5. **Average Directional Index (ADX):** Finally, the ADX is calculated as the rolling average of the DX over the period q:

$$
ADX_t^q = \frac{1}{q} \sum_{i=t-q+1}^{t} DX_i
\tag{12}
$$

The ADX measures the strength of a trend, regardless of its direction. Higher ADX values indicate a stronger trend, while lower values suggest a weak or non-trending market. The ADX is used with directional indicators to combine both the trend direction and its strength.

2.7 GARCH Volatility

The Generalized Autoregressive Conditional Heteroskedasticity (GARCH) model is used to capture the time-varying volatility observed in financial time series.

The GARCH(p, q) model is defined as follows:

$$
\sigma_t^2 = \omega + \sum_{i=1}^{p} \alpha_i \epsilon_{t-i}^2 + \sum_{j=1}^{q} \beta_j \sigma_{t-j}^2
\tag{13}
$$

where:

- σ_t^2 is the conditional variance (volatility) at time t,
- ω is a constant,
- α_i are the coefficients for the past squared residuals ϵ_{t-i}^2 (ARCH terms),
- β_j are the coefficients for the past conditional variances σ_{t-j}^2 (GARCH terms),
- ϵ_t is the residual (error) term at time t, typically modeled as:

$$\epsilon_t = \sigma_t z_t$$

where σ_t is the conditional volatility at time t and z_t is a standard normally distributed random variable, $z_t \sim N(0, 1)$.

This model forecasts future volatility based on past volatility and past shocks to the returns. The GARCH feature captures periods of increased volatility, making it an essential tool for financial risk management and forecasting.

The conditional volatility forecasted from the GARCH model is stored for each period as:

$$GARCH_t^q = \sigma_t \tag{14}$$

where σ_t is the square root of the conditional variance σ_t^2.

2.8 Bollinger Bands

Bollinger Bands are used to measure price volatility relative to a moving average. The bands consist of an upper band (UB) and a lower band (LB), calculated as:

$$\begin{aligned} UB_t^q &= MA_t^q + k \times \sigma_t^q \\ LB_t^q &= MA_t^q - k \times \sigma_t^q \end{aligned} \tag{15}$$

where σ_t^q is the standard deviation over period q and k is the number of standard deviations. The Bollinger relative B is computed as:

$$B_t^q = \frac{x_t - LB_t^q}{UB_t^q - LB_t^q} \tag{16}$$

This feature measures the position of the close price relative to the Bollinger Bands. B_t^q is used later as a feature for a data mining procedure.

2.9 Other Features

To include possible autocorrelation in the data, lagged returns, as well as relative high and low positions, are included as additional features to the opening price, day of the week in integer form with 0 being Monday and 6 Sunday and month in year as integer in interval 1–12. The last added feature is the volume of trades in the form of the percentage change.

Quantile Transformation. All presented indicators are transformed into quartiles to discretize continuous data. All explanatory variables are separated into 7 bins defined by 7 quantiles and related intervals between them. A showcase of input data is presented in Table 1. The observed instrument is Bitcoin, and data is derived from the longest possible period from the source of the data (September 2014) until September 2024. Data is collected from Yahoo Finance [1] which is the source for all the necessary data needed for the purpose of this paper. (close and open price, high and low position, volume of trades). As the majority of the features are affected by the length of a period q, 4 periods are taken into account for all the effected indicators, concretely 5 days (1 week), 10 days (2 weeks), 20 days (1 month) and 40 days (2 months). The number of lags used is 4.

Table 1. Showcase of transformed technical indicators.

Date	High	Low	Close	Volume	Day	Month
2014-12-05	4	3	5	4	4	12
2014-12-06	1	4	2	0	5	12
2014-12-07	1	1	3	2	6	12
2014-12-08	1	5	0	6	0	12
2014-12-09	1	6	0	6	1	12

Date	Noise 5	Gradient 5	High low 5	Noise 10	Gradient 10	High	Low 10
2014-12-05	3	2	2	3	3		3
2014-12-06	2	2	1	2	3		2
2014-12-07	3	2	2	2	3		3
2014-12-08	1	1	0	1	1		0
2014-12-09	0	1	0	0	1		0

Date	Noise 20	Gradient 20	High low 20	Noise 40	Gradient 40	High	Low 40
2014-12-05	3	3	4	3	3		3
2014-12-06	3	2	3	3	3		3
2014-12-07	3	2	4	3	3		3
2014-12-08	2	2	2	2	3		2
2014-12-09	1	1	0	2	3		1

Date	RSI 5	RSI 10	RSI 20	RSI 40	ATR 5	ATR 10
2014-12-05	2	3	3	3	1	2
2014-12-06	2	4	2	3	1	2
2014-12-07	2	3	2	3	1	1
2014-12-08	1	1	2	3	1	1
2014-12-09	1	0	1	3	2	1

Date	ATR 20	ATR 40	GARCH	ADX 5	ADX 10	ADX 20
2014-12-05	3	4	2	5	3	0
2014-12-06	3	4	2	5	3	0
2014-12-07	2	4	1	6	3	0
2014-12-08	3	4	1	6	4	1

(continued)

---

[1] Yahoo Finance: https://finance.yahoo.com/.

Table 1. (*continued*)

Date	High			Low		Close	Volume	Day	Month
2014-12-09	3			4		2	6	4	1

Date	ADX	40		Bollinger	5	Bollinger 10	Bollinger 20	Bollinger 40	Returns lag
2014-12-05	2			3		3	3	3	5
2014-12-06	2			2		2	3	3	2
2014-12-07	2			3		2	3	3	3
2014-12-08	2			0		0	1	2	0
2014-12-09	2			0		0	0	2	0

Date	Returns	Lag	1	High lag	1	Low lag 1	Returns lag 2	High lag 2	Low lag 2
2014-12-05	1			2		4	1	1	4
2014-12-06	5			4		3	1	2	4
2014-12-07	2			1		4	5	4	3
2014-12-08	3			1		1	2	1	4
2014-12-09	0			1		5	3	1	1

Date	Returns	Lag	3	High lag	3	Low lag 3	Returns lag 4	High lag 4	Low lag 4
2014-12-05	4			3		1	3	3	1
2014-12-06	1			1		4	4	3	1
2014-12-07	1			2		4	1	1	4
2014-12-08	5			4		3	1	2	4
2014-12-09	2			1		4	5	4	3

3 Results of Rules Mining

This section is dedicated to mining rules of Bitcoin price behavior based on features presented in the previous chapter. The used method is GUHA, concretely 4 ft-Miner and CF-Miner. Firstly, the focus will be on predicting the movement of the price of a Bitcoin the following day, respectively predicting if the return will be positive or negative. Then by a similar procedure, the high and low positions exceeding certain threshold predictions will be constructed. In the last part of this section, the CF-Miner will be used to find a rule, which sorts the prediction of a return into ordinal bins.

To simulate and analyze the effectiveness of the proposed method, the study employed the GUHA framework using the CleverMiner tool implemented in Python. The framework was applied to historical Bitcoin price data spanning from September 2014 to September 2024, with data sourced from Yahoo Finance. The dataset included key variables such as closing price, high/low positions, and trading volumes, which were transformed into quantile bins for rule mining.

The simulations focused on mining association rules using two primary methods:

- 4 ft-Miner: Utilized for identifying rules that predict next-day price movement, as well as high and low positions exceeding specific thresholds.

- CF-Miner: Applied to sort return probabilities into ordinal categories, highlighting conditions that increased the likelihood of positive returns.

Key parameters included:

- Periods: Indicators were calculated for four distinct periods: 5 days (1 week), 10 days (2 weeks), 20 days (1 month), and 40 days (2 months).
- Lagged Variables: Up to 4 lags were incorporated to account for historical dependencies.
- Quantile Transformation: All continuous explanatory variables were discretized into seven quantile bins to facilitate rule mining.

The computational experiments were performed on a machine with windows 11, running Python version 3.11, and leveraging the CleverMiner framework. Confidence, Above Average Difference (AAD), and Base metrics were used to evaluate the quality of mined rules, ensuring that the results were interpretable and statistically significant.

3.1 Using *4 ft-Miner* with Sequential Literals to Predict the Next Day Return

In this section, we apply the *4 ft-Miner* method within the GUHA framework to predict the next-day price movement of Bitcoin (successor Return_Prediction_Bin) based on technical indicators described in the previous section of this paper. This rule mining aims to identify conditions that significantly increase the probability of a price decrease or increase the following day. We focus on using sequential literals, because of the data transformation logic (splitting the indicators into 7 sequential quantile bins) to capture dependencies between consecutive quantile intervals and their impact on the next-day price movement. The minimal length of a sequence of literals is set to 1 and maximal is set to 2. This is because the total number of categories after transformation is only 7 as well as because of the computational complexity reasons. The use of sequential literal is particularly relevant in technical analysis, where consecutive low values of indicators such as price, gradient, or average directional index can signal a potential downward trend and opposite.

The rule with the highest confidence and relative occurrence higher than 2% found in the data is the following:

- Antecedents:

  - Close(0 1): The closing price is in the 0th or first quantile.
  - Gradient_5(0 1): The 5 days period gradient is in the 0th or first quantile.
  - Bollinger_5(0 1): The Bollinger Bands relative position is in the 0th or first quantile.
  - Returns_Lag_4(0): The return from 4 days ago is in the 0th quantile, indicating low returns.

- Successor:

  - Return_Prediction_Bin(False): This indicates that the next day's price will decrease if all of the conditions of Antecedents will hold.

The mined rule suggests that when the closing price, gradient, Bollinger Bands calculated in 5 days period relative position, and past returns are all in their lowest quantiles (0th or first quantile for the first three indicators and 0th quantile for returns), the next-day price is likely to decrease. The key metrics associated with this rule are summarized below:

- Base: The rule applies to 75 records in the dataset, representing 2.08% of the total data points.
- Confidence: The confidence of this rule is 80.65%, meaning that in 80.65% of cases where the conditions in the antecedent are met, the next day's price did decrease.
- Above Average Difference (AAD): The AAD score is 0.524, indicating that the probability of a price decrease is 52.4% higher when the antecedent conditions are met compared to the unconditional probability of a price decrease.

The results demonstrate that combining multiple low quantile indicators, Close, Gradient_5, Bollinger_Relative_5, and Returns_Lag_4, significantly increases the likelihood of predicting a next-day price decrease. The use of sequential cedents allowed us to capture important interrelationships inside these technical indicators, which may not have been as evident with a simple analysis as it is shown in the case of Bollinger_5, which is in this rule in 0th or first quantile.

The high **confidence** (80.65%) and **above-average difference** (0.524) suggest that this rule has significant predictive power. The four-fold contingency table (see Fig. 1) shows that this rule is applicable on the price decrease, because of the big proportion of true predicted decreases against false ones, but it does not tell anything about the price if this rule is not present (similar proportion of true and false in predicted increases).

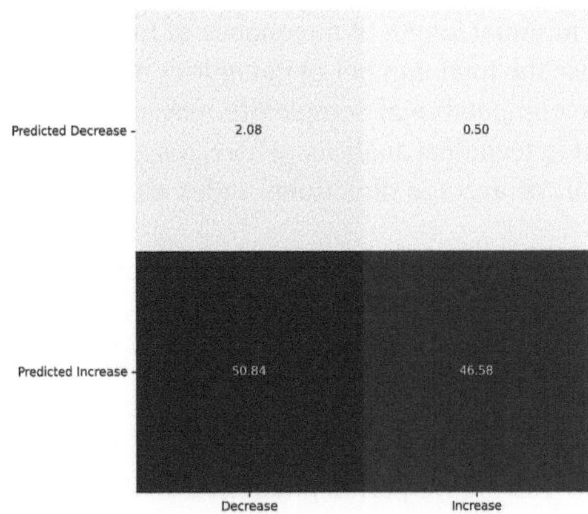

Fig. 1. Four-Fold Contingency Table for Predicting Next-Day Price Decrease.

The use of *CleverMiner* and the *4 ft-Miner* method proves to be an applicable tool in automating technical analysis for Bitcoin price prediction. It allows for the systematic discovery of patterns across multiple indicators, significantly improving the efficiency and accuracy of predictions. By leveraging sequential cedents and quantile transformations, we were able to uncover complex interactions that provide deeper insights into price behavior and trends.

3.2 *4 ft-Miner* **Procedure with Sequential and Right-Cuts Literals to Predict the Next Day High Position**

In this section, a showcase of a rule where the next day's high position exceeds 1% over the opening price (High_Prediction) is presented. In this rule-mining process, both sequential literals and right-cut literals for specific indicators are used. The right-cut literals are particularly important for indicators that are expected to take higher values in an upward trend, allowing us to capture cases where these indicators are in their upper quantiles. Variables used for creating right-cut literals are Percentage_ATR, Rel_High_Low, RSI and ADX.

The use of the right-cut method for these literals is motivated by the nature of the technical indicators being analyzed. For upward-trending market conditions, high values of volatility, directional movement, and trend strength indicators typically suggest a likely continuation of the trend. Therefore, allowing the literals for such indicators to extend into their higher quantiles helps capture the conditions that are most conducive to a high position exceeding 1%.

The rule with the highest confidence found is this:

- Antecedents:

  - Low (Kaur & Dharni, 2023): The lowest price of the day is in the highest quantile (sixth quantile).
  - Noise_10(0): The noise over a 10-period window is in the lowest quantile (0th quantile), indicating minimal deviation from the trend.
  - Gradient_10(0): The gradient of the 10-period moving average is in the lowest quantile (0th quantile), signaling a stable or slightly declining trend.
  - ATR_20 (Kaur & Dharni, 2023): The 20-period normalized average true range is in the highest quantile (6th quantile), indicating increased volatility.
  - ADX_5(6 5 4): The 5 days period average directional index is in the 4th, 5th, or 6th quantile, reflecting a strong trend in the market.

- Successor Cedent:

  - High_Prediction(True): This indicates that the next day's high position will exceed 1% above the opening price.

The mined rule demonstrates that when these conditions hold, there is a strong likelihood of the next day's high exceeding 1% over the opening price. The use of right-cut literals for indicators like ATR_20 and ADX_5 allows the rule to capture market conditions where high volatility and trend strength signal an upward trend.

Confidence and Rule Metrics The following key metrics are associated with this rule:

- Base: The rule applies to 75 records in the dataset, representing 2.08% of the total data points.
- Confidence: The confidence of this rule is 96.15%, meaning that in 96.15% of cases where the antecedent conditions are met, the next day's high exceeded 1%.

- Above Average Difference (AAD): The AAD score is 0.665, indicating that the probability of the next day's high exceeding 1% is 66.5% higher when the antecedent conditions are met compared to the unconditional probability of this event.

The results of this rule highlight that combining indicators like Low, Noise, Gradient, Percentage_ATR, and ADX creates a strong signal for predicting significant upward price movement. The high confidence (96.15%) and the AAD score of 0.665 suggest that this rule is highly reliable for forecasting when the next day's high will exceed 1%, particularly in volatile market conditions.

Figure 2 visually demonstrates the distribution of true positives, false positives, true negatives, and false negatives for the prediction of next-day high position exceeding 1%. As shown in the heatmap, the large proportion of true positives (2.08) relative to false positives (0.08) indicates that the rule effectively captures upward price movements when the specified technical indicators are in the highest or lowest quantiles. As in the previous example, when the rule does not hold, it does not tell anything about the next day's high position, but this is not the scope of the analysis.

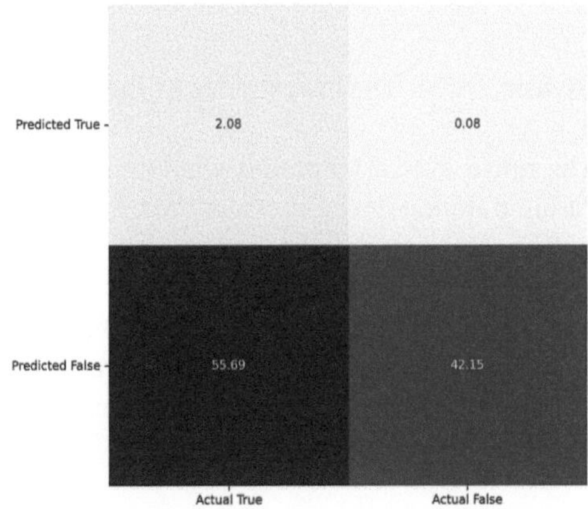

Fig. 2. Four-Fold Contingency Table for Predicting Next-Day High Position Exceeding 1%.

This example illustrates how using right-cut literals in conjunction with sequential indicators allows us to detect significant trends in technical analysis and predict high price movements. The application of the *4 ft-Miner* method and the use of *CleverMiner* proved highly effective in automatically identifying such patterns, contributing to the efficiency of technical analysis in Bitcoin price forecasting.

3.3 *4 ft-Miner* Procedure with Sequential and Left-Cuts Literals to Predict the Next Day Low Position

In this section, we explore the prediction of whether the next day's low position will exceed 1% below the opening price (Low_Prediction). The approach follows a similar methodology to the previous section, with a key difference: left cuts were used for cedents where technical indicators are expected to show low values in a downtrend scenario. The aim of using left cuts is to capture conditions where indicators are low, signaling potential downward pressure on the market. The maximal length of literals for these cuts is set to 3, allowing the rule-mining process to capture more complex relationships.

The highest confidence rule mined in this scenario is as follows:

- Antecedents:

  - Low (Kaur & Dharni, 2023): The Low value is in the highest quantile.
  - Gradient_5(0): The 5 days period gradient is in the lowest quantile.
  - High_Low_5(0 1): The relative high-low position over 5 periods is in the 0th or 1st quantile.
  - RSI_5(0 1 2): The 5 days period RSI is in the 0th, 1st, or 2nd quantile, indicating oversold conditions.
  - GARCH (Kaur & Dharni, 2023): The GARCH volatility is in the highest quantile, signaling high market uncertainty and potential for price drops.

- Successor Cedent:

  - Low_Prediction(True): This indicates that the next day's low position will exceed 1% below the opening price if all antecedent conditions are met.

This rule reveals that when the Low value, gradient, and other key technical indicators show low values, there is a high probability that the next day's low position will drop significantly below the opening price.

Confidence and Rule Metrics The mined rule shows strong predictive power for identifying cases where the next day's low position exceeds 1% below the opening price. Key performance metrics are as follows:

- Base: The rule applies to 75 records in the dataset, representing 2.08% of the total data points.
- Confidence: The confidence of this rule is 86.21%, meaning that in 86.21% of cases where the antecedent conditions are satisfied, the next day's low position exceeded 1% below the opening price.
- Above Average Difference (AAD): The AAD score is 0.550, indicating that the probability of a significant low price drop is 55.0% higher when the antecedent conditions are met compared to the unconditional probability.

Figure 3 displays the four-fold contingency table. Although the proportion is not as high as in the case of predicting a high position, the percentage of true positives (2.08) relative to false positives (0.33) supports the rule's effectiveness in predicting downward price movements.

The results demonstrate that left-cut cedents, applied to key technical indicators, offer a robust method for predicting significant downward price movements. The use of *CleverMiner* and the *4 ft-Miner* method with left-cut literals enables the discovery of

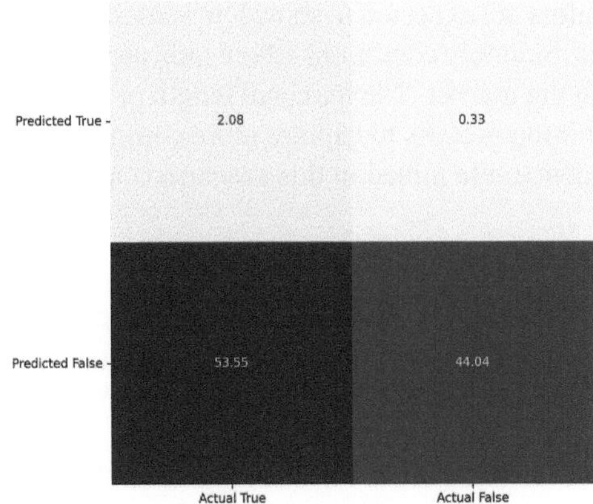

Fig. 3. Four-Fold Contingency Table for Predicting Next-Day Low Position Exceeding 1%.

complex interactions between multiple low-value indicators, providing valuable insights for trading strategies aimed at capitalizing on market downturns.

3.4 Sorting Probability of Return Categories Using *CF-Miner*

In this section, we explore the use of the *CF-Miner* method within the GUHA framework to find a rule that sorts the probability of price returns into three distinct categories: negative returns (below −0.3%), neutral returns (between −0.3% and 0.3%), and positive returns (above 0.3%). The goal is to identify a set of conditions (cedents) that result in an ascending order of occurrence probabilities, where negative returns have the lowest probability, neutral returns are more likely, and positive returns have the highest probability.

CF-Miner is particularly well-suited to this task because it focuses on the ordered nature of the target variable. Unlike the *4 ft-Miner*, which identifies association rules based on frequent patterns of antecedents and successors without considering order, *CF-Miner* aims to explicitly sort outcomes according to probabilities. This is crucial in scenarios where the goal is not just to predict an outcome but to ensure that outcomes are sorted in a desired order. In our case, the desired order reflects increasing probabilities of positive returns relative to neutral and negative returns.

As the focus is on getting a high probability of positive returns, the right cut method is used in variables specified in Sect. 3.2 with a maximal length of literal equal to 3. From the rest of the columns, only one category can be used to create the cedent.

Selected Rule and Metrics The rule that produces the highest proportion of positive returns relative to negative returns is as follows:

- Antecedents:

    - High_Low_10(5 4 3): The relative high-low position over the past 10 periods is in the 3rd, 4th, or 5th quantile.

- GARCH_Volatility(0): The GARCH model's volatility estimate is in the lowest quantile, indicating low market volatility.
- Low_Lag_3(0): The low price from 3 days ago is in the lowest quantile.

This rule suggests that when the market is characterized by a low high-low range, low volatility, and a historically low price from 3 days ago, the likelihood of positive returns exceeds that of neutral and negative returns.

The key strength of *CF-Miner* in this context lies in its ability to sort the outcomes based on their occurrence probabilities. In this case, the model effectively sorted the return categories (negative, neutral, positive) in ascending order of probability. Figure 4 compares the distribution of occurrences in the return categories with and without applying the rule. It is clearly visible that applying the rule significantly increases the relative occurrence of positive returns, while the occurrences of negative returns are reduced.

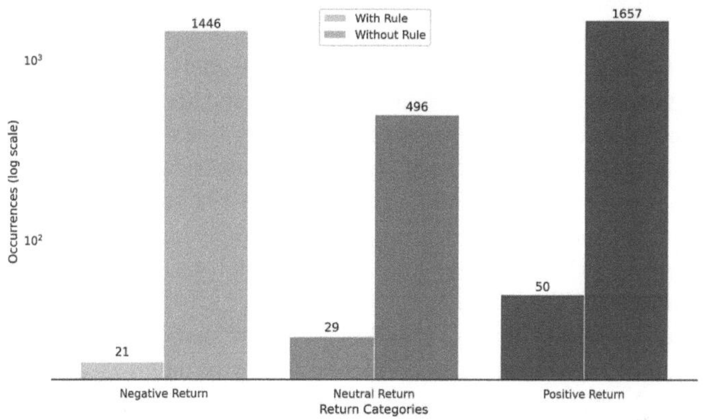

Fig. 4. Comparison of Return Category Occurrences with and Without Rule.

CF-Miner is distinct from the more basic 4 ft-Miner in that it actively seeks patterns that order outcomes in a desired way. While 4 ft-Miner would simply look for frequent patterns in the antecedents that are associated with a successor, it would not ensure that these patterns result in an increasing probability distribution across categories. Sometimes it is better to sort a probability of getting higher returns and that is where CF-Miner comes in handy.

The CF-Miner method proved to be good in identifying a rule that orders return probabilities in a desired ascending manner. By using the right cuts for the technical indicators, the method effectively captured conditions where the probability of positive returns significantly outweighs that of negative returns, making this rule highly actionable for trading strategies focused on positive price movements.

4 Conclusion

This paper explored the use of technical analysis indicators in predicting Bitcoin price movements through the application of the GUHA framework, specifically utilizing the *4 ft-Miner* and *CF-Miner* methods implemented in the *CleverMiner* tool. By focusing on key indicators such as the Relative Strength Index (RSI), Average Directional Index (ADX), and Bollinger Bands, the study demonstrated the potential of data mining techniques to extract association rules and uncover patterns in historical price data, providing actionable insights into the effects of various conditions on Bitcoin price movements.

The results indicate that combining sequential cedents and right/left-cut literals provides strong predictive power, particularly in capturing upward and downward trends in Bitcoin's price. Notably, the *4 ft-Miner* method identified rules with high confidence, such as predicting next-day price decreases or exceeding high positions based on specific indicator combinations. The usage of the *CF-Miner* allowed sorting return probabilities effectively, revealing conditions under which positive returns are more likely than neutral or negative ones.

While the findings highlight the utility of technical analysis indicators in automating trading strategies, this study has several limitations. First, the analysis focuses solely on Bitcoin and does not consider other asset classes or cryptocurrencies, limiting the generalizability of the results. Second, the study relies on historical data, which may not fully capture future market dynamics influenced by external factors such as macroeconomic events, regulatory changes, or technological advancements. Third, the use of one-day prediction windows, while effective for short-term analysis, does not address the needs of long-term investors or traders.

Future research should aim to expand the scope of analysis by applying the GUHA framework to other financial instruments and exploring the inclusion of additional technical and fundamental indicators. Incorporating external factors, such as sentiment analysis from social media or news, could enhance the robustness of the predictions. Additionally, exploring longer prediction horizons and optimizing the rule-mining process for scalability in real-time trading environments could further advance the applicability of this approach. Finally, integrating machine learning methods alongside data mining techniques could provide a hybrid framework that balances interpretability and predictive accuracy.

This paper serves as a proof of concept for automating the observation of technical indicators, demonstrating that systematic data mining can unlock valuable insights for trading strategies and financial risk management. By addressing the outlined limitations and expanding on the methods employed, future research can contribute to a more comprehensive understanding of the potential for data-driven automation in financial markets.

Acknowledgments. The research project was supported by Grant No. IGA F4/18/2024 of the Internal Grant Agency, Faculty of Informatics and Statistics, Prague University of Economics and Business.

References

Ansari Saleh, A., Pawan Kumar, S., Nguyen Van, T., Nguyen Viet, T., Vo Minh, H.: Prediction of BRIC stock price using ARIMA, SutteARIMA, and Holt-winters. Comput. Mate. Contin. **70**(1), 523–534 (2022)

Chang, T.H., Wang, N., Chuang, W.B.: Stock price prediction based on data mining combination model. J. Global Inform. Manage. **30**(7), 1–19 (2021)

Farid, S., Tashfeen, R., Mohsan, T., Burhan, A.: Forecasting stock prices using a data mining method: evidence from emerging market. Int. J. Fin. Econ. **28**(2), 1911–1917 (2023)

Fang, J., Qin, Y., Jacobsen, B.: Technical market indicators: an overview. J. Behav. Exp. Financ. **4**, 25–56 (2014). https://doi.org/10.1016/j.jbef.2014.09.001

Hájek, P., Havel, I., Chytil, M.: The GUHA method of automatic hypotheses determination. Computing. **1**, 293–308 (1966)

Kaur, J., Dharni, K.: Data mining–based stock price prediction using hybridization of technical and fundamental analysis. Data Technol. Applic. **57**(5), 780–800 (2023)

Leippold, M., Wang, Q., Zhou, W.: Machine learning in the Chinese stock market. J. Financ. Econ. **145**(2), 64–82 (2022)

Máša, P., Rauch, J.: A novel algorithm weighting different importance of classes in enhanced association rules. Knowl.-Based Syst. **294**, 111741 (2024). https://doi.org/10.1016/j.knosys.2024.111741

Rauch, J., Šimnek, M., Chudán, D., Máša, P.: Mechanizing hypothesis formation: principles and case studies. In: Chapter 15: Clever Miner GUHA and Python. CRC Press, London (2022)

Rezaei, H., Faaljou, H., Mansourfar, G.: Stock price prediction using deep learning and frequency decomposition. Expert Syst. Appl. **169**, 114332 (2021)

Rubio, L., Alba, K.: Forecasting selected Colombian shares using a hybrid ARIMA-SVR model. Mathematics. **10**(13), 2181 (2022)

Vuong, P.H., Dat, T.T., Mai, T.K., Uyen, P.H.: Stock-price forecasting based on XGBoost and LSTM. Comput. Syst. Sci. Eng. **40**(1), 237–246 (2022)

Yang, J., Wang, Y., Li, X.: Prediction of stock price direction using the LASSO-LSTM model combines technical indicators and financial sentiment analysis. PeerJ Comput. Sci. **8**, e1148 (2022)

Zolfaghari, M., Gholami, S.: A hybrid approach of adaptive wavelet transform, long short-term memory and ARIMA-GARCH family models for the stock index prediction. Expert Syst. Appl. **182**, 115149 (2021)

Data Science Techniques for Opinion Mining in Industrial Applications

Mehdi Belguith$^{(\boxtimes)}$ and Chafik ALoulou

ANLP-RG, MIRACL Lab., University of Sfax, Sfax, Tunisia
`belguith.mehdi2017@gmail.com, chafik.aloulou@fsegs.usf.tn`

Abstract. Opinion mining is a crucial research area in data science that focuses on the automatic analysis of textual data (such as reviews, customer comments, and feedback) to identify and evaluate sentiments expressed. In industrial contexts, it plays a vital role in enhancing customer feedback analysis, optimizing predictive maintenance, and improving process efficiency. In this research, we propose an innovative opinion mining method that combines Natural Language Processing (NLP), domain ontologies, and Machine Learning (ML) techniques to assist managers in making informed decisions about their products and services. Our method begins by constructing domain-specific ontologies, which play a key role in aspect detection within user comments. We then collect a raw dataset from social media platforms, such as YouTube and Facebook, focusing on four domains: Tunisian restaurants, smartphones, the 2019 Tunisian elections, and Tunisian TV programs. This dataset undergoes multiple pre-processing steps, including transliteration, stop-word removal, and vectorization, to prepare it for analysis. To classify sentiments, we employ two advanced deep learning models (i.e., LSTM and Bi-GRU), which are applied to analyze both the overall sentiment of each comment and the sentiment associated with specific aspects. This dual-layer analysis provides a detailed understanding of user opinions, identifying whether sentiments are positive, negative, or neutral at both the comment and aspect levels. This work not only advances the field of opinion mining through the integration of ontologies and deep learning but also offers valuable practical implications for decision-making in a wide range of industrial applications.

Keywords: Opinion Mining · Artificial Intelligence · Machine Learning · Ontologies

1 Introduction

With the exponential growth and accessibility of the internet, social media platforms such as Facebook and YouTube have become essential venues for individuals to share their opinions on a wide range of topics.

These platforms produce vast amounts of user-generated content, offering a valuable resource for extracting insights into opinions, preferences, and trends that can guide decision-making in industrial applications [1]. For instance, a consumer might write, "The delivery was super fast, but the product quality is disappointing.". This single

A. Mirzazadeh et al. (Eds.): ODSIE 2024, CCIS 2482, pp. 340–354, 2026.
https://doi.org/10.1007/978-3-031-93601-2_21

statement reflects both a positive sentiment about the delivery service and a negative sentiment about the product quality. Such nuanced opinions highlight the importance of advanced analysis techniques to capture the diverse sentiments expressed in consumer feedback.

In this work, we propose an innovative method for opinion mining that extracts sentiments (positive, neutral, or negative) from user comments on social media. The dataset used in this study comprises a collection of comments gathered from social media platforms, primarily YouTube and Facebook.

The goal is to automatically extract customer opinions from their comments on social media. These insights will help managers make better decisions regarding their products or services. For example, if customers find a product's price too high, the company should explore ways to make the product more affordable. Similarly, if customers complain that the service in a restaurant is too slow, the manager should work with the staff to address the issue and improve efficiency.

Our proposed method for opinion mining is based on natural language processing techniques, domain ontologies, and machine learning.

The structure of this paper is as follows: Sect. 2 provides a brief overview of the Tunisian dialect and highlights the associated processing challenges. Related works are reviewed in Sect. 3. Our proposed method for opinion mining is detailed in Sect. 4, followed by the evaluation results presented in Sect. 5. Finally, Sect. 6 concludes the paper and outlines promising directions for future research.

2 Tunisian Dialect: A Brief Description and Processing Challenges

In this section, we briefly describe Tunisian dialect in order to clarify its specificities and processing difficulties, compared to other languages.

2.1 Description

The Tunisian Dialect (TD), also known as "al-Tounsi" or "al-Darija," is the primary spoken language in Tunisia and belongs to the Maghrebi branch of Arabic dialects. Although Modern Standard Arabic (MSA) is the official language, TD is used in everyday conversation and local media. It combines Arabic with vocabulary from Amazigh (Berber), as well as foreign languages like French, Italian, Spanish, and Turkish, reflecting Tunisia's historical influences from the 18th and 19th centuries. Table 1 presents examples of foreign loanwords integrated into the Tunisian dialect.

TD has notable regional variations across Tunisia's 24 administrative areas, categorized into six main sub-dialects: the northeast dialect (Greater Tunis, Bizerte, Nabeul), coastal dialect (Sousse, Monastir, Mahdia), Sfax dialect (Sfax and surrounding areas), southwest dialect (closer to Standard Arabic), southeast dialect (similar to western Libyan), and northwest dialect (similar to eastern Algerian). Each region has subtle urban-rural differences.

Phonologically, TD retains most Arabic sounds, with some variations, such as the pronunciation of " القاف " (qāf) as " ڨ " (g) in certain areas, and the simplification of " الجيم " (jīm) to " ز " (z) when near another " ز " (z). Words in TD often omit middle consonants

Table 1. Examples of foreign words.

Tunisian word	Foreign word	English translation
bersane /برسانة	Italian/Persienne	Window
tIfwn / تلفون	French /Téléphone	Phone
qrnyt / قرنيط	Berber/Quarnit	Octopus
Kenastro /كنسترو	Spanish/Canastro	Basket
Bey/ بَيْ	Turkish/Bey	Sir/Lord

and may replace " الف" (alif) with " ياء" (yā') (e.g., " رأيت" (ra'aytu) meaning "I saw" becomes " رُيت" [rīt]). It also incorporates some Amazigh vocabulary and loanwords from various languages.

Distinctive grammar includes the use of " إنتِ" (inti) for both genders in the second-person singular, and common question words like " علاش" ('alāsh) meaning "why," " كيفاش" (kīfāsh) meaning "how," and " آش" (āsh) meaning "what." Negative verbs often end with " ش" (sh) (e.g., " مانحبش" (mā naḥibbish) meaning "I don't like").

Though widely spoken, TD is primarily a spoken language and is rarely used in formal writing or official contexts, reflecting Tunisia's rich cultural and linguistic heritage.

2.2 The Challenges of Processing Tunisian Dialect

Tunisian Dialect (TD) presents significant challenges for computational models due to its complex nature as described below.

Lack of Standardized Writing System The unique blend of languages and the absence of a standardized writing system make Tunisian Dialect difficult to process. Without a consistent orthographic system, computational models struggle to accurately interpret and generate TD text.

Regional Variability and Loanwords Variability between regional dialects, coupled with the significant influence of loanwords (French, Italian, etc.), further complicates the development of language models for TD.

Registers of Tunisian Dialect TD features two main registers: the colloquial register used in everyday and informal conversations, and the intellectualized register, which blends TD with Modern Standard Arabic (MSA) for formal contexts like public discussions or interviews. The switching between these registers adds complexity to TD processing.

2.3 The Tunisian Dialect on Social Media

The Tunisian dialect also faces challenges in digital spaces, especially on social media, where it is evolving rapidly. Young Tunisians often mix Tunisian Arabic with languages like French and English, creating a hybrid form known as "Franglish" or "Tunisian-English." Additionally, the emergence of "Arabizi" (the use of Latin characters and numbers to represent Arabic sounds) introduces further complexity.

For example, "7" is used to replace the Arabic letter "ح". The use of the Latin alphabet for phonetic transcription of Tunisian Arabic is widespread, making it more accessible but also more challenging for traditional NLP tools to process.". Table 2 shows some examples of Tunisian Dialect Adaptation on Social Media.

Table 2. Examples of Tunisian Dialect Adaptation on Social Media.

Feature	Example	Explanation
Franglish	Bahi, I'll go au marché.	Mixing Tunisian Arabic (Bahi), French (au marché), and English (I'll go).
Arabizi Phonetics	3amalt 7aga jdida.	"3" replaces the Arabic letter " ع", and "7" replaces " ح".
Code-Switching	Tawa I'm working sur mon projet.	Switching between Tunisian Arabic (Tawa), English (I'm working), and French (sur mon projet).
Latinized Tunisian Words	Chkoun bech yji ?	Tunisian Arabic written in Latin script; "Chkoun" means "Who" and "bech yji" means "will come".
Emoji and Abbreviations	LOL, chnowa hatha :)!	Incorporation of internet slang ('LOL') and emoticons alongside Arabizi Tunisian.

Furthermore, Tunisian Arabic suffers from a lack of linguistic resources, such as annotated corpora, lexicons, and tools specific to the dialect, which reduces the performance of NLP models. Resources available for Standard Arabic or other Arabic dialects are often not suitable for DT. This scarcity of resources makes it even more difficult for NLP systems to manage the complexities of Tunisian Arabic on social media, especially given its rapid evolution and diverse linguistic influences.

3 Related Work

In this section, we review related work for opinion mining in prior studies. We distinguish three main approaches, which are the knowledge-based approach, the machine learning-based approach, the deep learning-based approach, and the transfer learning approach.

In this section, we review previous studies on opinion mining and highlight relevant research. We identify four main approaches: the knowledge-based approach, the machine learning approach, the deep learning approach, and the transfer learning approach.

3.1 The Knowledge-Based Approach

This approach is based on predefined rules and dictionaries or lexicons that associate words with specific opinions (e.g., "good" represents a positive sentiment, 'bad' represents a negative sentiment). It works well when domain knowledge is available but struggles with context and complex sentences.

The earliest studies on aspect-based opinion mining relied on frequent nouns and noun phrases for aspect extraction.

Samha et al. [2] introduced a framework for analyzing user opinions by leveraging Natural Language Processing (NLP), an ontology, and frequency-based tag sets. Their approach demonstrated promising results, outperforming a prior reference model.

Building on this, Mowlaei et al. [3] proposed an extension of two lexicon generation techniques for opinion mining. Their method integrates statistical techniques with a genetic algorithm, combining dynamically generated lexicons with established static lexicons to enhance opinion classification in reviews. Experimental evaluations indicate that their approach surpasses traditional classification methods, achieving improvements of 6% in precision, 1% in recall, and 7.4% in F-measure.

3.2 Machine Learning Based Approach

Tamrakar et al. [4] compiled a dataset of Nepali text from various online sources and applied Part-of-Speech (POS) tagging to extract key opinion-related features. Their sentiment classification experiments, conducted using Support Vector Machine (SVM) and Naïve Bayes (NB), achieved accuracy rates of 76.8% and 77.5%, respectively.

Similarly, Mubarok et al. [5] introduced a three-phase approach for opinion analysis. Their methodology includes data preprocessing with POS tagging, feature selection using the Chi-Square test, and sentiment classification leveraging a Naïve Bayes model. The evaluation results indicate that their aspect-based opinion analysis system attained an F1-score of 78.12%, highlighting its effectiveness in sentiment classification.

3.3 Deep Learning Models

Al-Dabet et al. [6] introduced an approach for extracting opinion targets in Arabic text using attention-based neural models. Their model was evaluated on the SemEval-2016 annotated dataset in the hotel domain, demonstrating superior performance compared to both baseline and prior methods, with an F-measure of 72.83%.

Abdelgwad et al. [7] explored three deep learning architectures (Bidirectional Gated Recurrent Units (BGRU), Convolutional Neural Networks (CNN), and Conditional Random Fields (CRF)) which were integrated into a unified BGRU-CNN-CRF model for opinion target extraction. Their approach achieved an F1-score of 69.44%. For opinion polarity classification, they developed a deep learning model incorporating an Interactive Attention Network based on Bidirectional GRU (IAN-BGRU), which attained an accuracy of 83.98%.

Al-Smadi et al. [8] proposed a hybrid model for Opinion Target Expression (OTE) extraction, combining a character-level Bidirectional Long Short-Term Memory network

(Bi-LSTM) with a Conditional Random Field (CRF) classifier. Experimental results indicated a substantial 39% improvement over the baseline in aspect OTE extraction. Additionally, they introduced an LSTM-based model for aspect opinion polarity classification, yielding a 6% improvement compared to the baseline.

Mai and Le [9] developed a Joint Sentiment Analysis (JSA) approach that simultaneously performs Sentence-level Sentiment Analysis (SSA) and Aspect-level Sentiment Analysis (ASA). Their experiments demonstrated that the joint model outperformed the ASA model, achieving an F1-score of 81.78% compared to 81.06% for the separate approach. The findings suggest that JSA effectively captures the complementary advantages of SSA and ASA, with SSA benefiting more significantly from the joint framework.

3.4 Transfer Learning Models

Transfer learning models adapt pretrained models on large datasets for specific tasks with smaller datasets. Applying BERT to sentiment analysis exemplifies transfer learning, where the model is fine-tuned on task-specific datasets to enhance performance.

Zhao et al. [10] introduced a novel sentiment analysis approach that leverages multi-model fusion transfer learning to accurately assess public opinion sentiment.

Their method incorporates the ERNIE model for dynamic word vector representation and integrates TextCNN and BiGRU models to extract relevant features, enhancing the overall performance of sentiment analysis. They transfer the feature extraction layer parameters from a pre-trained model to the target model and incorporate an attention mechanism to highlight key sentiment features. The experimental results demonstrate that this approach outperforms traditional methods, achieving higher accuracy, precision, recall, and F1-measure.

Jahangir et al. [11] examined the efficacy of fine-tuning the BERT model for opinion mining tasks, comparing it with other machine learning algorithms to assess performance improvements. The authors fine-tune BERT on a sentiment analysis dataset and compare its performance with traditional machine learning algorithms. The results indicate that BERT fine-tuning significantly enhances sentiment analysis accuracy, outperforming other algorithms.

Li and Lee [12] integrated transfer learning into case-based reasoning to enhance new-product portfolio management by analyzing customer feedback. The authors propose a transfer-learning-based opinion mining framework that leverages case-based reasoning cycles to configure new-product portfolios. The framework effectively analyzes customer feedback, leading to improved product portfolio decisions.

Sekaran and Ragupathi [13] focused on summarizing opinions based on specific features, employing transfer learning techniques to improve the accuracy of sentiment analysis. The authors propose a feature-based opinion summarization model that utilizes transfer learning to adapt to new domains. The model effectively summarizes opinions, providing valuable insights for decision-making.

Best et al. [14] investigated the application of transfer learning to sentiment analysis in image-based data, demonstrating its effectiveness in handling complex datasets. The authors apply transfer learning to mine sentiment from image-based data, achieving improved sentiment analysis performance.

4 Proposed Method

In this research work, we propose first of all to build domain ontologies and then we will use these ontologies and machine learning techniques for opinion mining.

4.1 Ontology Building

A domain ontology is a structured framework that defines the key concepts, entities, and relationships within a specific field, enabling consistent knowledge representation and reasoning. It provides a shared vocabulary for both humans and machines, improving data interpretation, integration, and decision-making.

For example, in the restaurant domain, an ontology might include concepts such as menu items, ingredients, orders, and customers, with relationships defining how ingredients are used in dishes, how orders are processed, and how customers interact with the restaurant. This structured knowledge helps optimize operations like order management and recommendation systems, as well as enhances customer service through better understanding of preferences and dietary restrictions.

In this study, we focus on four domains which are: restaurant, election, smartphones and TV channels. For each domain we built an ontology [15]:

RestOnto. This ontology focuses on the restaurant domain, offering a deep semantic representation of the concepts and relationships associated with this field. It is based on a dataset of 2719 instances.

ElectOnto. This ontology is dedicated to the election sector, providing a formal and organized structure of relevant elements extracted from a dataset, containing 3076 instances that we collected in the context of this work.

PhoneOnto. This ontology is related to the smartphone domain. It is based on the dataset of Belguith et al. [16], which includes a total of 5813 instances.

ChannelOnto. This ontology is dedicated to television channels and is constructed based on the database of Medhaffar et al. [17] which contains 13667 instances.

The structure of our ontologies[1] is based on three main classes and a single object property called "hasAsentiment" that establishes the links between these three classes. The class "Sentiment" has three values (positive, negative, or neutral). The class "Sentiment-Expression" represents the textual expression of the sentiments and could be composed of subclasses. The class "Target" categorizes various aspects within a comment.

The primary goal is to analyze and determine the sentiment (i.e., whether positive, negative, or neutral) associated with each aspect. Figure 1 illustrates a segment of the "Food" ontology, which was developed using the "Protégé" tool.

4.2 Opinion Mining

We propose an opinion mining method based on artificial intelligence techniques. More specifically, we propose utilizing Natural Language Processing (NLP), domain-specific ontologies, and Machine Learning (ML) techniques to enhance analysis and

---

[1] We build our ontologies using Protégé 5.6.2 software, based on the four datasets.

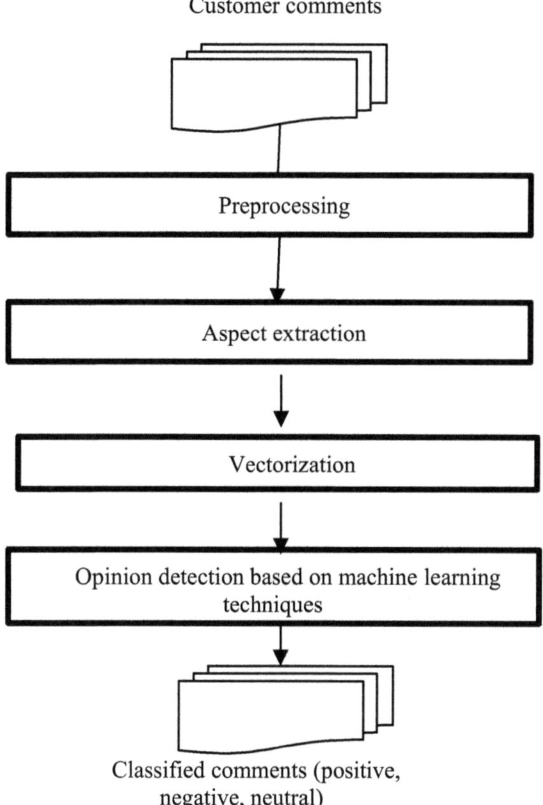

Fig. 1. Main steps of our opinion mining method.

classification. Initially, we gather a raw dataset from social media platforms, primarily YouTube and Facebook. This dataset comprises user comments related to our four interest domains. Next, we perform several pre-processing tasks (based on NLP techniques) on the raw dataset to prepare it for the learning phase, including transliteration, removal of stop words, and more. Subsequently, we convert the dataset into numerical vectors through a vectorization process. Lastly, we apply learning models to determine the opinion of each comment. Figure 1 shows these steps that we discuss in detail in the following sections. The ontologies are used for aspect detection inside the comments.

Dataset Collection and Pre-processing Our dataset primarily consists of user-generated comments in the Tunisian dialect, sourced from various social media platforms such as Facebook and YouTube. These comments span four distinct domains: smartphones, Tunisian restaurants, the 2019 Tunisian elections, and television programs.

Notably, the linguistic diversity within the dataset reflects different writing styles. Comments may appear in standard Arabic (using Arabic script), Tunisian Arabic (transcribed in Arabizi), or a mix of both through code-switching, where Arabic script and Arabizi are combined within the same text.

Data Pre-processing mainly consists of the following tasks: Transliteration, Light Stemming and Cleaning.

Transliteration. We begin by transliterating all Tunisian Arabizi comments into Tunisian Arabic to ensure uniformity in processing all the words within the comments. Table 3 showcases two examples from our dataset, where we apply this transliteration preprocessing step.

Table 3. Examples of comments before and after transliteration.

	Comment 1	Comment 2
Before translation	Ma3omri rit naib sabi9an ymthel nfsh mart o5ra fi eltinti5abat ba3ed el5ssart.	Alndhafh w elbena hiya elkol ya3tikom saht
After translation	ما عمري ريت نائبا سابقًا يمثل نفسه مرة أخرى في الانتخابات بعد الخسارة	النظافه و البنا هي الكل يعطيكم الف صحة
English translation	I have never seen a former deputy who applied again for the elections after failure.	The cleanliness and the good taste are the main key may God give you good health.

Light Stemming. Following the transliteration of the dataset, we perform a light stemming process, which involves removing both prefixes and suffixes from words [18, 19]. This ensures that words sharing the same root are treated as equivalent units (i.e., as the same word). Numerous studies in the literature have demonstrated that light stemming significantly enhances sentiment classification results for Arabic and its dialects [20]. Table 4 illustrates the comments from Table *3*, before and after the application of light stemming.

Table 4. Examples of light stemming.

	Comment 1	Comment 2
Before light stemming	ما عمري ريت نائبا سابقًا يمثل نفسه مرة أخرى في الانتخابات بعد الخسارة	النظافه و البنا هي الكل يعطيكم الف صحة
After light stemming	ما عمري ريت نائب سابق يمثل نفسه مرة أخرى في انتخاب بعد خسارة	نظافه و بنا هي كل يعطي الف صح
English translation	I have never seen a former deputy who **apply** again for **election** after failure	**cleanliness** and **good taste** are **main** key may God give you good health.

Cleaning. The objective of the data cleaning preprocessing step is to eliminate irrelevant words, symbols, and punctuation marks that do not contribute to sentiment analysis, while also reducing the time complexity of the learning process. To achieve this, we remove stop words (also known as empty words), URLs, user mentions, punctuation, and any duplicated or redundant characters within words. Table 5 provides an example of a comment before and after the cleaning process is applied.

Table 5. Example of comment cleaning.

	Before cleaning	After cleaning
Comment	الف‏‏|||||||||||||||| يعطيييييك⬜⬜ صحة	يعطيك الف صحة
English translation	may God give you ⬜⬜gooooood health	may God give you good health

Aspect extraction. The primary goal of this work is to determine the sentiment (positive, neutral, or negative) expressed in each comment. Our first task, therefore, is to identify the aspects (such as product or service features like price, taste, speed, and quality of service) that are referenced in the comments. It is important to note that a single comment may contain multiple aspects, as illustrated by the example in Table 6.

Table 6. Example of aspects.

	Comment	Aspects
Tunisian	اكل بنين بصراحه اما سوم ياسر غالي	اكل سوم
English translation	Delicious food but the price so expensive	Food price

The aspect extraction is based on the domain ontologies. Indeed, ontologies have proven to be an efficient means for extracting aspects for a specific domain [21–24]. Ontologies help in opinion mining by providing a structured framework to identify and categorize key aspects in text. They define specific aspects (like quality, price, service) and their relationships, enabling better extraction and classification of relevant terms. Ontologies also assist in disambiguating terms by linking synonyms to the same aspect and improve sentiment analysis by associating aspects with corresponding sentiments.

For example, in the restaurant domain, an ontology might categorize aspects like food quality, service, and atmosphere. Reviews containing phrases like "delicious food" (linked to food quality), "slow service" (linked to service), and "cozy ambiance" (linked to atmosphere) would be properly categorized, improving the accuracy of aspect-based sentiment analysis.

Vectorization Word embedding is a method of representing words as numerical vectors, where words with similar meanings and contexts are mapped to similar vectors. This approach has become a fundamental tool in sentiment analysis and various other Natural Language Processing (NLP) tasks. Word embeddings are also commonly used as input layers in deep learning models [25, 26]. In our case, we utilize the Keras layer, an integration layer within Keras designed to facilitate neural network operations on textual data.

Machine Learning As discussed in the previous section, we focus on two deep learning models: LSTM and Bi-GRU. We applied these models to the four datasets outlined in Table 7. This table provides a breakdown of the number of comments used for both training and testing in each domain. Additionally, it details the distribution of comments into positive, negative, and neutral categories. It is important to note that, for each domain, 80% of the data is allocated to the training set, while the remaining 20% is reserved for the test set.

Table 7. Train and test Dataset.

		#Positive	#Neutral	#Negative	#Aspects
Restaurants	Train	1150	1366	867	2068
	Test	288	341	216	515
Tunisian election 2019	Train	2063	3603	880	7 052
	Test	516	901	220	900
Tunisian TV programs	Train	1506	1913	526	2 037
	Test	377	478	132	503
Smartphones	Train	753	3685	307	4 745
	Test	188	921	77	1187

For our models, several parameters required tuning, particularly the number of epochs and batch size. We tested various values for these parameters and reported only the configurations that yielded the best results.

We set the number of epochs to 10 for all experiments. To regularize the neural networks and mitigate the risk of overfitting, we applied a dropout rate of 0.2. Additionally, we utilized the Adam optimizer, as outlined in Table 8.

Note that Dropout is an efficient regularization technique for avoiding overfitting in deep neural networks[2]. The number of units refers to the number of memory units assigned to each input in a sequential manner, where each unit processes the filtered information and passes it on to the next memory unit.

Word Embedding is a collection of statistical language modeling techniques that map words and phrases to vectors of real numbers. These vectors capture both the semantic and syntactic properties of words, facilitating more meaningful representations. The dimension of free embedding refers to the length of the vectors in the word embedding, determining the depth of the representation.

---

[2] https://towardsdatascience.com/multi-sample-dropout-in-keras-ea8b8a9bfd83

Table 8. Hyper parameters used to LSTM/ Bi-GRU model.

Hyper Parameter	Value/Type
Dropout	0.2
Number of LSTM/ BiGRU units	196
Word embedding	Keras embedding layer
Dimension of free embedding layer	100
Batch size	128
Epochs	10
Optimizer	Adam

The Batch size denotes the number of training examples processed in a single iteration. A larger batch size can speed up training, while a smaller batch size may provide more precise updates to the model.[3]

An epoch refers to one complete cycle of training in which the neural network processes the entire training dataset once. Multiple epochs are used to iteratively improve model performance[4].

We divided our collected dataset into two distinct parts. Approximately 80% of the total data is allocated for training, while the remaining 20% is reserved for testing, ensuring a balanced evaluation of the model's performance.

5 Result Analysis

For the four DataSets (i.e. Restaurant, Tunisian Election 2019, TV Program, Telephone) we experimented with two deep learning models (i.e. LSTM and Bi-GRU).

For the evaluation, we mainly used the following metrics:

- Precision (P): represents the percentage of targeted opinions found by the system that are correct.
- Recall (R): is the percentage of targeted annotated reviews found by the system that are also found by the system.
- F_measure: is an average of Precision and Recall

$$F\_measure = 2^*P^*R/(P + R) \tag{1}$$

Our deep neural networks have yielded promising results for opinion mining in the Tunisian dialect. The outcomes of all experiments are presented in Table 9.

From the results, we observe that the LSTM model consistently outperforms the Bi-GRU model across all datasets. The performance gap is particularly noticeable in

---

[3] https://radiopaedia.org/articles/batch-size-machine-learning#:~:text=Batch%20size%20is%20 20a%20term,iteration%20and%20epoch%20values%20equivalent

[4] https://www.baeldung.com/cs/epoch-neural-networks

Table 9. Experiments Results of sentiment model (% F-measure).

DataSet	LSTM	Bi-GRU
Restaurant	69%	65%
Tunisian election 2019	74%	69%
TV program	78%	74%
Smartphone	86%	76%

the Smartphone dataset, where LSTM achieves an F-measure of 86%, compared to 76% for Bi-GRU. This suggests that LSTM is more effective in capturing long-range dependencies in sentiment analysis for this dataset.

The TV Program and Tunisian Election 2019 datasets also show relatively high performance, with LSTM achieving 78% and 74%, respectively. These results indicate that deep learning models are effective for sentiment analysis in diverse domains of the Tunisian dialect. However, the lower F-measure for the Restaurant dataset (69% for LSTM and 65% for Bi-GRU) suggests that domain-specific challenges, such as ambiguous sentiments or informal language variations, may impact classification accuracy.

Overall, the findings highlight the potential of deep learning approaches for opinion mining in under-resourced languages like Tunisian Arabic.

6   Conclusion and Perspectives

In this paper, we have presented an innovative approach for opinion mining in the Tunisian dialect. Our method stands out by not only determining the overall opinion (positive, negative, or neutral) of a comment but also evaluating the sentiment for each specific aspect within the comment. For instance, while the overall opinion of a product may be negative, the sentiment regarding individual aspects (such as color or price) could be positive. This aim is to assist managers in making more informed decisions about their products, services, and other business operations.

Our proposed method is based on deep learning models (LSTM and Bi-GRU), natural language processing techniques and domain ontologies. It allows to (i) extract the aspects based on the ontologies and (ii) classify the opinion for each aspect into positive, neutral or negative based on the deep learning techniques. Several pre-processing tasks are applied to the raw dataset using NLP techniques to prepare it for the learning phase. These tasks include transliteration and the elimination of stop words.

We trained our models on four distinct datasets, each corresponding to a specific domain: smartphones, Tunisian restaurants, the Tunisian elections, and Tunisian TV programs. Experimental results show that best results are obtained by the LSTM model.

For future work, we plan to explore additional deep learning approaches, including transformer-based techniques [27]. Furthermore, we aim to expand our dataset to encompass a broader range of domains.

References

1. Seifian, A., Shokouhyar, S., Bahrami, M.: Exploring customers' purchasing behavior toward refurbished mobile phones: a cross-cultural opinion mining of amazon reviews. Environ. Dev. Sustain. **26**(11), 28131–28159 (2024)
2. Samha, A.K., Li, Y., Zhang, J.: Aspect-based opinion extraction from customer reviews. Comput. Sci. Inform. Technol., 149–160 (2014). https://doi.org/10.5121/csit.2014.4413
3. Mowlaei, M.E., Abadeh, M.S., Keshavarz, H.: Aspect-based sentiment analysis using adaptive aspect-based lexicons. Expert Syst. Appl. **148**, 113234 (2020). https://doi.org/10.1016/j.eswa.2020.113234
4. Tamrakar, S., Bal, B.K., Thapa, R.B.: Aspect based sentiment analysis of Nepali text using support vector machine and naive Bayes. Techn. J. **2**(1), 22–29 (2020). https://doi.org/10.3126/tj.v2i1.32824
5. Mubarok, M.S., Adiwijaya, Aldhi, M.D.: Aspect-based sentiment analysis to review products using Naïve Bayes. In: International Conference on Mathematics: Pure, Applied and Computation. AIP Publishing, Indonesia (2017). https://doi.org/10.1063/1.4994463
6. Al-Dabet, S., Tedmori, S., Al-Smadi, M.: Extracting opinion targets using attention-based neural model. SN Comput. Sci. **1**(5) (2020). https://doi.org/10.1007/s42979-020-00270-4
7. Abdelgwad, M.M., Soliman, T.H., Taloba, A.I., Farghaly, M.F.: Arabic aspect-based sentiment analysis using bidirectional GRU: based models. J. King Saud Univ. Comput. Inform. Sci. **34**(9), 6652–6662 (2022). https://doi.org/10.1016/j.jksuci.2021.08.030
8. Al-Smadi, M., et al.: Using long short-term memory deep neural networks for aspect-based sentiment analysis of Arabic reviews. Int. J. Mach. Learn. Cyber. **10**(8), 2163–2175 (2019). https://doi.org/10.1007/s13042-018-0799-4
9. Mai, L., Le, B.: Joint sentence and aspect-level sentiment analysis of product comments. Ann. Oper. Res. **300**(2), 493–513 (2021). https://doi.org/10.1007/s10479-020-03534-7
10. Zhao, Z., Liu, W., Wang, K.: Research on sentiment analysis method of opinion mining based on multi-model fusion transfer learning. J. Big Data. **10**(1) (2023). https://doi.org/10.1186/s40537-023-00837-x
11. Jahangir, M., Islam, M.A., Alam, M.S.: Transfer learning using BERT & comparative analysis of ML algorithms for opinion mining. In: IEEE International Conference on Artificial Intelligence and Knowledge Engineering (AIKE), pp. 1–5, Japan (2023). https://doi.org/10.1109/AIKE57629.2023.00007
12. Li, S.M., Lee, C.K.M.: Transfer-learning-based opinion mining for new-product portfolio configuration over the case-based reasoning cycle. Appl. Sci. **12**(23), 12477 (2022). https://doi.org/10.3390/app122312477
13. Sekaran, R., Ragupathi, A.: Feature-Based Opinion Summarization Using Transfer Learning. LAP Lambert Academic Publishing (2015)
14. Best, N., Ott, J., Linstead, E.: Exploring the efficacy of transfer learning in mining image-based software artifacts. J. Big Data. **7**(1), 1–10 (2020)
15. Belguith, M., Aloulou, C., Gargouri, B.: Aspect level sentiment analysis based on deep learning and ontologies. SN Comput. Sci. **5**(1), 58 (2024). https://doi.org/10.1007/s42979-023-02362-3
16. Belguith, M., Aloulou, C., Gargouri, B.: Building domain ontologies for Tunisian dialect: towards aspect sentiment analysis from social media. In: 3rd International Conference on Intelligent Systems and Patterns Recognition, pp. 252–267. Springer, Tunisia (2023)
17. Medhaffar, S., Bougares, F., Estève, Y., Hadrich Belguith, L.: Sentiment analysis of Tunisian dialects: linguistic resources and experiments. In: 3rd Arabic Natural Language Processing Workshop, pp. 55–61. Association for Computational Linguistics, Spain (2017)

18. Belguith, M., Azaiez, N., Aloulou, C., Gargouri, B.: Social media sentiment classification for Tunisian dialect: a deep learning approach. In: 2nd International Conference on Intelligent Systems and Patterns Recognition, pp. 377–393. Springer, Tunisia (2022)
19. Mekki, A., Zribi, I., Ellouze, M., Hadrich Belguith, L.: TTK: A toolkit for Tunisian linguistic analysis. Comput. Speech Lang. **86**(C), 101617 (2024)
20. Mekki, A., Zribi, I., Ellouze, M., Hadrich Belguith, L.: Treebank creation and parser generation for Tunisian social media text. In: 17th International Conference on Computer Systems and Applications (AICCSA), pp. 1–8. IEEE/ACS, Turkey (2020). https://doi.org/10.1109/AICCSA50499.2020.9316462
21. Sharma, S., Saraswat, M., Dubey, A.K.: Multi-aspect sentiment analysis using domain ontologies. In: Communications in Computer and Information Science, pp. 263–276. Springer (2022). https://doi.org/10.1007/978-3-031-21422-6_19
22. Carneiro, E., Guedes, G.P., Belloze, K. T.: A Review of Ontology-Based Approaches for Sentiment Analysis: Possible Improvements on the Brazilian Affective Computing Scenario (2022)
23. García-Díaz, J.A., Cánovas-García, M., Valencia-García, R.: Ontology-driven aspect-based sentiment analysis classification: an infodemiological case study regarding infectious diseases in Latin America. Futur. Gener. Comput. Syst. **112**, 641–657 (2020). https://doi.org/10.1016/j.future.2020.06.019
24. Pour, A.A.M., Jalili, S.: Aspects extraction for aspect level opinion analysis based on deep CNN. In: 26th International Computer Conference, Computer Society of Iran (CSICC), pp. 1–6. IEEE, Iran (2021)
25. Ellouze, M., Hadrich Belguith, L.: Interpreting the structure and results of a data warehouse model using ontology and machine learning techniques. Int. J. Hybrid Intell. Syst. **20**(4), 317–332 (2024)
26. Masmoudi, A., Hamdi, J., Hadrich Belguith, L.: Deep learning for sentiment analysis of Tunisian dialect. Computación y Sistemas. **25**(1) (2021). https://doi.org/10.13053/cys-25-1-3472
27. Du, J., Jiang, Y., Liang, Y.: Transformers in opinion mining: addressing semantic complexity and model challenges in NLP. Transactions on computational and scientific. Methods. **4**(10) (2024)

Investigating the Performance of Recently Proposed Metaheuristic Optimization Algorithms on Real-World Engineering Design Problems

Alper Buğra Polat[✉] ⓘD, Elif Varol-Altay ⓘD, and Osman Altay ⓘD

Department of Software Engineering, Manisa Celal Bayar University, 45400 Manisa, Turkey
alperpolat3573@gmail.com, {elif.altay,osman.altay}@cbu.edu.tr

Abstract. The obstacles in design engineering arise from the boundary conditions and objectives to be met within the constraints. These challenges range from machinery, energy, automotive, industrial, and even construction sectors. For these kinds of problems, metaheuristic optimization algorithms that search large search spaces at once while evading local optima are being adopted more often. This study approaches three engineering design problems—the tubular column, corrugated bulkhead, and cantilever beam—using three novel metaheuristic frameworks termed as Puma Optimizer (PO), Greylag Goose Optimization (GGO), and Human Evolutionary Optimization Algorithm (HEOA). Each of the algorithms was tried out on each of the problems for 30 trials per problem, with the maximum reliability achieved with proper documentation done during the trials. The results of the study proved that PO was superior to GGO and HEOA in every scenario tested. This type of comparative study offers quantitative data for researchers developing the subject as well as a strong starting point for the choice of appropriate methods for solving actual engineering problems.

Keywords: Real-world Engineering Design Problem · Metaheuristic Optimization Algorithm · Optimization

1 Introduction

The constraints of pre-established parameters must be taken into consideration while making tactical and important decisions during the design, manufacturing, and maintenance of any engineering system [1]. There are constraints, and in order to achieve a particular purpose, it is necessary to choose the best alternative among the possibilities; this is the essence of optimization [2]. It is the optimization process that allows either improving the effectiveness of this process or decreasing the time and quantity for the system management. For example, if we take system design cost as a factor of optimization, it is then possible to envisage it in a manner like the maximum of a variable that has various constraints on the conditions [3].

Constrained optimization can be effectively employed in industrial and technological problems of diverse nature. The most interesting abstract problems of this nature have a

© The Author(s), under exclusive license to Springer Nature Switzerland AG 2026
A. Mirzazadeh et al. (Eds.): ODSIE 2024, CCIS 2482, pp. 355–368, 2026.
https://doi.org/10.1007/978-3-031-93601-2_22

lot of real restrictions over the number of solutions to be entertained [4]. It has been possible to successfully deal with these restricted types of problems by many meta-heuristics recently [5]. Because there will be many functions to deal with and several constraints to guide regions, the focus of these approaches are approximation not exactitude. Indeed, there are several problems that can be addressed using typical tools of expansion and conventional mathematical optimization, covering many of the established approaches; however, they are clearly cost-ineffective.

Metaheuristic approaches can be noted as effective methods capable of achieving, with a desired approximate accuracy, the global or close to global optimal solution. Different techniques have been designed to prevent metaheuristic algorithms from becoming trapped within exploration or exploitation areas of the search space. It has been proven through "the no-free lunch" theorem that there is no universal algorithm for solving all problems [6]. Such algorithms can conduct the exploration and search of the solution space and are known to apply functions that need not be continuous [7]. Even though exact solutions cannot be found, acceptable solutions are obtained in practice, and the optimization algorithm allows flexibility of the complete search through stochastic transition mechanisms.

In this study, three different engineering design problems were solved with the help of three recently proposed metaheuristic algorithms. These algorithms are Greylag Goose Optimization Algorithm (GGO), Puma Optimizer (PO), and Human Evolutionary Optimization Algorithm (HEOA). The aim of the paper is the application of these new algorithms, comparing their performance and determining the areas where these algorithms can be utilized more effectively. This strategy hopes to assist in directing scientists who are practitioners in engineering and artificial intelligence to enhance findings in future studies.

This paper compares algorithms by testing them on the same real-world engineering design problem under equal conditions and analyzing the obtained results. The second part of the paper contains a literature review. The third part of the paper defines the definitions and objectives of the selected algorithm. Finally, in the fourth part of this report are presented descriptions of real-world engineering design problems and a detailed comparison of the test results. The fifth and final part of the paper presents the conclusions.

2 Literature Review

Optimization is the process of finding the best results at the minimum cost. Metaheuristic optimization is a different approach to optimization. Metaheuristic optimization algorithms are divided into eleven categories but can be examined under five general categories: physics-based, swarm-based, evolutionary-based, human-based, and game-based algorithms [8]. The movements of animals in groups are studied as an effort to develop swarm-based optimization algorithms. There are numerous swarm-based algorithms that have been developed. For instance, we can look at Tuna Swarm Optimization [9], Northern Goshawk Optimization [10], Swallow Swarm Optimization [11], and Cat and Mouse-Based Optimization [12].

Evolutionary-based meta-heuristic optimization algorithms can be classified as optimization algorithms that have been designed after key biological principles such as the processes of evolution and genetics. They include algorithms like genetic algorithms [13], differential evolution [14], partial reinforcement optimizer [15], tree growth algorithm [16], among others. Whereas, physics-based meta-heuristic algorithms are more focused on the physical world. They include Big Bang-Big Crunch [17], Archimedes Optimization Method [18], Multiverse Optimizer [19], and Thermal Exchange Optimization [20].

As the name suggests, game-based metaheuristic optimization algorithms are inspired by players and are based on the rules of games. Some examples include the Soccer Game Optimizer [21], Teamwork Optimization Algorithm [22], World Cup Optimization Algorithm [23], and Darts Game Optimizer [24]. Human-centered metaheuristic optimization algorithms are also constructed from the models of social and individual actors, connections, and activities. Some examples are Teaching-Learning-Based Optimization [25], Human Mental Search [26], Social Emotional Perception [27], and Brainstorming Optimization Algorithm [28].

Numerous real-world engineering design problems have been discussed in the literature [29]. Metaheuristic optimization approaches have been some of the methods used to solve such problems. The application of metaheuristic approaches include, but is not limited to, the following areas: pressure vessel design, three-bar truss design, speed reducer design, welded beam design, multiple disk clutch brake design, piston lever, corrugated bulkhead design, tubular column design, and cantilever beam design. These problems have been solved by some well-established metaheuristic optimization algorithms [30].

Although there are some studies available in the literature, systematic comparisons of new algorithms under the same conditions for different problems are still lacking. This paper tries to fill this gap by applying three newly proposed metaheuristic techniques, namely, Puma Optimizer (PO), Greylag Goose Optimization (GGO), and Human Evolutionary Optimization Algorithm (HEOA), to three engineering design problems: tubular column, corrugated bulkhead, and cantilever beam. In all the problems, each algorithm runs 30 times, and every detail has been documented to ensure its reliability. From these, the results have proven that PO performs better in all scenarios compared with GGO and HEOA. It offers a comparative analysis that can be useful quantitatively to the researchers and lays a solid foundation for selecting the best strategies toward real-world engineering challenges.

3 Metaheuristic Algorithms

3.1 Puma Optimizer

The PO is a hunting algorithm that uses both exploration and exploitation concepts, like the hunting style of pumas. The search area in this algorithm is conceptualized as a puma's territory; the optimal solution is a dominant puma, and the potential solutions are subordinate female pumas. During the exploration phase, the algorithm simulates a puma's search for food in its unfamiliar environment, making random movements and using various exploration strategies to search for alternatives beyond local optima. The exploitation phase, however, deals with the more specific task of simulating hunting in places that worked during exploration. One such improvement offered by the PO algorithm is an intelligent phase-change mechanism. This mechanism is based on the puma's decision-making and memory that continuously changes from exploration to exploitation in such a way that one does not overpower the other, preventing early convergence. Each phase applies specific operators that make sure that effective search is accomplished, thus enhancing the algorithm's capacity to solve complicated optimization problems [31].

3.2 Human Evolutionary Optimization

The HEOA revolves around human evolution focusing on two aims, i.e., expanding the horizon and enhancing the existing features: The HEOA uses chaos theory along with social roles to simulate evolution and search optimization intervals with reasonable efficiency. HEOA starts with logistic chaos mapping, which determines a level of randomness at the beginning of the above process; therefore, the mode is able to cover many possible options and does not become prematurely redundant. The algorithm splits into two distinct processes, which this modification refers to as the candidate solution refinement and search processes, respectively. The search process replicates the first experience of primitive man: the examination of unfamiliar terrains or resources that were non-manifold and fortuitous in some cases. This phase employs Levy flight mechanisms to provide more randomness in the distance traveled, thus allowing the algorithm to cover more ground within the search space. In the social phase, the algorithm was able to sort out the four basic types of candidate solutions available, namely the leaders, the explorers, the followers, and the losers. Using their prior knowledge, leaders concentrate on regions with high potentials, exploratory agents advance towards unfamiliar territories in search of plausible solutions, and followers copy the best-winning solutions, whereas the losers are weeded out and new solutions are produced in the more promising areas. This adaptive, role-based structure enables the algorithm to perform a trade-off between exploration and exploitation effectively [32].

3.3 Greylag Goose Optimization

The GGO algorithm is a type of metaheuristic that has been inspired by animal behavior. It solves complex optimization problems by, among other things, drawing on the social aspects of the graylag's dynamic activity and its behavioral characteristics. This GGO algorithm is designed based on the understanding that geese's cooperative use of their V-shaped flight pattern during migration allows for the group to go farther than an individual because of the reduced wind resistance when the entire group flies together. This type of characteristic is reflected in the two-dimensional aspect of the proposed algorithm, as it helps in balancing between exploration and exploitation. A performance paradigm shift occurs whereby, in ideal scenarios, situations mimic scouting geese that actively seek new zones where they do local searches in the neighborhood, performing random wholesome jumps to imposing areas while systematically searching for attractive ones. The best identified solutions from exploration are employed, promising solutions taking lead. Roles switch during a hunt, with each leader performing core tasks to facilitate energy conservation, similar to how goose engages in a flight. The switching of roles in what task sharing guarantees is to move from one task to another, and it assuages the risks of premature convergence. The GGO also uses information sharing updates in solution processes that focus on geese's social commitment and care. This dynamic interaction among candidate solutions is like the way in which geese look after one another and cooperate to achieve the group's survival. The algorithm is flexible; the gander's strategy is to use landmarks in their environment that allow for effective searching in a space of high dimensions while enabling an optimal solution [33].

4   Engineering Design Problems and Experimental Results

In this section, the previously mentioned metaheuristic optimization approaches shall be implemented and demonstrated on selected engineering design problems. These engineering design problems are the ones that require formulation, modeling, and solving of technical management tasks that have some set constraints [30]. The mentioned metaheuristic optimization methods will be tested with the tubular column design problem, corrugated bulkhead design problem, and cantilever beam problem. These problems were chosen as they represent diverse challenges in engineering optimization: cantilever beam focuses on structural strength and material efficiency [34], tubular column highlights stability under compressive loads [35], and corrugated bulkhead addresses weight minimization in complex geometries [36]. Their explicit statements are given in Table 1. In these problems, D stands for the decision variable set, h indicates the number of equality restrictions, and g is the total number of constraints that are inequalities. f_{min} stands for the precise reply expected from the optimization procedure on the case.

Table 1. Details of used real-world engineering problems.

Name	D	g	h	f_{min}
Cantilever beam	5	1	0	1.3399576
Tubular column design	2	6	0	26.486361473
Corrugated bulkhead design	4	6	0	6.8429580100808

All the methods were tested in an identical way on an Intel(R) Core (TM) i7-10750H CPU @ 2,60GHz PC that deployed the Windows 10 Home Version 22H2 Operating System, and MATLAB R2021a as the programming environment. Algorithm parameters are default values found in the literature and are shown in Table 2. Each method was allowed to perform a maximum of 1000 iterations with 30 search agents and 30 runs. The most optimum convergence results that emanated from the best method over the 30 runs were recorded. The results appeared in the form of these convergence curves and were presented in graphical form as tables based on their best, mean and standard deviation values (SD).

Table 2. Parameters of used algorithms

	Parameter settings
PO	$Mega\_Explor = 0.99,\ Mega\_Exploit = 0.99$
HEOA	$A = 0.6,\ LN = 0.4,\ EN = 0.4,\ FN = 0.1$
GGO	–

4.1 Cantilever Beam Problem

The cantilever beam problem, also known as the cantilever beam optimization problem, is a problem commonly encountered by engineers. There are five variables in this case; x_1, x_2, x_3, x_4 and x_5; and the objective is to identify the minimum values at these variable coordinates. The variables are all integer type, and the variables have been defined limits [37]. The outcomes of the PO, HEOA, and GGO algorithms' solutions to the cantilever beam problem are displayed in Table 3. The decision variables in the solutions are displayed in Table 4. The Fig. 1. provides a schematic of the problem. The mathematical formulations of the problem are given in the equations that follow.

Table 3. Results of cantilever beam problem

	Best	Mean	SD
PO	**1.339956925**	**1.339964869**	**1.04304E-05**
HEOA	1.343324902	1.400264597	0.031274146
GGO	3.803248136	6.11433987	0.914114524

Table 4. Decision variables of the best solution for cantilever beam problem

	Parameters Values					f_{min}
	x_1	x_2	x_3	x_4	x_5	
PO	59.05507961	88.16206437	46.06581546	9.293125861	59.79564374	**1.339956925**
HEOA	90.65449855	59.42304977	91.73289517	51.26896635	74.79054761	1.343324902
GGO	17.14870521	11.00173095	13.01756604	9.490422651	10.29106451	3.803248136

Minimize:

$$f(x) = 0.6224(x_1 + x_2 + x_3 + x_4 + x_5) \tag{1}$$

Subject to:

$$g(x) = \frac{60}{x_1^3} + \frac{27}{x_2^3} + \frac{19}{x_3^3} + \frac{7}{x_4^3} + \frac{1}{x_5^3} - 1 \leq 0 \tag{2}$$

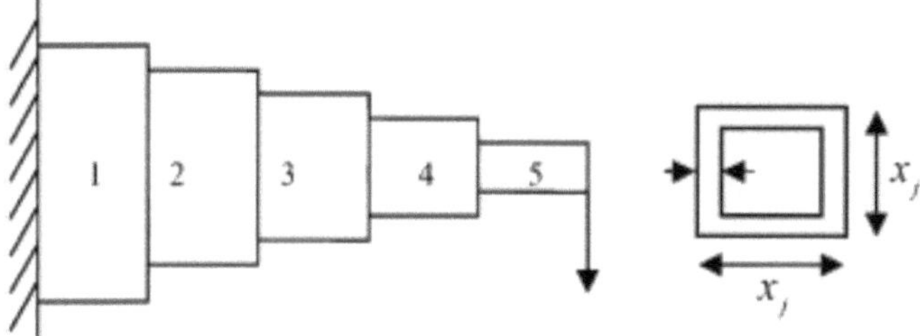

Fig. 1. Cantilever beam

with bounds:

$$0.01 \leq x_1, x_2, x_3, x_4, x_5 \leq 100 \tag{3}$$

As it is shown in Table 3. the best result for the cantilever beam problem was obtained by PO. The characteristics of the results and decision variables are summarized in Table 4.

The best convergence curves found on each data set between 30 runs for comparison are plotted in Fig. 2.

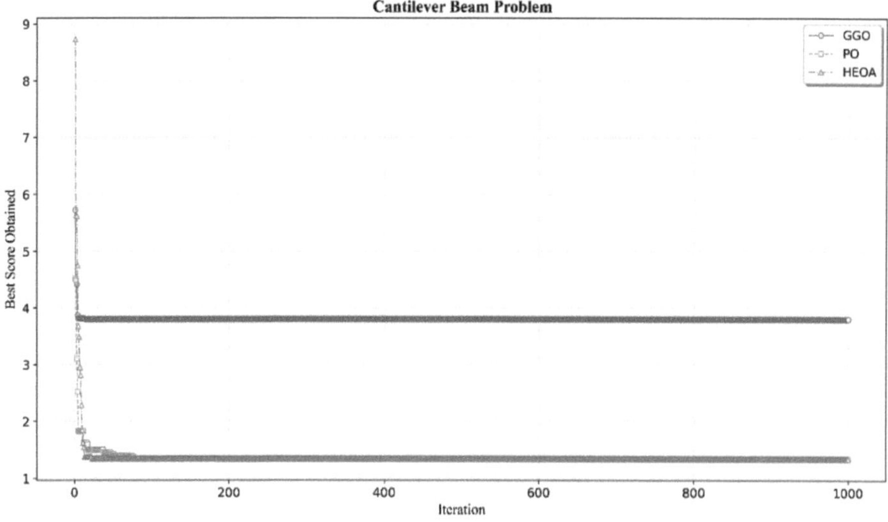

Fig. 2. Convergence curve of the methods used on cantilever beam problem

4.2 Tubular Column Design Problem

The tubular column design problem is an engineering design type problem whose focus is to come up with reinforced structural designs of tubular columns that can effectively retain their shape and true form after being subjected to compressive loading while at the same time achieving least material and cost use. These columns have been employed widely on structures such as construction columns, bridge columns, and various industrial applications due to their high bending, shear and torsional resistance [1]. The outcomes of the PO, HEOA, and GGO algorithms' solutions to the tubular column design problem are displayed in Table 5. The decision variables in the solutions are displayed in Table 6. The Fig. 3. illustrates the type of structure that a uniform tubular column should have in order to be able to bear a compressive force and at the same time remote the stress with the best cost to install the structure. The mathematical formulations of the problem are given in the equations that follow.

Table 5. Results of tubular column design problem

	Best	Mean	SD
PO	**26.48636047**	**26.48636047**	**1.44538E-14**
HEOA	26.50889451	27.80401586	1.943380072
GGO	26.48636048	27.46379482	1.175567656

Table 6. Decision variables of the best solution for tubular column design problem

	Parameters Values		f_{min}
	x_1	x_2	
PO	6.118462014	0.278846652	**26.48636047**
HEOA	5.14254875	0.255276939	26.50889451
GGO	5.452181346	0.291626332	26.48636048

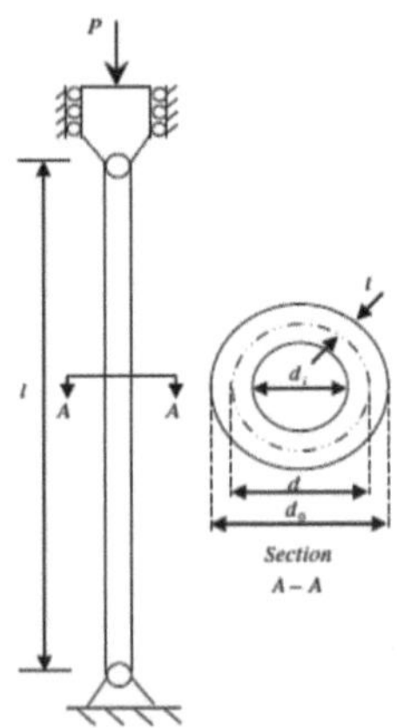

Fig. 3. Tubular column design

Minimize:

$$f(d, t) = 9.8dt + 2d \tag{4}$$

Subject to:

$$g_1 = \frac{P}{\pi dt\sigma_y} - 1 \le 0 \tag{5}$$

$$g_2 = \frac{8PL^2}{\pi^3 Edt(d^2 + t^2)} - 1 \le 0 \tag{6}$$

$$g_3 = \frac{2.0}{d} - 1 \le 0 \tag{7}$$

$$g_4 = \frac{d}{14} - 1 \le 0 \tag{8}$$

$$g_5 = \frac{0.2}{t} - 1 \le 0 \tag{9}$$

$$g_6 = \frac{t}{0.8} - 1 \le 0 \tag{10}$$

As it is shown in Table 5. the best result for the tubular column design problem was obtained by PO. The characteristics of the results and decision variables are summarized in Table 6.

The best convergence curves found on each data set between 30 runs for comparison are plotted in Fig. 4.

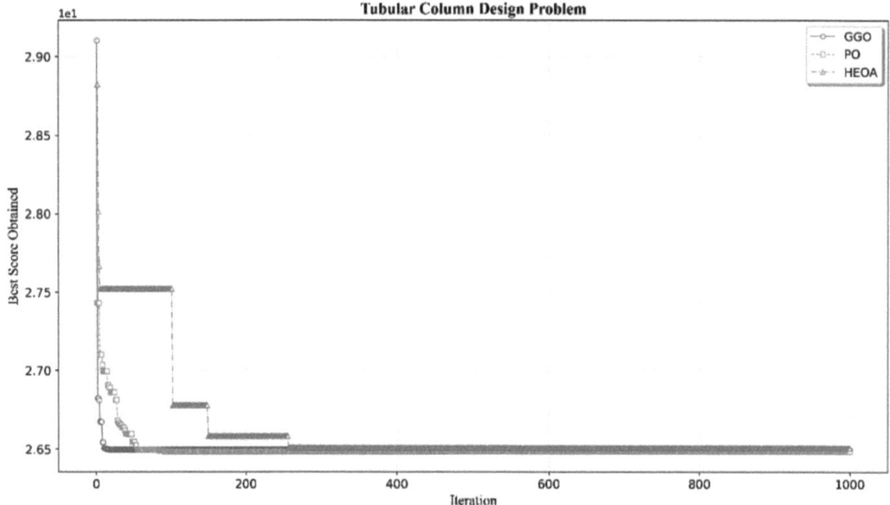

Fig. 4. Convergence curve of the methods used on tubular column design problem

4.3 Corrugated Bulkhead Design Problem

The corrugated bulkhead design is aimed for use in chemical tankers and product tankers, in order to enhance efficiency in the cleaning of the cargo tank in the loading pool. The bulwarks of a tanker must be as thin as it gets [38]. The outcomes of the PO, HEOA, and GGO algorithms' solutions to the corrugated bulkhead design problem are displayed in Table 7. The decision variables in the solutions are displayed in Table 8. This problem has four dimensions: width b, depth h, length l, and plate thickness t. The mathematical formulations of the problem are given in the equations that follow.

Table 7. Results of corrugated bulkhead design problem

	Best	Mean	SD
PO	**6.842957472**	**6.842957472**	**3.61345E-15**
HEOA	6.984966962	7.429865251	0.443646202
GGO	7.384374868	8.584305964	1.077131039

Table 8. Decision variables of the best solution for corrugated bulkhead design problem

	Parameters Values				f_{min}
	x_1	x_2	x_3	x_4	
PO	70.16497311	21.15064668	11.92520754	2.065655324	**6.842957472**
HEOA	98.62828621	21.0477967	63.00824254	0.505248025	6.984966962
GGO	36.1302396	37.39697081	58.82085146	1.069636088	7.384374868

Minimize:

$$f(b, h, l, t) = \frac{5.885t(b + l)}{b + \sqrt{(l^2 - h^2)}} \tag{11}$$

Subject to:

$$g_1 = th\left(0.4b + \frac{1}{6}\right) - 8.94\left(b + \sqrt{(l^2 - h^2)}\right) \geq 0 \tag{12}$$

$$g_2 = th^2\left(0.2b + \frac{1}{12}\right) - 2.2\left(8.94\left(b + \sqrt{(l^2 - h^2)}\right)\right)^{\frac{4}{3}} \geq 0 \tag{13}$$

$$g_3 = t - 0.0156b - 0.15 \geq 0 \tag{14}$$

$$g_4 = t - 0.0156l - 0.15 \geq 0 \tag{15}$$

$$g_5 = t - 1.15 \geq 0 \tag{16}$$

$$g_6 = l - h \geq 0 \tag{17}$$

with bounds:

$$0 \leq b, h, l \leq 100 \; and \; 0 \leq t \leq 5 \tag{18}$$

As it is shown in Table 7. the best result for the corrugated bulkhead design problem was obtained by PO. The characteristics of the results and decision variables are summarized in Table 8.

The best convergence curves found on each data set between 30 runs for comparison are plotted in Fig. 5.

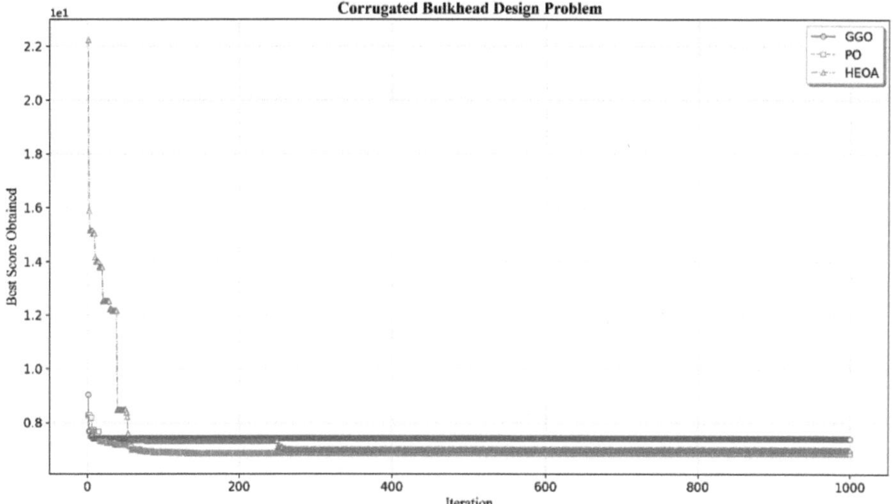

Fig. 5. Convergence curve of the methods used on corrugated bulkhead design problem

5   Conclusion

In recent decades, metaheuristic algorithms have proven to be one of the most effective approaches in solving complex and ill-defined engineering design tasks. Such optimization techniques have been implemented in domestic and foreign practice in such branches as mechanical, structural, and industrial designing, which require sophisticated balances between search and refinement of new solutions, respectively. This investigation puts to the test the newly designed Puma Optimizer (PO), Human Evolutionary Optimization Algorithm (HEOA), and Greylag Goose Optimization (GGO) algorithms in relation to three problems: cantilever beam, tubular column, and corrugated bulkhead design problems.

According to the findings from solutions of the cantilever beam, tubular column, and corrugated bulkhead problems, which were solved using the newly created algorithms PO, HEOA, and the GGO metaheuristic, it can be stated that PO performed the best in the overall evaluation. PO secured the optimal solutions to all three issues on a consistent basis and topped the other algorithms in terms of finding more global optimums and reliability of reaching the optimal solutions. While HEOA offered the second-best solutions to one of the three problems, two out of three problems were solved by GGO to the second-best solution. These findings factors also highlight the ability of PO to combine exploration and exploitation in an effective manner in terms of convergence and speed. HEOA and GGO had some ability in these areas, algorithms may be investigated in different optimization landscapes with the objective of increasing their efficiency in diverse applications by adjusting their parameters. But PO proved to be reliable and robust in the face of complex engineering design problems. These algorithms may be investigated in different optimization landscapes with the objective of increasing their efficiency in diverse applications by adjusting their parameters.

In the future, studies may aim to examine how these algorithms perform on larger-scale or multi-objective problems to establish the scope of applicability and efficiency. Combining different algorithms' strengths in hybridization techniques should also be an area worth looking into regarding improvement. In addition, utilizing dynamic parameter control mechanisms, which modify certain parameters during the optimization procedure, might grant better efficiency to these methods as well. Ultimately, the application of these algorithms to the prevailing ones in uncertain or changing environments would broaden the scope with respect to their practical relevance and assist in modifying them for higher industrial and engineering purposes.

References

1. Martins, J.R., Ning, A.: Engineering Design Optimization. Cambridge University Press (2021)
2. Altay, O.: Chaotic slime mould optimization algorithm for global optimization. Artif. Intell. Rev. **55**(5), 3979–4040 (2022)
3. Deb, K.: Optimization for Engineering Design: Algorithms and Examples. PHI Learning Pvt, Ltd (2012)
4. Altay, O., Altay, E.V.: A novel chaotic transient search optimization algorithm for global optimization, real-world engineering problems and feature selection. PeerJ Comput. Sci. **9**, e1526 (2023)
5. Altay, E.V.: Gerçek dünya mühendislik tasarım problemlerinin çözümünde kullanılan metasezgisel optimizasyon algoritmalarının performanslarının incelenmesi. Int. J. Innovat. Eng. Applic. **6**(1), 65–74 (2022)
6. Altay, E.V., Altay, O.: Güncel metasezgisel optimizasyon algoritmalarının CEC2020 test fonksiyonları ile karşılaştırılması. Dicle Üniversitesi Mühendislik Fakültesi Mühendislik Dergisi. **12**(5), 729–741 (2021)
7. Yang, X.S.: Metaheuristic optimization: algorithm analysis and open problems. In: International Symposium on Experimental Algorithms, pp. 21–32. Springer, Berlin, Heidelberg (2011, May)
8. Abdel-Basset, M., Abdel-Fatah, L., Sangaiah, A.K.: Metaheuristic algorithms: a comprehensive review. In: Computational Intelligence for Multimedia Big Data on the Cloud with Engineering Applications, pp. 185–231 (2018)

9. Xie, L., Han, T., Zhou, H., Zhang, Z.R., Han, B., Tang, A.: Tuna swarm optimization: a novel swarm-based metaheuristic algorithm for global optimization. Comput. Intell. Neurosci. **2021**(1), 9210050 (2021)

10. Dehghani, M., Hubálovský, Š., Trojovský, P.: Northern goshawk optimization: a new swarm-based algorithm for solving optimization problems. IEEE Access. **9**, 162059–162080 (2021)

11. Neshat, M., Sepidnam, G., Sargolzaei, M.: Swallow swarm optimization algorithm: a new method to optimization. Neural Comput. & Applic. **23**(2), 429–454 (2013)

12. Dehghani, M., Hubálovský, Š., Trojovský, P.: Cat and mouse based optimizer: a new nature-inspired optimization algorithm. Sensors. **21**(15), 5214 (2021)

13. Holland, J.H.: Genetic algorithms. Sci. Am. **267**(1), 66–73 (1992)

14. Storn, R., Price, K.: Differential evolution–a simple and efficient heuristic for global optimization over continuous spaces. J. Glob. Optim. **11**, 341–359 (1997)

15. Taheri, A., RahimiZadeh, K., Beheshti, A., Baumbach, J., Rao, R.V., Mirjalili, S., Gandomi, A.H.: Partial reinforcement optimizer: an evolutionary optimization algorithm. Expert Syst. Appl. **238**, 122070 (2024)

16. Cheraghalipour, A., Hajiaghaei-Keshteli, M., Paydar, M.M.: Tree growth algorithm (TGA): a novel approach for solving optimization problems. Eng. Appl. Artif. Intell. **72**, 393–414 (2018)

17. Erol, O.K., Eksin, I.: A new optimization method: big bang–big crunch. Adv. Eng. Softw. **37**(2), 106–111 (2006)

18. Hashim, F.A., Hussain, K., Houssein, E.H., Mabrouk, M.S., Al-Atabany, W.: Archimedes optimization algorithm: a new metaheuristic algorithm for solving optimization problems. Appl. Intell. **51**, 1531–1551 (2021)

19. Ghannadi, P., Kourehli, S.S.: Multiverse optimizer for structural damage detection: numerical study and experimental validation. Struct. Design Tall Spec. Build. **29**(13), e1777 (2020)

20. Kaveh, A., Dadras, A.: A novel meta-heuristic optimization algorithm: thermal exchange optimization. Adv. Eng. Softw. **110**, 69–84 (2017)

21. Purnomo, H.D., Wee, H.M.: Soccer game optimization: an innovative integration of evolutionary algorithm and swarm intelligence algorithm. In: Meta-Heuristics Optimization Algorithms in Engineering, Business, Economics, and Finance, pp. 386–420. Igi Global (2013)

22. Dehghani, M., Trojovský, P.: Teamwork optimization algorithm: a new optimization approach for function minimization/maximization. Sensors. **21**(13), 4567 (2021)

23. Razmjooy, N., Sheykhahmad, F.R., Ghadimi, N.: A hybrid neural network–world cup optimization algorithm for melanoma detection. Open Med. **13**(1), 9–16 (2018)

24. Dehghani, M., Montazeri, Z., Givi, H., Guerrero, J.M., Dhiman, G.: Darts game optimizer: a new optimization technique based on darts game. Int. J. Intell. Eng. Syst. **13**(5), 286–294 (2020)

25. Rao, R.V., Savsani, V.J., Vakharia, D.P.: Teaching–learning-based optimization: a novel method for constrained mechanical design optimization problems. Comput. Aided Des. **43**(3), 303–315 (2011)

26. Mousavirad, S.J., Ebrahimpour-Komleh, H.: Human mental search: a new population-based metaheuristic optimization algorithm. Appl. Intell. **47**, 850–887 (2017)

27. Xu, Y., Cui, Z., Zeng, J.: Social emotional optimization algorithm for nonlinear constrained optimization problems. In: Swarm, Evolutionary, and Memetic Computing: First International Conference on Swarm, Evolutionary, and Memetic Computing, SEMCCO 2010, Chennai, India, December 16–18, 2010. Proceedings 1, pp. 583–590. Springer, Berlin, Heidelberg (2010)

28. Shi, Y.: An optimization algorithm based on brainstorming process. In: Emerging Research on Swarm Intelligence and Algorithm Optimization, pp. 1–35. IGI Global (2015)

29. Abualigah, L., Elaziz, M.A., Khasawneh, A.M., Alshinwan, M., Ibrahim, R.A., Al-Qaness, M.A., et al.: Meta-heuristic optimization algorithms for solving real-world mechanical engineering design problems: a comprehensive survey, applications, comparative analysis, and results. Neural Comput. Applic. **34**, 4081–4110 (2022)
30. Altay, E.V., Altay, O., Özçevik, Y.: A comparative study of metaheuristic optimization algorithms for solving real-world engineering design problems. CMES-Comput. Model. Eng. Sci. **139**(1), 1039–1094 (2024)
31. Abdollahzadeh, B., Khodadadi, N., Barshandeh, S., Trojovský, P., Gharehchopogh, F.S., El-kenawy, E.S.M., et al.: Puma optimizer (PO): A novel metaheuristic optimization algorithm and its application in machine learning. Clust. Comput. **27**, 5235–5283 (2024)
32. Lian, J., Hui, G.: Human evolutionary optimization algorithm. Expert Syst. Appl. **241**, 122638 (2024)
33. El-Kenawy, E.S.M., Khodadadi, N., Mirjalili, S., Abdelhamid, A.A., Eid, M.M., Ibrahim, A.: Greylag goose optimization: nature-inspired optimization algorithm. Expert Syst. Appl. **238**, 122147 (2024)
34. Zhao, Q., Liu, Y., Wang, L., Yang, H., Cao, D.: Design method for piezoelectric cantilever beam structure under low frequency condition. Int. J. Pavem. Res. Technol. **11**(2), 153–159 (2018)
35. Kvocak, V., Kanishchev, R., Platko, P., Hodovanets, E., Al Ali, M.: Stability of steel columns with concrete-filled thin-walled rectangular profiles. Sustain. For. **15**(23), 16217 (2023)
36. Ko, D.E., Shin, S.H.: A study on the optimum design of corrugated bulkhead for product carrier. In IOP Conference Series: Materials Science and Engineering (Vol. 269, 1, p. 012084). IOP Publishing (2017, November)
37. Yucel, M., Bekdas, G., Nigdeli, S.M., Sevgen, S.: Artificial neural network model for optimum design of tubular columns. Int. J. Theor. Appl. Mech. **3**, 82–86 (2018)
38. Shin, S.H., Ko, D.E.: A study on minimum weight design of vertical corrugated bulkheads for chemical tankers. Int. J. Naval Archit. Ocean Eng. **10**(2), 180–187 (2018)

Comparison of Feasible Initial Solution Methods for the Knapsack Problem

Halil İbrahim Ayaz[1]([⊠]) [iD], Belkız Torğul[2] [iD], and Turan Paksoy[1] [iD]

[1] Necmettin Erbakan University, Konya, Turkey
hiayaz@erbakan.edu.tr
[2] Fırat University, Elazığ, Turkey

Abstract. The Knapsack Problem (KP) is a combinatorial optimization challenge that involves selecting a subset of items with specific weights and values to maximize an objective without exceeding a weight limit. As an NP-hard problem, it has theoretical significance and real-world applications like resource allocation and project selection. Solving KP often relies on metaheuristic algorithms, which benefit from high-quality initial solutions generated using heuristic approaches such as greedy algorithms, random selection with feasibility checks, or hybrid strategies. However, generating feasible initial solutions becomes increasingly difficult in large-scale cases with numerous alternatives and budget constraints. This study systematically compares initial solution generation methods for KP under such constraints. Using a tailored dataset for project selection, we evaluate the convergence performance of the artificial bee colony algorithm with different initialization strategies. The study examines how initial solutions influence convergence rates and final results. Key findings show that the Random method works well for small datasets (10 projects) but struggles with larger ones, failing at 100+ projects. The Random with Feasibility Check method consistently produces near-optimal solutions. The Repair method ensures feasibility but increases computation time, while the MaxMin method offers greater efficiency with fewer iterations and reduced computation time. These insights are valuable for tackling large-scale, budget-constrained optimization problems.

Keywords: Initial Solution Generation · Project Selection · Knapsack Problem · Metaheuristic Methods

1 Introduction

The Knapsack Problem is one of the most extensively researched combinatorial problems. There are numerous variations of the problem, as well as a multitude of real-world applications. The objective of the knapsack problem is to select a subset of available items with the maximal total weight while ensuring that the sum of the weights does not exceed a given capacity [1]. The knapsack problem is one of the most frequently studied problems in operations research. Knapsack problems are used in various ways to solve real-world problems in multiple fields, such as economics, engineering, and business, including cryptography and applied mathematics [2].

A. Mirzazadeh et al. (Eds.): ODSIE 2024, CCIS 2482, pp. 369–388, 2026.
https://doi.org/10.1007/978-3-031-93601-2_23

The classical knapsack is the 0–1 knapsack, also known as the binary knapsack. Nevertheless, the family of knapsack problems is extensive, encompassing a multitude of variations. For example, in economics, a real-world application might involve evaluating a set of projects, each consisting of multiple components, under a set of constraints—such as available resources. The objective is to identify a subset of the n projects that maximizes profit while remaining within the constraints of the corresponding resource limits. This variant of the knapsack problem has been the subject of extensive research, leading to the development of both exact algorithms and heuristic methods for its resolution. Exact algorithms encompass dynamic programming techniques, branch and bound approaches, and systematic strategies. In large-scale instances, a range of heuristic optimization techniques are employed to attain optimal solutions, including simulated annealing, tabu search, and genetic algorithms [1].

A heuristic approach to a problem generally exhibits four characteristics. First, by definition, the heuristic does not guarantee that an optimal solution will be generated. Second, a heuristic finds "good" solutions relatively quickly. Third, the quality of the solution cannot be measured without a given or internally generated lower/upper bound. Finally, a time, iteration, or quality limit is usually imposed to terminate the heuristic process [3].

The efficacy of the heuristic algorithms is evaluated based on the expected value of the optimal solution that can be attained within a specified computational time frame. The computational results show that using initial solutions generated by a construction heuristic, rather than random initialization, often improves the performance of edge exchange heuristics. However, this improvement is contingent upon the specific edge exchange heuristic employed, the characteristics of the problem, and the available computing time. The feasible initial solution may be generated through the application of a construction heuristic or it can be randomly generated [4].

The initial basic feasible solution represents a significant step towards achieving an optimal solution to the problem. However, existing methods for generating the initial basic feasible solution do not always produce a feasible solution that minimizes the number of iterations needed to reach the optimal solution. The value of the initial basic feasible solution directly impacts the optimal solution; therefore, an improved method for generating the initial basic feasible solution is crucial and highly preferred. Consequently, there is still a need to develop a more effective method of the initial basic feasible solution [5, 6].

In particular project selection problems, when there is a large number of potential projects and a limited budget, the random generation of the initial solution is prone to difficulties in producing suitable solutions. In this study, we compare the initial solution generation methods (random, random method with feasibility check, MaxMin, and base approaches) in knapsack problems where the budget is low and the number of alternative projects is high. We also discuss the convergence speeds of the compared methods to the optimum value and the effect of the first solutions obtained with different methods on the quality of the results.

This paper is organized as follows: Sect. 2, give a brief literature about artificial bee colony algorithm, knapsack problem, and initial solutions. Materials and methods are given in Sect. 3. The experimental results are discussed in Sect. 4. The conclusion of the paper is presented in Sect. 5.

2 Literature Review

The knapsack problem is a well-established combinatorial optimization problem with significant implications in various fields, particularly project selection and resource allocation. The essence of the knapsack problem lies in its objective to maximize the total value of selected items without exceeding a specified capacity. This problem has been extensively studied, leading to the development of various algorithms and methodologies aimed at finding optimal or near-optimal solutions.

One of the notable approaches to solving the knapsack problem is through heuristic algorithms, such as the Artificial Bee Colony (ABC) algorithm. Hasoon's work highlights the application of the ABC algorithm for project selection, demonstrating its effectiveness in maximizing profits while adhering to budget constraints. The results indicated that the ABC algorithm outperformed traditional methods, such as genetic algorithms, in terms of precision and speed, making it a viable option for financial applications in project selection [7]. This aligns with the broader trend in optimization literature, where heuristic methods are increasingly favored for their efficiency in handling NP-hard problems like the knapsack problem.

In addition to heuristic approaches, variations of the knapsack problem have been explored, such as the Set-Union Knapsack Problem (SUKP). Wei and Hao introduced an iterated two-phase local search algorithm designed explicitly for the SUKP, which generalizes the classical 0–1 knapsack problem. Their findings suggest that this approach effectively maximizes profit while managing weight constraints, thereby providing a robust framework for project selection scenarios where multiple items are grouped into subsets [8]. This reflects the versatility of knapsack problem formulations in addressing complex decision-making scenarios.

Moreover, the dynamic version of the knapsack problem has garnered attention for its applicability in real-time decision-making contexts. The work by Dizdar et al. emphasizes revenue maximization in dynamic settings, where items arrive sequentially, and decisions must be made on the fly. This dynamic aspect is crucial for project selection in environments characterized by uncertainty and variability, as it allows for adaptive strategies that can respond to changing conditions [9].

The integration of graph theory with the knapsack problem has also been explored, particularly in resource allocation contexts. Köse and Özbek's research on device-to-device communications illustrates how maximal independent sets can be utilized within the framework of the knapsack problem to optimize resource allocation in cellular systems. This innovative approach underscores the potential for interdisciplinary applications of the knapsack problem, extending its relevance beyond traditional optimization contexts [10].

Furthermore, the Multiple Knapsack Problem (MKP) has been recognized for its practical implications in project selection and resource management. Setiawan et al. noted that MKP can model a variety of real-world scenarios, including capital budgeting and cargo loading, thereby reinforcing its significance in operational research [11]. The

MKP's adaptability to various constraints and objectives makes it a powerful tool in decision-making processes across different sectors.

In summary, the knapsack problem remains a fundamental focus in optimization research, with broad applicability in fields like project selection and resource allocation. The evolution of algorithms, from heuristic methods to dynamic programming and graph-based approaches, reflects the ongoing efforts to address the complexities of this NP-hard problem. As researchers continue exploring innovative solutions, the knapsack problem will undoubtedly maintain its relevance in theoretical and practical domains.

2.1 Project Selection as Knapsack Problem

The selection and scheduling of projects within transport infrastructure and other sectors have garnered significant attention in recent literature, particularly regarding the methodologies employed to optimize these processes. Ječmen's work highlights the variability in transport infrastructure development across different countries despite similar funding levels, suggesting that strategic allocation of resources plays a crucial role in project outcomes [12]. This observation aligns with the broader discourse on project selection, where the balance between efficiency and fairness is a recurring theme. Naldi emphasizes the importance of developing models that address this trade-off, proposing that effective project selection must consider both profit maximization and equitable resource distribution [13].

In the context of project portfolio management, Schiffels provides insights into the behavioral aspects influencing decision-making. His findings indicate that decision-makers often prioritize projects based on perceived value and resource requirements, which can lead to suboptimal selections if not managed properly [14]. This is particularly relevant in environments where cooperation among stakeholders is essential, as the quality of collaboration can significantly impact the decision-making process. Furthermore, Schiffels' experimental studies reveal that many organizations lack structured selection processes, which can hinder their ability to optimize project portfolios effectively [14].

The methodologies for project selection have evolved, with various approaches being proposed to enhance decision-making. For instance, Bakirli introduces a fuzzy multi-objective model that employs meta-heuristics to solve complex project selection problems, demonstrating that traditional methods may not always yield optimal results [15]. This is echoed in Harrison's survey of defense sector applications, which reveals a predominance of exact solvers like integer linear programming, despite the NP-hard nature of these optimization problems [16]. The need for robust methodologies is further underscored by Mavrotas, who discusses integrating robustness analysis into multi-objective optimization frameworks, thereby enhancing the reliability of project selection outcomes [17].

The knapsack problem (KP) is a foundational model in project selection, with various adaptations being explored in the literature. Ječmen notes that both the multiple knapsack problem (MKP) and the multidimensional knapsack problem share similar solving algorithms, which can be leveraged across different applications [12]. This versatility is crucial, as it allows for developing tailored solutions that can address specific project selection challenges. Additionally, Chang's work highlights the limitations of traditional knapsack formulations, suggesting that they may not fully account for the ranking of projects based on their objective values, thus necessitating more nuanced approaches [18].

In the realm of project portfolio selection (PPS), integrating fairness and equity considerations has gained significant attention. The literature reveals various methodologies and criteria that aim to balance project selection with equitable resource allocation. One prominent approach is the Maximin criterion, employed as a fairness indicator in budget allocation scenarios. This criterion seeks to minimize the maximum loss among all projects, promoting a more equitable distribution of resources across departments or categories. The application of the Maximin criterion is particularly relevant in contexts where perceived fairness can influence operational outcomes and employee satisfaction [19].

Moreover, the Gini coefficient has become a valuable metric for assessing income disparities and equity in project selection. Its application extends beyond economic analysis to include resource allocation in various domains, such as telecommunications and environmental management. The Gini coefficient's versatility in measuring equity underscores its importance in project portfolio selection, where the goal is to ensure that resources are allocated in a manner that minimizes disparities among competing projects [19].

In addition to fairness metrics, the literature also discusses the role of expert evaluations in project selection. The process often involves aggregating assessments from multiple experts to form a comprehensive utility function that reflects the collective judgment on project viability [20]. This aggregation can be performed using various statistical methods, such as the geometric or arithmetic mean, which help in deriving a compromise solution that aligns with the preferences of all stakeholders involved [20]. The complexity of this process is further compounded by the need to consider interdependencies among projects, which can significantly affect the overall portfolio performance [19].

The binary (KP) is a foundational model for understanding project selection dynamics. In this context, the challenge lies in selecting a subset of projects that maximizes overall profit while adhering to resource constraints [21]. The KP framework has been adapted to address multi-agent scenarios, where different stakeholders vie for limited resources, necessitating a balance between individual and collective interests [19]. This adaptation highlights the intricate interplay between optimization and fairness in project portfolio selection.

The literature reveals a complex interplay between efficiency, fairness, and decision-making in project selection. Integrating advanced methodologies, such as genetic algorithms and meta-heuristics, alongside behavioral insights, provides a comprehensive framework for optimizing project portfolios. As organizations continue to navigate resource allocation challenges, the ongoing development of robust and equitable models will be essential for achieving sustainable outcomes in project management.

2.2 Initial Solution Generation Methods

Heuristic algorithms require an initial feasible solution to begin solving the problem. Several solution techniques have been proposed in the literature to generate this initial solution. [3] developed a specialized branch-and-bound algorithm using linear programming cuts, feasible-solution generators, Lagrangean relaxation, and subgradient optimization. Computational results showed it effectively solved "hard" problems with up to 3000 binary variables. An unanticipated benefit of this algorithm was its ability to generate high-quality feasible solutions early, often outperforming two recently published heuristics in both solution quality and computation time. [5] developed a better polynomial-time heuristic solution technique to obtain an improved initial feasible solution to the transportation problem. This heuristic was implemented using C++ programming. Comparative studies were conducted to evaluate its performance against the best available methods in the literature, using results from several numerical problems. The findings show that the proposed heuristic consistently leads to the minimal total cost solution in most cases, achieving optimal results in 88.89% of the studied problems. [22] proposed two initial solution algorithms for use with a metaheuristic technique (simulated annealing) and tested them against other published algorithms. The algorithms were developed as modules for flexibility in using various metaheuristics. The proposed algorithms, tested on a 37-bus distribution network with distributed resources such as electric vehicles, achieved results within 0.1% of the optimal solution. In comparison, a deterministic technique took around 26 hours to find the optimal solution, while simulated annealing achieved results in about 1 minute. [23] explored initial solution choices for fitting a phase-type distribution (PH) using the expectation maximization (EM) method. Tested with four-state PHs, numerical results showed that the EM method converges faster from various initial solution structures.

[24] identified limitations and computational errors in Vogel's Approximation Method (VAM) and developed an Improved Vogel's Approximation Method (IVAM) to address these issues. IVAM was compared to VAM on a numerical example, resulting in a 2.27% reduction in total transportation cost. Compared with twelve other methods, IVAM produced optimal solutions in seven cases, better solutions in four, and the same solution in one. A statistical analysis of 1500 randomly generated transportation problems showed that IVAM outperformed VAM in 71.8% of cases, particularly for larger problems, by providing better initial feasible solutions to the transportation problem. [25] introduced the variable-sized bin packing problem with time windows, a real-world logistics issue. The problem was first formulated as an integer programming model, followed by the proposal of three independent heuristics: a modified best-fit heuristic for an initial solution, an iterative local search using the shortest path decoder, and a primal heuristic based on column generation. Two categories of instances were used to test the approach, and the results show that the proposed methods effectively solve the problem, yielding near-optimal solutions in a short time. [26] developed a mathematical model and proposed a hybrid algorithm combining gray wolf optimization and invasive weed optimization for the flexible job shop scheduling problem, aiming to minimize makespan. The invasive weed optimization algorithm is integrated into the gray wolf algorithm to enhance both global and local search capabilities. An effective initial population generation mechanism is also introduced to improve the initial solution quality. Additionally,

a variable neighborhood search structure is proposed, adapting dynamically during the search. Experimental results demonstrate that the proposed algorithm outperforms other improved gray wolf algorithms in terms of effectiveness and efficiency across various benchmark instances.

[6] proposed a new method for obtaining the initial basic feasible solution of the transportation problem, called the Bilqis Chastine Erma (BCE) method. They evaluated the performance of this proposed method using thirty-five numerical examples and compared it with other existing methods. The experimental results indicated that BCE achieved a lower total minimal cost than VAM, TDM1, TOCM-MT, and JHM, while also reaching the fastest solving time. BCE achieved the optimal solution for thirty-one out of the thirty-five examples (88.57%). [27] proposed initialization and diversification strategies to enhance the local search for the weighted partial maximum satisfiability problem. The initialization strategy uses a novel structural entropy definition to generate a solution close to a high-quality feasible one. The diversification strategy selects variables based on the best benefits or by focusing on clauses with the greatest penalty, choosing variables probabilistically. They developed a local search solver, , and a hybrid solver, -. Experimental results on recent MaxSAT Evaluations show that these solvers perform better or similarly to state-of-the-art competitors. [28] reviewed research on the Multi-Trip Vehicle Routing Problem with Time Windows (MTVRPTW) and proposed a two-stage heuristic algorithm. This approach generates multiple high-quality initial solutions using greedy and ILS algorithms in the first stage and then optimizes the total distance using the VNS algorithm in the second stage. [29] proposed a Hybrid Evolution Strategies-Simulated Annealing (HES-SA) algorithm to minimize makespan in job shop scheduling. The Evolution Strategies (ES) algorithm generates an initial solution with multi-mutation operators to improve performance. To avoid local minima and enhance local search, the ES algorithm is combined with Simulated Annealing (SA). The HES-SA algorithm was tested on benchmark problems and compared with other techniques, showing superior performance in terms of makespan values.

3 Materials and Methods

In this study, the effects of the initial solution for the artificial bee algorithm developed by Karaboga [30] were measured. Four different initial solution generation algorithms were used; these methods are given in detail below. The material and methods of the artificial bee colony algorithm used as a metaheuristic algorithm are given in detail in the following section.

3.1 Artificial Bee Colony

The Artificial Bee Colony (ABC) algorithm is designed to tackle intricate optimization challenges by mimicking the behaviors of worker and scout bees found in honey bee colonies. The process begins with randomly placing employed bees throughout the search space. Each iteration of the algorithm encompasses three distinct phases: the worker bee phase, the scout bee phase, and the foraging phase. In the employed bee phase, individual worker bees explore potential new solutions near their current solutions. During this exploration, a bee evaluates its source alongside a randomly chosen source from other known solutions, generating a new candidate solution between the existing and the new sources. If this new solution proves superior, the bee adopts it. The subsequent phase involves scout bees, which utilize a probabilistic method to assess the quality of various sources, favoring those that yield better solutions. Each scout bee then investigates its selected source, determining if it is an improvement over its previous source. The scout bee will switch to this new source if a better source is found. If no improvement is observed over a set period, an explorer bee will abandon its current source and randomly select a new one. These three phases are iteratively repeated for a predetermined number of iterations or until a specified CPU time is reached.

After identifying the random food sources necessary for the initial population, the scout bees return to the hive to relay critical information regarding the nectar's quantity, quality, and the geographical coordinates of the visited food sources to the worker bees through a behavior known as the waggle dance. During this dance, the duration of the waggle indicates to the worker bees that a longer performance correlates with a more fruitful food source. The process by which a employed bee discovers a new food source in proximity to the current one can be mathematically represented by Eq. (1).

$$v_{ij} = x_{ij} + \phi_{ij}\left(x_{ij} - x_{kj}\right) \tag{1}$$

Each worker bee is designated to explore several food sources, returning to the hive after investigating only one of these sources for nearby neighbors. The collective knowledge regarding the identified food source is then shared among all employed bees and scout bees. The scout bee selects a food source based on a probability that is directly proportional to the quantity of nectar available. This probability is determined using a specific formula outlined in Eq. (2)

$$p_i = \frac{fit_i}{\sum_{n=1}^{SN} fit_n} \tag{2}$$

Pseudo-code for the artificial bee colony algorithm is provided in Algorithm 1. Different initial solutions approach are used for the first step of the algorithm. These various approaches are given in the next section.

Algorithm 1. Pseudo-code for Artificial Bee Colony Optimization

```
Initialize the population of solutions (food sources) randomly
Evaluate the fitness of each solution
Set cycle = 1
While cycle <= Maximum_Cycles:
# Employed Bee Phase
For each employed bee:
    Generate a new solution by modifying the current solution
    Evaluate the new solution
    Apply greedy selection between the new and old solutions
# Onlooker Bee Phase
Calculate the probability of selecting each food source based on its
fitness
For each onlooker bee:
    Select a solution based on the probability
    Generate a new solution by modifying the selected solution
    Evaluate the new solution
    Apply greedy selection
# Scout Bee Phase
If any solution is not improved for a certain limit:
    Replace it with a new random solution
    Memorize the best solution found so far
    cycle = cycle + 1
Return the best solution found
```

3.2 Initial Solution Generation Methods

For metaheuristic algorithms, the quality of the initial solution is essential when the budget constraint is low and the number of alternative projects is high. In this section, four different initial solution generation methods are described in detail.

The Random Method. The Random Method is a simple but effective for starting from any point in the solution space of the knapsack problem. In this approach, a subset of the available items is randomly selected and checked to see if this subset fits the capacity of the knapsack. The random method allows the algorithm to initially explore different regions of the solution space, which allows for a broader search in various metaheuristics. However, a disadvantage of the random method is that the quality of the initial solution cannot be guaranteed, so the solution is often treated as raw data that needs to be optimized. The random method is beneficial when the solution space is large, and it is impossible to rely on specific initial knowledge. Metaheuristic algorithms aim to improve this random solution in an iterative process to arrive at better solutions.

Random Method with Feasibility Check. Projects are selected one by one. Here, a project is chosen randomly from the list of projects and added to the portfolio. After each selection process, it is checked whether the cost of the total selected projects exceeds the budget constraint. If an added project exceeds the budget constraint, the last added project is removed, and a feasible initial solution is obtained. Pseudo code of this method is given Algorithm 2.

Algorithm 2. Pseudo-code for Random Method with Feasibility Check

```
While cost(selected_project)<budget
    Choose random project from list
    Add this project to selected_project
End
Remove last added project
```

Repair Method. The initial solutions required for the metaheuristic algorithm are generated randomly. Then, it is checked whether the generated solution exceeds the budget constraint. A random item in the portfolio is removed from the selected problem set if it does. This process is continued until the budget constraint is met. The resulting feasible solution is given to the metaheuristic algorithm to be developed. Pseudo code of this method is given Algorithm 3.

Algorithm 3. Pseudo-code for Repair Method

```
While cost(selected_project)>budget
    Choose random project from list
    Remove this project to selected_project
End
```

MaxMin Method. The number of projects contained in the initial randomly generated solutions is generated within certain lower and upper limits, considering the projects in the list. While determining the lower limit of the number of projects to be included in the solution, the projects with the highest cost from the projects ranked descendingly according to the project cost are added to the portfolio in order. When the budget limit is exceeded, the last project added is removed and the number of projects included determines the lower limit. When determining the upper limit, the least costly project is added to the portfolio. In the same way, projects are continued to be added until the budget limit is exceeded. When the budget limit is exceeded, the last project added is removed, and the total number of projects in the portfolio is used to meet the upper limit. Pseudo code of this method is given Algorithm 4.

Algorithm 4. Preudo-code for MaxMin Method

```
Determine upper limit considering low cost projects
Determine lower limit considering high cost projects
Generate random number between upper and lower limit
Generate a initial solution considering generated random
number
```

Comparison of Methods. The random method generates initial solutions randomly without any additional steps. It requires less time compared to other methods. However, as the number of alternative projects increases, the number of infeasible solutions increases. The random method with feasibility check and the repair method guarantee to produce a feasible solution in the end. However, additional steps are required to obtain a feasible solution. This may extend the running time of the algorithm. Finally, according to the MaxMin method, the feasibility rate of the initial solutions produced within certain limits is high. It can work faster than the random method and repair method with conformity control.

4 Experimental Evaluation

In this study, four different methods are used to generate initial solutions for metaheuristic algorithms. The dataset used in the study was randomly generated, and the information for the dataset is given in the next section. The artificial bee colony was used as the metaheuristic algorithm. The parameters used in the artificial bee colony are given in detail below.

4.1 About Experiences

A dataset of 1000 projects is generated, each with randomly assigned costs and profits.

Profits are generated from a uniform distribution between 10,000 and 50,000, while costs are set to be 60% of the profit. The total budget available for selecting these projects is capped at 100,000 units. To evaluate different scenarios, feasible solutions are selected by considering subsets of varying sizes, including 10, 20, 30, 40, 50, 100, 250, 500, and all 1000 projects. The selection process aims to maximize profitability while ensuring that the total cost of chosen projects remains within the budget limit. This approach allows for analysis of trade-offs between the number of projects considered and the overall profitability under budget constraints.

The Artificial Bee Colony (ABC) algorithm is a swarm-based optimization technique inspired by the foraging behavior of honey bees. The algorithm parameters are crucial for its performance. In this setup, the maximum number of iterations, **MaxIt**, is set to 250, which defines how long the algorithm will run. The **nScoutBee** parameter specifies 25 scout bees responsible for exploring new solutions. Among the potential sites, **nEliteSite** (4 sites) represents the best-performing locations, while **nBestSite** (20 sites) includes additional promising locations. For exploitation, **nEliteSiteBee** allocates 300 recruited bees to each elite site, while **nBestSiteBee** assigns 100 recruited bees to each of the best sites. This balance of exploration and exploitation aims to ensure efficient convergence to optimal solutions.

Evaluation metrics play a crucial role in assessing the performance of optimization algorithms. Two key metrics are the *Convergence Rate* and *Solution Quality*. The *Convergence Rate* measures the speed at which the algorithm approaches the optimal solution, providing insights into its efficiency and time required to achieve results. Meanwhile, *Solution Quality* evaluates the value of the final solution relative to the optimal solution, highlighting the algorithm's effectiveness and accuracy. Together, these metrics comprehensively understand an algorithm's overall performance and suitability for specific problem domains. These experiments were performed on a computer with Intel(R) Core(TM) i5–2400 CPU @ 3.10GHz 3.10 GHz processor and 16 Gb RAM. This computer has windows 10 operating system.

4.2 Computational Results

In this study, 1000 projects with random cost and profit values were generated. A problem set was formed by selecting the first ten projects from the projects. Then, problem sets were formed by selecting the first 20, 30, 40, 50, 100, 250, 500, and all projects. For each problem set, the optimum values considering cost and profit are calculated with GAMS optimization software and given in Table 1.

Table 1. Optimum Results of Problems.

# of available projects	10	20	30	40	50	100	250	500	1000
Optimum value	66,500	66,672	66,673	66,674	66,675	96,973	96,977	96,977	110,064

The problem sets were solved by using the same metaheuristic algorithm with four different initial solution algorithms considering different number of projects. For the evaluation of the results, the number of iterations to reach the final solution, the final solution, and the running time of the algorithm were recorded.

The results of the random method are given in Table 2. For the dataset containing ten projects, optimal solutions were obtained. As the number of projects increased, the optimal solution was not obtained in all trials in the problem sets where the number of projects was 30, 40, and 50. No solution was obtained for problem sets with 100 or more projects. The reason for this is that as the number of projects increased, the first solutions were unsuitable.

Table 2. Random Method.

# of available project	Iteration to the final result	Final result	Average time
10	2,17	66,500	7,73
20	38,38	66,666,12	7,66
30	53,81	31,224,58	7,53
40	90	66,650	7,46
50	2,17	66,500	7,73
100	No solution found		
250	No solution found		
500	No solution found		
1000	No solution found		

The random method with feasibility check, it was ensured that the first solutions were feasible. For all problem sets, values close to the optimal value were generated and solutions were obtained in all iterations. The results of this method are presented in Table 3.

Table 3. Random Method with Feasibility Check.

# of available project	Iteration to final result	Final result	Average time
10	1,14	66,500	7,74
20	11,17	66,666,78	7,73
30	9	66,668,11	7,71
40	18,7	66,668,34	7,73
50	27,35	66,669,37	7,79
100	53,46	96,963,12	7,72
250	61,53	96,963,64	7,88
500	82,09	96,964,1	9,28
1000	169,9	101,681,69	10,77

With the Repair method, infeasible solutions were made suitable by removing the projects one by one from the selected problem set at each step. When the number of projects is high, the running time of the algorithm may increase since it is necessary to remove a large number of projects. For example, when the number of projects is 1000, the average number of randomly selected projects is 500. In this case, 490 projects must be removed on average for the solution to be feasible. In cases where the number of projects is much higher, this process will significantly impact the running time of the algorithm. The results of the method are presented in Table 4. As in the previous method, near-optimal values were obtained, but the results' quality is relatively lower than the previous method. At the same time, the running time of the method is slightly higher.

In the MaxMin method, the number of projects to be selected in randomly generated solutions is restricted. In this way, while increasing the number of feasible solutions, the process of adding or subtracting one by one, as in the two previously mentioned methods, is avoided. Here, the aim is to increase the number of feasible solutions while shortening the time needed to generate the initial solutions. As can be seen from Table 5, values close to the optimal value were obtained. The number of steps required to reach the final value has increased relatively. However, when it is considered for the time, it is possible to talk about a relative decrease.

4.3 Discussion on Results

The results of the study demonstrate the distinct performance characteristics of the tested methods across different dataset sizes. The Random method, while effective for smaller datasets, such as the 10-project problem set where optimal solutions were consistently

Table 4. Repair Method.

# of available project	Iteration to final result	Final result	Average time
10	1,66	66,500	7,61
20	11,13	66,666,21	7,57
30	8,7	66,667,55	7,57
40	15,58	66,668,62	7,59
50	22,23	66,669,28	7,66
100	67,3	96,962,75	7,88
250	70,52	96,963,8	8,47
500	78,36	96,964,06	9,53
1000	162,43	101,638,11	11,69

Table 5. MaxMin Method.

# of available project	Iteration to final result	Final result	Average time
10	2,68	66,500	7,46
20	17,78	66,665,67	7,60
30	13,95	66,664,86	7,59
40	20,59	66,668,08	7,60
50	24,99	66,668,05	7,63
100	86,57	96,963,97	8,67
250	74,68	96,964	8,32
500	98,57	96,963,73	9,33
1000	166,81	101,382,49	11,54

found, exhibited significant performance degradation with increasing dataset size. As the problem scale grew to 30, 40, and 50 projects, the method struggled to achieve consistent optimal solutions. Beyond 100 projects, it failed to provide any valid solutions, indicating its limitations when applied to more complex, large-scale problems.

In contrast, the Random method with Feasibility Check displayed enhanced performance and robustness across all tested dataset sizes. By integrating a feasibility check into the solution generation process, this method was able to consistently produce near-optimal solutions in every iteration. This improved reliability highlights the importance of incorporating feasibility checks to maintain performance regardless of dataset size.

The Repair method introduced a systematic approach to ensuring solution feasibility by iteratively removing projects from the initial set. This process proved effective in achieving near-optimal solutions; however, it introduced considerable computational overhead, particularly for larger datasets. For instance, in the 1000-project dataset, the method required the removal of an average of 490 projects, significantly increasing runtime and reducing overall efficiency. This indicates that while the method ensures feasible and high-quality solutions, its computational cost may render it impractical for very large datasets.

The MaxMin method provided a more efficient alternative by limiting the initial solution size, thus reducing the need for extensive adjustments. This method successfully maintained near-optimal solution quality with lower computational requirements compared to the Repair method. However, achieving the final solution required more iterations, suggesting a trade-off between runtime efficiency and convergence speed.

Overall, the findings indicate that methods incorporating feasibility considerations, particularly the Random method with Feasibility Check and the MaxMin method, offer superior performance for larger datasets. These methods balance solution quality with computational efficiency, demonstrating their potential for scalable applications. Future work could explore hybrid approaches that combine the strengths of these methods to further enhance performance for large-scale optimization problems.

In order to compare the methods, the number of iterations until the final result of each method was measured. The results are presented in Fig. 1. As can be seen from Fig. 1, the mathematical method that generates the optimal results has the highest number of steps. It is followed by the repair method, where too many projects are extracted to obtain the optimal solution. The repair method was followed by Random method with feasibility Check. The method that reaches the final result with the least number of iterations is MaxMin.

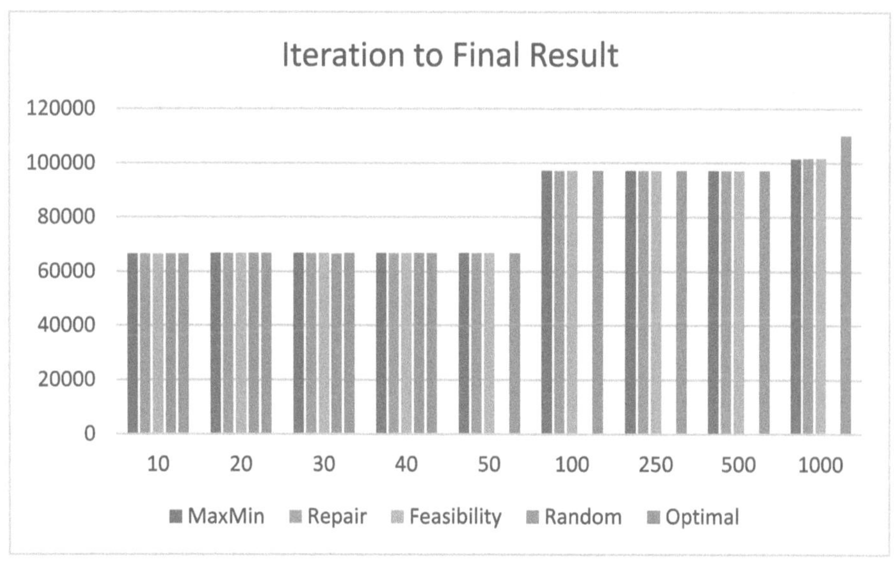

Fig. 1. Iteration to Final Result.

The running times of the methods are given in Fig. 2. As the number of projects increases, the running time of the repair method increases. This is because as the number of projects increases, the number of projects that need to be removed from the random solution obtained to obtain a feasible solution increases. In cases where the number of projects is higher, the relative difference will be higher. MaxMin method followed the repair method. When the running times of the methods were compared, the best results were obtained with Random with Feasibility Check.

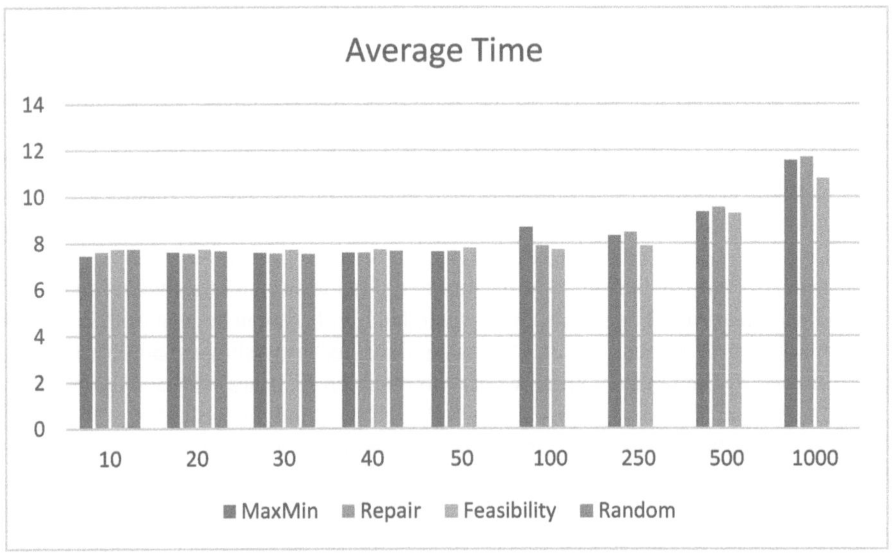

Fig. 2. Running Times of Methods.

5   Conclusions

This study has provided a detailed comparison of four initial solution generation methods—Random, Random with Feasibility Check, Repair, and MaxMin—for the knapsack problem under constrained budgets and a high number of project alternatives. The findings emphasize the critical role of initial solution strategies in the performance of meta-heuristic algorithms, particularly in scenarios where resource constraints and large-scale datasets pose significant challenges.

The Random method, while straightforward and computationally efficient, exhibited notable limitations as the complexity of the problem increased. Its ability to generate optimal solutions was restricted to smaller datasets, such as the 10-project problem set. Beyond this, the method struggled to consistently produce feasible solutions for larger datasets, often resulting in suboptimal or infeasible outcomes. This limitation highlights the trade-off between simplicity and reliability in solution quality.

In contrast, the Random with Feasibility Check method demonstrated consistent and robust performance across all dataset sizes. By incorporating a feasibility check during the initial solution generation phase, this method effectively eliminated infeasibility issues. The results showed that it not only achieved near-optimal solutions but also ensured solution feasibility across iterations, making it a highly reliable option for tackling larger and more complex knapsack problems. This method balances computational simplicity and solution quality, making it particularly appealing for real-world applications with limited computational resources.

The Repair method introduced a systematic approach to adjust infeasible solutions by iteratively removing excess projects until feasibility was achieved. This approach yielded near-optimal results, demonstrating its effectiveness in improving raw initial solutions. However, the method's performance was hindered by its high computational demand, especially for large datasets. For instance, when applied to the 1000-project dataset, the method required substantial computational effort to remove an average of 490 projects to meet budget constraints. While effective, this computational overhead makes the Repair method less suitable for time-sensitive applications.

The MaxMin method addressed some of the computational inefficiencies associated with the Repair method by limiting the number of projects in the initial solutions. By imposing predefined lower and upper bounds on the number of selected projects, the MaxMin method reduced the need for extensive feasibility adjustments. This approach resulted in feasible initial solutions that were more quickly implemented, achieving comparable solution quality to the Repair method with significantly lower runtime. However, the method required a higher number of iterations to converge to the final solution, indicating a trade-off between runtime efficiency and convergence speed.

When comparing all four methods, the Random with Feasibility Check and MaxMin methods emerged as the most practical and effective approaches for solving large-scale knapsack problems under constrained budgets. The Random with Feasibility Check method is particularly advantageous in scenarios where high-quality initial solutions are essential for efficient convergence. Meanwhile, the MaxMin method offers a balance between computational efficiency and solution feasibility, making it a viable choice for applications with stricter time constraints.

This study underscores the importance of carefully selecting initial solution strategies in metaheuristic optimization. The choice of method should be guided by the specific requirements of the problem, including dataset size, computational resources, and desired solution quality. Future research could build on these findings by exploring hybrid approaches that combine the strengths of multiple methods. For example, a hybrid method could leverage the feasibility guarantee of the Random with Feasibility Check method while incorporating the efficiency gains of the MaxMin approach. Additionally, integrating machine learning techniques to predict and enhance initial solutions based on problem characteristics could further improve the effectiveness of metaheuristic algorithms in solving complex optimization problems.

The data used in this study is randomly generated at specific intervals; however, since it is fixed, it can be reapplied. This enables the results to be easily reproduced by others, ensuring their reliability can be verified. Furthermore, these results can be compared by applying different methods. To enhance the study's applicability to more realistic scenarios, increasing the dataset size will improve the model's generalizability by introducing greater diversity in the data. While heuristic algorithms typically perform well on small to medium-sized datasets, performance issues can arise when applied to larger datasets. As the dataset size increases, the processing times of heuristic algorithms generally increase linearly, leading to longer analysis times, potential memory overloads, and higher processing power requirements. As future exploration strategies, integrating advanced techniques such as deep learning with heuristic methods could yield more powerful and effective results through hybrid approaches.

In conclusion, the study highlights the nuanced trade-offs between different initial solution generation methods. By addressing the specific challenges associated with budget constraints and large-scale datasets, these methods pave the way for more efficient and effective solutions to the knapsack problem, reinforcing their relevance in both theoretical research and practical applications, particularly in project selection and resource allocation scenarios.

Disclosure of Interests. The authors declare that they have no competing financial or non-financial interests that could have influenced the work presented in this article. This research was conducted independently and was not influenced by any commercial, governmental, or non-governmental funding sources.

References

1. Assi, M., Haraty, R.A.: A survey of the knapsack problem. In: 2018 Interna-Tional Arab Conference on Information Technology (ACIT), pp. 1–6. IEEE (2018). https://doi.org/10.1109/ACIT.2018.8672677
2. Hançer, E.: An ABC algorithm inspired by Boolean operators for knapsack and lot sizing problems. Acad. Plat. J. Eng. Sci. **6**, 142–152 (2018). https://doi.org/10.21541/apjes.337415
3. Nauss, R.M.: Solving the generalized assignment problem: an optimizing and heuristic approach. INFORMS J. Comput. **15**, 249–266 (2003). https://doi.org/10.1287/ijoc.15.3.249.16075
4. Perttunen, J.: On the Significance of the Initial Solution in Travelling Sales-man Heuristics. (1994).
5. Juman, Z.A.M.S., Hoque, M.A.: An efficient heuristic to obtain a better ini-tial feasible solution to the transportation problem. Appl. Soft Comput. J. **34**, 813–826 (2015). https://doi.org/10.1016/j.asoc.2015.05.009
6. Amaliah, B., Fatichah, C., Suryani, E.: A new heuristic method of finding the initial basic feasible solution to solve the transportation problem. J. King Saud Univ. Comput. Inform. Sci. **34**, 2298–2307 (2022). https://doi.org/10.1016/j.jksuci.2020.07.007
7. Armaneesa Naaman Hasoon: Projects selection in knapsack problem by using artificial bee Colony algorithm. Tikrit J. Pure Sci. **23**, 137–142 (2023). https://doi.org/10.25130/tjps.v23i2.662
8. Wei, Z., Hao, J.K.: Iterated two-phase local search for the set-union knap-sack problem. Futur. Gener. Comput. Syst. **101**, 1005–1017 (2019). https://doi.org/10.1016/j.future.2019.07.062
9. Dizdar, D., Gershkov, A., Moldovanu, B.: Revenue maximization in the dy-namic knapsack problem. Theor. Econ. **6**, 157–184 (2011). https://doi.org/10.3982/te700
10. Kose, A., Ozbek, B.: Resource allocation for underlaying device-to-device communications using maximal independent sets and knapsack algorithm. In: 2018 IEEE 29th Annual International Symposium on Personal, Indoor and Mobile Radio Communications (PIMRC), pp. 1–5. IEEE (2018). https://doi.org/10.1109/PIMRC.2018.8580784
11. Setiawan, F., Sadiyoko, A., Setiardjo, C.: Application of pigeon inspired optimization for multidimensional knapsack problem. Int. J. Ind. Eng. Eng. Manage. **2**, 45–56 (2020). https://doi.org/10.24002/ijieem.v2i1.3841

12. Ječmen, K., Mocková, D., Teichmann, D.: Solving transport infrastructure investment project selection and scheduling using genetic algorithms. Mathematics. **12** (2024). https://doi.org/10.3390/math12193056

13. Naldi, M., Nicosia, G., Pacifici, A., Pferschy, U., Leder, B.: A Simulation study of Fairness-Profit Trade-off in Project Selection based on HHI and Knapsack Models (2016). https://doi.org/10.1109/EMS.2016.24

14. Schiffels, S., Fliedner, T., Kolisch, R.: Human behavior in project portfolio selection: insights from an experimental study. Decis. Sci. **49**, 1061–1087 (2018). https://doi.org/10.1111/deci.12310

15. Bakirli, B.B., Gencer, C., Aydoğan, E.K.: A combined approach for fuzzy multi-objective multiple knapsack problems for defence project selection. J. Oper. Res. Soc. **65**, 1001–1016 (2014). https://doi.org/10.1057/jors.2013.36

16. Harrison, K.R., Elsayed, S., Weir, T., Garanovich, I.L., Galister, M., Boswell, S., Taylor, R., Sarker, R.: Multi-period project selection and scheduling for Defence capability-based planning. In: 2020 IEEE International Conference on Systems, Man, and Cybernetics (SMC), pp. 4044–4050. IEEE (2020). https://doi.org/10.1109/SMC42975.2020.9283334

17. Mavrotas, G., Figueira, J.R., Siskos, E.: Robustness analysis methodology for multi-objective combinatorial optimization problems and application to project selection. Omega (United Kingdom). **52**, 142–155 (2015). https://doi.org/10.1016/j.omega.2014.11.005

18. Chang, P.T., Lee, J.H.: A fuzzy DEA and knapsack formulation integrated model for project selection. Comput. Oper. Res. **39**, 112–125 (2012). https://doi.org/10.1016/j.cor.2010.10.021

19. Naldi, M., Nicosia, G., Pacifici, A., Pferschy, U.: Profit-fairness trade-off in project selection. Socio Econ. Plan. Sci. **67**, 133–146 (2019). https://doi.org/10.1016/j.seps.2018.10.007

20. Roland, J., Figueira, J.R., De Smet, Y.: Finding compromise solutions in project portfolio selection with multiple experts by inverse optimization. Comput. Oper. Res. **66**, 12–19 (2016). https://doi.org/10.1016/j.cor.2015.07.006

21. Gómez-Herrera, F., Ramirez-Valenzuela, R.A., Ortiz-Bayliss, J.C., Amaya, I., Terashima-Marín, H.: A quartile-based hyper-heuristic for solving the 0/1 knapsack problem. In: Lecture Notes in Computer Science (Including Subseries Lecture Notes in Artificial Intelligence and Lecture Notes in Bioinfor-matics), pp. 118–128. Springer Verlag (2018). https://doi.org/10.1007/978-3-030-02837-4_10

22. Sousa, T., Morais, H., Castro, R., Vale, Z.: Evaluation of different initial solution algorithms to be used in the heuristics optimization to solve the en-ergy resource scheduling in smart grids. Appl. Soft Comput. **48**, 491–506 (2016). https://doi.org/10.1016/J.ASOC.2016.07.028

23. Bražėnas, M., Valakevičius, E.: On structured initial solution generation for phase type fitting with EM method. Inform. Technol. Control. **47**, 197–208 (2018). https://doi.org/10.5755/J01.ITC.47.2.18169

24. Babu, M.A., Hoque, M.A., Uddin, M.S.: A heuristic for obtaining better ini-tial feasible solution to the transportation problem. Opsearch. **57**, 221–245 (2020). https://doi.org/10.1007/S12597-019-00429-5

25. Liu, Q., Cheng, H., Tian, T., Wang, Y., Leng, J., Zhao, R., Zhang, H., Wei, L.: Algorithms for the variable-sized bin packing problem with time win-dows. Comput. Ind. Eng. **155**, 107175 (2021). https://doi.org/10.1016/J.CIE.2021.107175

26. Lu, Y., Huang, Z., Cao, L., Schmid, M., Braun, S., Wang, Y., Song, Y., Zou, Y., Lei, Q., Wang, X.: A hybrid gray wolf weed algorithm for flexible job-shop scheduling problem. J. Phys. Conf. Ser. **1828**, 012162 (2021). https://doi.org/10.1088/1742-6596/1828/1/012162

27. Krzywanski, J., Grabowska, K., Skrobek, D., Moen Uddin, G., Gao, Y., Zyl-ka, A., Kulakowska, A., El Fil, B., Zhang, Z., Zhou, J., Wang, X., Yang, H., Fan, Y.: Initial solution generation and diversified variable picking in local search for (weighted) partial MaxSAT. Entropy. **24**, 1846 (2022). https://doi.org/10.3390/E24121846

28. Lu, Y.: A two-stage heuristic algorithm for high-quality initial solu-tion generation in MTVRPTW. SPIE. **12790**, 127901S (2023). https://doi.org/10.1117/12.2689564
29. Khurshid, B., Maqsood, S.: A hybrid evolution strategies-simulated annealing algorithm for job shop scheduling problems. Eng. Appl. Artif. Intell. **133**, 108016 (2024). https://doi.org/10.1016/J.ENGAPPAI.2024.108016
30. Karaboga, D., Basturk, B.: A powerful and efficient algorithm for numerical function optimization: artificial bee colony (ABC) algorithm. J. Glob. Optim. **39**, 459–471 (2007). https://doi.org/10.1007/s10898-007-9149-x

A Comparative Approach to Multi-response Optimization for AWJM of Monel 400 Alloys Using Hybrid Metaheuristic-MCDM Techniques

Ayyappan Solaiyappan[1]([⊠]) [iD] and Sivakumar Mahalingam[2] [iD]

[1] Department of Mechanical Engineering, Government College of Technology,
Coimbatore 641013, Tamilnadu, India
`ayyappansola@gmail.com`

[2] Department of Mechanical Engineering, SRM TRP Engineering College,
Tiruchirappalli 641013, Tamilnadu, India

Abstract. This investigation aimed to determine the optimal parameters for Abrasive Water Jet Machining (AWJM) of Monel 400 alloys. To enhance key performance metrics of Surface Roughness (SR) and Material Removal Rate (MRR), critical AWJM process variables namely Transverse Rate (TR), Stand-off Distance (SD), and Abrasive Flow Rate (FR) were considered. Various metaheuristic algorithms, such as Multiverse Optimization (MVO), Antlion Optimization (ALO), Grey Wolf Optimization (GWO), and the Grasshopper Algorithm (GHO), were employed to optimize the machining parameters. To obtain Pareto-optimal solutions, these metaheuristic techniques were integrated with Multi-Criteria Decision-Making (MCDM) methods, including TOPSIS, Deng's method and EDAS, using the Hypervolume (HV) indicator. Experimental validation demonstrated an error margin of less than 3% between predicted and actual results, confirming the effectiveness of the proposed optimization framework. The optimal AWJM machining parameters of TR = 250 mm/min, SD = 1 mm, and FR = 238 g/min were identified using the GWO-EDAS approach, yielding an MRR of 0.449 g/s and SR of 2.43 μm.

Keywords: Abrasive Water Jet Machining · Multi-Criteria Decision-Making · Hypervolume · Material Removal Rate · Surface Roughness

1 Introduction

Monel 400 alloys are extensively used in numerous industrial applications, such as nuclear reactors, heat exchangers, marine applications, chemical processing equipment, sensors, pumps, and valves [1]. This alloy exhibits high strength and exceptional corrosion resistance across a broad range of acidic and alkaline environments [2]. In conventional machining, cutting forces induce shear deformation along the shear plane, making Monel 400 particularly challenging to machine due to its rapid work hardening compared to steel. The conventional machining of Monel 400 leads to increased manufacturing costs due to excessive tool wear and compromised product quality resulting

A. Mirzazadeh et al. (Eds.): ODSIE 2024, CCIS 2482, pp. 389–418, 2026.
https://doi.org/10.1007/978-3-031-93601-2_24

from residual stresses. Consequently, non-conventional machining techniques are preferred for processing such superalloys. Abrasive Water Jet Machining (AWJM) is particularly advantageous due to its ability to achieve high-quality surface finishes, making it a preferred method in industries such as automotive and aerospace manufacturing, surgical equipment production, and the defence sector [3]. In AWJM, material removal and surface quality enhancement are achieved through the mechanical energy imparted by high-velocity abrasive particles combined with water. This process generates minimal heat in the cutting zone, making it suitable for machining hard materials. Additionally, AWJM is a highly versatile cutting process, and can machine composites, ceramic materials, metals, and non-metals. Compared to electrochemical machining and wire-electrical discharge machining, AWJM offers higher material removal rates, while providing superior surface integrity in comparison to the plasma-arc cutting process. Furthermore, AWJM can effectively cut components up to 250 mm thickness while maintaining high dimensional accuracy [4]. Although machining Monel 400 alloys using AWJM presents significant potential advantages, research on its feasibility remains limited [5-7]. Therefore, further comprehensive investigation is necessary to explore its effectiveness for the machining of Monel 400 alloys and optimize machining parameters.

The investment, maintenance, operational, and tooling costs associated with AWJM are generally high. Therefore, to maximize the efficiency and cost-effectiveness of these machines, they must operate under optimal or near-optimal process conditions [8, 9]. Traditionally, the selection of AWJM process parameters relies on the machinist's expertise or information available in machining handbooks. However, parameter selection based solely on experience rarely ensures optimal machining efficiency and quality. Furthermore, machining handbooks address only a limited number of applications, and the selected parameters are often suboptimal, preventing the full utilization of AWJM capabilities. A systematic, data-driven approach is essential to achieve consistent and high-performance results. Without appropriate optimization techniques, determining the best AWJM process parameters requires extensive experimentation, which is costly, time-intensive, and laborious. Additionally, solving complex machining problems using classical optimization methods is often impractical due to the significant computational effort required. While Response Surface Methodology (RSM) has been widely employed as an optimization tool for AWJM, metaheuristic algorithms have emerged as a more effective approach for solving complex real-world problems, offering enhanced solution accuracy and robustness [10, 11]. Metaheuristic algorithms are inspired by the social behaviours of birds, animals and, insects, enabling enhanced search patterns within the solution space by improving convergence speed and computational efficiency. Among the recently developed nature-inspired algorithms, the Ant Lion Optimizer (ALO), Grasshopper Optimization (GHO), Grey Wolf Optimizer (GWO), and Multiverse Optimization (MVO) algorithms have demonstrated significant potential for addressing real-time machining optimization challenges [11].

Jai Rajesh et al. (2024) employed various metaheuristic algorithms, including the Firefly Algorithm, Cuckoo Search Algorithm, Particle Swarm Optimization (PSO), GWO algorithm, Multi-Objective Teaching Learning-Based Optimization (MOTLBO), and Salp Swarm Algorithm, to optimize AWJM parameters to improve MRR and depth of cut [12]. Additionally, hybrid optimization techniques, such as the Adaptive Decreasing

Method Multi-Objective Jaya (ADM-MO-Jaya) algorithm, have demonstrated significant effectiveness in optimizing the AWJM process [13]. Process optimization of AWJM has also been performed using the Taguchi DEAR methodology, as reported by Manoj et al. (2024) [14]. Furthermore, Arun et al. (2024) applied the Taguchi method alone to determine the optimal AWJM cutting parameters for Monel 400 alloys, without incorporating any metaheuristic techniques. A review of the existing literature indicates that the application of nature-inspired metaheuristic algorithms such as the ALO, GWO, GHO, and MVO in the optimization of AWJM parameters for Monel 400 alloys remains relatively unexplored. This gap highlights the need for further research to enhance machining performance and process efficiency through advanced optimization approaches.

The reviewed literature highlights the increasing sophistication of intelligent optimization techniques and underscores their potential for broader applications. Despite their demonstrated effectiveness, there remains significant scope for further exploration and implementation in diverse industrial contexts. In addressing complex industrial optimization challenges, Multi-Criteria Decision-Making (MCDM) techniques such as 'Evaluation based on Distance from Average Solution (EDAS)', the 'Technique for Order of Preference by Similarity to Ideal Solution (TOPSIS)', and Deng's Method are widely utilized. These methods provide systematic frameworks for decision-making by evaluating multiple conflicting criteria. To enhance the optimization of AWJM parameters for Monel 400 alloys, this study integrates MCDM techniques with metaheuristic algorithms, including the ALO, GHO, GWO, and MVO. A comparative analysis is carried out to assess the performance of these approaches concerning convergence rate, computational efficiency, and the accuracy of the Pareto optimal front. This evaluation aims to establish a more robust and effective optimization framework for AWJM parameter selection.

Section 2 presents a comprehensive explanation of the experimental work conducted on AWJM process. The research methodology adopted for this investigation is detailed in Sect. 3. Section 4 covers the development of mathematical models along with an Analysis of Variance (ANOVA) study. In Sect. 5, the various metaheuristic algorithms and MCDM techniques employed in this study are discussed. Section 6 provides an in-depth analysis of the results, highlighting the influence of AWJM process parameters and the optimization outcomes. Finally, the conclusions drawn from this research are summarized in Sect. 7.

2 Methodology of the Investigation

The investigation is planned to be conducted in four phases. In Phase I, a comprehensive experimental study is carried out to analyze the effects of AWJM process parameters namelyTraverse Rate (TR), Stand-off Distance (SD), and Abrasive Flow Rate (AF) on Material Removal Rate (MRR) and Surface Roughness (SR) while machining Monel 400 alloys. Phase II focuses on developing regression models and performing ANOVA analysis. In Phase III, surface topography is examined using Scanning Electron Microscopy (SEM), along with a detailed discussion of the results. Finally, Phase IV involves optimizing AWJM process parameters using ALO, GWO, GHO, and MVO algorithms, integrated with EDAS, TOPSIS, and Deng's method. Figure 1 illustrates the step-by-step process flow of this investigation.

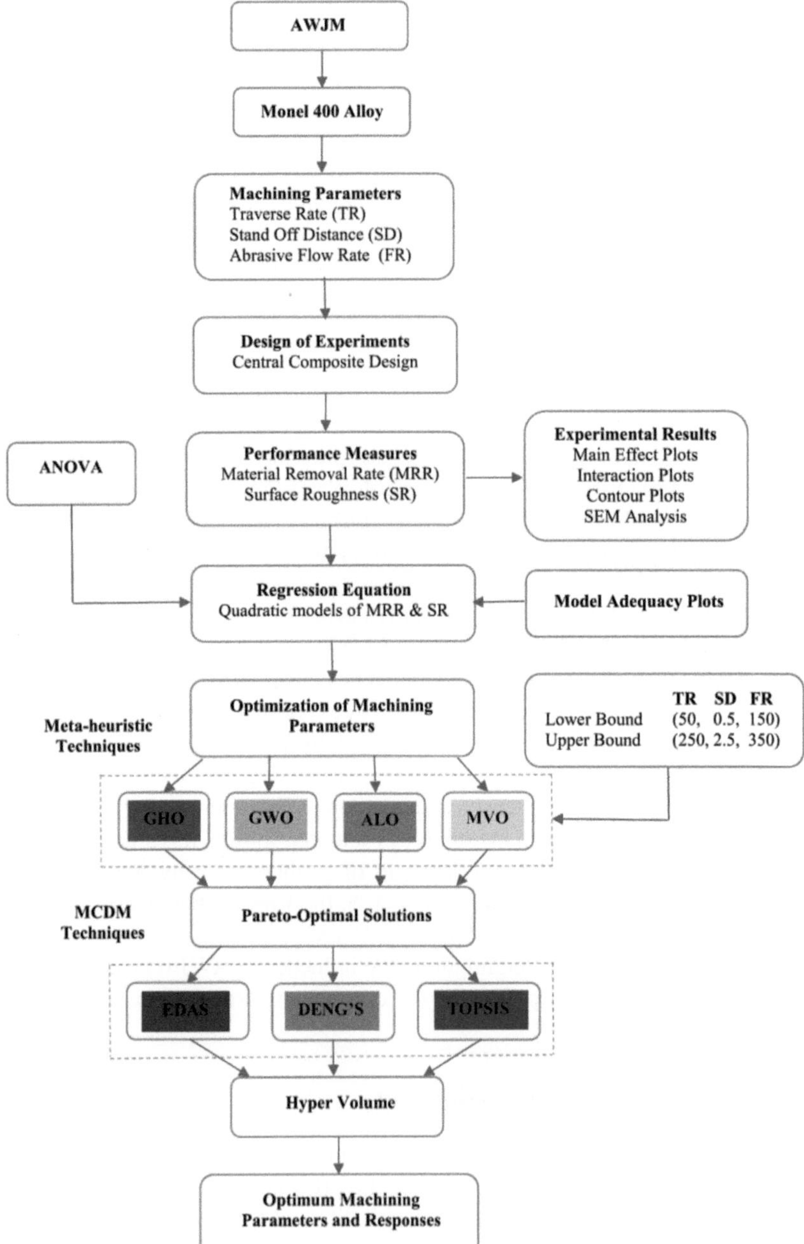

Fig. 1. Flow chart of the proposed methodology.

3   Experimentation

Figure 2 presents a photograph of the AWJM setup. The system comprises several key components, including a numerical control system, hydraulic pump system, abrasive supply system, water supply system, mixing chamber, wooden plate, hammer, nozzle, workpiece, and fixture. The cutting head operates along the X, Y, and Z axes, guided by a CNC program input into the control panel. Additionally, the numerical control system is equipped with control knobs that allow for precise adjustment of machining parameters, ensuring optimal performance in the AWJM process.

Fig. 2. Abrasive Water Jet Machining (AWJM) Setup.

Initially, water is pressurized using a hydraulic pump driven by an electric motor. The flow direction of the water is regulated by direction control valves to ensure precise operation. An intensifier within the system further increases the water pressure from a lower to a significantly higher level. As the pressure rises, fluctuations may occur within the system, requiring stabilization. The high-pressure water is then directed through an orifice into the mixing chamber, where it combines with abrasive particles supplied by the abrasive feed system. This mixture of high-pressure water and abrasives generates a high-velocity jet, which is then directed onto the workpiece to facilitate material removal through erosion. The operational parameters of the AWJM machine are summarized in Table 1.

In this study, GMA garnet abrasive with a mesh size of 80 is used as the abrasive medium. The workpiece selected for experimentation is a 152 × 152 mm square plate of Monel 400 alloy with a thickness of 5 mm. The chemical composition of Monel 400 alloy comprises 67.209% nickel (Ni), 29.24% copper (Cu), 1.66% iron (Fe), and 1.03% manganese (Mn), with trace amounts of cobalt (Co), aluminum (Al), silicon (Si), chromium (Cr), tungsten (W), and molybdenum (Mo) present in minor proportions [15]. Anish and Kumanan (2017) investigated AWJM characteristics for Inconel 617, considering process parameters such as traverse speed (50–80 mm/min), stand-off distance

Table 1. Operating Conditions of AWJM.

Power	50 HP
Table area	3 × 1.5 m
Table thickness	1.5 m
Water pressure	4000 Bar
Cutting pressure	3500 Bar
Abrasive pressure	6 Bar
Taper	0.5 mm
Impingement angle	90°
Nozzle length	80 mm
Nozzle diameter	1.016 mm
Orifice diameter	0.33 mm
Abrasive quality	GMA Garnet # 80

(1.5 4.5 mm), abrasive flow rate (0.24 0.44 kg/min), and water pressure (180 275 MPa) [16]. Their findings indicate that traverse speed has the most significant influence on machining performance, followed by abrasive mass flow rate [17].

The investigation of AWJM for Inconel 618 reveals that stand-off distance (2 4 mm) significantly influences performance, along with traverse rate (0.22 0.42 kg/min) and water pressure [18]. High water pressure is always preferred for cutting nickel-based alloys. The critical AWJM process parameters, including Traverse Rate (TR), Stand-off Distance (SD), and Abrasive Flow Rate (AF), were selected as control factors to systematically assess key performance indicators, such as Material Removal Rate (MRR) and Surface Roughness (SR) represented by R_a values. These selections were based on literature findings and machine constraints. Table 2 presents the actual and coded values of the chosen process parameters used in the experiment. The step value (ΔX_i) of uncoded parameters can be determined using the given equation.

$$\Delta X_i = \frac{X_{i,,\max} - X_{i,,\min}}{nl - 1} \tag{1}$$

where, $X_{i,,max}$—The upper limit of the parameters, $X_{i,,min}$—The lower limit of the parameters and nl—Number of levels.

The Central Composite Design (CCD) of RSM was utilized to design and analyze the experiment. Parameters such as water pressure, abrasive pressure, cutting pressure, nozzle diameter, impingement angle, and orifice diameter were set to predefined values. The control knobs were carefully adjusted to manually regulate parameters like Abrasive Flow Rate (AF), Stand-off Distance (SD), and Traverse Rate (TR). In this study, the Material Removal Rate (MRR) was determined by calculating the ratio of weight loss to machining time, as described in Eq. (2).

$$MRR = \frac{Weight\ Loss}{Machining\ Time} \tag{2}$$

Table 2. Process Factors and Level Setting.

S. No.	Variables		Levels				
	Uncoded	Units	−2	−1	0	1	2
1	Traverse rate (TR)	mm/min	50	100	150	200	250
2	Stand-off distance (SD)	mm	0.5	1	1.5	2	2.5
3	Abrasive flow rate (FR)	g\min	150	200	250	300	350

Weight loss is calculated by subtracting the specimen's post-machining weight from its pre-machining weight. Another key performance criterion, surface roughness, is measured using Martalk measurement equipment. Figure 3 illustrates the sample surface roughness graph for the R_a measurement. To evaluate surface roughness, the average of three measurements is taken as the performance metric.

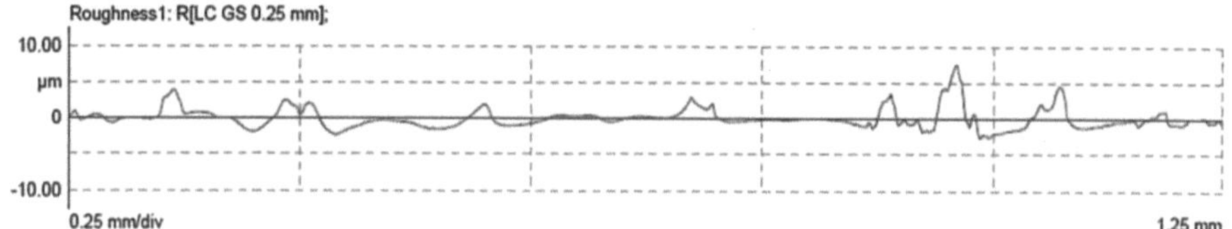

Fig. 3. Sample Surface Roughness measurement graph—R_a value.

Table 3 shows the experimental design and the observed performance measures.

Table 3. Experimental Observation.

Exp. No	Traverse Rate (TR)		Stand-off Distance (SD)		Abrasive Flow Rate (FR)		MRR (g\s)	SR (µm)
	X_1	mm\min	X_2	mm	X_3	g\min		
1	2	250	0	1.5	0	250	0.234	2.78
2	1	200	−1	1	−1	200	0.167	2.27
3	0	150	0	1.5	2	350	0.196	2.06
4	−1	100	1	2	1	300	0.163	2.25
5	−1	100	−1	1	1	300	0.155	2.23
6	−1	100	1	2	−1	200	0.135	2.74
7	1	200	−1	1	1	300	0.197	2.40
8	0	150	0	1.5	0	250	0.152	2.62
9	0	150	2	2.5	0	250	0.143	2.86

(continued)

Table 3. (*continued*)

Exp. No	Traverse Rate (TR)		Stand-off Distance (SD)		Abrasive Flow Rate (FR)		MRR (g\s)	SR (μm)
	X_1	mm\min	X_2	mm	X_3	g\min		
10	1	200	1	2	1	300	0.226	2.66
11	−2	50	0	1.5	0	250	0.128	2.48
12	0	150	0	1.5	−2	150	0.149	2.83
13	0	150	0	1.5	0	250	0.153	2.38
14	0	150	0	1.5	0	250	0.152	2.36
15	0	150	0	1.5	0	250	0.152	2.31
16	0	150	−2	0.5	0	250	0.161	2.13
17	0	150	0	1.5	0	250	0.153	2.42
18	1	200	1	2	−1	200	0.192	2.70
19	−1	100	−1	1	−1	200	0.175	2.65
20	0	150	0	1.5	0	250	0.173	2.40

4 Mathematical Modelling and Analysis of Variance

Regression models for MRR and SR outputs are developed as in Eq. (3) using Minitab 19 software to characterize the AWJM experiments.

$$Y_u = b_0 + \sum\nolimits_{i=1}^{n} b_i\, X_{iu} + \sum\nolimits_{i=1}^{n} b_{ii}\, X_{iu}^{2} + \sum\nolimits_{j>i}^{n} b_{ij} X_{iu} X_{ju} + e_u \qquad (3)$$

where Y_u—The predicted value of the response (u), b_0, b_i, b_{ii}, b_{ij}—Coefficients of the variables, X_{iu}—Independent variables for the response (u) and e_u—Regression error for the response (u).

The regression equation is evaluated using Fisher's test (F-ratio) and the ANOVA approach. A model is considered significant if the probability value is below 0.05, with a confidence level of 95%. Probability values exceeding 0.1 indicate terms that are not statistically significant. Table 4 presents the ANOVA results for the MRR and SR models in the machining of Monel 400 alloys by AWJM.

Model adequacy is tested by the coefficient of determination values shown in Table 5.

The F-value of 9.61 for the MRR model demonstrates its statistical significance. In this case, TR and FR are identified as key contributing factors within the model. The coefficient of determination (R^2) for the model is 0.8964, demonstrating a strong fit. Therefore, the regression equation for MRR, as given in Eq. (4), is validated as suitable for predicting MRR in further analysis.

$$MRR = 0.534 - 0.001762 * TR - 0.1317 * SD - 0.001634 * FR + 0.000003 * TR^2 + 0.00014 * SD^2 + 0.000002$$
$$(4)$$
$$SR = 5.48 - 0.02225 * TR - 0.083 * SD - 0.00999 * FR + 0.000021 * TR^2 + 0.0768 * SD^2 + 0.000003 * FR^2 +$$
$$(5)$$

Table 4. ANOVA results for Regression Models of MRR and SR.

Source	DF		Adj SS		Adj MS		F-Value		P-Value	
	MRR	SR	MRR	SR	MRR	SR	MRR	SR	MRR	SR
Model	9	9	0.0136	0.9508	0.0015	0.1056	9.61	8.07	0.001	0.002
Linear	3	3	0.0101	0.7034	0.0034	0.2345	21.49	17.92	0.000	0.000
TR	1	1	0.0084	0.0361	0.0084	0.0361	53.41	2.76	0.000	0.128
SD	1	1	0.0000	0.3192	0.0000	0.3192	0.08	24.39	0.786	0.001
FR	1	1	0.0017	0.3481	0.0017	0.3481	10.99	26.60	0.008	0.000
Square	3	3	0.0018	0.0731	0.0006	0.0244	3.83	1.86	0.046	0.200
TR*TR	1	1	0.0013	0.0705	0.0013	0.0705	8.51	5.39	0.015	0.043
SD*SD	1	1	0.0000	0.0093	0.0000	0.0093	0.00	0.71	0.989	0.420
FR*FR	1	1	0.0007	0.0011	0.0007	0.0011	4.27	0.09	0.066	0.775
2-way interaction	3	3	0.0016	0.1742	0.0006	0.0581	3.52	4.44	0.057	0.031
TR*SD	1	1	0.0009	0.0421	0.0009	0.0421	5.90	3.21	0.036	0.103
TR*FR	1	1	0.0004	0.1250	0.0004	0.1250	2.50	9.55	0.145	0.011
SD*FR	1	1	0.0003	0.0072	0.0004	0.0072	2.16	0.55	0.173	0.475
Error	10	10	0.0016	0.1309	0.0002	0.0131				
Lack-of-fit	5	5	0.0012	0.0733	0.0002	0.0147	3.42	1.27	0.102	0.399
Pure error	5	5	0.0004	0.0576	0.0001	0.0115				
Total	19	19	0.0151	1.0817						

Table 5. Model Summary.

	S	R-sq	R-sq (adj)	R-sq (pred)
MRR	0.012520	89.64%	80.32%	33.61%
SR	0.114394	87.90%	77.01%	35.81%

The SR model's F-value of 8.07 indicates its significance. In this model, SD, FR, and the TR*FR interaction are identified as important terms. The developed regression equation for SR, as presented in Eq. (5), is well-fitted and can be used to predict SR performance in further investigations, as evidenced by the model's coefficient of determination (R^2) of 0.879.

5 Meta-Heuristics and MCDM Procedures

Shahrzad et al. (2017) introduced the Grasshopper Optimization Algorithm (GHO), inspired by the swarming behaviour of grasshoppers in search of food [19]. The mathematical model of GHO, which simulates this swarming behaviour, is presented in Eq. (6).

$$X_i = S_i + G_i + A_i \tag{6}$$

The equation can be written as $X_i = r_1 S_i + r_2 G_i + r_3 A_i$ where r_1, r_2, r_3—random numbers in [0,1]. The social interaction component (S_i) is determined by Eq. (7).

$$S_i = \sum_{\substack{j=1 \\ j \neq 1}}^{N} s(d_{ij})\hat{d}_{ij} \tag{7}$$

The G (gravity force on the ith grasshopper) component is calculated as in Eq. (8).

$$G_i = -g\hat{e}_g \tag{8}$$

The A component (wind advection) is calculated as in Eq. (9).

$$A_i = u\hat{e}_w \tag{9}$$

Table 6 provides a summary of the variables and notations used in the GHO algorithm. Figure 4(a) presents the pseudo-code for the GHO algorithm.

Table 6. The Symbols and Notations for GHO Algorithm.

Symbols	Explanation
X_i	The position of the ith grasshopper
S_i	The social interaction
G_i	The gravity force on the ith grasshopper
A_i	The wind advection
N	Number of grasshoppers
d_{ij}	Distance between the ith and the jth grasshopper
s	Function to define the strength of social forces
$\hat{d}_{ij}$	Unit vector from the ith grasshopper to the jth grasshopper
f	Intensity of attraction
l	Attractive length scale
g	Gravitational constant

(continued)

Table 6. (*continued*)

Symbols	Explanation
$\hat{e}_g$	Unity vector towards the centre of the earth
u	Constant drift
$\hat{e}_w$	Unity vector in the direction of the wind

(a) Pseudo-code for GHO Algorithm	(b) Pseudo-code for GWO Algorithm
1: Initialize the swarm X_i ($i = 1, 2, 3,...., n$) 2: Initialize C_{max}, C_{min} and ite_{max} 3: Calculate the fitness of each agent 4: $T =$ The best search agent 5: **While** ($I < ite_{max}$) 6: update C 7: **For** each search agent 8: Normalize the distance between the grasshoppers 9: Update the position of the current search agent 10: Bring the current search agent back if it goes outside the boundaries 11: **End for** 12: Update T if there is a better solution 13: *I = I + 1* 14: **End While** 15: Display the Pareto optimal solutions	1: Initialize no. of the grey wolf X_{ij} ($i = 1,2,..nv$ and $j = 1, 2,..nd$) 2: While ($it < nitr$) 3: Determine the fitness function F_{ik} ($k = 1, 2,..nf$) of each wolf 4: Calculate Pareto optimal distance f_i 5: Sort f_i in descending order and set as sf_i and store the first wolf's data as X_{it}. and F_{it}. 6: Using the sorted data, assign $X_a. = X_1.$, $X_b. = X_2.$ and $X_d. = X_3.$ 7: Compute $a = 2 - it * (\frac{2}{nitr})$ 8: For each wolf Update the position using 9: $A1 = 2 * a * rand() - a$ and $C1 = 2 * rand()$ and $D_a. = abs(C1 * X_a. - X_i.)$ and $X1. = X_a. - A1 * D_a.$ 10: $A2 = 2 * a * rand() - a$ and $C2 = 2 * rand()$ and $D_b. = abs(C1 * X_b. - X_i.)$ and $X2. = X_b. - A2 * D_b.$ 11: $A3 = 2 * a * rand() - a$ and $C3 = 2 * rand()$ and $D_d. = abs(C1 * X_d. - X_i.)$ and $X3. = X_d. - A3 * D_d.$ 12: $X_i. = \left(\frac{X1.+X2.+X3.}{3}\right)$ 13: Check $X_i.$ within bounds 14: End

Fig. 4. Pseudo-code for GHO and GWO algorithms.

Mirjalili et al. (2014) developed the Grey Wolf Optimizer (GWO), inspired by the natural hunting behaviours of grey wolves [20]. GWO involves the mathematical modelling of grey wolves' hunting strategies to solve complex engineering problems. In this approach, the alpha (α) wolf represents the optimal solution, while the prey signifies the best possible outcome for the problem. The beta (β) and delta (δ) wolves represent the second and third most optimal alternatives, respectively, whereas all other solutions are categorized as omega (ω) wolves. Equations (10) and (11) illustrate how the prey is encircled during the optimization process.

$$\vec{D} = \left| \vec{C} * \vec{X_p}(t) - \vec{X}(t) \right| \tag{10}$$

$$\vec{X}(t+1) = \vec{X_p}(t) - \vec{A} * \vec{D} \tag{11}$$

$$Coefficient\ vector\ \vec{A} = 2\vec{a} * \vec{r_1} - \vec{a} \tag{12}$$

$$Coefficient\ vector\ \vec{C} = 2 * \vec{r_2} \tag{13}$$

The distance between the prey and the positions of the alpha (α), beta (β), and delta (δ) wolves are represented in Eqs. (14), (15), and (16), respectively.

$$\vec{D_\alpha} = \left| \vec{C_1} * \vec{X_\alpha} - \vec{X} \right| \tag{14}$$

$$\vec{D_\beta} = \left| \vec{C_1} * \vec{X_\beta} - \vec{X} \right| \tag{15}$$

$$\vec{D_\delta} = \left| \vec{C_1} * \vec{X_\delta} - \vec{X} \right| \tag{16}$$

With the use of mathematical Eqs. (17) and (19), these wolves direct the hunting process. Random updates are made to the positions of the other wolves (ω).

$$\vec{X_1} = \vec{X_\alpha} - \vec{A_1} * \vec{D_\alpha} \tag{17}$$

$$\vec{X_2} = \vec{X_\beta} - \vec{A_2} * \vec{D_\beta} \tag{18}$$

$$\vec{X_3} = \vec{X_\delta} - \vec{A_3} * \vec{D_\delta} \tag{19}$$

Using the equation below, a grey wolf adjusts its position to $\vec{X}$ $(t+1)$ from the position of $\vec{X}$ based on the location of its prey $\vec{X_p}$. The position is updated by refining $\vec{A}$ and $\vec{C}$ values.

$$\vec{X}\ (t+1) = \frac{\vec{X_1} + \vec{X_2} + \vec{X_3}}{3} \tag{20}$$

Table 7 presents the key notations and symbols used in the GWO algorithm. Figure 4(b) illustrates the pseudo-code of the GWO algorithm.

Table 7. The Symbols and Notations for GWO Algorithm.

Symbols	Explanation
$\vec{D}$	Distance between the position of prey and grey wolf
t	Current iteration
$\vec{X_p}$	Position of prey
$\vec{X}$	Position of a randomly chosen grey wolf
$\vec{X_\alpha}$	Position of α wolf
$\vec{X_\beta}$	Position of β wolf
$\vec{X_\delta}$	Position of δ wolf
$\vec{D_\alpha}$	Distance between the position of prey and α wolf
$\vec{D_\beta}$	Distance between the position of prey and β wolf
$\vec{D_\delta}$	Distance between the position of prey and δ wolf

(continued)

Table 7. (*continued*)

Symbols	Explanation
$\vec{a}$	Decrease from 2 to 0 linearly over iterations.
$\vec{r_1}$, $\vec{r_2}$	Random vectors in [0, 1]
X(t)	Position of ants

Seyadali (2015) introduced the Antlion Optimizer (ALO), a nature-inspired optimization algorithm that mimics the unique predatory behaviour of antlions in their natural environment [21]. The ALO algorithm begins by mathematically modelling the interaction between antlions and ants within the cone-shaped trap. To replicate the foraging behaviour of ants, a random walk model is utilized. Table 8 presents the symbols and the parameters of the ALO algorithm.

Table 8. The Symbols and Notations for ALO Algorithm.

Symbols	Explanation
cumsum	Cumulative sum
m	Maximum number of iterations
t_s	Step of random walk (iteration)
$r(t)$	Stochastic function
M_{Ant}	Matrix for saving the position of each ant
$A_{i,j}$	Value of the jth variable (dimension) of ith ant
n	Number of ants
d	Number of variables
M_{OA}	Matrix for saving the fitness of each ant
$A_{i,j}$	Value of jth dimension of ith ant
f	Objective function
$M_{Antlion}$	Matrix for saving the position of each antlion
$AL_{i,j}$	j^{th} dimension's value of ith antlion
z	Number of antlions
M_{OAL}	Matrix for saving the fitness of each antlion
X_i^t	Updated position of ants
a_i	Minimum of random walk of ith variable
c_i^t	Minimum of ith variable at t^{th} iteration

(*continued*)

Table 8. (*continued*)

Symbols	Explanation
d_i^t	Maximum of ith variable at t^{th} iteration
c^t	Minimum of all variables at t^{th} iteration
c_j^t	Minimum of all variables for t^{th} ant
d_i^t	Maximum of all variables for i^{th} ant
I	Minimum of all variables at t^{th} iteration
d^t	Vector includes the maximum of all variables at t^{th} iteration
t	Current iteration
T	Maximum number of iterations
w	Constant
$Antlion_j^t$	Position of selected j^{th} antlion at t^{th} iteration
Ant_j^t	Position of ith ant at t^{th} iteration
R_A^t	Random walk around the antlion selected by the roulette wheel at t^{th} iteration
R_E^t	Random walk around the elite at t^{th} iteration

Equation (21) describes this random walk behaviour.

$$X(t) = \left[0, \, cumsum\big(2r(t_{s,1}) - 1\big), \ldots, cumsum\big(2r(t_{s,m}) - 1\big)\right] \tag{21}$$

The positions of ants after the random walk are saved in the following matrix.

$$M_{Ant} = \begin{bmatrix} A_{1,1} & \cdots & A_{1,d} \\ \vdots & \ddots & \vdots \\ A_{n,1} & \cdots & A_{n,d} \end{bmatrix} \tag{22}$$

An objective function is evaluated for each ant's position, and all ants' values are stored in the following matrix.

$$M_{OA} = \begin{bmatrix} f\big([A_{1,1}, A_{1,2}, \ldots, A_{1,d}]\big) \\ \vdots \\ f\big([A_{n,1}, A_{n,2}, \ldots, A_{n,d}]\big) \end{bmatrix} \tag{23}$$

The hidden positions of antlions within the search space are stored in the following matrix.

$$M_{Antlion} = \begin{bmatrix} AL_{1,1} & \cdots & AL_{1,d} \\ \vdots & \ddots & \vdots \\ AL_{z,1} & \cdots & AL_{z,d} \end{bmatrix} \tag{24}$$

The fitness values of each antlion are recorded in the matrix below.

$$M_{OAL} = \begin{bmatrix} f\left([AL_{1,1}, AL_{1,2}, \ldots, AL_{1,d}]\right) \\ \vdots \\ f\left([AL_{z,1}, AL_{z,2}, \ldots, AL_{z,d}]\right) \end{bmatrix} \tag{25}$$

The random walks of ants, derived from Eq. (21), were normalized using the following min-max normalization method to ensure that the values remain within the search space.

$$X_i^t = \frac{\left(X_i^t - a_i\right)\left(d_i - c_i^t\right)}{\left(d_i^t - a_i\right)} + c_i \tag{26}$$

Equations (27) and (28) describe the random walk of ants within a hypersphere surrounding a selected antlion, defined by the vectors, c_j^t and d_j^t.

$$c_j^t = Antlion_j^t + c^t \tag{27}$$

$$d_j^t = Antlion_j^t + d^t \tag{28}$$

This algorithm incorporates a roulette wheel selection mechanism to probabilistically select antlions based on their fitness values during the optimization process. Moreover, the radius of the ants' random walks within the hyper-sphere is dynamically adjusted according to the equations presented below.

$$c^t = \frac{c^t}{I} \tag{29}$$

$$d^t = \frac{d^t}{I} \tag{30}$$

where $I = 10^w \frac{t}{T}$

An antlion updates its position to match the most recent position of the hunted ant, as defined by Eq. (31).

$$Antlion_j^t = Ant_j^t \ \ if \ f\left(Ant_j^t\right) > f\left(Antlion_j^t\right) \tag{31}$$

To preserve the elitism characteristic, each ant is assumed to perform a random walk around a selected antlion, chosen via the roulette wheel mechanism, as well as the elite, following the equation below.

$$Ant_i^t = \frac{R_A^t + R_E^t}{2} \tag{32}$$

Figure 5 presents the pseudo-code for ALO Algorithm.

The Multiverse Optimizer (MVO) is inspired by the principles of multiverse theory and cosmology [22]. To attain a stable state, multiple universes engage in interactions through white, black, and wormholes. This phenomenon is systematically conceptualized and mathematically formulated as a meta-heuristic algorithm tailored for global

```
Pseudo-code for ALO Algorithm
```

1: Initialize the Population of ants (P_{ij}) and antlions (AP_{ij}) $i = 1,2,3\ldots na$ and $j = 1,2,3\ldots np$
2: For each ant and antlions, $i = 1$ to na
3:     Calculate the fitness function values of ants as mrr_i and sr_i
4:     Calculate the fitness function's values of antlions as $amrr_i$ and asr_i
5: End
6: While ($it <= max\_it$)
7:         Compute non-dominated antlions
8:         Compute the best antlion BP_{it} (elite) based on crowding distance (F_{itj})
9:         Update the archive within its maximum size
10:       For each ant $i = 1$ to na
11:               Select an antlion (k) using the roulette wheel selection method
12:               Compute $I = 10^W \left(\frac{it}{max\_it}\right)$ where $W = 2,3,4,5,6$ based on it value
13:               Compute $l = \frac{lb}{I}$ and $u = \frac{ub}{I}$
14:               If rand$(0,1) < 0.5$ $l = l + AP_k$
15:               Else $l = AP_k - l$
16:               End
17:               If rand$(0,1) \geq 0.5$ $u = u + AP_k$
18:               Else $u = AP_k - u$
19:               End
20:               For each $j = 1$ to np
21:                       Compute $Q = [0\ cumsum(2 * rand(max\_it, 1) - 1)]$
22:                       Calculate $a1 = \min(Q), b1 = \max(Q), c1 = l_j$ and $I = u_j$
23:                       Compute $RP_j = \frac{(Q - a1)(d1 - c1)}{(b1 - a1)} + c1$
24:               End
25:               If rand$(0,1) < 0.5$ $l = l + BP_{it}$
26:               Else $l = BP_{it} - l$
27:               End
28:               If rand$(0,1) \geq 0.5$ $u = u + BP_{it}$
29:               Else $u = BP_{it} - u$
30:               End
31:               For each $j = 1$ to np
32:                       Compute $Q = [0\ cumsum(2 * rand(max\_it, 1) - 1)]$
33:                       Calculate $a1 = \min(Q), b1 = \max(Q), c1 = l_j$ and $i = u_j$
34:                       Compute $SP_j = \frac{(Q - a1)(d1 - c1)}{(b1 - a1)} + c1$
35:               End
36:               Compute the position of the ant $P_i = \frac{RP_i + SP_i}{2}$
37:       End
38:       For each ant, $i = 1$ to na
39:               Calculate the fitness function values of ants as mrr_i and sr_i
40: End
41:       Replace antlion (AP_{ij}) if it is better than ant (P_{ij})
52: End
53: Display the Pareto optimal solutions

Fig. 5. Pseudo-code for ALO algorithm.

optimization problems. Within this framework, each solution is regarded as a distinct universe, while each variable within a solution corresponds to an entity within that universe. To maintain a balance between exploration and exploitation, every universe is assumed to contain wormholes that facilitate the random movement of objects across different universes. These wormholes modify objects irrespective of inflation rates, ensuring diversity and effective search capability. Furthermore, wormhole tunnels are always presumed to exist between any given universe and the best universe discovered thus far, enabling local refinements and high inflation rates. Table 9 provides the symbols and notations used in the MVO algorithm.

Table 9. The Symbols and Notations for MVO Algorithm.

Symbols	Explanation
n	Number of universes (solutions)
x_i^j	j^{th} parameter of ith universe
Ui	i^{th} universe
$NI\,(Ui)$	Normalized inflation rate of the ith universe
$r1$	Random number [0, 1]
x_k^j	j^{th} parameter of kth universe selected by a roulette wheel selection
X_j	j^{th} parameter of the best universe formed so far
TDR	Travelling distance rate, a coefficient
WEP	Wormhole existence probability, a coefficient
lb_j	Lower bound of jth variable
ub_j	Upper bound of jth variable
x_i^j	j^{th} parameter of ith universe
r_2, r_3, r_4	Random numbers [0, 1]
p	Exploitation accuracy over the iterations

The mathematical representation of this process is provided below.

$$
x_i^j = \begin{cases} \begin{cases} X_j + TDR\left(\left(ub_j - lb_j\right)r_4 + lb_j\right) & r_3 < 0.5 \\ X_j - TDR\left(\left(ub_j - lb_j\right)r_4 + lb_j\right) & r_3 \geq 0.5 \end{cases} & r_2 < WEP \\ x_i^j & r_2 \geq WEP \end{cases}
\tag{33}
$$

The Wormhole Existence Probability (WEP) defines the probability of a wormhole's existence in universes. As shown in Eq. (34), WEP increases linearly over iterations to enhance the exploitation phase of the optimization process.

$$
WEP = \min + l\left(\frac{\max - \min}{L}\right)
\tag{34}
$$

$$
TDR = 1 - \frac{l^{\frac{1}{p}}}{L^{\frac{1}{p}}}
\tag{35}
$$

Pseudo-code for MVO Algorithm is presented in Fig. 6.

Keshavarz et al. (2015) introduced the 'Evaluation based on Distance from Average Solution (EDAS)' method, a MCDM approach designed to address conflicting objectives [23]. This method evaluates alternatives based on their desirability using two key metrics: Positive Distance from Average (PDA) and Negative Distance from Average (NDA). These metrics quantify how each alternative (solution) deviates from the average solution, providing a structured means of comparative assessment.

Pseudo-code for MVO Algorithm

1: Initialize the Population of the random universe (P_{ij}) $i = 1,2,3\ldots nu$ and $j = 1,2,3\ldots np$
2: Set $W_m = 0.2$ and $W_x = 1$
3: While $(it <= max\_it)$
4:          Compute $W = W_m + it\left(\frac{W_x - W_n}{max\_it}\right)$ and $T = 1 - \frac{it^{\frac{1}{6}}}{max\_it^{\frac{1}{6}}}$
5:          For each universe $i = 1$ to nu
6:                    Calculate the fitness function's values as mrr_i and sr_i
7:      End
8:          Compute non-dominated universes
9:          Compute the best universe based on crowding distance (F_{itj})
10:          Update the archive within its maximum size
11:          Sort the universe and set it as SP_{ij}. Normalise the inflation rate and set it as NF_i
12:          Store the first sorted universe and its fitness function as SP_{ij}, mrr_{it} and sr_{it}
13:          For each Universe, $i = 2$ to nu
14:                    For each variable $j = 1$ to np
15:                              If rand $(0,1) < NF_i$
16: $n =$ select a universe based on the roulette wheel selection method
17: $P_{ij} = SP_{nj}$
18:                    End
19:                              If rand$(0,1) < W$
20:   If rand $(0,1) < 0.5$
21: $P_{ij} = P_{itj} + T\left[(ub_j - lb_j)rand(0,1) + lb_j\right]$
22:   Else
23: $P_{ij} = P_{itj} - T\left[(ub_j - lb_j)rand(0,1) + lb_j\right]$
24:   End
25:                              End
26:                    End
27:      End
28: End
29: Display the positions and the fitness values of the universe from the archive i.e. Pareto optimal solutions

---

Fig. 6. Pseudo-code for MVO algorithm.

In the EDAS method, alternatives are evaluated based on higher PDA values and lower NDA values. Figure 7(a) presents the pseudo-code for the EDAS method. Additionally, Deng's method, proposed by Deng in 2007, was developed to identify the optimal solution for MCDM problems [24]. This approach determines the best alternative by considering both the magnitude of alternatives and their conflicts with the ideal solution. The selection process is based on ranking conflicting indices, ensuring a structured evaluation. Figure 7(b) illustrates the pseudo-code for Deng's method.

The TOPSIS method is a widely used approach in decision-making optimization strategies for consolidating multiple objectives into a single objective [25, 26]. The core principle of the TOPSIS is to determine the alternative that maximizes its geometric distance from the negative ideal solution while minimizing its geometric distance from the positive ideal solution. This approach enables the integration of multiple objectives into a single objective value, thereby enhancing structured decision-making. Figure 8 illustrates the pseudo-code for the TOPSIS method.

The hypervolume (HV) indicator is a set-based metric used to evaluate the performance of Evolutionary Multi-Objective Algorithms (EMOAs) and to guide the search process. It is widely preferred in multi-objective optimization due to its Pareto compliance [27]. A higher hypervolume value indicates a superior solution set, signifying

(a) Pseudo-code for EDAS Method	(b) Pseudo-code for Deng's Method
1: Read objectives matrix – O_{ij} with weights (W_j) and type of objectives (OT)	1: Read objectives matrix – O_{ij} with weights (W_j) and type of objectives (OT)
2: For each Response j = 1 to nr	2: For each Alternate i = na
3: Determine average value (AV$_j$)	3: For each Response j = nr
4: End	4: Compute the Normalized value of O_{ij} (N_{ij})
5: For each Alternate i = na	5: Calculate Weighted Normalized value (A_{ij})
6: For each Response j = nr	6: End
7: Compute positive and negative distance from average (PDA_{ij} and NDA_{ij})	7: End
8: Calculate Weighted Normalized value (A_{ij})	8: For each Response j = nr
9: End	9: Determine positive ideal (P_j) and negative ideal solution (M_j)
10: End	10: End
11: For each Alternate i = na	11: For each Alternate i = na
12: Calculate the weighted sum of PDA and NDA (SP$_i$ and SN$_i$)	12: Calculate the degree of conflicts ($\cos\Theta_i^+$) and ($\cos\Theta_i^-$)
13: Compute the normalized value of SP$_i$ and SN$_i$ (NSP$_i$ and NSN$_i$)	13: Determine the degree of similarity (S_i^+) and (S_i^-)
14: Determine the appraisal score (AS$_i$)	14: Compute Overall performance index (R_i)
15: End	15: End
16: Arrange alternatives in descending order based on AS_i	16: Arrange alternatives in descending order based on R_i
17: Display the alternate which has the highest AS_i value	17: Display the alternate which has the highest R_i value.

Fig. 7. Pseudo-code for EDAS and Deng's Methods.

Pseudo-code for TOPSIS method

1: Read alternate and objectives matrix – O_{ij} with weights (W_j) and types of objectives (ot_j)

2: For each alternative (i = 1,2,3…m) and objective (j = 1,2,3…n)

3:    Compute the normalized value of O_{ij} using $N_{ij} = \dfrac{O_{ij}}{\sqrt{\Sigma_{i=1}^{m} O_{ij}^2}}$

4:    Calculate the Performancee matrix (A_{ij}) using $A_{ij} = N_{ij} * W_j$

5: End

6: For each objective (j=1,2,3…n)

7: Determine positive ideal (P_j) and negative ideal solution (M_j)

8:    For minimization objective – $P_j = \min_{1<i<m} (A_{ij})$ and $M_j = \max_{1<i<m} (A_{ij})$

9:    For maximization objective – $P_j = \max_{1<i<m} (A_{ij})$ and $M_j = \min_{1<i<m} (A_{ij})$

10: End

11: For each alternative (i=1,2,3…m)

12:    Calculated Ideal (SP_i) and negative ideal separation (SM_i)

13:    $SP_i = \sqrt{\Sigma_{j=1}^{n} (A_{ij} - P_j)^2}$ $SM_i = \sqrt{\Sigma_{j=1}^{n} (A_{ij} - M_j)^2}$

14:    Determine relative closeness (R_i)

15: End

16: Rank alternatives w.r.t. R_i in descending order.

Fig. 8. Pseudo-code for TOPSIS Method.

better algorithm performance. The HV indicator measures the volume or area covered by the Pareto optimal set points relative to a given reference point. The computation of HV is defined in Eq. (36).

$$HV\,(r,p) = \mathcal{L}\left(\bigcup_{p^n \in\, p} \left[f_1\left(p^n\right), r_1 \right] * \cdots * \left[f_m\left(p^n\right), r_m \right] \right) \tag{36}$$

where $HV\,(r, p)$—The volume of the space covered by an approximation set p.

$r \in R$—The chosen reference point

$\mathcal{L}$ (.)—Lebesgue measure

p—Set of Pareto optimal points

R—Set of reference points

f_1, f_2—Objective functions

This concept is illustrated in Fig. 9, which depicts the objective functions with a population consisting of three Pareto optimal solution points. Let p^1, p^2 and p^3 represent these solution points, while r serves as the reference point.

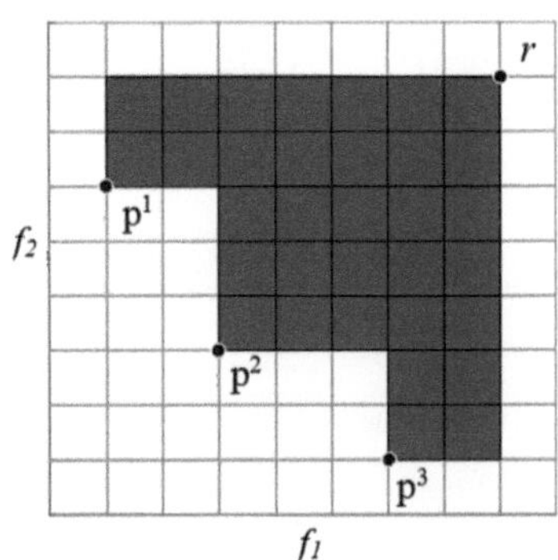

Fig. 9. Hypervolume Indicator.

6   Results and Discussion

6.1   Influence Process Parameters

Figure 10(a) presents the contour plot of the response surface for MRR. The plot clearly illustrates that MRR varies directly with the traverse rate. As mechanical force increases, a significant rise in the erosion rate is observed. Additionally, as the traverse rate increases from its minimum to the maximum value, MRR exhibits a gradual upward trend. The third contour plot in Fig. 10(a) illustrates the influence of SD & AF on MRR. In this plot, MRR is independent of SD.

As shown in the first contour plot of Fig. 10(b), at a traverse rate (TR) of 150 mm/min, roughness values increase as SD rises from its minimum to maximum value. This observation indicates a direct proportional relationship between SD and surface roughness. As the SD increases, the water jet expands further and becomes more susceptible to external drag from the surroundings. With a higher SD, the jet's width increases while its kinetic energy decreases, resulting in a rougher surface. Conversely, a lower SD is preferable, as it allows the jet to create a smoother surface. Additionally, as the flow rate of abrasive particles increases, particle collisions become more frequent, leading to a loss of kinetic energy upon impact with the surface. This phenomenon contributes to smoother surface formation. Moreover, in ductile materials, surface roughness is significantly influenced by particle velocity [28, 29]. Particles with higher kinetic energy

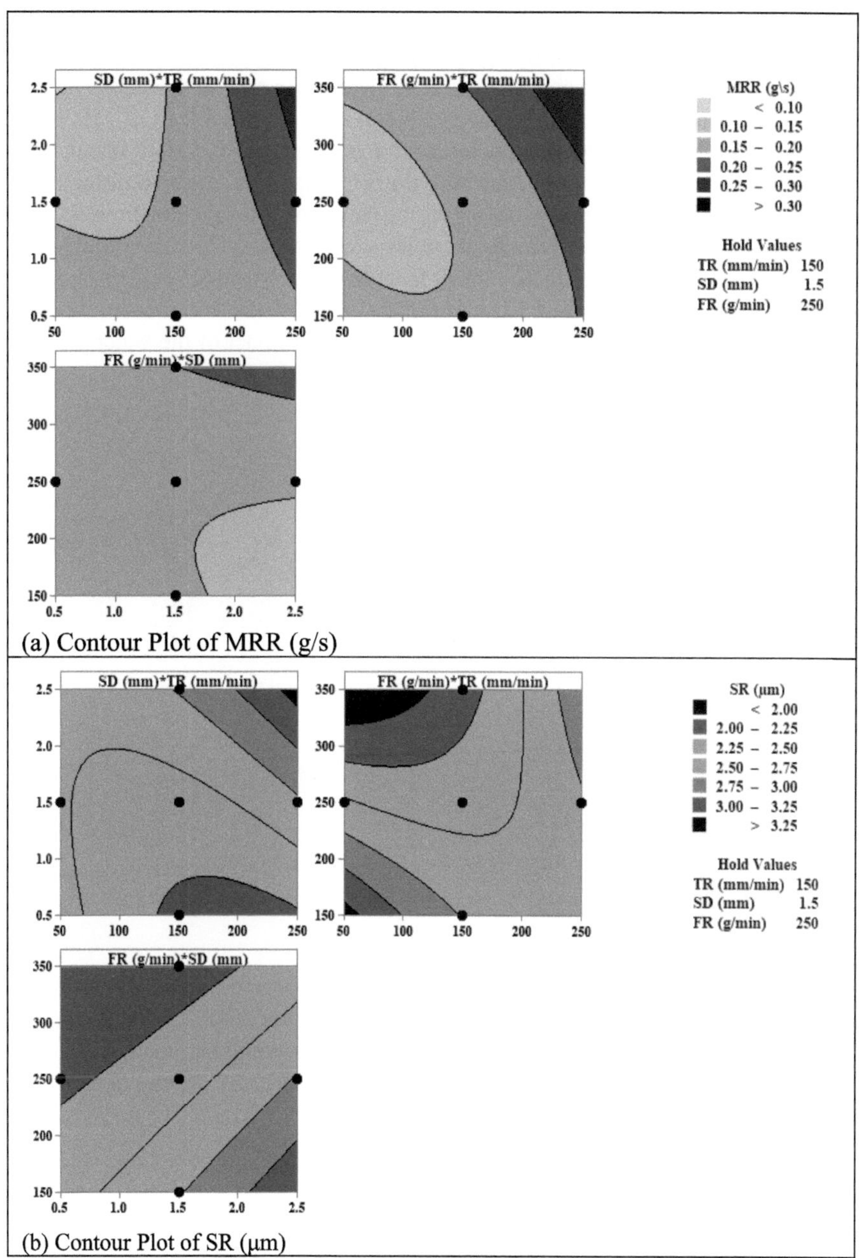

Fig. 10. Contour Plots for MRR and SR.

generate larger chips and result in rougher surfaces [30]. Cracks can be observed in the SEM image in Fig. 11 when AWJM is operated at a TR of 250 mm/min. At lower abrasive particle flow rates, SR increases due to the inconsistent distribution of abrasive particles within the water jet during AWJM [31]. To achieve a smoother surface, the FR must be maximized. The combined effect of FR and TR (ranging from 150 to 250 mm/min) significantly reduces SR. When TR decreases, the erosion rate also declines due to the reduced mechanical force acting on the machined surface in AWJM [32]. However, at TR values exceeding 200 mm/min, the interaction between SD and MRR becomes more pronounced. Furthermore, Fig. 11 illustrates the formation of a crater, resulting from the substantial removal of material by hard abrasive particles. This phenomenon occurs due to the repeated impact of high-velocity particles at an increased TR. The ploughing-like score marks observed on the surface result from hard abrasive particles sliding against the specimen, as illustrated in Fig. 12. Additionally, abrasive particles become embedded in the machined surface due to the high kinetic energy impact of the water jet and the sharp cutting edges of garnet abrasives [32].

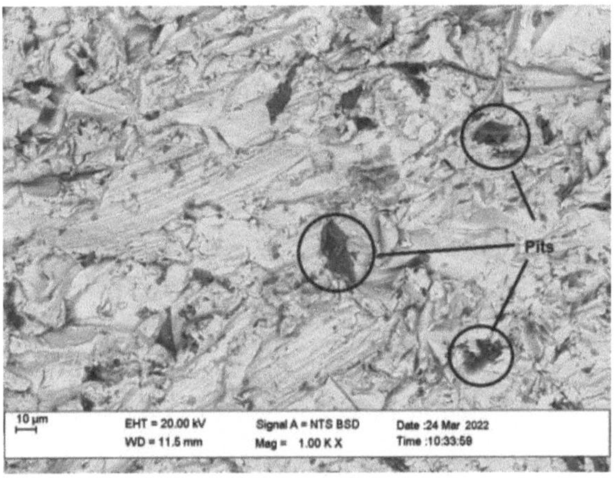

Fig. 11. SEM image of specimens at TR = 250 mm/min, SD = 1.5 mm, FR = 250 g/min.

Energy Dispersive X-ray (EDAX) analysis was conducted on the Monel 400 alloy specimen under operating conditions of TR = 150 mm/min, SD = 0.5 mm, and FR = 250 g/min. The peaks shown in Fig. 13 highlight the essential elements present in the specimen, including Cu, Ni, and Si. Furthermore, Mg elements are also detected in the analysis, attributed to the embedded garnet abrasives.

6.2 Optimization of Process Parameters

To optimize the process parameters, the GHO, GWO, ALO, and MVO algorithms were utilized to achieve the dual objectives of maximizing Material Removal Rate (MRR) while minimizing Surface Roughness (SR) simultaneously. Given that these objectives are conflicting, a non-dominated Pareto optimal solution was implemented instead of converting the multi-objective problem into a single-objective one using weighted coefficients. Using the regression Eqs. (4) and (5), the response values for MRR and SR were computed for each metaheuristic algorithm. The respective algorithmic parameters, as

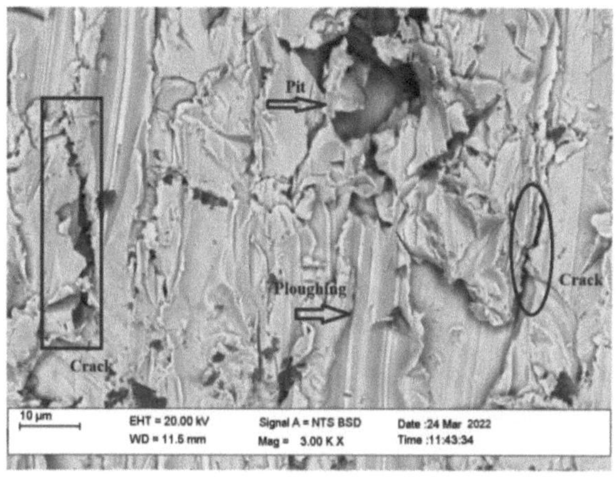

Fig. 12. SEM image of specimens at TR = 150 mm/min, SD = 0.5 mm and FR = 250 g/min.

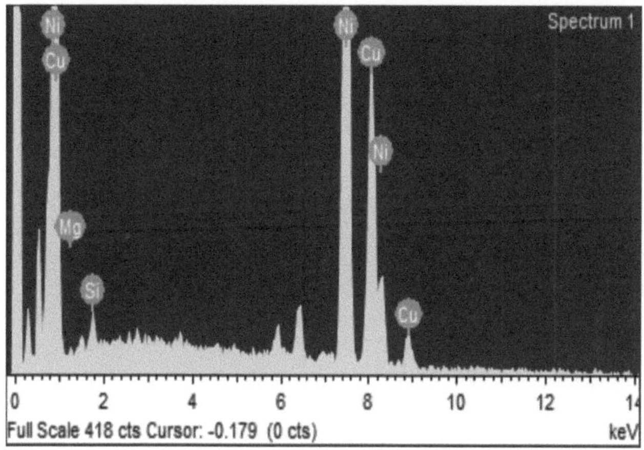

Fig. 13. EDAX image of the specimen at TR = 150 mm/min, SD = 0.5 mm and FR = 250 g/min.

shown in Table 10, were applied across various metaheuristic approaches to determine the best performance measures. Each algorithm was executed for 100 iterations.

Figures 14(a) and (b) illustrate the convergence graphs of MRR and SR for all metaheuristic algorithms i.e. GHO, GWO, ALO, and MVO. Each algorithm was executed 30 times, and the resulting solutions were recorded. To transform the conflicting dual objectives into a single objective, the EDAS, Deng's, and TOPSIS methods were integrated with each metaheuristic algorithm. Figure 15 presents the Pareto optimal solutions obtained over 30 runs using different MCDM techniques.

Table 10. Algorithm parameters for GHO, GWO, ALO and MVO algorithms

Parameters	Value
MVO algorithm	
Minimum wormhole existence probability (WEP_min)	0.2
Maximum wormhole existence probability (WEP_max)	1
Exploitation accuracy (p)	6
Number of universe	100
Maximum time (number of iterations)	100
Archive size	100
ALO algorithm	
Number of ant lion (na)	100
Number of iterations ($nitr$)	100
Accuracy level of exploitation	3 to 6 (based on % of no. of iterations)
Archieve size	100
GWO algorithm	
Scale factor (SF)	2
No. of Grey wolf / population size (np)	100
No. of iteration ($nitr$)	100
GHO algorithm	
No. of grasshopper (ng)	100
No. of iterations ($nitr$)	100
Factors to control exploration and exploitation (C_{min} & C_{max})	0.00004 and 1
Archieve size	100

Table 11 displays the HV values computed using Eq. (36) for the four different MCDM-metaheuristic algorithm combinations employed in this study. The algorithm with the highest HV value is considered the most efficient. Among all tested approaches, the GWO combined with EDAS demonstrated superior performance, achieving the highest hypervolume value of 0.076. Based on the HV values obtained for different algorithmic combinations, the corresponding optimal process parameters and performance measures are summarized in Table 12.

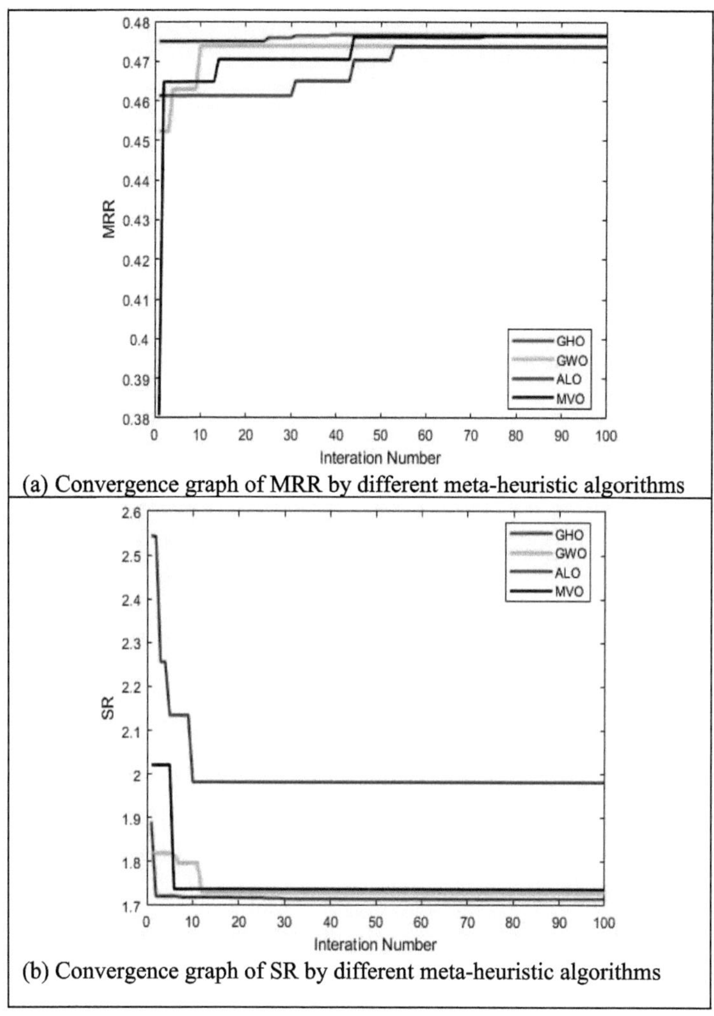

(a) Convergence graph of MRR by different meta-heuristic algorithms

(b) Convergence graph of SR by different meta-heuristic algorithms

Fig. 14. Comparison of convergence plot by various algorithms.

The confirmatory tests are performed after rounding off certain optimal process parameter values obtained using Hybrid Metaheuristics and MCDM approaches, considering the practical applicability of the AWJM process. As shown in Table 13, the confirmatory test results indicate that the percentage error for the chosen performance characteristics is less than 3% for the outputs produced by the respective algorithm combinations. For Monel 400 alloys to achieve improved quality, these process parameter values can be easily adjusted within the AWJM setup.

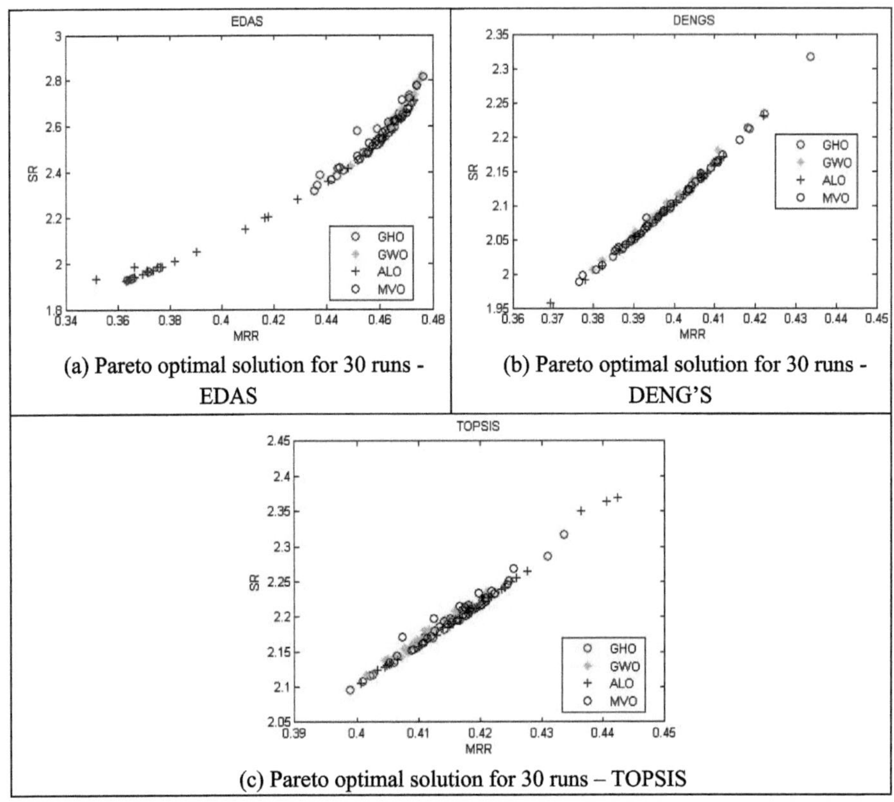

(a) Pareto optimal solution for 30 runs - EDAS

(b) Pareto optimal solution for 30 runs - DENG'S

(c) Pareto optimal solution for 30 runs – TOPSIS

Fig. 15. Pareto optimal solutions for 30 runs (100 Iterations).

Table 11. Hypervolume for different MCDM-Metaheuristics.

MCDM/Algorithm	GWO	GHO	ALO	MVO
EDAS	0.076	0.006	0.071	0.012
DENG'S	0.007	0.004	0.005	0.004
TOPSIS	0.021	0.009	0.014	0.008

Table 12. Optimum parameters by various algorithms.

Algorithm/MCDM	Parameters and Responses	EDAS	DENG'S	TOPSIS
GHO	TR (mm/min)	250.0000	250.0000	250.0000
	SD (mm)	0.8276	0.6073	0.6375
	FR (g/min)	231.51435	191.1261	199.3914

(*continued*)

Table 12. (*continued*)

Algorithm/MCDM	Parameters and Responses	EDAS	DENG'S	TOPSIS
	MRR (g/s)	0.43541	0.3989	0.4059
	SR (μm)	2.31811	2.0959	2.1343
GWO	TR (mm/min)	249.8823	249.9809	249.9626
	SD (mm)	1.0186	0.5625	0.6790
	FR (g/min)	237.5948	183.5510	194.8178
	MRR (g/s)	0.4488	0.3903	0.4072
	SR (μm)	2.4299	2.0519	2.1426
ALO	TR (mm/min)	250.0000	250.0000	250.0000
	SD (mm)	0.5000	0.5718	0.6945
	FR (g/min)	150.0000	189.5171	197.4201
	MRR (g/s)	0.3632	0.3946	0.4099
	SR (μm)	1.9264	2.0735	2.1577
MVO	TR (mm/min)	249.9999	249.9775	249.9593
	SD (mm)	0.8526	0.5619	0.6894
	FR (g/min)	244.4401	186.3224	200.2413
	MRR (g/s)	0.4417	0.3918	0.4108
	SR (μm)	2.3662	2.0594	2.1629

Table 13. Confirmatory Results.

Hybrid Techniques	AMJM Process Parameters			Predicted Performance Results		Experimental Results		Error (%)	
	TR (mm/min)	SD (mm)	FR (g/min)	MRR (g/s)	SR (μm)	MRR (g/s)	SR (μm)	MRR	SR
GHO-DENG'S	250	0.6	191	0.3989	2.0959	0.409	2.15	2.54	2.58
GWO-EDAS	250	1.0	238	0.4488	2.4299	0.439	2.39	2.19	1.64
ALO-EDAS	250	0.5	150	0.3632	1.9264	0.369	1.96	1.61	1.75
MVO-TOPSIS	250	0.7	200	0.4108	2.1629	0.418	2.10	1.76	2.91

7 Conclusion

In this study, Abrasive Water Jet Machining (AWJM) operations were performed on Monel 400 alloys by varying Traverse Rate (TR), Stand-off Distance (SD), and Feed Rate (FR) of abrasives. The primary effects of these parameters on Material Removal Rate (MRR) and Surface Roughness (SR) were analyzed using main effect plots, contour plots, interaction plots, and Scanning Electron Microscopy (SEM) images. Furthermore, metaheuristic algorithms, including the Grasshopper Optimization Algorithm (GHO), Grey Wolf Optimizer (GWO), Ant Lion Optimizer (ALO), and Multiverse Optimizer (MVO), were employed in conjunction with Multi-Criteria Decision-Making (MCDM) techniques such as EDAS, Deng's method, and TOPSIS. The key findings of this study are summarized below.

The investigation confirms that Traverse Rate (TR) enhances the performance of the AWJM process for Monel 400 alloys, with higher TR values preferred to achieve a high removal rate.

1. To obtain a smoother surface, lower TR, lower Stand-off Distance (SD), and higher abrasive Feed Rate (FR) are recommended. SEM images validate the influence of these parameters on the specimen surface.
2. The GWO-EDAS combination was found to be the most effective, yielding the Pareto optimal solutions with the highest hypervolume indicator of 0.076 among 12 different metaheuristic-MCDM combinations.
3. The GWO-EDAS approach identified the optimal AWJM machining parameters as TR = 250 mm/min, SD = 1 mm, and FR = 238 g/min, resulting in an MRR of 0.449 g/s and an SR of 2.43 μm.
4. Confirmatory experiments for selecting optimal solutions demonstrated practical applicability, with an observed error margin of less than 3%.

For future research, Machine Learning (ML) algorithms can be employed to develop more advanced predictive models for the abrasive water jet machining of Monel 400 alloys. Additionally, key AWJM characteristics, such as radial overcut and dimensional accuracy, can be analyzed to further evaluate process performance advantages. Instead of utilizing the hypervolume indicator to consolidate multiple objectives, Spacing Metrics, a quality assessment tool, may be considered as an alternative approach. By achieving precise parameter values, the findings of this study offer an optimized metaheuristic-MCDM framework for enhancing the AWJM process.

References

1. Ventrella, V.A., Berretta, J.R., De Rossi, W.: Micro welding of Ni-based alloy Monel 400 thin foil by pulsed Nd: YAG laser. Phys. Procedia. **12**(B), 347–354 (2011). https://doi.org/10.1016/j.phpro.2011.03.143
2. Tayal, A., Kalsi, N.S., Gupta, M.K., Garcia-Collado, A., Sarikaya, M.: Reliability and economic analysis in sustainable machining of Monel 400 alloy. Proc. Inst. Mech. Eng. C J. Mech. Eng. Sci. **235**(21), 1–17 (2021). https://doi.org/10.1177/0954406220986818
3. Shukla, R., Singh, D.: Experimentation investigation of abrasive water jet machining parameters using Taguchi and evolutionary optimization techniques. Swarm Evolut. Comput. **32**, 167 183 (2017). https://doi.org/10.1016/j.swevo.2016.07.002
4. Natarajan, Y., Murugesan, P.K., Mohan, M., Liyakath Ali Khan, S.A.: Abrasive water jet machining process: a state of art of review. J. Manuf. Process. **49**, 271 322 (2020). https://doi.org/10.1016/j.jmapro.2019.11.030
5. Arun Raja, A.C., Senkathir, S., Geethapriyan, T., Nanditta, R.V., Goyal, K.: Performance characteristics of abrasive water-jet machining on monel 400 using two different abrasive grain size. AIP Conf. Proc. **2395**, 040006 (2021). https://doi.org/10.1063/5.0068390

6. Arun, A., Rajkumar, K., Sasidharan, S., Balasubramaniyan, C.: Process parameters optimization in machining of monel 400 alloy using abrasive water jet machining. Mater. TodayProc. **98**, 28 32 (2024). https://doi.org/10.1016/j.matpr.2023.08.308

7. Rajesh, N., Lokanadham, R.: Optimization of machining parameters & studies on characteristics of Monel k400 alloy using abrasive water jet machining using ANFIS. Mater. Today Proc. **98**, 46 40 (2024). https://doi.org/10.1016/j.matpr.2023.08.376

8. Chakraborty, S., Bhattacharyya, B., Diyaley, S.: Applications of optimization techniques for parametric analysis of non-traditional machining processes: a review. Manage. Sci. Lett. **9**(3), 467 494 (2019). https://doi.org/10.5267/j.msl.2018.12.004

9. Abdelaoui, F.Z.E., Jabri, A., Barkany, A.E.: Optimization techniques for energy efficiency in machining processes—a review. Int. J. Adv. Manuf. Technol. **125**, 2967 3001 (2023). https://doi.org/10.1007/s00170-023-10927-y

10. Perec, A.: Multiple response optimization of abrasive water jet cutting process using response surface methodology (RSM). Procedia Comput. Sci. **192**, 931 940 (2021). https://doi.org/10.1016/j.procs.2021.08.096

11. Okwu, M.O., Tartibu, L.K.: Ant lion optimization algorithm. In: Metaheuristic Optimization: Nature-Inspired Algorithms Swarm and Computational Intelligence, Theory and Applications. Studies in Computational Intelligence 927, pp. 33 41. Springer, Cham (2021). https://doi.org/10.1007/978-3-030-61111-8_9

12. Jai Rajesh, P., Balambica, V., Achudhan, M.: Advanced machining performance through optimization of Awjm parameters using metaheuristic techniques, 012097. J. Phys. Conf. Ser. **2837** (2024). https://doi.org/10.1088/1742-6596/2837/1/012097

13. Wan, L., Liu, J., Qian, Y., Wang, X., Wu, S., Du, H., Li, D.: Analytical modeling and multi-objective optimization algorithm for abrasive waterjet milling Ti6Al4V. Int. J. Adv. Manuf. Technol. **123**, 4367 4384 (2022). https://doi.org/10.1007/s00170-022-10396-9

14. Manoj, M., Jinu, G.R., Kumar, J.S.: Process optimization of abrasive water jet machining of aluminum hybrid composite using Taguchi DEAR methodology. Int. J. Met. **18**, 933 943 (2024). https://doi.org/10.1007/s40962-023-01071-0

15. Vengatajalapathi, N., Ayyappan, S., Rajasekar, V.: Eco-friendly filtration of Nickel from the sludge during electrochemical machining of Monel 400 alloys. Global NEST J. **24**(2), 203 211 (2022). https://doi.org/10.30955/gnj.004226

16. Nair, A., Kumanan, S.: Multi-performance optimization of abrasive water jet machining of Inconel 617 using WPCA. Mater. Manuf. Process. **32**(6), 693 699 (2017). https://doi.org/10.1080/10426914.2016.1244844

17. Llanto, J.M., Tolouei-Rad, M., Vafadar, A., Aamir, M.: Recent Progress trend on abrasive waterjet cutting of metallic materials: a review. Appl. Sci. **11**(8), 3344 (2021). https://doi.org/10.3390/app11083344

18. Samson, R.M., Rajak, S., Kannan, T.D.B., Sampreet, K.R.: Optimization of process parameters in abrasive water jet machining of Inconel 718 using VIKOR method. J. Inst. Eng. Ser. C. **101**, 579 585 (2020). https://doi.org/10.1007/s40032-020-00569-4

19. Saremi, S., Mirjalili, S.: Andrew Lewis.: grasshopper optimisation algorithm: theory and application. Adv. Eng. Softw. **105**, 30 47 (2017). https://doi.org/10.1016/j.advengsoft.2017.01.004

20. Mirjalili, S., Mirjalili, S.M., Lewis, A.: Grey wolf optimizer. Adv. Eng. Softw. **69**, 46 61 (2014). https://doi.org/10.1016/j.advengsoft.2013.12.007

21. Mirjalili, S.: The ant lion optimizer. Adv. Eng. Softw. **83**, 80 98 (2015). https://doi.org/10.1016/j.advengsoft.2015.01.010

22. Mirjalili, S., Mirjalili, S.M., Hatamlou, A.: Multi-verse Optimizer: a nature-inspired algorithm for global optimization. Neural Comput. Applic. **27**, 495 513 (2016). https://doi.org/10.1007/s00521-015-1870-7

23. Keshavarz Ghorabaee, M., Zavadskas, E.K., Olfat, L., Turskis, Z.: Multi-criteria inventory classification using a new method of evaluation based on distance from average solution (EDAS). Informatica. **26**(3), 435 451 (2015). https://doi.org/10.15388/Informatica.2015.57
24. Deng, H.: A similarity-based approach to ranking multicriteria alternatives. In: Huang, D.S., Heutte, L., Loog, M. (eds.) Advanced Intelligent Computing Theories and Applications. With Aspects of Artificial Intelligence. ICIC 2007 Lecture Notes in Computer Science, vol. 4682, pp. 253 262. Springer, Berlin, Heidelberg (2007). https://doi.org/10.1007/978-3-540-74205-0_28
25. Ananthakumar, K., Rajamani, D., Balasubramanian, E., Paulo Davim, J.: Measurement and optimization of multi-response characteristics in plasma arc cutting of Monel 400™ using RSM and TOPSIS. Measurement. **135**, 725 737 (2019). https://doi.org/10.1016/j.measurement.2018.12.010
26. Nagarajan, V., Solaiyappan, A., Mahalingam, S.K., Nagarajan, L., Salunkhe, S., Nasr, E.A., Shanmugam, R., Hussein, H.M.A.M.: Meta-heuristic technique-based parametric optimization for electrochemical machining of Monel 400 alloys to investigate the material removal rate and the sludge. Appl. Sci. **12**(6), 2793 (2022). https://doi.org/10.3390/app12062793
27. Shang, K., Ishibuchi, H., He, L., Pang, L.M.: A survey on the hypervolume indicator in evolutionary multiobjective optimization. IEEE Trans. Evolut. Comput. **25**(1), 1 20 (2021). https://doi.org/10.1109/TEVC.2020.3013290
28. Begic-Hajdarevic, D., Cekic, A., Mehmedovic, M.: Almina Djelmic.: experimental study on surface roughness in abrasive water jet cutting. Procedia Eng. **100**, 394 399 (2015). https://doi.org/10.1016/j.proeng.2015.01.383
29. Ahmed, D.H., Naser, J., Deam, R.T.: Particles impact characteristics on cutting surface during the abrasive water jet machining: numerical study. J. Mater. Process. Technol. **232**, 116 130 (2016). https://doi.org/10.1016/J.JMATPROTEC.2016.01.032
30. Haj Mohammad Jafar, R., Nouraei, H., Emamifar, M., Papini, M., Spelt, J.K.: Erosion modeling in abrasive slurry jet micro-machining of brittle materials. J. Manuf. Process. **17**, 127 140 (2015). https://doi.org/10.1016/j.jmapro.2014.08.006
31. Haghbin, N., Spelt, J.K., Papini, M.: Abrasive waterjet micro-machining of channels in metals: comparison between machining in air and submerged in water. Int J Mach Tool Manu. **88**, 108 117 (2015). https://doi.org/10.1016/j.ijmachtools.2014.09.012
32. Uthayakumar, M., Khan, M.A., Thirumalai, K.S., Slota, A., Zajac, J.: Machinability of nickel based superalloy by abrasive water jet machining. Mater. Manuf. Process. **31**(13), 1733 1739 (2015). https://doi.org/10.1080/10426914.2015.1103859

Internet and Mobile Technologies

Developing Users' Acceptance Framework towards Mobile Commerce: An Exploratory Study

Arshan Kler[1]([✉]), Bhupinder Preet Bedi[1], and Simerjeet Singh Bawa[2]

[1] Department of Management, Chandigarh School of Business, Chandigarh Group of Colleges, Jhanjeri, Mohali 140307, Punjab, India
{arshan.j1053,bhupinder.j3141}@cgc.ac.in
[2] Chitkara Business School, Chitkara University, Rajpura, Punjab, India

Abstract. By integrating millions of individuals; mobile phones are helping to close the digital divide in India. Mobile phones are revolutionizing how businesses operate and how customers interact with them as a result of the quick rise in internet access and advancements in wireless and mobile technologies. The emergence of a new business model known as Mobile Commerce (M-Commerce), which uses the internet on mobile devices, is the result of the internet's amazing commercial opportunities for e-commerce. Since m-commerce is still in its earliest stages in India and a large portion of the public has not had the chance to use this new technology, there are very few academic researches on the subject. Moreover, despite the strong potential of m-commerce, the actual level of m-commerce activities remains relatively low. The current study purposes to determine the primary variables impacting the perception of Mobile users towards Mobile Commerce. A survey of 593 mobile internet users from urban areas in Punjab and Chandigarh Region was conducted, and was examined using SEM technique. The findings indicate that four elements including perceived usefulness, perceived ease of use, social influence, and perceived enjoyment greatly impact mobile users' perceptions. The m-commerce adoption was not significantly inclined by variables: perceived risk and facilitating conditions. These results will help service providers to create appropriate marketing strategies to attract mobile users to accept m-commerce.

Keywords: Mobile Commerce · Perception · Mobile Users · Perceived Usefulness · Perceived Ease of Use

1 Introduction

The world has evolved into a global information society due to the internet, where any individual can obtain any information at any time from any location. The internet has developed into a useful supplement to the conventional means of information access. Using electronic markets to contact clients and establish lasting connections has greatly benefited businesses.

India is currently experiencing the Second Information Technology Revolution. Since the 1990s, there has been rapid growth in number of internet users. According to

A. Mirzazadeh et al. (Eds.): ODSIE 2024, CCIS 2482, pp. 421–436, 2026.
https://doi.org/10.1007/978-3-031-93601-2_25

data from the Internet and Mobile Association of India (IAMAI) and KANTAR IMRB report, as of December 2023 [1], 821 million people in India were online, making it the country with the second-highest internet user base behind China. This massive rise in mobile internet users was found to be driven by better quality data offerings at reasonable pricing. In terms of mobile internet usage activities, communication, social networking, and entertainment were found to be the most popular in urban India, while entertainment was the most popular in rural India, followed by social networking and communication. Future internet access will mostly come through mobile devices; therefore, ecosystem growth will require collaborative efforts from all stakeholders.

Mobile phones are revolutionizing corporate business practices and customer-business interactions in India due to the country's rapid internet penetration and advancements in wireless and mobile technology. From a business perspective, every company seeks to develop and thrive in a distinctive manner. By targeting a new customer base—mobile users—and offering superior goods and services, mobile commerce offers a special means of growing a firm. Telecommunication and the internet are becoming more and more commonplace in everyone's everyday life (M-commerce—Reports, Statistics & Marketing Trends, 2024) [2]. With the advent of mobile internet usage, a new type of business model has evolved, and the internet has produced incredible commercial potential for e-commerce.

The definition of mobile commerce means using internet for doing every transaction to buy product or services via mobile devices. Modern smartphones, running on advanced mobile operating systems, offer sophisticated computing capabilities and seamless connectivity. The mobile devices help in accessing internet at anytime and anywhere due to which the users of m-commerce have been increasing drastically in the recent years.

The growing adoption of smartphones and affordable internet access is driving the surge in mobile commerce in developing countries. Specifically, post Covid-19, the scenario has changed, people are now preferring online shopping instead of going to the marketplace. The Majority of the studies that are conducted in India regarding m-commerce are theoretical studies or had considered the application of m-commerce by focusing on the attitude or intention of mobile users. But before identifying the key factors which affect attitude of users, it is important to discover the variables that influences the perception of the towards M-commerce. This will assist service providers in developing suitable marketing plans to draw in a sizable user base of mobile users for m-commerce.

By creating a research model that pinpoints the variables influencing mobile consumers' perceptions of m-commerce, the current study closes these gaps. This model's uniqueness stems from its incorporation of the essential elements of mobile commerce adoption and its ability to discern how mobile users perceive the industry, which in turn influences their intention and actual behaviour. The current study is conducted in three major sections, firstly, variables of Mobile commerce which affects the mobile users in acceptance are identified. Secondly the factors identified are then validated using confirmatory factor analysis through structural equation modelling. Lastly the structural model is utilized to identify the impact of all variables and their significance on perception towards mobile commerce.

2 Review of Literature

2.1 Variables of Mobile Commerce

Perceived Usefulness. Perceived usefulness remains a crucial driver of m-commerce adoption [3–5]. Consumers are more likely to adopt m-commerce when they perceive tangible benefits and value [6]. This is particularly relevant in the context of increasing mobile penetration and the demand for convenient shopping experiences. Recent trends suggest that the integration of personalized recommendations and targeted advertising within m-commerce platforms can further enhance perceived usefulness.

Perceived Ease of Use. The increasing sophistication of mobile interfaces and the simplification of transaction processes contribute to the perceived ease of use of m-commerce applications [4] User-friendly design, intuitive navigation, and seamless checkout experiences are essential for attracting and retaining m-commerce users. The rise of mobile wallets and one-click purchasing options further enhances the convenience and ease of use associated with m-commerce.

Perceived Enjoyment. Perceived enjoyment continues to be a significant factor in m-commerce adoption, particularly for entertainment-related applications and interactive shopping experiences [5]. Gamification, personalized content, and social shopping features can enhance the enjoyment associated with m-commerce. The integration of augmented reality and virtual reality technologies is also creating more immersive and engaging m-commerce experiences.

Perceived Risk. Despite advancements in security measures, perceived risk remains a barrier to m-commerce adoption [4, 7]. Concerns about data breaches, fraud, and privacy issues can deter consumers from engaging in m-commerce transactions. Building trust and transparency through robust security protocols, clear privacy policies, and secure payment gateways is crucial for mitigating perceived risk and fostering m-commerce adoption. The increasing adoption of biometric authentication and secure payment technologies can contribute to enhancing consumer trust.

Social Influence. Social influence continues to shape m-commerce adoption, particularly among younger demographics [5, 8, 9]. Social media marketing, influencer endorsements, and peer recommendations can significantly impact consumer behaviour. The integration of social commerce features within m-commerce platforms allows users to share their purchases and experiences, further amplifying social influence.

Facilitating Conditions. Facilitating conditions, such as reliable mobile networks, secure payment systems, and supportive infrastructure, are essential for m-commerce to thrive [10]. The expansion of 5G networks and the increasing availability of high-speed internet access are creating a more conducive environment for m-commerce growth. Furthermore, the development of interoperable payment systems and the increasing adoption of mobile wallets are facilitating seamless and secure m-commerce transactions [11]. Government initiatives and regulatory frameworks also play a crucial role in creating a supportive environment for m-commerce development [12]. As m-commerce continues to evolve, addressing the challenges related to security, privacy, and infrastructure will be crucial for realizing its full potential [2]. The growth of m-commerce is expected to continue in the coming years, driven by increasing mobile penetration, technological advancements, and changing consumer behaviour.

2.2 Perception Toward Using Mobile Commerce

In order to engage with the world, perception is essential. We are inundated with stimuli all the time, both at work and in our personal life. Our reality is shaped by the way we understand and use this information. The deliberate process of choosing, arranging, and interpreting sensory data to derive a meaningful understanding of our environment is called perception. Once we have an understanding of the circumstance, we can act appropriately [13]. Positive consumer impressions of the digitalization will be helpful in forecasting future behaviour for the purposes of new technology adoption study. Researchers and practitioners who want to apply the TAM model to these upcoming technologies will find this to be significant.

2.3 Theoretical Framework

Varied scholars have developed modified and extended versions of TAM over time for varied levels of technology adoption. Given the characteristics of m-commerce and the market environment, it is crucial to take into account a variety of elements while researching how mobile users' perceptions affect the uptake of m-commerce. Therefore, Following Model has been developed to achieve this objective and hypotheses has been developed respectively (Fig. 1).

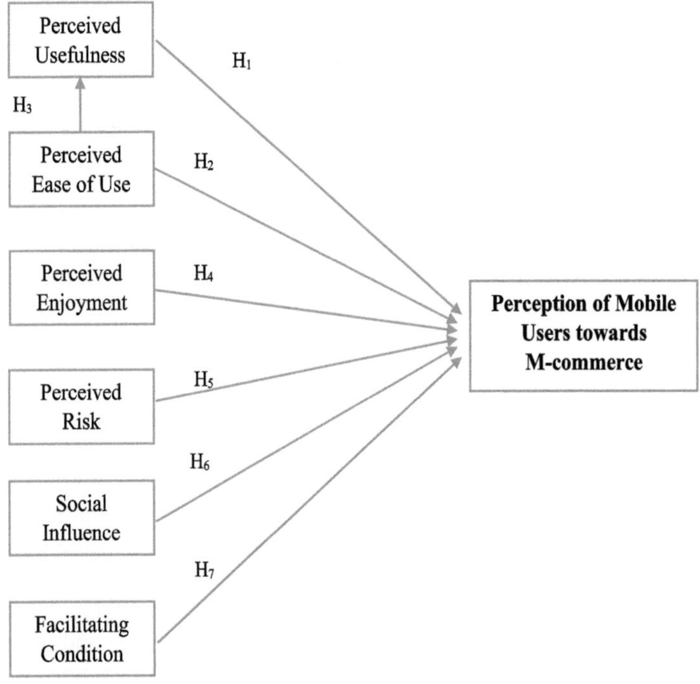

Source: Author's Compilation

Fig. 1. Theoretical Model. Source: Author's Compilation

2.4 Hypotheses Development

To achieve the first objective, appropriate statistical tools and techniques were used. For extracting the factors related to perception, Exploratory Factor Analysis (EFA) was utilized. Then Structural Equation Modelling (SEM) method is applied to study impact of the extracted factors on the perception. The hypotheses for the conceptual framework are shown below in Table 1.

Table 1. Hypotheses Development.

Hypotheses name	Hypotheses statement
	Factors influencing perception towards m-commerce
H_1	Perceived usefulness (PU) positively impacts mobile users' perception of Mobile Commerce
H_2	Perceived ease of use (PEOU) positively impacts perception of mobile users towards Mobile Commerce
H_3	Perceived ease of use (PEOU) has indirectly positive impact through Perceived Usefulness (PU) on perception of mobile users towards Mobile Commerce
H_4	Perceived Enjoyment (PE) positively impacts perception of mobile users towards Mobile Commerce
H_5	Perceived Risk (PR) negatively impacts perception of mobile users towards Mobile Commerce
H_6	Social Influence (SI) positively impacts perception of mobile users towards Mobile Commerce
H_7	Facilitating Condition (FC) positively impacts perception of mobile users towards Mobile Commerce

Source: Author's Compilation

3 Research Methodology

3.1 Research Design

A structure or framework that guides the gathering and examination of study data is called a research design. The creation of a design like this makes research as effective as possible and produces the most information. The study aims to accomplish its stated objectives through the implementation of both descriptive and exploratory research methods.

3.2 Sampling Design

A universe consists of all the items in any field of inquiry. A sampling design helps in framing a plan for defining a sample from a given universe. Sample for this study will be taken from internet users of Punjab and Chandigarh Region, residing in urban areas. Punjab comprises of twenty-two districts in all with a total population of 2.77 crore and Chandigarh have population of 10.55 lakhs. Out of this population, Punjab comprises of total 1.03 crore urban population and Chandigarh having 10.26 lakh population [14]. For this study, only those respondents will be considered who are of age 21 to 40 years, as it was found in the Census report [14] that majority of educated working people lie in this age category. More than three fourth of population of Punjab state is concentrated in Amritsar and Ludhiana. For the purpose of sample selection, study is restricted to four districts of Punjab and Chandigarh city.

For the purpose of collecting the primary data, stratified proportionate sampling technique was used. Under this technique, the final sample is split according to the proportion of each stratum after the overall population is split up into various subgroups or strata. Therefore, for the present study the population was categorized by creating strata on the basis of the number of populations that are between 21 to 40 years in urbanized districts of Punjab.

3.3 Sample Size

600 sample size for the present study has been calculated using a formula given by American Marketing Association [15]. In the formula, the confidence level is considered at 95% and the confidence interval at 4%.

For choosing the sample from the defined districts of Punjab and UT Chandigarh, Purposive Judgmental sampling technique has been used. Those individuals were chosen as the sample for the study whose input will satisfy the purpose of the study based on the judgment of the researcher.

3.4 Data Collection

The structured questionnaire was employed to assemble the main information for the investigation. It is developed from extensive literature review and interviews conducted using open ended questions among the experts in the field of mobile commerce such as mobile application developers, digital marketers, professors etc. the questionnaire was then tested by conducting pilot survey that helped in validating the final questionnaire.

4 Data Analysis

4.1 Reliability Analysis

Before being used for the study's execution, the questionnaire must be reviewed for accuracy, correctness, and efficacy. Therefore, in order to evaluate the questionnaire's reliability and gather data for the current study in accordance with the sample plan, a pilot study comprising of one hundred respondents was carried out.

Based on review of literature, six factors were extracted regarding the adoption of mobile commerce. By applying Cronbach's alpha method, the consistency of the factors was checked. The values of Cronbach's alpha coefficients for all the six variables and one dependent variable i.e., Perception was more than 0.7, indicating that all construct variables are trustworthy in accordance with [16] recommendations (Table 2).

Table 2. Reliability Results.

Factors	Cronbach's Alpha
Perceived Usefulness (PU)	0.812
Perceived Ease of Use (PEOU)	0.952
Perceived Enjoyment (PE)	0.874
Perceived Risk (PR)	0.818
Social Influence (SI)	0.877
Facilitating Conditions (FC)	0.801
Perception (PRCP)	0.822

Source: SPSS 21 Output

4.2 EFA-Exploratory Factor Analysis

The variables were computed by means of Principal Component Analysis (PCA) in conjunction with Exploratory Factor Analysis. As recommended by [17] on the basis of Eigen value of 1 or higher and factor loading of at least 0.5 or more was obtained using Varimax rotation (Table 3).

Table 3. KMO and Bartlett's test of Sphericity Results.

Kaiser-Meyer-Olkin Measure of Sampling Adequacy.		0.806
Bartlett's Test of Sphericity	Approx. Chi-Square	10732.792
	Df	351
	Sig.	.000

Source: SPSS 21 Output

The sampling adequacy, or whether the sample is large enough for doing research, is measured by the KMO. Bartlett's test of sphericity is applied to determine whether the variables are multicollinear. Bartlett's Test results should have a number that is less than 0.05 [18]. The sample data's KMO value is determined to be 0.806, and the Bartlett's Test yields a significant result ($p < 0.001$). A decent number for optimal factor analysis results is 76.597, which is the total variance explained by the thirteen retrieved factors. Only variables with Eigen values greater than one were taken into account (Table 4).

4.3 SEM-Structural Equation Modelling

The construct validity of the data is investigated using a multidimensional data analysis technique known as structural equation modelling (SEM). It is used in regression modelling to validate the theoretical model developed to get the desired outcome. A statistical approach called CFA is used to confirm the factors that were identified through

Table 4. Factor Loadings, Communalities and Cronbach's Alpha.

Factors	Name	Item	Loading	Communalities
Factor 1	Perceived Usefulness (PU)	PU1	0.925	0.885
		PU3	0.913	.659
		PU4	0.821	0.851
		PU2	0.788	0.693
Factor 2	Perceived Enjoyment (PE)	PE3	0.880	0.757
		PE1	0.852	0.752
		PE2	0.843	0.810
		PE4	0.836	0.732
Factor 3	Perception (PRCP)	PRCP1	0.894	0.851
		PRCP3	0.861	0.694
		PRCP4	0.807	0.825
		PRCP2	0.776	0.704
Factor 4	Facilitating Condition (FC)	FC2	0.918	0.724
		FC1	0.848	0.860
		FC4	0.834	0.699
		FC3	0.826	0.710
Factor 5	Social Influence (SI)	SI3	0.916	0.753
		SI2	0.841	0.733
		SI1	0.837	0.885
		SI4	0.751	0.623
Factor 6	Perceived Risk (PR)	PR2	0.902	0.617
		PR3	0.825	0.832
		PR4	0.810	0.700
		PR1	0.761	0.696
Factor 7	Perceived Ease of Use (PEOU)	PEOU2	0.942	0.877
		PEOU1	0.918	0.920
		PEOU3	0.899	0.842

Source: SPSS 21 Output

the use of EFA. It is a method for confirming the set of observed variables' structural integrity. It assesses how accurately the number of evolved constructs is represented by the observed variable.

Goodness of fit Results. The goodness of fit results for the measurement model are displayed in the table below. It is clear that every model fit index falls within the suggested threshold and the range provided by [16] (Fig. 2).

Structural Model: Impact of Key factors on User's Perception towards M-commerce. The model was created in direction to highlight the elements influencing how mobile consumers view mobile commerce. For this purpose, the factors confirmed in the CFA were considered (Fig. 3).

Because the study's sample size is substantial, the chi-square value was determined to be meaningful. According to [16, 19], when the sample size is greater than 200, the

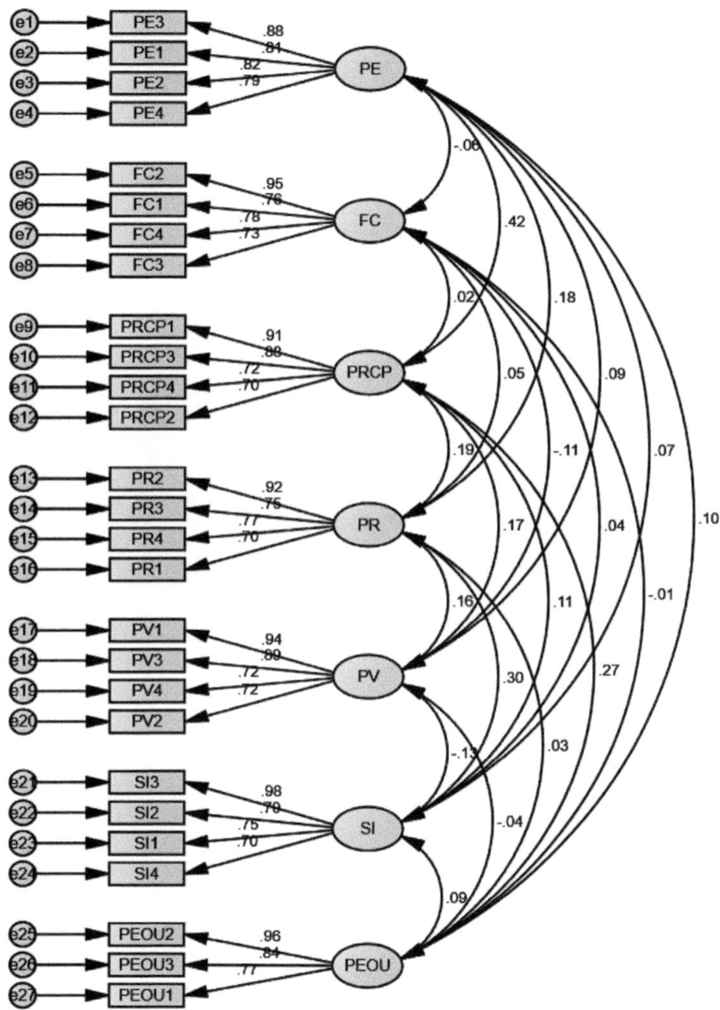

Source: SPSS AMOS 21 Output

Fig. 2. Measurement Model. Source: SPSS AMOS 21 Output

chi-square value is sensitive to the sample size. However, the measurement model's chi-square/degree of freedom value was 2.613, falling between [16, 20] recommended threshold limit of less than 3. With a p-value >0.05, the model was determined to be significant. It was also discovered that the goodness of fit (GIF) value was higher than the 0.90 upper bound. The measurement model is fit for the study since its CFI value of 0.953 indicates that it is within the minimal limit of 0.95, which is required for the model to be fit. It was determined that every goodness of fit index was meaningful and appropriate for the research (Table 5).

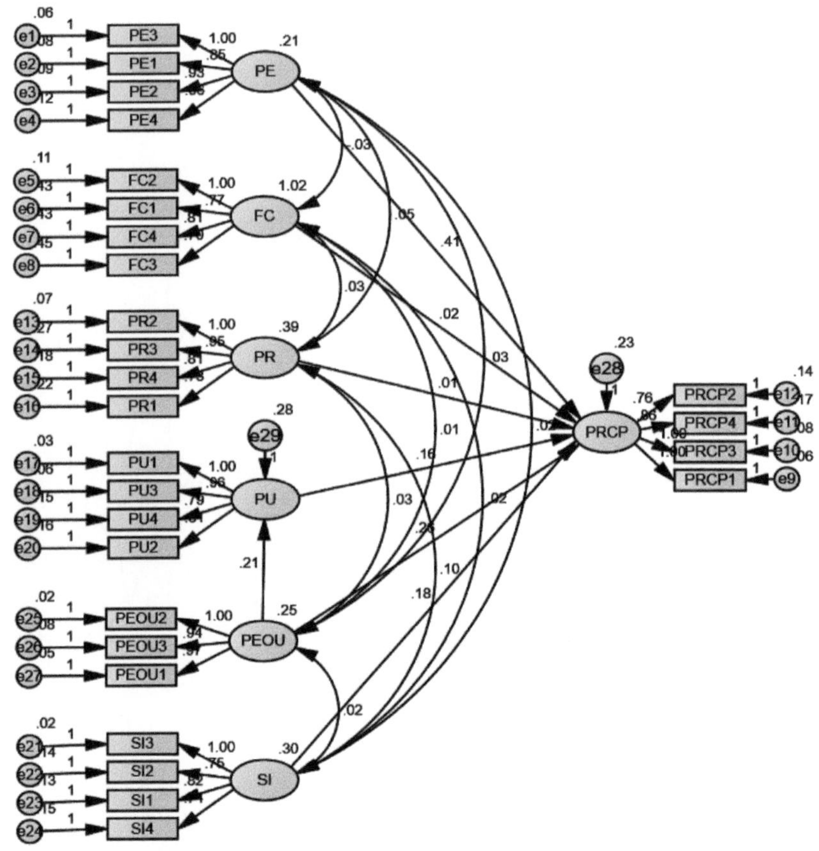

Source: SPSS AMOS 21 Output

Fig. 3. Structural Model. Source: SPSS AMOS 21 Output

Hypotheses Testing Results. Perceived Ease of Use (PEOU), Perceived Usefulness (PU), Perceived Enjoyment (PE), Social Influence (SI), Perceived Risk (PR), and Facilitating Condition (FC) were the six independent or exogenous factors, and Perception (PRCP) was the one dependent or endogenous variable. The SEM analysis results show both standardized and unstandardized estimates but for the current research purpose only standardized estimates were considered because with unstandardized weights, it is difficult to estimate the regression weights and beta coefficients for controlled variables.

Above is the structural model of the theoretically presented framework. It includes all the independent variables and its relationship with one dependent variable is studied. Former research has confirmed the association of the PEOU and PU, and the indirect effect of PEOU on perception through PU. So, for the current analysis, the effect of PEOU on PU and its indirect impact on perception is also considered (Table 6).

The table above represents the results of hypotheses testing conducted using SEM analysis. Based on the p-value of less than 0.01 for each of the seven hypotheses, the

Table 5. Measurement Model Fit Indices.

Model Fit Indices	Recommended Criterion (Hair et al., 2010)	Measurement Model
Chi-Square	–	802.098
Probability	> 0.05	0.000
(CMIN/DF)	< 3	2.613
Goodness of Fit (GFI)	>0.9	0.910
RMR	0.03–0.08	0.032
RMSEA	0.03–0.08	0.052
CFI	>0.95	0.953
NFI	>0.90	0.927
TLI	>0.90	0.946

Source: Author's Compilation

outcomes show five of the hypotheses are accepted, while the remaining two were rejected due to insignificance.

The PEOU had indirect effect on perception through PU ((β value $= 0.258$ and $p < 0.01$), which was also found to be significant. This shows that both these original factors of the original TAM given by [21] are significantly important variables that influence the perception of Indian mobile users. Among the other factors which were taken into consideration, Perceived Enjoyment (β value $= 0.409$ and $p < 0.01$) and social influence (β value $= 0.180$ and $p < 0.01$) were found positively significant factors that influence mobile users' perception towards mobile commerce.

It was discovered that two variables had little effects on how mobile consumers perceived m-commerce. These were, facilitating condition and perceived risk. Both these factors have significant value of above 0.05, which shows that they were not significant for the mobile users' perception. Figure 4 shows the structural equation modelling estimations for the structural model in which all the path estimates are shown.

5 Discussion

The present study was set out to determine the variables that affect how mobile users in the post-COVID scenario perceive mobile commerce. This goal was accomplished in two sections. First, a detailed investigation of the literature was done to regulate the elements that were found to have a major impact on how mobile users perceived m-commerce. A defined questionnaire was created as per the literature analysis, and data was gathered from 593 respondents. The model was developed in direction to pinpoint the elements that, in the post-COVID era, significantly affect how mobile users perceive themselves.

The perception of mobile users towards mobile commerce can be inclined by many variables. The results showed that out of six variables obtained, four factors were found

Table 6. Summary of Hypotheses Testing.

Hypothesis Name	Hypothesis statement	Path Coefficient	P-value	Conclusion
H_1	Perceived usefulness (PU) positively impacts mobile users' perception of mobile commerce	0.163	***	Accepted
H_2	Perceived ease of use (PEOU) positively impacts perception of mobile users towards Mobile Commerce	0.258	***	Accepted
H_3	Perceived ease of use (PEOU) has indirectly positive impact through Perceived Usefulness (PU) on perception of mobile users towards Mobile Commerce	0.214	***	Accepted
H_4	Perceived Enjoyment (PE) positively impacts perception of mobile users towards Mobile Commerce	0.409	***	Accepted
H_5	Perceived Risk (PR) negatively impacts perception of mobile users towards Mobile Commerce	0.009	0.819	Rejected
H_6	Social Influence (SI) positively impacts perception of mobile users towards Mobile Commerce	0.180	***	Accepted
H_7	Facilitating Condition (FC) positively impacts perception of mobile users towards Mobile Commerce	0.018	0.417	Rejected

Note: ***denotes for significant at 1% significance level
Source: Author's Compilation

to be significant for perception. Sensationally it was discovered that the most important and powerful factor influencing how mobile users view m-commerce is enjoyment. It suggests that consumers of mobile devices are drawn to the entertainment value they offer, which enhances their opinion of m-commerce. These results are consistent with those of [22] who found that users mostly utilize m-commerce for social networking, games, and movies, particularly in covid scenarios. Another author like [23] revealed that the mobile users will have positive behaviour towards m-commerce when they feel good and excited.

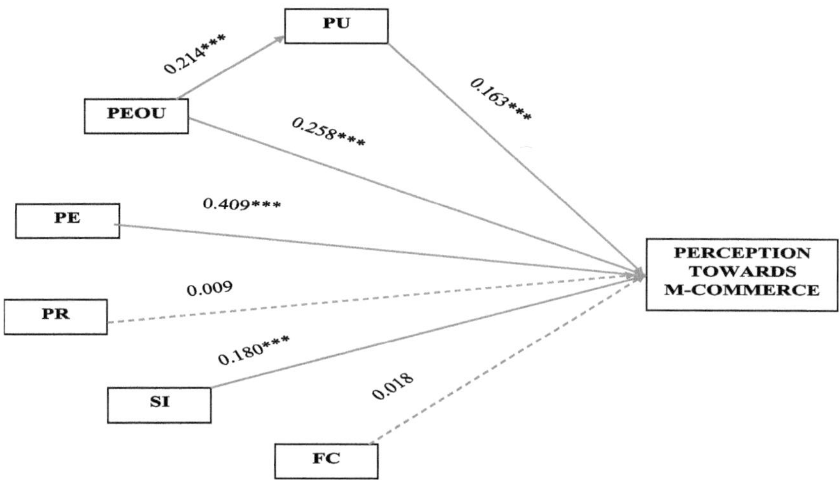

Fig. 4. Users' Acceptance Framework towards Mobile Commerce. Note: *** denotes for significant at 1% significance level. Source: Author's Compilation

PEOU and PU was also discovered to be a crucial element influencing how mobile users see m-commerce. This implies that the mobile users perceive the easiness of using a technology very important which affects their perception. The users will perceive positive behaviour, if they can find what they want easily and with minimum efforts through m-commerce. These results are in consistent with the results drawn by various authors in the past [6, 23, 24]. The outcome of PEOU on PU tested and found significant in this study because in original TAM, which is given by Davis [21] it was found significant.

Family, friends, neighbors, the media, and other elements of society all have a big impact on how others view mobile users. The findings demonstrated that mobile consumers' perceptions of m-commerce are positively and significantly impacted by social influence. This suggests that a mobile user's opinion of using m-commerce is influenced by the society in which they reside. These results coincide with other investigations executed through different authors, such as [25]. who discovered that SI impact a critical role in predicting how mobile users will behave when it comes to m-commerce.

It was discovered that mobile users' perceptions of m-commerce were not significantly impacted by perceived risk or enabling conditions. The study's findings suggest clients are unlikely to accept mobile commerce technologies if they believe there are significant hazards involved. Customers' reluctance to adopt electronic systems is largely due to the threats they believe using internet technologies will present. In the past various studies have found mixed results regarding Facilitating Conditions, since some found facilitating condition as significant factor that affects the adoption of m-commerce [26–28] and some studies have found it as not significant for the adoption of m-commerce [22, 29, 30].

So, it can be concluded that the perception of mobile users in India towards m-commerce are affected by various factors which ultimately leads to the acceptance of using internet on mobile devices. Indians consider mobile phones as very important part of their life and they use it for various purposes apart from communication. It played a significant role in the times of Covid-19 when people were stuck at their homes during lockdown. This could only be possible with the emergence of mobile commerce, which enabled users to use internet on their mobile devices and now Indian people are accepting m-commerce for using it in their daily life. The marketers and service providers in India need to consider all these factors which were originate as noteworthy in this study for developing their application or market strategies accordingly.

6 Conclusion

This study investigated the factors influencing mobile users' perceptions of m-commerce in the post-COVID-19 era. The findings reveal that perceived enjoyment, perceived ease of use, perceived usefulness, and social influence are significant factors positively affecting mobile users' perception of m-commerce. Surprisingly, perceived enjoyment emerged as the most influential factor, highlighting the importance of entertainment value in attracting mobile users. This aligns with broader trends indicating the increasing use of mobile devices for entertainment purposes. The significance of perceived ease of use and perceived usefulness underscores the importance of user-friendly interfaces and tangible benefits in driving m-commerce adoption. The positive impact of social influence suggests that social interactions and recommendations play a crucial role in shaping mobile users' perceptions of m-commerce. Contrary to expectations, perceived risk and facilitating conditions were not found to be significant factors in this study. This suggests that users may be less concerned about security risks or infrastructure limitations than previously thought. Overall, the study provides valuable insights for marketers and service providers seeking to enhance m-commerce adoption in India. By focusing on the key factors identified in this study, businesses can develop targeted strategies to improve user experience and promote the adoption of m-commerce services.

7 Limitations and Future Scope of Research

This study has certain limitations that should be acknowledged. The sample was limited to mobile users in India, which may limit the generalizability of the findings to other contexts. Future research could explore cross-cultural comparisons to understand how these factors vary across different countries and demographics. The study relied on self-reported data, which may be subject to biases. Future research could incorporate objective measures of m-commerce usage to complement self-reported data. The study focused on a specific time period, the post-COVID-19 era. Future research could investigate how these factors evolve over time as technology and user behaviour change. Additionally, the study did not delve into specific types of m-commerce applications. Future research could explore the factors influencing adoption for different m-commerce categories, such as mobile shopping, mobile banking, and mobile entertainment. Furthermore, the study did not examine the impact of emerging technologies, such as 5G and augmented reality, on m-commerce adoption. Future research could investigate how these technologies shape user perceptions and behaviour. Finally, future research could explore the interplay between the identified factors and other relevant variables, such as consumer trust, privacy concerns, and government regulations [7, 10, 12]. By addressing these limitations and exploring these future research directions, we can gain a more comprehensive understanding of the factors driving m-commerce adoption and develop more effective strategies to promote its growth.

References

1. IAMAI and KANTAR IMRB Report. ICUBE 2023, Internet and Mobile Association of India (IAMAI) and KANTAR IMRB International [online] (2023). Retrieved from https://images.assettype.com/afaqs/2021-06/b9a3220f-ae2f-43db-a0b436a372b243c4/KANTAR_ICUBE_2023_Report_C1.pdf, Last Accessed on 24th Nov 2024
2. M-commerce—Reports, Statistics & Marketing Trends (2024). Retrieved from https://www.emarketer .com /topics/industry/mcommerce, Last Accessed on 15th Oct 2024)
3. Hsu, C.-W., Yeh, C.-C.: Understanding the critical factors for successful M-commerce adoption. Int. J. Mob. Commun. **16**(1), 50–62 (2018)
4. Nassar, M.A., Abdou, H.A., Youssef, M.A.: M-commerce adoption decisions in Egypt: the effect of perceived risk on behavioral intention. J. Bus. Retail Manage. Res. **9**(1), 26–39 (2014)
5. Song, J., Mort, G.S., Hyun, Y.J.: Examining the role of intrinsic motivators in m-commerce acceptance. Int. J. Mob. Commun. **5**(3), 203–222 (2007)
6. Pipitwanichakarn, T., Wongtada, N.: Mobile commerce adoption among the bottom of the pyramid: a case of street vendors in Thailand. Int. J. Mob. Commun. **16**(1), 65–85 (2018)
7. Almonte, R., Tang, Z., Smith, J.: Security concerns and trust in the adoption of m-commerce. J. Electron. Commer. Res. **21**(3), 201–218 (2020)
8. Khalifa, M., Shen, N.: Explaining the adoption of transactional B2C mobile commerce. J. Enterp. Inf. Manag. **21**(2), 110–124 (2008)
9. Wei, R., Marthandan, G., Chong, A.Y.L., Ooi, K.B., Arumugam, S.: What drives Malaysian m-commerce adoption? An empirical analysis. Ind. Manag. Data Syst. **109**(3), 370–388 (2009)
10. Smith, A.D.: Online payment service providers and customer relationship management. Int. J. Electron. Fin. **2**(3), 257–283 (2008)
11. Moghavvemi, S., Salleh, N.A.M., Zhao, W., Mattila, M.: The entrepreneur's perception on the adoption of information system innovation. J. Small Bus. Manag. **58**(3), 447–475 (2020)
12. Gitau, L., Nzuki, D.: Analysis of determinants of m-commerce adoption by online consumers. Int. J. Bus. Human. Technol. **4**(3), 88–94 (2014)
13. Agyekum, C.K., Haifeng, H., Agyeiwaa, A., Agyekum, C.K., Haifeng, H., Agyeiwaa, A.: Consumer perception of product quality. Microecon. Macroecon. **3**(2), 25–29 (2015)
14. Census of India 2011. Report on Post Enumeration Survey, Registrar General & Census Commissioner, India. Retrieved from www.censusindia.gov.in/2011Census/pes/Pesreport.pdf
15. Godden, B.: (2004). http://williamgodden.com/sscicalc.html, last accessed 10th August 2024
16. Hair, J.F., Anderson, R. E., Babin, B.J., Black, W.C.: Multivariate data analysis: A global perspective (Vol. 7) (2010)
17. Hair, J.J.F., Black, W.C., Babin, B.J., Anderson, R., Tatham, R.: Multivariate data analysis, 6th edn. Prentice-Hall, Englewood Cliffs, NJ (2006)
18. Field, A.: Discovering Statistics Using SPSS: Introducing Statistical Method, 3rd edn. Sage Publications, Thousand Oaks, CA (2009)
19. Schumacker, R.E., Lomax, R.G.: A Beginner's Guide to Structural Equation Modeling, 3rd edn. Routledge, New York, NY (2010)
20. Straub, D.W.: Validating instruments in MIS research. MIS Q. **13**)2, 147–166 (1989)

21. Davis, F.D., Bagozzi, R.P., Warshaw, P.R.: User acceptance of computer technology: a comparison of two theoretical models. Manag. Sci. **35**(8), 982–1003 (1989)
22. Herero, A., Martin, H.S., Salmones, M.G.: Explaining the adoption of social networks sites for sharing user-generated content: a revision of the UTAUT2. Comput. Hum. Behav. **71**, 209–217 (2017)
23. Zhang, L., Zhu, J., Liu, Q.: A meta-analysis of mobile commerce adoption and the moderating effect of culture. Comput. Hum. Behav. **28**(5), 1902–1911 (2012)
24. Ghazali, E.M., Mutum, D.S., Chong, J.H., Nguyen, B.: Do consumers want mobile commerce? A closer look at M-shopping and technology adoption in Malaysia. Asia Pac. J. Mark. Logist. **30**(4), 1064–1086 (2018)
25. Pedersen, P.E.: An adoption framework for mobile commerce. In: Towards the E-Society, pp. 643–655. Springer, Boston, MA (2001)
26. Lai, I.K., Lai, D.C.: User acceptance of mobile commerce: an empirical study in Macau. Int. J. Syst. Sci. **45**(6), 1321–1331 (2014)
27. Lallmahomed, M.Z., Lallmahomed, N., Lallmahomed, G.M.: Factors influencing the adoption of e-Government services in Mauritius. Telematics Inf. **34**(4), 57–72 (2017)
28. Verkijika, S.F.: Factors influencing the adoption of mobile commerce applications in Cameroon. Telematics Inform. **35**(6), 1665–1674 (2018)
29. Oliveira, T., Thomas, M., Baptista, G., Campos, F.: Mobile payment: understanding the determinants of customer adoption and intention to recommend the technology. Comput. Hum. Behav. **61**, 404–414 (2016)
30. Chong, A.Y.L.: A two-staged SEM-neural network approach for understanding and predicting the determinants of m-commerce adoption. Expert Syst. Appl. **40**(4), 1240–1247 (2013)

A Smartphone-Based Solution for Enhancing Interactive Learning Experiences in Early Childhood Education Using Near Field Communication and Holographic Modules

Maria Seraphina Astriani[1]([✉]), Bayu Prakoso Dirgantoro[2],
Jude Joseph Lamug Martinez[1], and Lee Huey Yi[3]

[1] Computer Science Department, School of Computing and Creative Arts, Bina Nusantara
University, Jakarta, Indonesia 11480
`seraphina@binus.ac.id`
[2] Graphic Design and New Media Program, School of Computing and Creative Arts, Bina
Nusantara University, Jakarta, Indonesia 11480
[3] Water Lily, 75200 Melaka, Malaysia

Abstract. Near Field Communication (NFC) tags offer dynamic information storage and retrieval capabilities when paired with an NFC reader. Despite its vast applicability across sectors, including education, the adoption of NFC technology in enhancing the student learning experience remains limited. This is largely due to the perceived necessity for specialized, high-computational-power devices that are not typically found in everyday life. This research explores the use of portable devices, specifically smartphones, that can compute like desktop computers by utilizing NFC technology. A novel solution is developed to support interactive learning for early childhood education, focusing on sea animal creatures. The approach involves attaching NFC tags to small sea animal figures; utilizing smartphones as NFC readers and projecting the retrieved information into a hologram to create an engaging and immersive environment. This innovative solution aims to broaden the educational applications of NFC technology and inspire further research at the intersection of NFC, holograms, and pedagogy, ultimately enriching the learning experience in early childhood education.

Keywords: Near Field Communication · Hologram · Smartphone-based ·
Interactive Learning · Portable

1 Introduction

Near Field Communication (NFC) is a short-range wireless technology which allows devices to communicate with one another at distances of a few centimeters [1, 2]. By simply placing devices close together (NFC tag and the reader), users can quickly exchange or transfer the data. This seamless interaction improves daily tasks, making NFC a vital component of modern mobile technology. Its simplicity of use has led to its implementation in a variety fields.

A. Mirzazadeh et al. (Eds.): ODSIE 2024, CCIS 2482, pp. 437–448, 2026.
https://doi.org/10.1007/978-3-031-93601-2_26

Education is one of the areas which utilizes information technology (IT) to help students learn [3]. The use of this technology usually involves the use of tools that must be purchased specifically and are rarely used in everyday life, that requires a special budget allocation. Because the main goal of education area is to teach the students, if a school requires to spend their funds for IT purposes, the fund usually will be compressed as much as possible.

When students are learning new material, the use of NFC may contribute to making the learning process more interactive, which is especially suitable for early childhood education students [4–6]. To use NFC technology, it requires an NFC tag and a tool that is capable of reading and writing data to the NFC tag. To purchase the NFC tag may not require a large budget because the unit price of the tag is affordable. However, an NFC reader device that is capable to reading and writing data needs to be purchased. If this device is not planned to be reused in the future because the school do not have any plan to utilize the device; it may become electronic waste and this condition is not ideal. Researchers are challenged to explore the potential of tools that are often found in daily life so that their functions can be optimized to solve this problem.

Based on literature reviews and previous research, a smartphone is a suitable device to use in this study because it is commonly used in daily life and can read and write digital data on an NFC tag [7, 8]. Other advantages of smartphones incorporate are the ability to process data, the ability to connect to a network or the internet, and do not require separate tools for input (touch screen) and output (screen). The smartphone's small size and portability make it ideal for solutions that do not require a large installation space.

The researchers attempt to combine the benefits of NFC technology and smartphones to create interactive learning for early childhood education for studying sea animal creatures. Children enjoy toys, and small sea animal figures are used as one of the media in this solution to attract their attention. To provide a WOW effect to the students who use the solution, researchers are challenged to incorporate additional elements, specifically hologram special effects. According to research that have been done, this effect is able to make people be amazed and are expected to be able to remember the material that has been taught better [8–10]. Other research also bring the effectiveness of using holographic technology to improve the performance of students while they are learning the material [11].

NFC with holographic module using a smartphone is specially developed to enhance interactive learning experiences in early childhood education. The solution was intentionally designed so that students can use it by bringing the small sea animal creature figures close to a smartphone that functions as an NFC reader and seeing the results in the form of a special effect hologram. Based on the scalability and adaptability gap that been mentioned on the previous research [12], the installation and application to produce a special effect hologram was upgraded on this research solution by using a transparent pyramid-shaped panel to accommodate up to four students at the same time. The results projected on the panel is the learning materials in the form of multimedia, such as static images (with audio) or videos of sea creature animals. This paper discussed the literature review, method, results, discussion, conclusion, and recommendation of the solution. It is hoped that by using this interactive solution, students will gain more experience, making them more interested in learning compared to the traditional learning methods.

2 Method

This study presents an advancement of the method derived from the prior solution [8, 10, 12]. This study additionally emphasizes the enhancement of user experience to ensure improved comfort for students while using the solution. This enhancement can be accomplished by focusing on the improvement of the two main aspects of the solution: installation and application (web-based application). The general architecture is shown in Fig. 1.

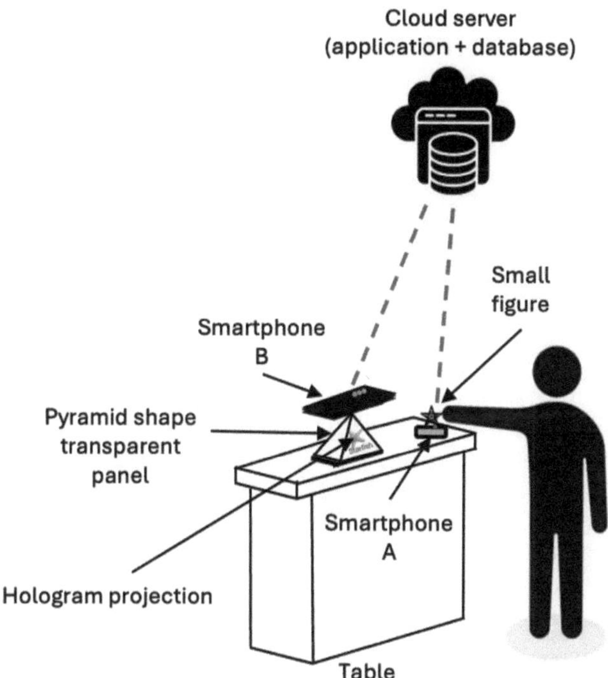

Fig. 1. Architecture of the solution.

2.1 Installation

Installation in this study refers to the arrangement or setting up of devices or physical objects in accordance with the architecture (Fig. 1). The installation is specifically designed to be portable to make it easy to install. Previous studies used only one panel to display a hologram special effect [8, 10, 12]. However, in this study, a transparent panel with a pyramid shape must be provided to accommodate students to see the hologram results up to a maximum of four different sides, with the assumption that one side is seen by each student. The small figures also do not use small sea creature figures that are commonly sold, but rather are custom made using a 3D printer with a tag compartment as a base for the figure. This tag compartment is located at the bottom of the small sea animal creature figure and is designed to store the NFC tag to ensure it cannot be easily removed when used.

The tag specification that been used is ISO 14443-3A with NFC forum type 2 data format [8]. This type of tag is widely used because it is capable of reading and writing data, and its price is quite affordable. This tag operates at 13.56 MHz radio frequency, but its data transmission range is limited to about 4 cm or less.

Any smartphone brand and operating system (OS) can be used in this study if it has the ability to write and read data from an NFC tag. There are two smartphones used (named Smartphone A and Smartphone B). However, one smartphone (Smartphone B) can use one without the write and read NFC tag feature because it will be placed on a transparent pyramid-shaped panel to project the output. Smartphones with the write and read NFC tag feature (Smartphone A) can be configured with an existing application to write the NFC ID for the sake of convenience. The NFC ID can be named using alphanumeric characters, underscores, and/or dashes. Other symbols and spaces are not permitted because the application may fail to process data from the NFC tag.

2.2 Web-Based Application

In general, if a mobile application is chosen to be used, the application must first be installed on each device, and it is quite dependent on the hardware specifications. Other issues may arise if the solution created is quite a lot so it will be requiring many devices and the application to be installed one by one. To facilitate this issue and to ensure that the application is not dependent on the specifications of the smartphone device being used, the application need to be made based on the website. This application is built with PHP (Hypertext Preprocessor) technology with Laravel framework, and MySQL database. This web-based application can be accessed on smartphones by using a browser. Any browser can be used, but researchers prefer Google Chrome because it is compatible with iOS and Android operating systems.

The application's features developed in this study are the enhancements over previous research based on users' feedback and researchers' experience during development until user testing. The enhancements carried out in this study can be seen in Table 1.

The web-based application has the main feature to manage learning materials. This solution also includes new supporting features such as Viewer Settings, Manage Course, and Manage Learning Materials. These features can be used independently by teachers, and the user interfaces were designed to be as simple as possible to make it easier for them to use the application. The Manage Learning Materials feature allows admin/teacher to add current topics as well as other topic to be learned by students in the future. Manage Course maintains course names, and Viewer Setting allows the user to set the display on smartphone's screen. If the value is set to 1, the output can only be seen from one side of the pyramid's transparent panel. If the hologram special effect display can be seen by two students at the same time, the teacher can set the viewer display to 2. The maximum limit value that can be chosen is 4, because the transparent panel used is in the shape of a pyramid with four sides. If the Viewer Setting feature is not used, the output will only be displayed on one side. Figure 2 shows an example of a display based on the value selected in the Viewer Settings feature.

Table 1. Comparison of the enhancement.

Feature	Previous research [8, 10, 12]	Enhancement
Installation: Transparent panel	1 side	4 sides (using pyramid shape)
Installation: Figure	Not customized	Customized with tag compartment
Installation: NFC tag	Paste under the figure	Inside the tag compartment
Installation: Reader	NFC reader or smartphone	Smartphone
Application: Viewer setting	Not available	Available
Application: Manage course	Not available	Available
Application: Manage learning materials > materials column adjustment	Topic and material group	Topic and sub-topic
Application: Manage learning materials > custom file	Not available	Available

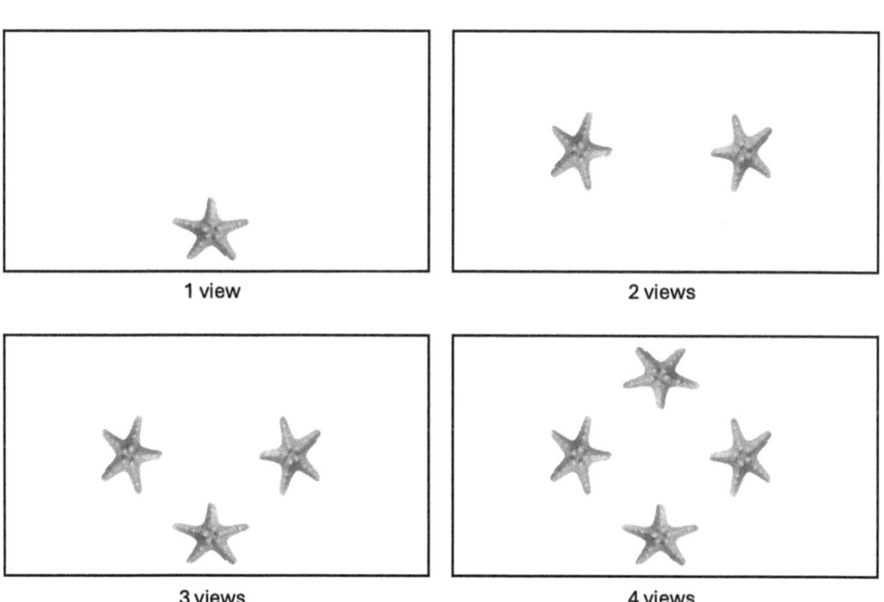

Fig. 2. Output based on the value set on Viewer Setting.

2.3 How It Works

The solution in this research begins with the process that occurs on Smartphone A, which contains a module for reading NFC tag data located on the small figure of sea a creature. If the NFC tag is detected, the application reads the NFC ID data and sends it to a web application server in the cloud. The server runs on the Google Cloud Platform. The application will then check the data to database. If the NFC ID data is found in the database, the material's file will be retrieved, and the results will be displayed on the screen of Smartphone A through a browser. The resulting display on Smartphone A will be transmitted to Smartphone B by using screen mirroring. Communication on Smartphone A and B is done using a wireless network connection (Wireless Fidelity/Wi-Fi). After the output of Smartphone A screen is successfully displayed on Smartphone B's screen, the output on the Smartphone B screen will be reflected on a transparent panel, creating an immersive display of hologram special effect. Figure 3 illustrates a flow diagram of how this solution works.

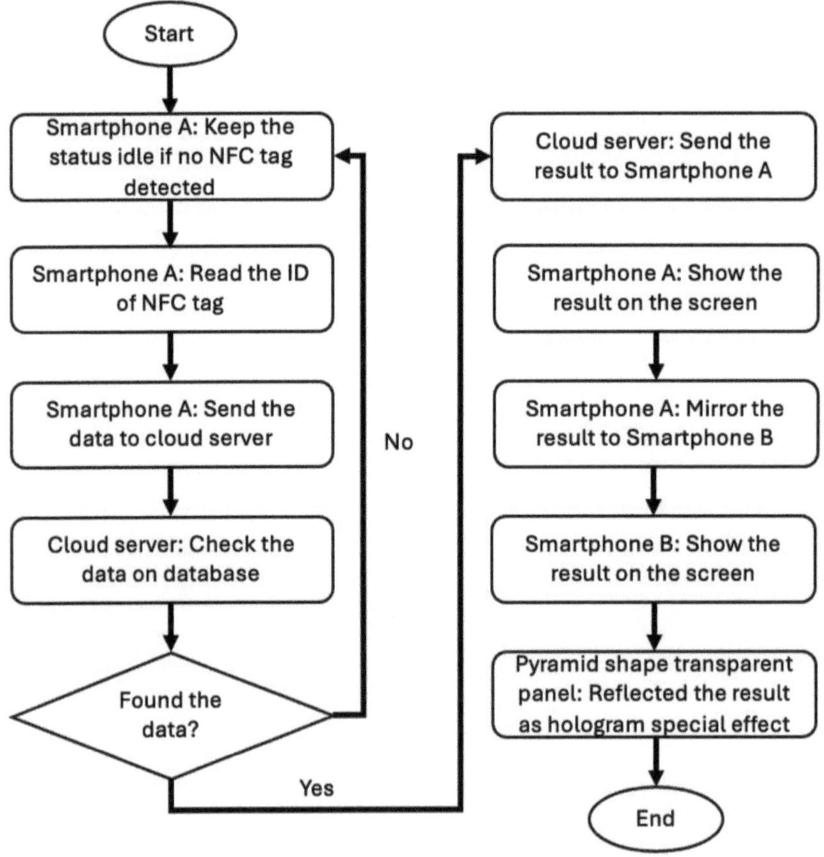

Fig. 3. Flow diagram of the solution.

3   Results and Discussion

3.1   Installation Preparation

Small figures of sea animal creatures such as whale, starfish, and sea turtle need to be made and prepared in advance by using a 3D printer before the installation process is carried out. The process of creating small figures begins with 3D modeling, followed by 3D printing preparation, bed adjustment, printing setup, printing process, and 3D printing results.

In the 3D modeling stage, the existing design is based on a freely available 3D animal gallery, which serves as the foundation of the model. To enhance its functionality, we integrated a dedicated compartment specifically designed to hold an NFC tag beneath the animal figure. This addition allows for interactive capabilities, such as scanning or communication. The NFC tag compartment and the 3D animal model were then meticulously combined using Blender, a professional 3D modeling software. Through this process, we seamlessly merged the animal figure with the NFC tag compartment, resulting in a unified, functional model.

Once the model was completed in Blender, in preparing for 3D process, researchers exported the file in .stl format, preparing it for 3D printing. To carry out the printing process, researchers used the XYZ Print R2 software, which is directly linked to the 3D printer hardware, specifically the Da Vinci Jr. Before printing, we carefully adjusted the model's position and scale to ensure it properly accommodated the NFC tag. Additionally, we applied specific settings, including rectilinear supports and a honeycomb base, to increase the model's structural integrity and improve the chances of a successful print. The estimated printing time is 5:49:00. The process bed adjustment, printing setup, printing process, and 3D printing result can be seen in Figs. 4, 5, and 6. The NFC tag that has been inserted into the compartment (please have a look on the white part) can be seen in the small sea turtle figure on the right in Fig. 6.

Fig. 4. Bed adjustments and printing setup.

A pyramid-shaped transparent panel is prepared with a size of 9.5 cm of the perimeter and 5 cm of height (Fig. 7). This transparent panel has an opening at the bottom. The material used can be Polyethylene Terephthalate (PET) plastic or acrylic. In addition to being made by hand, this pyramid-shaped transparent panel can be purchased ready-made from the marketplace. To strengthen the structure of this pyramid, the top can be

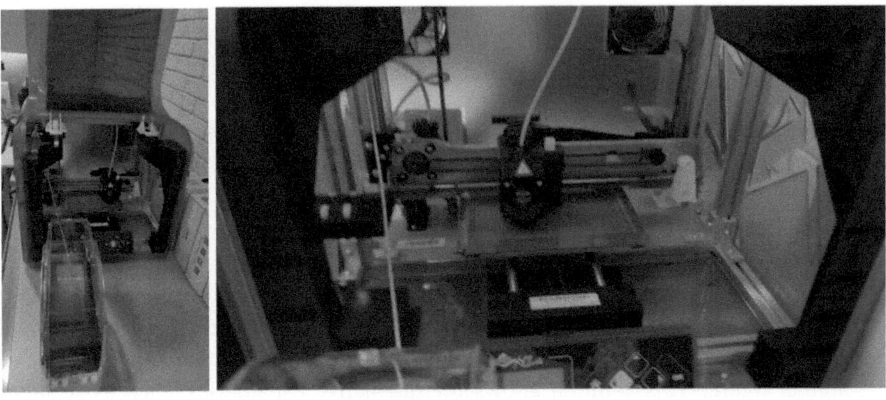

Fig. 5. Printing process.

Fig. 6. 3D printing result.

added with a flat panel to place Smartphone B and a base on the bottom part to make the pyramid sturdier.

Fig. 7. Pyramid-shape transparent panel.

3.2 Solution

The web application was developed by a small team and used the Integrated Development Environment (IDE) Visual Studio Code. The application consists of two parts: students' section (NFC tags are read until the special effect hologram appears) and admin/teacher section. The admin/teacher section was developed to help them to set up and manage the materials being taught. The custom file feature is a highlighted feature in this web application because it allows users to freely manage the existing display using the Rich Text Editor (RTE) and the results will be saved to the MySQL database in the form of LONGTEXT data type. This feature allows teachers to use a various types of data file, including video, images, and audio, if the format is supported and can be displayed in the browser. Figure 8 illustrates the admin/teacher section of the Manage Learning Materials list feature.

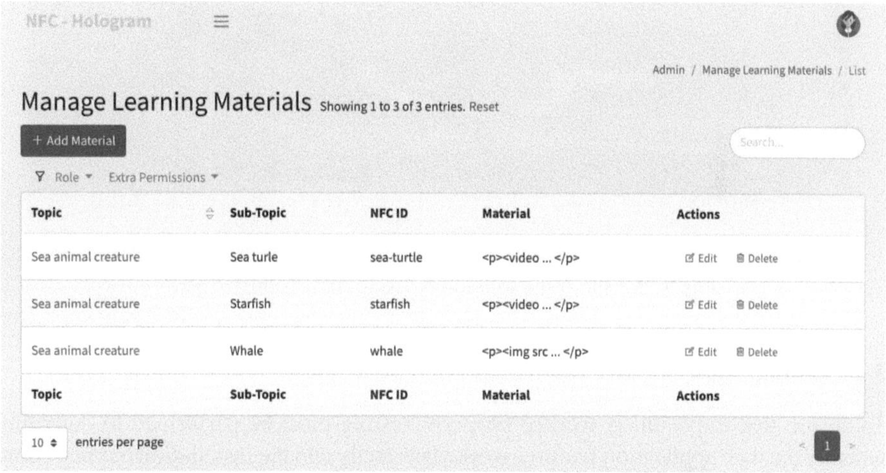

Fig. 8. Web application—Manage Learning Materials list.

After the application has been developed and the data of teaching material related to sea animal creatures has been entered by admin/teacher, all devices must be arranged in accordance with the installation architecture shown in Fig. 1. NFC functionality on Smartphone A must be enabled, and both smartphones must have Wi-Fi turned on. Each NFC tag also needs to have the same NFC ID as the one entered in the application, because this ID will point to the Uniform Resource Locator (URL) where the application is located. The result of this solution when using the starfish small figure can be seen in Fig. 9. If students want to learn the information about other sea anime creatures, such as whales, they can replace the small figure with a whale and place it on Smartphone A (screen facing down), and the results of the special effect hologram will be visible on the pyramid, providing an engaging and immersive experience. To make it easier to detect NFC tags, researchers recommend placing small figures in the center of the smartphone.

This solution does not store any personal information related with the students. So, it will not raise any data privacy or security issue. The personal information that been stored in the database is the username and password to let the administrators or teachers login to the web-based application to manage the learning materials.

Fig. 9. Result of the solution by using starfish small figure.

3.3 Testing, Analysis, and Implementation Strategies

To ensure that the solution worked properly, testing must be performed to determine whether the web application features worked perfectly and the installation has been done properly. Before the test is carried out, it is necessary to ensure that the web application has been opened in the browser on Smartphone A.

Testing is divided into two stages: Alpha and Beta testing. Testing was conducted ten times, emphasizing usability and functionality within the testing environment for both Alpha and Beta testing. Beta testing involving users, specifically students aged approximately 5–6 years old and adults serving as administrators or teachers. There were four test scenarios conducted to be measure. The test scenarios included reading the NFC tag with the one on the small sea animal creature figures: sea turtle, sending the NFC ID data to the cloud server and processing it on the server, displaying the output from the server on Smartphone A and mirroring it on Smartphone B, and successfully reflecting the special effect hologram on the transparent pyramid-shaped panel. The test result qualifies as passed if the test scenario is carried out successfully. All tests that have been conducted have good results. Alpha and Beta tests have passed the four test scenarios (100% successful rate), and these mean the usability and functionality has worked and able to produce the results as expected. However, the users need to be briefed first to let them know how to operate it because this solution is not common existed in daily life.

Students are eager to try this solution, based on the testing that was carried out. Seeing the holographic special effect may ignite student' interest and encourage the student to try this interactive solution independently. The projected hologram results can help student learns about sea animal creatures, which provide multimedia information about their shapes and lives.

The preparation required for the implementation of this solution is the same as in the testing stage. This involves verifying that the web-based application is accessible on Smartphone A's browser and that the devices installation has been installed properly. Teachers need to be trained first to know how to use this solution. And before students use it, teachers need to provide examples of how to use small sea animal creature figures to display special effect holograms.

4 Conclusion and Recommendation

It is possible to utilize smartphones as NFC readers to projecting the retrieved information taken from a cloud server into a hologram special effect to create an engaging and immersive environment. To create this solution, it requires two smartphones (one smartphone must have the ability to read and write NFC tags), NFC tags, small figures (created by a 3D printer), web-based application (built with PHP and the Laravel framework) for students and admins/teachers, MySQL database, cloud server, and transparent pyramid-shaped panels. Students can use the small sea animal creature figure that have NFC tag inserted into the compartment by bringing them close to the smartphone (Smartphone A). The NFC tag ID that has been read then processed by the web-based application to let the students can see the learning material in the form of an immersive hologram reflected on the pyramid. The research has the potential to be expanded and used for other learning topics because it is intentionally designed to be flexible and not only limited to learning about sea animal creatures.

The solution presented in this paper has limitation, only allow NFC tags to be read one at a time. If multiple NFC tags can be read at the same time, it can increase interactivity and lead the way for future development of this NFC technology. It is hoped that this solution will broaden the application of NFC technology and provide inspiration for further research. The future to-do-list that can be done is to develop research using multiple NFC tags to be read simultaneously by enhancing the reader and the web-based application.

Acknowledgement. This work is supported by Bina Nusantara University as a part of Bina Nusantara University's BINUS International Research—Applied entitled "Learning with Interactive Smart Toys and Hologram Projection for Early Childhood Education" with contract number: 069C/VRRTT/III/2024 and contract date: March 18, 2024.

Data Availability. Data is collected by researchers in this study (researchers' data file). Data collection is conducted to assess the system's performance.

Authorship Statement. Maria Seraphina Astriani: Writing and finalizing the manuscript, literature review, and arranging all of the manuscript. She was responsible for conducting the research. Bayu Prakoso Dirgantoro, Jude Joseph Lamug Martinez, and Lee Huey Yi: formulating the methodology, analyzing the data, and contributed significantly to drafting the manuscript.

References

1. Nikitina, K., Melnikova, M., Biliatdinov, K.: Analysis of NFC technology evolution. Int. J. Open Inform. Technol. **12**(2), 49–54 (2024)
2. Chandrasekar, P., Dutta, A.: Recent developments in near field communication: a study. Wirel. Pers. Commun. **116**, 2913–2932 (2021)
3. Aizawa, S., Yoshimura, M.: Development of English learning system by using NFC tag. J. Rob. Netw. Artif. Life. **8**(2), 145–149 (2021)
4. Kıyanç, S., Abi, M.: Technological step in history education: NFC. Res. Exp. J. **4**(2), 77–87 (2019)
5. Zomer, N.R.: Technology use in early childhood education: a review of the literature. J. Educ. Inform. **1**(1), 1–25 (2018)
6. Grøver, V., Snow, C.E., Evans, L., Strømme, H.: Overlooked advantages of interactive book reading in early childhood? A systematic review and research agenda. Acta Psychol. **239**, 103997 (2023)
7. Chiang, T.W., Yang, C.Y., Chiou, G.J., Lin, F.Y.S., Lin, Y.N., Shen, V.R., Juang, T.T.Y., Lin, C.Y.: Development and evaluation of an attendance tracking system using smartphones with GPS and NFC. Appl. Artif. Intell. **36**(1), 2083796 (2022)
8. Astriani, M.S.: The prospect of combining NFC technology with hologram projection for early childhood education. In: 2024 2nd International Conference on Technology Innovation and Its Applications
9. Ramirez-Lopez, C.V., Castano, L., Aldape, P., Tejeda, S.: Telepresence with hologram effect: technological ecosystem for distance education. Sustain. For. **13**(24), 14006 (2021)
10. Astriani, M.S., Bahana, R., Ahmad, A.P.S., Kurniawan, A., Yi, L.H.: Augmented reality immersive world with hologram special effect in early childhood education. In: International Conference on Science, Engineering Management and Information Technology, pp. 176–186. Springer Nature Switzerland, Cham (2023)
11. Yu, Q., Li, B.M., Wang, Q.Y.: The effectiveness of 3D holographic technology on students' learning performance: a meta-analysis. Interact. Learn. Environ. **32**(5), 1629–1641 (2024)
12. Astriani, M.S., Bahana, R., Ahmad, A.P.S., Yi, L.H.: Portable augmented reality installation with hologram special effect for kids' education to learn everyday objects. In: E3S Web of Conferences, vol. 426, p. 02021 (2023)

Author Index

A

Abdulrahman, H. S.   232
Abu Taleb, Anas   36
Abu Al-Haija, Qasem S.   36
Aggoune-Mtalaa, Wassila   281
Alami, Salaheddine Kammouri   153
Albdairi, Mustafa   232
Alhajahjeh, Tareq   36
Almusawi, Ali   232
ALoulou, Chafik   340
Altay, Osman   355
Anas, Muhammad   51
Astriani, Maria Seraphina   437
Ayari, Mariem   281
Ayaz, Halil İbrahim   369

B

Bahri, Afef   17
Bawa, Simerjeet Singh   69, 91, 421
Bedi, Bhupinder Preet   421
Belguith, Mehdi   340
Bouziri, Hend   281

C

Caliskan Demir, Melisa   245
Chafi, Anas   153

D

Dandis, Khaled   245
Dirgantoro, Bayu Prakoso   437
Dobreva, Albena   106
Durak, Hatice Yıldız   3

E

Eğin, Figen   3
Epcim, Dursun Emre   179
Ergün, Serap   197, 217

G

Govindaraj, Vasanthi   169

H

Hatamzadeh, Saeed   303

J

Jouini, Eya   137

K

Kabarcik, Ahmet   232
Kler, Arshan   421

L

Lienkov, Serhii   265
Limam, Hela   137
Lytvynenko, Nataliia   265

M

Mahalingam, Sivakumar   389
Martinez, Jude Joseph Lamug   437
Masouri, Zahra   303
Mete, Süleyman   179
Mirkamali, SeyedSaeid   119
Mjadri, Houyem   17
Myasischev, Alexander   265

N

Nahavandi, Saeid   51
Nasri, Sonia   281
Neugebauer, Jakub   323

O

Odeh, Ammar   36
Oghbaei, Maryam   119
Onan, Aytuğ   3
Oueslati, Wided   17

Ovcharuk, Vadym   265
Özceylan, Eren   179

P
Paksoy, Turan   369
Pervaiz, Kashif   69, 91
Polat, Alper Buğra   355
Periyasamy, Prakash   169

Q
Qadri, Syed Shah Sultan Mohiuddin   232

R
Rahmani Hosseinabadi, Ali Asghar   119

S
Sagiroglu, Aslihan   245
Salameh, Walid   36

Seker, Ezgi Zehra   245
Sieliukov, Oleksandr   265
Solaiyappan, Ayyappan   389

T
Tarhonskyi, Vitalii   265
Taskin, Alev   245
Torğul, Belkız   369

V
Varol-Altay, Elif   355

Y
Yi, Lee Huey   437
Yüksel, Zeynep   179

Z
Zabraoui, Oussama   153
Zhang, Jingxin   51

MIX
Papier aus verantwortungsvollen Quellen
Paper from responsible sources
FSC® C105338

If you have any concerns about our products,
you can contact us on
ProductSafety@springernature.com

In case Publisher is established outside the EU,
the EU authorized representative is:
**Springer Nature Customer Service Center GmbH
Europaplatz 3, 69115 Heidelberg, Germany**

Printed by Libri Plureos GmbH
in Hamburg, Germany